Virology

Second Edition

Renato Dulbecco, M.D.

Distinguished Research Professor,
The Salk Institute, San Diego;
Senior Clayton Foundation Investigator and Professor Emeritus,
Departments of Pathology and Medicine,
University of California, San Diego, School of Medicine,
La Jolla, California

Harold S. Ginsberg, M.D.

Higgins Professor of Microbiology and Medicine,
Columbia University College of Physicians and Surgeons,
New York, New York

J. B. LIPPINCOTT COMPANY
Philadelphia

London   Mexico City   New York   St. Louis   São Paulo   Sydney

Virology, 2nd ed, is also published as a section of Microbiology, 4th ed,
by Davis, Dulbecco, Eisen, and Ginsberg, published by J. B. Lippincott
Company.

Acquisitions Editor: Lisa A. Biello
Sponsoring Editor: Richard Winters
Manuscript Editor: Lee Henderson
Indexer: Betty Herr Hallinger
Design Coordinator: Caren Erlichman
Designer: Arlene Putterman
Cover Designer: Anthony Frizano
Production Manager: Carol A. Florence
Production Coordinator: Kathryn Rule
Compositor: Bi-Comp, Inc.
Printer/Binder: The Murray Printing Company
Cover Printer: New England Book Components

Second Edition

Copyright © 1988, by J. B. Lippincott Company
Copyright © 1980, by Harper & Row, Publishers, Inc.
All rights reserved. No part of this book may be used or reproduced in any manner whatsoever without written permission except for brief quotations embodied in critical articles and reviews. Printed in the United States of America. For information write J. B. Lippincott Company, East Washington Square, Philadelphia, Pennsylvania 19105.

6 5 4

Library of Congress Cataloging-in-Publication Data

Dulbecco, Renato, DATE
   Virology.
   "Also published as a section of Microbiology, 4th ed.,
by Davis, Dulbecco, Eisen, and Ginsberg."
   Includes bibliographies and index.
   1. Medical virology.   I. Ginsberg, Harold S.,
DATE.   II. Title.   [DNLM: 1. Viruses.
QW 160 D879v]
QR360.D84   1988   616'.0194   87-21353
ISBN 0-397-50905-7

Cover photograph from Choppin PW et al: J Exp Med 112:945, 1960, courtesy of Rockefeller University Press.

Preface

Virology originated as a discipline to study diseases of major medical or economic importance, such as smallpox, rabies, foot-and-mouth disease, or tobacco mosaic disease. Subsequently, viruses were used as tools for studies of genetics, biochemistry, and molecular biology; as a result, the properties of many viruses are now understood at the molecular level. Viruses are therefore considered in two parallel ways in this book—as very interesting organisms and as agents of disease. Thus, *Virology* discusses both the biological and pathogenic properties of viruses.

The basic properties of the viruses are examined in depth, beginning with bacteriophages, which still allow the deepest insight into the biology of viruses, and following with animal viruses. This information serves as the basis for analyzing the mechanisms by which viruses interact with human or animal cells and organisms and produce disease. Indeed, this material permits an approach to understanding the mechanisms of viral pathogenesis at a molecular level. The properties of the animal cells relevant to this interaction are examined in a separate chapter, and then the viruses infecting humans are examined systematically. Two chapters review the principles of viral oncogenesis and the role of viruses in human cancer. The virus causing AIDS (the human immunodeficiency virus, or HIV), a retrovirus, is considered among them.

Unlike other virology texts, *Virology* applies the principles of basic virology to each viral family. For each family it discusses viral structure, mode of replication, genetic properties, and mechanisms of interaction with the cells. On this basis, problems of pathogenesis, immunology, epidemiology, and control of major viruses can be approached rationally.

Renato Dulbecco, M.D.
Harold S. Ginsberg, M.D.

Contents

Virology

44

Renato Dulbecco

The Nature of Viruses

Viruses, as infectious agents responsible for many diseases in humans, animals, and plants are of great medical and economic importance. They were recognized at the end of past century as infectious agents smaller than bacteria ("filterable agents"). Transmission by a cell-free filtrate was demonstrated in 1898 for foot-and-mouth disease, for fowl leukosis in 1908, and for chicken sarcoma in 1911. The discovery of viruses affecting bacteria, made in 1917, made available an important model system for investigations of basic virology.

Distinctive Properties

Passage through the usual bacterial filters, and multiplication only as obligatory parasites in living cells, proved inadequate to distinguish viruses from the smallest bacteria (e.g., rickettsiae). Viruses are distinguished from other microbes in more fundamental ways: their simple organization and their characteristic mode of replication. In addition, animal viruses produce characteristic effects on host cells: death, fusion, or transformation into cancer cells.

The free viral particles, called **virions,** are made up of two essential constituents: a **genome,** which can be DNA or RNA, associated with proteins or polyamines; and a protein coat (**the capsid**), sometimes surrounded by a membranous **envelope.** In addition, some virions have enzymes that are needed in the initial steps of replication of the genome. They may also contain other minor constituents (see below). So a virion is relatively simple: it is little more than a block of genetic material enclosed in a coat. Capsid and envelope protect the genome from the nucleases present in the environment, and they facilitate

1

its attachment and penetration into the cell in which it will replicate.

Viruses have a unique method of multiplication. Whereas in the replication of other microbes all the constituents are made within the cell envelope, finally causing the microbe to undergo binary fission, virions lack the machinery for using and transforming energy and for making the proteins specified by the viral genes. Accordingly, after the viral genome is released from the coat, it uses the machinery of the host cell to make the constituents of viruses in the cell's cytoplasm or nucleus. Progeny virions are then assembled from these constituents. Although all viruses have this basic mechanism of replication in common, they differ considerably in such characteristics as size and shape, chemical composition of the genome (Fig. 44–1), and the type of cells they infect.

HOST RANGE

Viruses are subdivided into **animal viruses, bacterial viruses (bacteriophages),** and **plant viruses.** Within a class each virus is able to infect only cells of a certain species or of a certain type. The host range is determined in part by the specificity of attachment to the cells, which depends on properties of both the virion's coat and specific receptors on the cell surface. It also depends on the availability of cellular factors required for the replication or transcription of the genome. The host range is broader in **transfection:** infection by the naked nucleic acid, the entry of which does not depend on specific receptors. Limitations determined by intracellular factors, however, persist.

Are Viruses Alive?

When Stanley crystallized tobacco mosaic virus in 1935, there followed extensive debates on whether it was a living being or merely a nucleoprotein molecule. As Pirie pointed out, these discussions showed only that some scientists had a more teleologic than operational view of the meaning of the word *life.* Life can be viewed as a complex set of processes resulting from the actuation of the instructions encoded in the genes; those of viral genes are actuated after the viral genome has entered a susceptible cell; hence, viruses may be considered alive when they replicate in cells. Outside cells, virions are metabolically inert chemicals. Thus, depending on the context, viruses may be regarded both as exceptionally simple microbes and as complex chemicals.

Viruses are then not organisms in the usual sense: they are parasitic genomes, related to plasmids (see Chap. 8). Moreover, some viral genomes, like certain plasmids, become integrated into the DNA of their host cells,

exercising the same form of parasitism displayed by movable DNA elements and by certain repeated sequences abundant in the DNA of eukaryotic cells.

The Viral Particles

GENERAL MORPHOLOGY

Viruses of different families have virions of different morphologies, which can be readily distinguished by electron microscopy. This relationship is useful for diagnosing viral diseases and especially for recognizing new viral agents of infection. For instance, the wheel-shaped virions of rotaviruses in feces of infants with diarrhea could readily be distinguished from other viruses also present in feces, and recognition of paramyxovirus nucleocapsids in thin sections of the brains of patients helped to reveal the viral origin of subacute sclerosing panencephalitis (SSPE).

However, different classes of viruses within the same family have virions of similar morphology. The identification can be refined by the binding of specific antibodies to virions, which is also recognizable by electron microscopy (see Immuno-Electron Microscopy in Chap. 50).

Virions belong to several morphological types (Fig. 44–2 and Table 44–1).

1. **Icosahedral virions** resemble small crystals. Extensive studies, especially by Klug and Caspar, have shown that these virions have an **icosahedral** protein shell (**the capsid**) surrounding a core of nucleic acid and proteins. The capsid and the core form the **nucleocapsid.** Examples are picornaviruses, adenoviruses, papovaviruses, and bacteriophage ϕX174 (Fig. 44–3A).

2. **Helical virions,** of which tobacco mosaic virus (see Fig. 44–3B) and bacteriophage M13 are examples, form long rods. The nucleic acid is surrounded by a **cylindrical capsid,** in which a helical structure is revealed by high-resolution electron microscopy.

3. **Enveloped virions** contain lipids. In most cases the nucleocapsid—in some viruses icosahedral, in others helical—is surrounded by a membranous **envelope.** Most enveloped virions are roughly spherical but highly **pleomorphic** (i.e., of varying shapes) because the envelope is not rigid. Herpesviruses and togaviruses are examples of **enveloped icosahedral** viruses (see Fig. 44–3C). In **enveloped helical** viruses, such as orthomyxoviruses (see Fig. 44–3D), the nucleocapsid is coiled within the envelope.

4. **Complex virion** structures belong to two groups. Those illustrated by poxviruses (see Fig. 44–3E) do not possess a clearly identifiable capsid but have several coats around the nucleic acid, while certain bacteriophages (see Fig. 44–3F) have a capsid to which additional structures are appended.

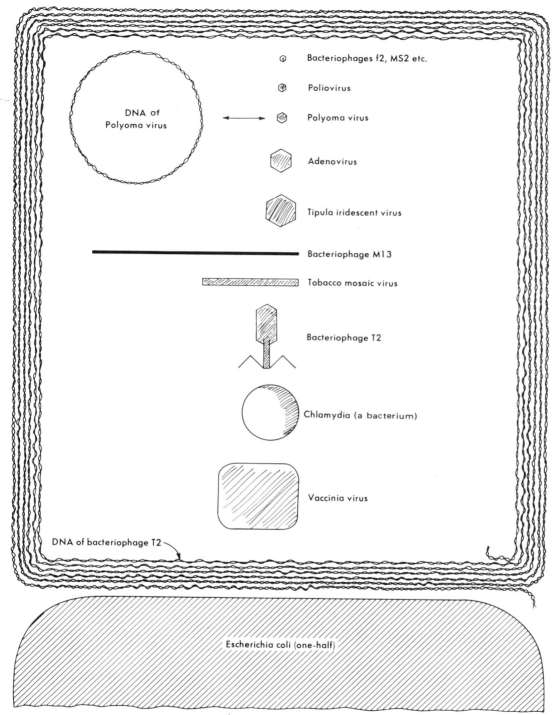

Figure 44–1. Comparative sizes of virions, their nucleic acids, and bacteria. The profiles and the lengths of the DNA molecules are all reproduced on the same scale.

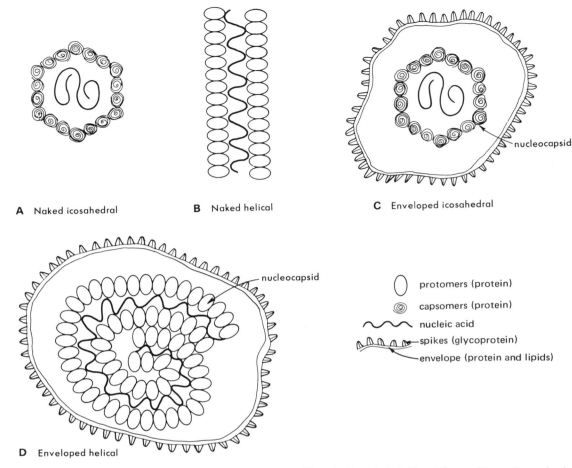

A Naked icosahedral   **B** Naked helical   **C** Enveloped icosahedral

D Enveloped helical

protomers (protein)
capsomers (protein)
nucleic acid
spikes (glycoprotein)
envelope (protein and lipids)

Figure 44—2. Simple forms of virions and of their components. The naked icosahedral virions (*A*) resemble small crystals; the naked helical virions (*B*) resemble rods with a fine regular helical pattern in their surface. The enveloped icosahedral virions (*C*) are made up of icosahedral nucleocapsids surrounded by the envelope; the enveloped helical virions (*D*) are helical nucleocapsids bent to form a coarse, often irregular coil within the envelope.

VIRAL GENOMES

In a given virus the genome may consist of either DNA or RNA, which may be either single or double stranded. The amount of genetic information per virion varies from about 3 to 300 kilobases (Kb). If 1 Kb is taken as the size of an average gene, small viruses contain perhaps three or four genes, and large viruses several hundred. The diversity of virus-specific proteins synthesized in the infected cells varies accordingly. With the exception of retroviruses (see Chap. 65), virions contain only a single copy of the genome; that is, they are **haploid.** The virions of some plant viruses contain only a fraction of the genome, and several virions, collectively containing the whole genome, must enter the same cell for viral multiplication to take place.

Double-Stranded Viral DNA

Table 44–2 gives the lengths of various viral DNAs, obtained in great part by cloning and sequencing. The base composition and the codon usage vary considerably: some viruses even contain **abnormal bases.** For instance, cytosine is replaced by 5-hydroxymethyl-cytosine in T-even coliphages (see Chap. 45), and substitutions for thymine are found in *Bacillus* and *Pseudomonas* phages. These differences from the host cells allow the viral DNA to escape the action of cellular nucleases or to be selectively recognized by virus-specified enzymes.

Many viral DNAs have **special features** related to their methods of replication. These features avoid the difficulty of complete replication of the ends of a linear molecule from an internal initiation (see Fig. 45–10). To

TABLE 44–1. Characteristics of Viruses

Morphological Class	Nucleic Acid*	Example Virus Family	Example Virus	Size of Capsid (nm)	No. of Capsomers	Size of Virions (Enveloped Viruses) (nm)	Special Features
Helical capsid							
Naked	DNA	Coliphage f1, M13		5 × 800			Single-stranded cyclic DNA
	RNA	Many plant viruses	Tobacco mosaic	17.5 × 300			
			Beet yellow	10 × 1200			
Enveloped	RNA	Orthomyxoviruses	Influenza	9 (diameter)		90–100	Segmented RNA
		Paramyxoviruses	Newcastle disease	18 (diameter)		125–250	
		Rhabdoviruses	Vesicular stomatitis			68 × 175	Bullet shaped
Icosahedral capsid							
Naked	DNA	Parvoviruses	Adeno-associated	20	12		Single-stranded linear DNA
		Coliphage φx174		22	12		Single-stranded cyclic DNA
		Papovaviruses	Polyoma	45	72		Cyclic DNA
			Papilloma	55	72		Cyclic DNA
		Adenoviruses		60–90	252		
	RNA	Coliphage F2 and others		20–25			
		Picornaviruses	Polio	28	32		
		Many plant viruses	Turnip yellow	28	32		
		Reoviruses		70	92		Segmented; double-stranded RNA
Enveloped	DNA	Herpesviruses	Herpes simplex	100	162	180–200	
		Hepadnaviruses	Hepatitis B	27	42		
Capsids of binal symmetry (i.e., some components icosahedral, others helical)							
Naked	DNA	Large bacteriophages	T2,T4,T6	Modified icosahedral head: 95 × 65; helical tail: 17 × 115			
Complex virions	DNA	Poxviruses	Variola } Vaccinia }			250 × 300	Brick shaped
			Contagious pustular dermatitis of sheep			160 × 260	

* DNA double stranded, RNA single stranded, unless specified in last column

this purpose some viral DNA are **cyclic,** and therefore without ends, whereas others are made to become cyclic after entering cells. Others have **terminal redundancies,** which enable incompletely replicated molecules to complete each other by recombination. Some viral DNAs have **palindromes** or **terminal proteins** at the ends, which act as primers during replication. These characteristics and their roles will be considered in greater detail together with DNA replication in Chapter 50 and in the Chapters on specific viruses (45, 46, 52–65).

Some viral DNAs have features that show their **relatedness to transposons** (see Chap. 8), such as **terminal repeats.** The DNAs of some herpesviruses are made up of two unequal transposons joined together, each with its own terminal repeats (see Chap. 53); each transposon undergoes frequent inversion independently of the other, so a population of virions contains four, equally frequent, kinds of DNA: ⟶ →, ⟵ →, ⟶ ←, ⟵ ←. Some DNAs have single-strand nicks at characteristic places, which define special genomic segments during

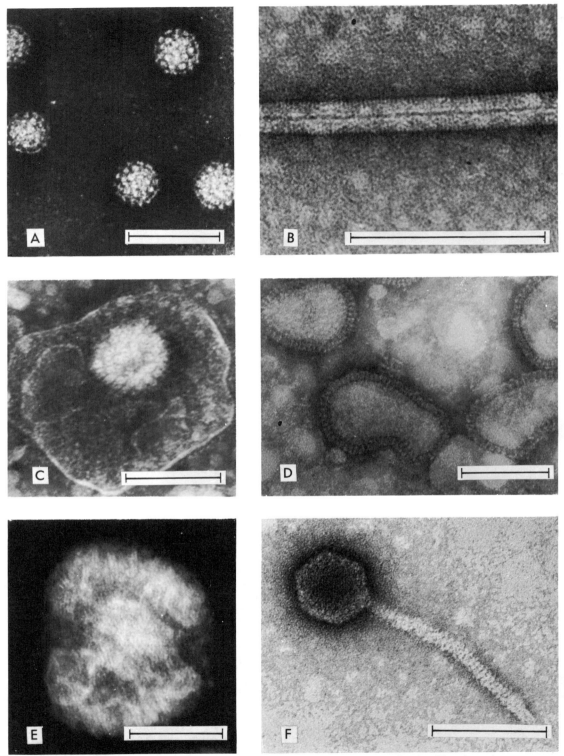

Figure 44–3. Electron micrographs of representative virions were obtained by negative staining; that is, suspending the virions in an electron-opaque salt solution so that structures are transparent on a dark background. Markers under each micrograph are 100 nm. (*A*) Naked icosahedral: human wart virus (papovavirus, Chap. 65). (*B*) Naked helical: segment of tobacco mosaic virus. (*C*) Enveloped icosahedral: herpes simplex virus (herpesvirus, Chap. 53). (*D*) Enveloped helical: influenza virus (orthomyxovirus, Chap. 56). (*E*) Complex virus: vaccinia virus (poxvirus, Chap. 54). (*F*) Coliphage λ (Chap. 46). (*A*, Noyes WF: Virology 23:65, 1964; *B*, Finch JT: J Mol Biol 8:872, 1964. Copyright by Academic Press, Inc. [London] Ltd.; *C*, courtesy of P. Wildy; *D*, Choppin PW, Stockenius W: Virology 22:482, 1964; *E*, courtesy of R. W. Horne; *F*, courtesy of F. A. Eiserling)

TABLE 44–2. Characteristics of Viral Nucleic Acids

Type of Nucleic Acid	Representative Virus	Mol. wt. (in 10⁶ daltons)	Kilobases per Strand*	No. of Segments	Polarity
DNA, DOUBLE STRANDED					
Hepatitis B (cyclic)		1.6	2.5		
Papovavirus (cyclic)	Polyoma	3.5	5.0		
	Papilloma	6	9		
Pseudomonas phage PMS2 (cyclic)		6	9		
Adenovirus	Types 12,18	21	32		
	Types 2,5	23	35		
Coliphages T3,T7		25	38		
Coliphage Mu		26	39		
Coliphage λ		31	47		
Coliphage T5		77	117		
Herpesvirus	Herpes simplex	100	151		
Coliphages T2,T4,T6		110	167		
Bacillus subtilis phage SP8		130	197		
Poxvirus	Vaccinia	160	242		
DNA, SINGLE STRANDED					
Parvovirus	Adeno-associated†	1.5	4.5		
Coliphage φx174 (cyclic)		1.7	5.2		
Coliphage M13 (cyclic)		2.4	7.3		
RNA, DOUBLE STRANDED					
Rotaviruses		15³	23	10	
Rice dwarf virus		15³	23	10	
Cytoplasmic polyhedrosis of silkworms		15³	23	10	
RNA, SINGLE STRANDED					
Satellite necrosis virus†		0.4	1.2	1	
Coliphage R17		1.3	4	1	+
Tobacco mosaic virus		2	6	1	+
Turnip yellow mosaic virus		2	6	1	+
Picornavirus	Polio	2.5	7.5	1	+
Bunyavirus		3‡	9	3	−
Retrovirus§	Rous sarcoma virus	3.5	10.5	1	+
Alphavirus	Sindbis	4	13	1	+
Rhabdovirus	Vesicular stomatitis virus	4	13	1	−
Orthomyxovirus	Influenza	6‡	18	8	−
Paramyxovirus	Newcastle disease	6	18	1	−

+, positive; −, negative

* A kilobase (1000 bases) corresponds to a molecular weight of about 700,000 for double-stranded and 350,000 for single-stranded nucleic acid; it can specify about 33,000 daltons of protein. The number of genes is approximately equal to the number of kilobases; φx174 has fewer kilobases because **some genes overlap each other** (see Chap. 45).

† These viruses are defective and multiply only in cells infected by a helper virus (adenovirus or tobacco necrosis virus, respectively). They probably specify only their own capsid, perhaps with another small protein.

‡ This value, as for other virions with segmented genomes, is the aggregate of all fragments.

§ Retroviruses have diploid virions.

entry into cells (Phage T5; see Regulation of Transcription in Chap. 45).

Control Elements of DNA Viruses

Genes contained in viral DNAs are controlled essentially like the genes of the host cells, and they have the corresponding characteristic sequences. DNAs of bacterial viruses, like bacterial genes, have promoters, operators, and ribosome-binding sites. DNAs of eukaryotic viruses have control regions comparable to those of eukaryotic genes, enhancers (see Chap. 64), and a TATA box for locating the exact initiation of the transcripts; they also have poly(A) addition sites at their 3′ ends where the messengers terminate. In poxviruses, however, the transcription signals do not conform to those of the host cells and are recognized by enzymes specified by viral genes. The structure of the genome is suitable for polycistronic transcription in bacterial viruses and for mono-cistronic transcription in eukaryotic viruses. The DNA of eukaryotic viruses encodes intervening sequences, whereas those of bacterial viruses usually do not. (For an exception in bacteriophage T4, see Posttranscriptional Regulation in Chap. 45.)

Single-Stranded Viral DNA

The DNA is single-stranded and cyclic in some very small bacteriophages (the icosahedral φX174 and the helical f1 and M13) and in one family of animal viruses (parvoviruses). The phages have DNA molecules of the same polarity in all virions; parvoviruses have strands of both polarities, but in different virions. Parvoviruses have also inverted terminal repeats that can form hairpins, important for replication.

Viral RNAs

Some RNA viral genomes are **double stranded** (in reovirus, in a phage, and in some viruses of lower animals, insects, yeasts, and plants). Other genomes are **single stranded.** Single-stranded genomes belong to two classes: **positive-strand** genomes that, on entering the cells, can directly act as messengers for protein synthesis; and **negative-strand** genomes that are not of messenger polarity and must be transcribed into messengers. Positive-strand RNAs of eukaryotic cells have the general organization of eukaryotic mRNAs: they have a cap at the 5′ end, and they end with a poly(A) chain at the other end. Picornaviruses are an exception: they do not have a cap but have a small protein covalently linked to the 5′ terminal uridylate. Negative-strand RNAs do not have caps, but each is terminated at the 5′ end with a nucleoside triphosphate.

Of these viruses, the retroviruses are closely related to transposons: their genome is flanked by two repeats, and it integrates into the cellular DNA where it is flanked by two short repeated cellular sequences. It is propagated by reverse transcriptions (RNA → DNA), like transposons of Drosophila and yeast.

Segmentation of the Genome

The genomes of double-stranded RNA viruses, and those of some negative-strand viruses, have a peculiarity: they are made of separate segments. For instance, the double-stranded reoviruses have ten segments, and the negative-stranded orthomyxoviruses have eight. The segmentation of the genome is probably a mechanism for avoiding polycistronic messengers, because eukaryotic cells rarely initiate protein synthesis internally in a messenger. A segment, however, may specify two proteins.

Control Sequences for RNA Viruses

Control sequences in RNA genomes have the function of interacting with the replication or transcription apparatus, with initiation factors for protein synthesis and ribosomes, and with the capsid. Short sequences with these functions, present near the ends of genomes and at the ends of genes, are recognized because they are conserved among related viruses.

Exceptional arrangements are found in some viral genomes. For instance, the positive-strand phage MS2 lacks a ribosome entry site for a lysis gene. This gene uses ribosomes that translate the upstream coat gene, with which it partly overlaps but in a different phase. Ribosomes that accidentally go out of phase in the coat gene can enter the lysis gene. This arrangement ensures that the lysis protein is made in much smaller amounts than the coat protein, so that the cells lyse only after enough virus is made. The polymerase gene of retroviruses has a similar arrangement. These are examples of gene overlap with regulatory function.

Origin of Viral Genomes

Because viral genomes could not have evolved readily except by replication within cells, it is logical to assume that viral genomes are ultimately derived from the genomes of host cells. A step in this evolution may be the incorporation of cellular genes into the genomes of transducing bacteriophages (see Chap. 46) and retroviruses (see Chap. 65). The evolutionary separation, however, is long, for homologies between viral genomes and cellular genes are rare. One is found in vaccinia virus, which has a gene with some homology with a cellular gene for a growth factor.

More stringent evidence for a cellular origin is found in positive-strand RNA viruses, the genomes of which resemble cellular messengers. The RNAs of some plant viruses terminate at the 3′ end with sequences that fold into the tertiary structure of tRNAs and can be aminoacylated by specific tRNA aminoacylases. At its 5′ position this cistron is connected to a poly(A) sequence from a cistron with all the features of a cellular mRNA. These genomes were evidently derived from the recombination of a cellular mRNA with a tRNA-like polynucleotide.

THE VIRION'S COATS
The Capsid

The study of the organization of the capsid is important because it uncovers the principles by which biological macromolecules assemble into complex structures. The capsid encloses the genome and gives the virions their characteristic shapes. It accounts for a large part of the virion's mass and is made up of protein molecules, which are specified by viral genes. Since viruses have small genomes, they cannot afford too many genes for specifying capsid proteins; hence, the capsid must be formed by the association of many protein units of a single kind or of relatively few kinds (**protomers**). For instance, poliovirus RNA (7 Kb) can specify at most 250,000 daltons of protein altogether; some of the proteins must be used for replication. Yet the poliovirus capsid weighs about 6×10^6 daltons. In fact, it contains only four unique proteins. The shape and dimensions of the capsid depend on characteristics of the constituent protomers and, for helical capsids, on the length of the viral nucleic acid.

The repeated protomers forming the capsid must be arranged in a regular architecture that utilizes bonds between the same pairs of chemical groups. This goal is attained in different ways in the icosahedral and helical capsids, as shown by extensive x-ray crystallographic studies.

ICOSAHEDRAL CAPSIDS. The icosahedral shape of many viruses is of considerable interest because the only closed shell that can be made with identical protomers is icosahedral. The simplest icosahedron is a regular solid with 12 vertices and 20 triangular faces (Fig. 44–4). To

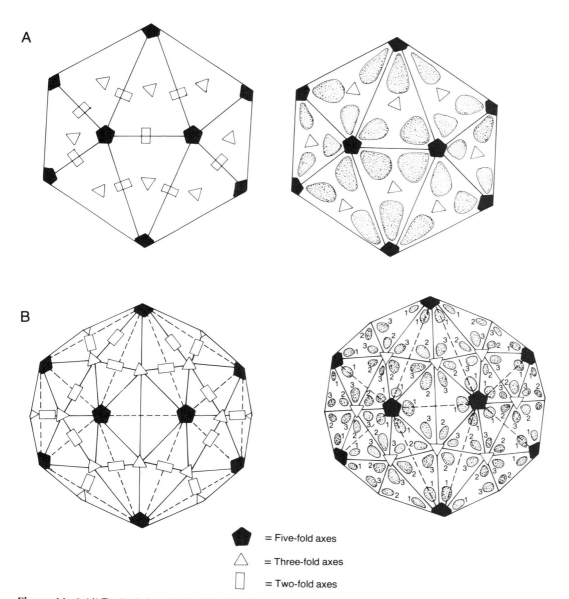

⬟ = Five-fold axes

△ = Three-fold axes

▯ = Two-fold axes

Figure 44–4. (A) The basic icosahedron. The drawing at left shows the triangular faces, with pentons at each vertex; the drawing at right shows the positions of the monomers around the fivefold axis. (B) A derived icosahedron (T = 3; see Appendix). Each triangular face of the icosahedron shown in A is subdivided into six half-triangles. The corners of the inscribed faces are solid lines; those of the basic faces are dashed lines. Monomers are arranged in pentons around the fivefold axes and in hexons around the threefold axes. 1, 2, and 3 indicate quasi-equivalent monomers.

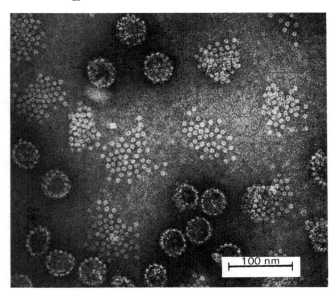

Figure 44–5. Preparation of rabbit papilloma virus (papovavirus) containing mostly empty capsids, i.e., devoid of nucleic acid. Some of the capsids have disintegrated during the preparation of the specimen for electron microscopy, each producing a small puddle of capsomers (some of the capsomers of the original capsid have been lost). The angular polygonal shape of the capsomers is evident, but it is not possible to differentiate hexamers from pentamers. (Breedis C et al: Virology 17:84, 1962)

make a shell there must be sixty protomers—three per face, each located at one of the vertices (see Fig. 44–4*A*) and all connected to one another in the same way. The five protomers around each vertex together form knobs recognizable by electron microscopy, known as **capsomers,** and, more specifically, **pentons.** The bonds between protomers in a capsomer are usually more stable than those between capsomers, so the capsomers tend to persist after the capsid is disrupted under mild conditions (Fig. 44–5).

To enclose space, asymmetric units with identical bonding must be related to one another by rotational symmetry. This is easily seen by considering the two-dimensional case, as shown in Figure 44–6. Accordingly, the icosahedron has **only rotational symmetry;** that is, it is brought to coincide with itself after rotation of an appropriate angle around certain axes (see Fig. 44–4*A*). The icosahedron has axes of three kinds: fivefold axes through the vertices (coincidence is achieved five times in a full turn), threefold axes through the centers of the faces (three coincidences), and twofold axes through the middle of each corner (two coincidences). For this reason it is said to have a 2–3–5 rotational symmetry. In a perfect icosahedron all protomers are therefore related to one another in exactly the same way.

THE SIMPLEST VIRIONS. Only the smallest and simplest virions have capsids made up of 60 identical protomers. One is that of the satellite tobacco necrosis virus. Enclosed in its capsid is a short RNA (about 1600 bases) with just one gene, that for the protomer of the capsid. Because of this simplicity this virus multiplies only in cells infected by the tobacco necrosis virus, which provides the proteins needed for its replication (hence the name satellite). Under suitable conditions the protomers of this virus can spontaneously assemble first into pentons, which then join to form the icosahedral capsids (**self-assembly**).

QUASI-EQUIVALENCE. Most other icosahedral viruses have more than 60 protomers per virion. They cannot form rigorous icosahedrons but form approximate ones, in the following way. In an icosahedron the triangular faces can be subdivided into a number of smaller triangles, generating **modified icosahedrons.** The smallest possible number of inscribed triangles per face is 3, followed by 4, 7, 9, 12, and so on (see Appendix). These figures are known as the **triangulation numbers (T)** of the various modified icosahedrons. Because each inscribed triangle must again have three protomers, their number in each modified icosahedron is $60 \times T$.

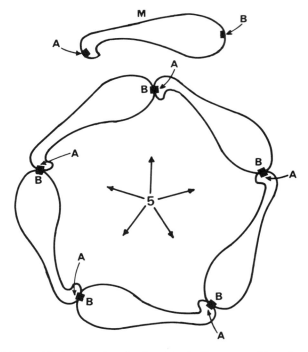

Figure 44–6. Formation of a closed ring by five asymmetric protomers (*M*), in which group B can form a bond with group A. Since the distance between successive A–B bonds and the angle of the A–B bonds to the axis of the monomer are constant, a closed ring having fivefold rotational symmetry around an axis through its center is formed.

For the basic icosahedron with 20 faces, T = 1. The simplest of the more complex viruses are alphaviruses and some plant viruses, for which T = 3; this icosahedron has 60 faces (see Fig. 44–4B). These capsids contain 180 protomers made up of identical polypeptide chains but with three **different configurations** (indicated by 1, 2, and 3 in Fig. 44–4B) that depend on the location of each protomer with respect to the axes of the icosahedrons. In all protomers the polypeptide chains have a tight boxlike structure, but with tails that establish connections between them. In Figure 44–4B all protomers No. 1 form **pentons** around the fivefold axes, whereas Nos. 2 and 3 protomers are regularly arranged in groups of six around the threefold axes (**hexons**). The relationships among protomers in hexons and pentons must be different, but the differences are small and can be taken care of by configurational differences of the same peptide chain (principle of quasi-equivalence).

HIGHER-ORDER CAPSIDS. In more complex viruses the capsids have a higher number of protomers and capsomers. Then the two types of capsomer are made up of different proteins, with arrangements that at times deviate from the basic icosahedral scheme. For instance, in adenoviruses (T = 25; Fig. 44–7), hexons contain three instead of six protomers, still retaining the ability to make contacts each with six other capsomers. This arrangement is compatible with icosahedral symmetry, because the axes going through the hexons have a threefold, not sixfold, symmetry (see Fig. 44–4B). In the construction of these capsids the hexons, which are the most abundant capsomers, often join together to form **complex structural units** with a closely packed hexagonal lattice. These units then assemble into capsids. The configuration of the polypeptide chains must still be different, depending on each hexon in the final assembly, whether at the center of a face of the basic icosahedron, at a corner, or adjacent to a penton.

These complex capsids cannot form by self-assembly. In principle, the hexons are capable of making, by self-assembly, flat sheets or cylinders; cylindrical forms are indeed found in cells infected by some viral mutants that are unable to make a complete capsid (Fig. 44–8). In con-

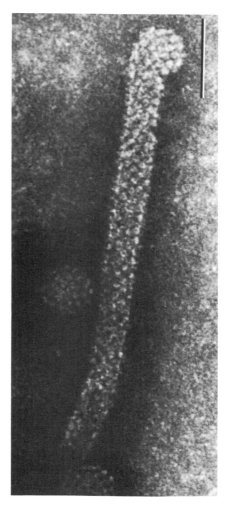

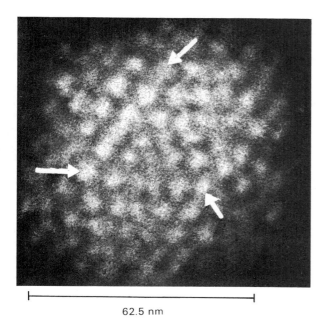

62.5 nm

Figure 44–7. Electron micrograph of GAL virus (chicken adenovirus) by negative staining, showing the capsomer structure. The arrowed capsomers are situated on the fivefold axis. (Wildly P, Watson JD: Cold Spring Harbor Symp Quant Biol 27:25, 1962)

Figure 44–8. A very long filament of human wart virus (papovavirus) hexons, together with a regular virion. The filament, of a diameter close to that of the virions, is made up of hexagonal capsomers. (Noyes WF: Virology 23:65, 1964)

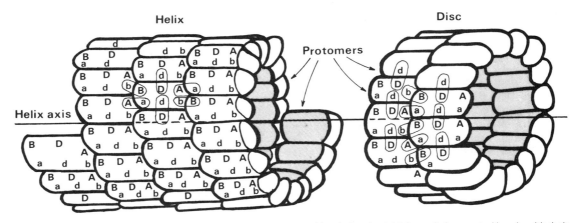

Figure 44–9. Constitution of the helical capsid. All protomers are identical and establish regularly repeated bonds with their neighbors between chemical groups indicated by letters. Since each protomer is staggered with respect to its lateral neighbors, it forms bonds with the two of them on each side along the helix axes. This confers considerable stability on the capsid. The protomers assemble first to constitute flat nonhelical discs containing two rings of protomers, with the staggered arrangement found in the finished helix. Under physiologic conditions, when the protomers of the discs associate with the RNA, they shift slightly to produce a helix. The helix grows in length by the addition and assimilation of discs.

trast, pentons make vertices; in combination with hexons, which make the flat faces, they allow the formation of a closed shell, approximately icosahedral, with the pentons at the axes of fivefold symmetry and the hexons in between. Large viruses, such as phage T4 or adenovirus, use **scaffolding proteins** as a mold for the capsid. The mold determines the dimensions of the capsid and the number of hexons, whereas the number of pentons remains fixed at 12. This principle explains how oblong capsids of variable length can be assembled.

An unusual arrangement is found in the **geminiviruses,** which cause important diseases in plants. The capsid is formed by two incomplete icosahedrons (T = 1) attached to each other, and it contains two different cyclic single-stranded DNAs, which are both required for infectivity.

HELICAL CAPSIDS. The roughly cylindrical helical capsid (Fig. 44–9) has the simplest organization. The identical protomers are bound end-to-end by identical bonds to form a ribbonlike structure, which is wound around the axis of the helix. The protomers in two successive turns of the ribbon each form bonds with two protomers of adjacent turns. This pattern confers great stability on the structure. The straight line in the center of the cylinder is an axis of rotational symmetry.

The diameter of the helical capsid is determined by the characteristics of its protomers, while the length is determined by the length of the nucleic acid it encloses. The tail of some bacteriophages (see below) is also helical but does not contain nucleic acid: its length is determined by the length of a filiform **tape protein** around which it assembles.

The capsids of naked helical viruses (e.g., tobacco mosaic virus) are very tight (see Fig. 44–3*B*). In contrast, the capsids of enveloped helical viruses (such as paramyxoviruses) are very flexible, because the whole capsid—and not only the nucleic acid—has to coil within the envelope. Often the turns of the helices are visible in electron micrographs (Fig. 44–10), suggesting a loose structure; in these viruses the envelope rather than the capsid may provide the required barrier to nucleases.

ORGANIZATION OF THE GENOME WITHIN THE CAPSID. In **icosahedral capsids** the nucleic acid is tightly packed, with attendant important topologic and thermodynamic problems, especially for stiff double-stranded DNA. In conjunction with certain basic proteins or with polyamines, DNA often forms a central **core** of parallel loops; folding is accompanied by some denaturation of the double helix. In some animal viruses, such as polyomaviruses, DNA folds tightly around cellular histones to form a **chromatinlike structure,** much like cellular chromatin in condensed chromosomes. Neither DNA nor RNA has a highly regular arrangement within the capsid, as deduced from absence of an orderly pattern in x-ray crystallography.

The highest level of organization is that of adenovirus DNA, which, in conjunction with virus-specified proteins, forms twelve equal balls, each located under one of the vertices of the icosahedral capsid. This arrangement is secondary to that of the capsid, because the DNA enters a preformed capsid.

In **helical capsids** the RNA is located in a helical groove between the protomers (Fig. 44–11), to which it is attached by multiple weak bonds, irrespective of sequences.

Capsids without nucleic acid—**empty capsids**—are found in most preparations of the icosahedral viruses. Electron microscopy reveals an external capsomeric

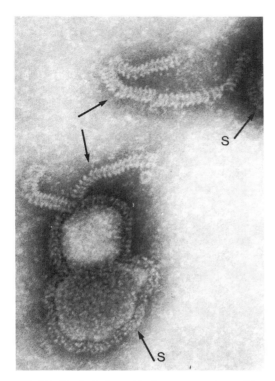

Figure 44–10. The helical capsid of a paramyxovirus with negative staining. Two particles of the simian paramyxovirus SV5 are seen: segments of the helical capsid protrude from both particles (*arrows*), probably owing to rupture of the envelope. Note the loose arrangement of the protomers and the hollow along the axis of the helix. The envelopes are covered by the characteristic spikes (*S*). (Choppin PW, Stockenius W: Virology 23:195, 1964)

Figure 44–11. Drawing of a segment of tobacco mosaic virus showing the helical nucleocapsid. In the upper part of the figure two rows of protein monomers have been removed to reveal the RNA. This drawing is based on results of x-ray diffraction studies. (Klug A, Caspar DLD: Adv Virus Res 7:225, 1960)

configuration similar to that of normal virions (Fig. 44–12). However, these capsids may lack some capsid proteins. Empty capsids are less stable than the capsids of virions, and during preparation of specimens for electron microscopy they disintegrate more readily (see Fig.

44–5). Their existence shows that the nucleic acid is not essential for the assembly of many kinds of capsids.

The Envelope

Envelopes, like cellular membranes, contain lipid bilayers and proteins with special functions. The presence of lipids makes enveloped viruses sensitive to disinfection or damage by lipid solvents, such as ether. The proteins of the envelope are of two kinds: one or more **glycoproteins** and a **matrix** protein.

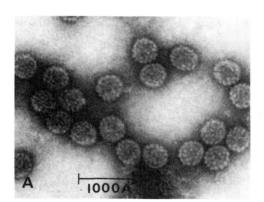

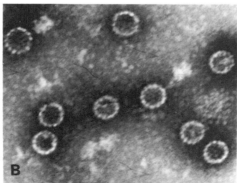

Figure 44–12. Electron micrographs with negative staining of purified polyoma virus, a papovavirus. (*A*) The full virions are not penetrated by the stain and show only the pattern on the surface of the capsid. (*B*) The empty capsids are penetrated by the stain.

GLYCOPROTEINS. Individual glycoprotein molecules are **transmembrane** proteins, with a large external domain and a small cytoplasmic domain. The two domains are connected to each other by a stretch of hydrophobic amino acids (**anchor sequence**), usually close to the carboxyl terminus, which anchors the protein to the lipid bilayer of the membrane. Some glycoproteins (e.g., the influenza hemagglutinins) are anchored at both the carboxyl and the amino ends, forming a loop. In these cases the hydrophobic sequence at the amino end corresponds to the **signal sequence,** which is usually cut off after it has guided the protein through the membrane.

Glycoproteins have oligosaccharide residues, in the form of either N- or O-glycosylation. In many viruses glycoproteins have one or two molecules of a **fatty acid** (palmitate or myristilate), of uncertain function, attached to the carboxyl end in the cytoplasmic domain.

Small groups of glycoproteins (up to four) form the **spikes** or other knoblike structures protruding at the surface of enveloped virions (see Figs. 44–3D and 44–10). Each virion has several hundred spikes; their number generally does not correspond to that of the capsid protomers. Glycoproteins, although functionally essential, may not be required for the formation of the envelope: some paramyxovirus mutants, which lack a glycoprotein, form virions with otherwise normal morphology, though they are noninfectious.

Glycoproteins perform important functions: they cause the attachment of virions to cell surfaces and the penetration of their genomes into cells. The rabies virus glycoprotein may also mediate the neurotoxic effect of the virus by binding to the acetylcholine receptors at the neuromuscular junction. These functions of the glycoproteins in attachment and penetration are reflected in the ability of virions of several viral families to cause **hemagglutination,** that is, the bridging of red blood cells. Some glycoproteins cause the **fusion** of the viral membrane with the cell's membrane; they may also cause **hemolysis.**

With orthomyxoviruses and paramyxoviruses hemagglutination is carried out by **hemagglutinins** that bind to the terminal N-acetylneuraminic (sialic) acid present on oligosaccharides of the cellular surface, which act as **receptors** for the viruses. Fusion is carried out by **fusion proteins** with a hydrophobic amino acid sequence at its amino terminus; this sequence penetrates into the cell's lipid bilayer, destabilizing it and causing it to form a single membrane with the virion's envelope.

Some viruses (e.g., influenza and parainfluenza) also possess glycoproteins with **neuraminidase** activity on their surfaces, which cleaves the terminal sialic acid from cellular oligosaccharides. This function is important when progeny virus is released from the cells, to free it from entrapment by oligosaccharides.

MATRIX PROTEINS (M PROTEINS). Matrix proteins are usually not glycosylated. Some are transmembrane proteins with multiple stretches of hydrophobic amino acids, which traverse the membrane; others are held to the inner side of the membrane by hydrophobic amino acids. These proteins reinforce the envelope, connect the nucleocapsid to the glycoproteins, and perform a crucial function during the formation of the virions (see Chap. 48).

In many enveloped viruses the liquid state of the lipid bilayer, and the absence of connections among the proteins of the envelope, prevent the formation of a rigid structure, leading to pronounced **pleomorphism** of the virions. However, in other viruses a firm connection between envelope and nucleocapsid confers on the virions characteristic shapes. Thus, the alphavirus virions are icosahedral, but, surprisingly, the symmetries of the envelope (T = 4) and that of the capsid (T = 3) do not coincide; the glycoproteins form different contacts with proteins of the capsid. Rhabdoviruses are **bullet shaped,** with the helical nucleocapsids coiled under the outer layer (Fig. 44–13).

The major proteins of the envelope are, like those of the capsid, specified by viral genes. They constitute the major **antigens** of the virions, which are important in the immune response and in virion identification. Some virions contain also, as a minor component, glycoproteins derived from the surface of the cells in which they are produced. This is typical for vesicular stomatitis virions, which contain 10% to 20% of such adventitious protein molecules. The proteins thus incorporated do not represent a random sample of those present at the cell surface; preferentially incorporated are glycoproteins of other enveloped viruses infecting the same cell.

Complex Virions

POXVIRUSES. The poxvirus virions, brick-shaped or ovoid, hold the viral DNA, associated with protein, in a **nucleoid** shaped like a biconcave disc and surrounded by several lipoprotein layers. A layer of coarse fibrils near the outer surface gives the virions a characteristically striated appearance in negatively stained preparations (see Fig. 44–3E; see also Chap. 54).

LARGE BACTERIOPHAGES. Some bacteriophages, such as coliphages T2, T4, and T6 (**T-even coliphages**) have very complex structures (see Fig. 44–3F), including an icosahedral head and a helical tail. They are said to have **binal symmetry** because they have components with different kinds of symmetry within the same virion.

Other Virion Components

In addition to the structural proteins found in the coats or in association with the nucleic acids, some virions contain **enzymes,** which perform functions essential to

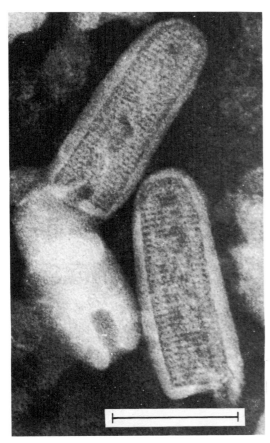

Figure 44–13. Virions of vesicular stomatitis virus (a rhabdovirus) with negative staining. The helical filament, present in a deeper layer, is visible in two particles. (Howatson AF, Whitmore GF: Virology 16:466, 1962)

the initial stages of viral multiplication (Table 44–3). Typical examples are enzymes transcribing the virion RNA into mRNA (in virions with a negative-strand RNA genome), those transcribing double-stranded RNA (in virions with such a genome), enzymes transcribing RNA into DNA (in retroviruses), and those that generate capped leaders for viral messengers by transferring them from cellular messengers (in orthomyxoviruses). Enzymes of these various types are not present in cells; they are specified by viral genes and are brought into the cells by the virions themselves. Virions of some viruses contain a variety of other enzymes, which often duplicate functions performed by the cellular enzymes. Their presence in virions may reflect specific needs of the virus. For instance, the vaccinia virions, which replicate in the cytoplasm, contain enzymes for the transcription of double-stranded DNA that recognize initiation and termination signals specific to vaccinia virus DNA.

DEFECTIVE VIRUSES

Viruses incapable of autonomous replication can arise by mutation of regular viruses; for multiplication they require a wild-type **helper virus,** coinfecting the same cells to provide the defective function. **Satellite viruses** are by nature defective; that is, they do not have a related replication-competent virus. These viruses are present only in cells infected by another unrelated virus, which acts as helper. Among satellite viruses the adeno-associated viruses, with a single-stranded DNA genome, require coinfecting adenovirus as helper. The satellite genome encodes only a few proteins, among which is the protein of its capsid. An important satellite virus associated with hepatitis B virus, the **δ agent** increases the severity of the disease. It has a cyclic RNA genome, resembling that of viroids (see below), and forms particles surrounded by hepatitis B surface antigen.

VIRUS-RELATED AGENTS
Viroids

Viroids, which are responsible for serious diseases of many plants, share with viruses some fundamental properties, such as a simple organization. However they are naked RNA, neither containing nor coding for any protein. Each viroid particle is a **cyclic single-stranded RNA molecule** containing between 250 and 400 nucleotides, depending on the strain. They resist enzymatic destruction because they have no free ends and because they have a very tight secondary structure (owing to self-complementary sequences), which makes them resemble small, compact rods.

All viroid strains, in spite of the different lengths of the RNAs, have similar characteristics. Their genome can be considered a double-stranded RNA, with many unpaired short "bubbles." The most striking feature is the lack of initiation codons for protein synthesis (AUG), or of their complements (in case the RNA is of the negative-strand type); there is no evidence that these RNAs are translated. They are replicated in the nucleus of infected cells by host enzymes through oligomeric double-stranded intermediates. Replication is blocked by alpha-amanitin, which inhibits RNA polymerase II, the enzyme that generates the transcripts of cellular genes destined to become mRNAs.

The base sequences of viroids have repeats, both direct and inverted, which suggest a relatedness to transposing elements (see Chap. 8). Moreover, they possess a sequence similar to that used by retroviruses—which are also closely related to transposing elements (see Chap. 65)—for initiating reverse transcription. Unlike retroviruses, however, viroids are not transcribed into DNA, and no sequences homologous to viroids are found in the DNA of the infected cells. DNA complementary to

TABLE 44–3. Characteristics of Virion Enzymes

Enzyme	Virus	Product of Function
ENZYMES AFFECTING INTERACTION OF VIRIONS WITH THE HOST CELL SURFACE		
Neuraminidase	Orthomyxovirus, paramyxovirus	Splits off NANA from surface polysaccharides
Endoglycosidase	*E. coli* K bacteriophages	Breaks down surface polysaccharides
Fusion factor*	Paramyxovirus	Alters lipid bilayer
ENZYMES TRANSCRIBING THE VIRAL GENOME INTO MESSENGER RNA		
DNA-dependent RNA polymerase	Poxvirus, polyhedrosis viruses of frogs, bacteriophages N4, SP02	Single-stranded mRNA
Double-stranded RNA transcriptase	Viruses with double-stranded RNA	Single-stranded mRNA
Single-stranded RNA transcriptase	Viruses with single-stranded RNA (negative strand)	Single-stranded mRNA (positive strand)
ENZYMES ADDING SPECIFIC TERMINAL GROUPS TO VIRAL mRNA MADE IN VIRIONS		
Nucleotide phosphohydrolase	Viruses synthesizing mRNA in virions (e.g., poxviruses, reoviruses)	Converts terminal 5'-triphosphate to diphosphate as prelude to guanylylation
Guanylyl transferase	Viruses synthesizing mRNA in virions (e.g., poxviruses, reoviruses)	Adds guanylyl residue to 5'-end diphosphate in mRNA
RNA methylases	Viruses synthesizing mRNA in virions (e.g., poxviruses, reoviruses)	Methylate guanylyl residue at 5'-end of mRNA and some riboses in 2' position
Poly(A) polymerase	Viruses synthesizing mRNA in virions (e.g., poxviruses, reoviruses)	Synthesizes poly(A) tail at 3' end of mRNA
ENZYMES INVOLVED IN COPYING VIRION RNA INTO DNA		
RNA-dependent DNA polymerase (reverse transcriptase)	Retroviruses	DNA–RNA hybrids; double-stranded DNA
RNase H (in association with reverse transcriptase)	Retroviruses	Breaks down RNA strand in RNA–DNA hybrids
Polynucleotide ligase	Retroviruses	Closes single-strand breaks in double-stranded DNA
ENZYMES FOR NUCLEIC ACID REPLICATION OR PROCESSING		
DNA-dependent DNA polymerase	Hepatitis B	Synthesizes double-stranded DNA
Deoxyribonucleases (exo- and endo-)	Poxviruses, retrovirus	Break DNA chains and crosslinks
Endoribonuclease	Viruses with single-stranded mRNA (e.g., poxvirus)	Processing of mRNA
OTHER ENZYMES		
Protein kinases	Hepatitis B	Phosphorylate proteins

NANA, N-acetylneuraminic acid

* No enzymatic activity known

viroid RNA is also infectious, and in cells it is transcribed into regular infectious viroid molecules.

A striking feature of viroid RNA is the presence of sequences highly homologous to some of the small nuclear RNAs, U_1 and U_3, which are involved in the splicing of introns in animal cells, and presumably also in plant cells. This finding suggests that viroids originated from introns and that their pathogenicity might be due to interference with the normal splicing of introns in the cell. Related to viroids are the **virusoids,** which are satellites of certain plant viruses and are encapsidated with their helper RNAs in the virions. A candidate for a viroidlike organism in humans is the δ agent (see Defective Viruses, above), which is much larger (1678 base pairs) and surrounded by a coat.

Agents of Slow Infections

The etiology of several transmissible slow diseases of humans (such as Creutzfeldt-Jakob disease and Kuru) or animals (scrapie) has defied characterization. The agents are like viruses in size and infectivity; no virus, however, has been isolated from the infected tissues. These contain a characteristic protein, which is also present in normal tissues. It has been suggested that the agent is of a novel type, not containing nucleic acid (**prion**); the nature of the agent, however, remains obscure (see Chap. 51).

Assay of Viruses

The methods used for the assay of viruses reflect their dual nature as both complex chemicals and living microorganisms. Viruses can be assayed by chemical and physical methods, by immunologic techniques, or by the consequences of their interaction with living host cells, i.e., their **infectivity.** Assays carried out by different techniques can differ vastly in their significance.

CHEMICAL AND PHYSICAL DETERMINATIONS
Counts of Physical Particles

Virions can be clearly recognized in the electron microscope; if a sample contains only virions of a single type, their number can be determined unambiguously. Virions are counted by mixing the sample with a known number of polystyrene latex particles, viewing droplets of the mixture in the electron microscope, and counting the two types of particles present in the same droplet. Simple arithmetic then yields the number of virions in the total sample (Fig. 44–14). This technique does not distinguish between infectious and noninfectious particles.

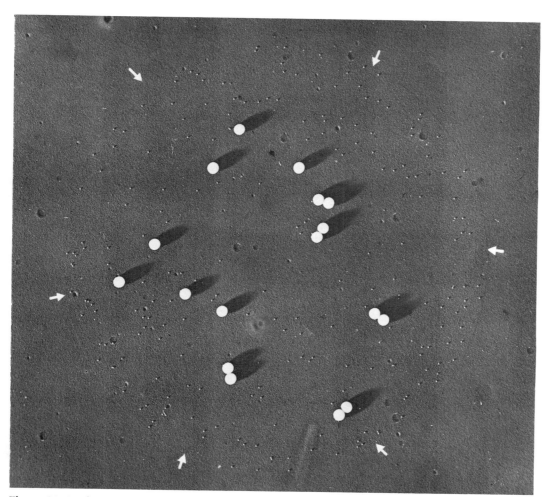

Figure 44–14. Counting of particles of poliovirus (a picornavirus) mixed with polystyrene latex particles. The mixture was sprayed in droplets on the supporting membrane, dried, and shadowed. The micrograph shows a droplet, the outline of which is partly visible (*arrows*). The small particles are virus, the large ones latex. There are 220 viral and 17 latex particles in the droplet. Since the latex concentration in the sample was 3.2×10^{10} particles/ml, the concentration of viral particles is $220/17 \times 3.2 \times 10^{10} = 4.1 \times 10^{11}$/ml. The precision of the assay based on this one droplet is only about ±50% (see Appendix), owing to the small number of latex particles counted. To obtain a greater precision, pooled counts from many similar drops would have to be used. (Courtesy of the Virus Laboratory, University of California, Berkeley)

Hemagglutination

Many viruses, both small and large, can agglutinate red blood cells (RBCs). This important property, discovered independently for influenza virus by Hirst and by McClelland and Hare in 1941, affords a simple, rapid method for viral titration. Hemagglutination is usually caused by the virions themselves; in some cases, however, as with poxviruses, it is caused by lipid hemagglutinins produced during viral multiplication.

Although the spectrum of red cell species that are agglutinated and the required conditions differ for different viruses, the phenomenon is basically similar in all cases: a virion or a hemagglutinin attaches simultaneously to two RBCs and bridges them, and at sufficiently high viral concentrations multiple bridging yields large **aggregates.**

HEMAGGLUTINATION ASSAY. The formation of aggregates can be detected in a number of ways. In the simplest, the **pattern method** (Fig. 44–15), the suspension of RBCs and virus is left undisturbed in small wells in a plastic plate for several hours. Nonaggregated cells sediment to the round bottom of the well and then roll toward the center, where they form a small, sharply outlined, round pellet. Aggregates, however, sediment to the bottom but do not roll; they form a thin film, which has a characteristic serrated edge. This method is used for **endpoint assays.** Serial twofold dilutions of the virus sample are each mixed with a standard suspension of RBCs (usually 10^7/ml). The last dilution showing complete hemagglutination is taken as the endpoint. This titer has an inherent imprecision at least as large as the dilution step used (usually twofold).

A more refined method is to determine the proportion of aggregated cells by observing their sedimentation in a photoelectric colorimeter, since aggregated RBCs sediment faster than nonaggregated ones and can be measured separately. The titer obtained either way is expressed in **hemagglutinating units.** The photometric assay is the more sensitive and permits the demonstration that a single influenza virus particle agglutinates two RBCs.

The presence of neuraminidases on the surface of orthomyxovirus and paramyxovirus virions affects the course of hemagglutination. At 37°C the viral neuraminidase ultimately dissociates the viruses from RBCs by splitting N-acetyl neuraminic acid from receptors; the virus then spontaneously elutes off the RBCs, which disaggregate. At 0°C, in contrast, the enzyme is much less active, and the virion–RBC union is stable. After the virus has eluted, cells cannot be agglutinated again by a new batch of virus, since they have lost the receptors; these cells are said to be **stabilized.** (The eluted virus, on the contrary, retains all its activities.)

However, cells stabilized by a given orthomyxovirus or paramyxovirus can sometimes be agglutinated by another virus of these families. Indeed, it is possible to arrange the viruses in a series (called a **receptor gradient**) such that any virus will exhaust the receptors for itself and the viruses preceding it in the gradient, but not for those that follow it. This result indicates that viruses differ in the precise specificity of their neuraminidases.

Heating the virus inactivates its neuraminidase with-

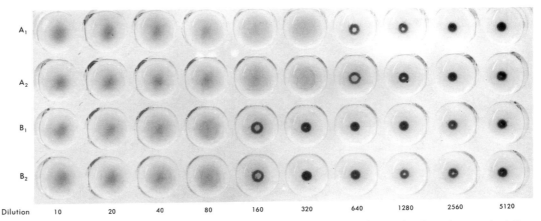

Figure 44–15. Results of a hemagglutination assay by the pattern method with influenza virus (an orthomyxovirus). Two samples, A and B, were diluted serially in steps: (1) 0.5 ml of each dilution was mixed with an equal volume of a red cell suspension, and (2) each mixture was placed in a cup drilled into a clear plastic plate and left for 30 minutes at room temperature. Each assay was made in duplicate. Sample A causes complete hemagglutination until dilution 320; sample B until dilution 80. In either case the subsequent dilution still shows partial hemagglutination. The hemagglutinating titer of A is 320; that of B, 80.

out destroying hemagglutinating activity. This **indicator** virus is useful for studying the union with receptors and mucoproteins without the complication of their enzymatic inactivation.

Assays Based on Antigenic Properties

Complement fixation (CF) or precipitation with antiserum can be used to measure amounts of virus. These methods have relatively low sensitivity and are used only for special purposes (see Chap. 50).

ASSAYS OF INFECTIVITY
Plaque Method

The plaque method is the fundamental assay method in virologic research, and it is also of great value in diagnosis: it combines simplicity, accuracy, and high reproducibility. First used with bacteriophages, this method was a key factor in the spectacular advances of research on phage and later also on animal viruses.

Bacteriophages are assayed in the following way. A phage-containing sample is mixed with a drop of a dense liquid culture of suitable bacteria and a few milliliters of melted soft agar at 44°C; the mixture is then poured over the surface of a plate (Petri dish) containing a layer of hard nutrient agar. The soft agar spreads in a thin layer and sets, and the bacteriophages diffuse through it until each meets and infects a bacterium. After 20 to 30 minutes the bacterium lyses, releasing several hundred progeny virions. These, in turn, infect neighboring bacteria, which again lyse and release new virus. The uninfected bacteria, in the meantime, grow to form a dense, opaque lawn. After a day's incubation the lysed areas stand out as transparent **plaques** against the dense background (Fig. 44–16A). The soft agar permits diffusion of phage to nearby cells but prevents convection to other regions of the plate; hence, secondary centers of infection cannot form.

With **animal viruses** a similar method is possible, the bacteria being replaced by a monolayer of cells growing on a solid support (see Chap. 47), and the nutrient medium is replaced by a solution containing the viral sample. Within an hour or so most of the virions attach to cells. Soft nutrient agar or some other gelling mixture is

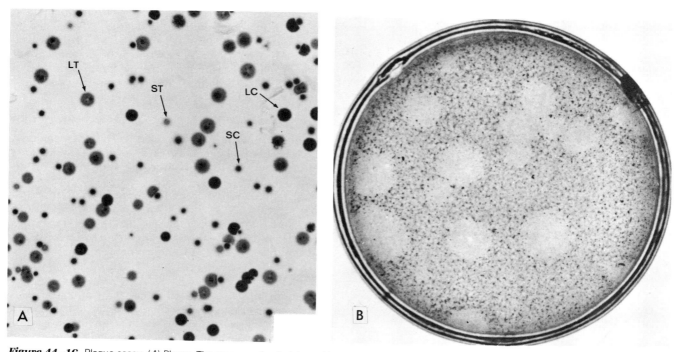

Figure 44–16. Plaque assay. (*A*) Phage. The progeny of cells infected by two phage types was diluted by a factor of 10^7; 0.1 ml of the diluted virus was assayed. The plate was counted 18 hours after plating. Four different plaque types, differing in plaque size and turbidity—large clear (*LC*), large turbid (*LT*), small clear (*SC*), and small turbid (*ST*)—can be distinguished, showing the great utility of plaque formation for genetic work with bacteriophages. Part of the plate is reproduced; a total of 407 plaques could be counted on the whole plate. The titer is 4.07×10^{10}/ml in the undiluted sample. The accuracy is ±10%. (*B*) Poliovirus (picornavirus). A sample of poliovirus type 1 was diluted by a factor of 2×10^5, and 0.1 ml was assayed on a monolayer culture of rhesus monkey kidney cells, with an agar overlay containing neutral red. The culture was incubated for 3 days at 37°C in an atmosphere containing 7% CO_2, which constitutes a buffer with the bicarbonate present in the overlay. Some of the plaques show partial confluence, but they can still be identified as separate plaques; 17 plaques can be counted on the photograph. The corresponding titer is 3.4×10^7/ml in the undiluted sample. The accuracy is ±50%.

poured over the cell layer. Plaques develop after 1 day to 3 weeks of incubation, depending on the virus (see Fig. 44–16*B*).

Plaques are detected in a variety of ways.

1. The virus often kills the infected cells, i.e., produces a **cytopathic effect**; the plaques are then detected by staining the cell layer with a dye that stains only the live cells (e.g., neutral red) or only the dead cells (trypan blue).

2. With certain viruses the cells in the plaques are not killed but acquire the ability to adsorb RBCs (see Maturation and Release of Animal Viruses—Enveloped Viruses, Chap. 48). The plaques are revealed by **hemadsorption,** i.e., by flooding the cell layer with a suspension of RBCs, then washing out those not attached to infected cells.

3. The infected cells may fuse with neighboring uninfected cells to form **polykaryocytes** (i.e., multinucleated cells), which are microscopically detectable (**syncytial plaques).**

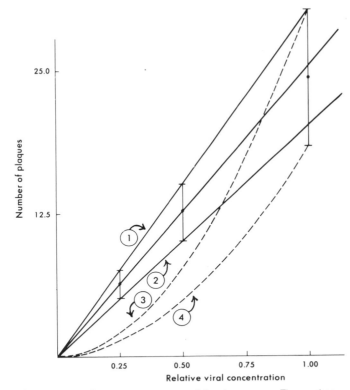

Figure 44–17. Dose–response curve of the plaque assay. The number of plaques produced by a sample of poliovirus type 1 at various concentrations was plotted versus the relative concentration of the virus; the accuracy of the assay (±2σ) is indicated for each point. The data are in agreement with a linear dose–response curve that falls between lines 1 and 2, and therefore with the notion that a single particle is sufficient to give rise to a plaque. Curves 3 and 4 give the range of data that would be obtained if at least two viral particles were required to initiate a plaque; the deviation is such that the hypothesis is ruled out (see Appendix).

4. Often the cells of the plaques contain large amounts of viral Ags, which can be detected by **immunofluorescence.**

The titer of the viral preparation is directly calculated from the number of plaques and the dilution of the sample, as shown in Figure 44–16*B*. As discussed in the appendix to this chapter, the accuracy of the assay depends on the number of plaques counted. An assay estimated from n plaques will be within $2/\sqrt{n}$ of the true value (e.g., with 100 plaques the range will be ±20%).

THE DOSE–RESPONSE CURVE OF THE PLAQUE ASSAY. The number of plaques in plates infected with different dilutions of the same viral sample is proportional to the concentration of the virus; i.e., the dose–response curve is linear (Fig. 44–17). A single virion is therefore sufficient to infect a cell. It follows that the viral population contained in a plaque is the progeny of a single virion, i.e., a clone, representing a genetically pure line. Plaques also provide useful genetic markers through their visible characteristics such as size, shape, and turbidity.

Pock Counting

When the chorionic epithelium of the chorioallantoic membrane of a chick embryo (see Fig. 48–1) is infected by vaccinia or herpes simplex virus, characteristic lesions (**pocks**) appear. They may be white or hemorrhagic; viral mutants may be distinguishable by the appearance of the pocks. This method, important initially, is now largely superseded by the plaque method.

Other Local Lesions

Tumor-producing viruses, such as the Rous sarcoma virus (see Chap. 65), can be assayed on monolayer cell cultures; they produce **proliferative foci,** each initiated by a single viral particle.

Many **plant viruses** can be titrated by counting the lesions produced on leaves rubbed with a mixture of virus and an abrasive material. The virus penetrates through ruptures of the cell walls caused by the abrasive, and the progeny spread to neighboring cells.

Endpoint Method

The endpoint method, used for assaying animal viruses before the advent of the plaque method, is still employed for certain diagnostic assays and for quantitating virulence or host resistance. The virus is serially diluted, and a constant volume of each dilution is inoculated into a number of similar **test units,** such as mice, chick embryos, or cell cultures. At each dilution the proportion of infected test units (**infectivity ratio**) is scored: for example, by (1) death or disease of an animal or embryo, (2) degeneration of a tissue culture, or (3) recognition of progeny virus *in vitro* (e.g., by hemagglutination).

The lower dilutions of the virus infect most of the test

units, and the highest dilutions infect none. A rough idea of the viral titer is given by the intermediate dilutions that produce signs of infection in only a fraction of the test units. The transition is not sharp, however, and only by combining the data from several dilutions is it possible to calculate the precise endpoint at which 50% of the test units are infected. At this dilution each sample contains on the average one **ID$_{50}$**, i.e., one **infectious dose for 50**% of the test units. One ID$_{50}$ can be shown mathematically to correspond to 0.7 plaque-forming units (see Appendix).

The interpolation to obtain the ID$_{50}$ can be carried out in a variety of ways. The method of Reed and Muench, though not mathematically derived, yields results in fair agreement with more rigorous methods. In this method (see Appendix) the dilution containing one ID$_{50}$ is obtained by interpolation between the two dilutions that straddle the 50% value of the infectivity ratio. The interpolation assumes that in the proximity of the ID$_{50}$ the infectivity ratio varies linearly with the log dilution. Usually the accuracy of the method is low, since the number of test units used at each dilution is small. When, for instance, six test units are employed at each tenfold dilution, as is common in diagnostic titrations, the titration is useful only to ascertain large differences in viral titer (50-fold or more) between two samples, which is adequate for many routine diagnostic procedures.

Viral titers obtained by the endpoint method are expressed in various equivalents of the ID$_{50}$: LD$_{50}$ (lethal dose) if the criterion is death; PD$_{50}$ (paralysis dose) if the criterion is paralysis; TC$_{50}$ (tissue culture dose) if the criterion is degeneration of a culture.

COMPARISON OF DIFFERENT TYPES OF ASSAYS

The focal assay methods (plaques, foci, and pocks) are most satisfactory for their high efficiency combined with simplicity, reproducibility, and economy. For example, to match the precision obtained by counting 100 plaques on a single culture, on would require more than 100 test units per decimal dilution in an endpoint titration. The precision of any type of assay is adversely affected by a variability in the response of the cells or organisms used in the assay; the variabilities can be very large in the pock assay and even larger in endpoint assays using animals.

The various methods of assay have different sensitivities and measure different properties. Assays based on infectivity are as much as a millionfold more sensitive than those based on chemical and physical properties. Chemical and physical techniques, moreover, titrate not only infectious but also noninfectious virions (empty capsids, particles with a damaged nucleic acid). These methods can therefore be useful for studies requiring measurement of the total number of viral particles. Hemagglutination or immunologic methods can also ti-

TABLE 44–4. Ratio of Viral Particles to Infectious Units

Virus	Ratio
ANIMAL VIRUSES	
Picornaviruses	
Poliovirus	30–1000
Foot-and-mouth disease virus	33–1600
Papovaviruses	
Polyoma virus	38–50
SV40	100–200
Papilloma virus	~10^4
Reoviruses	10
Alphaviruses	
Semliki Forest virus	1
Orthomyxoviruses	
Influenza virus	7–10
Herpesviruses	
Herpes simplex virus	10
Poxviruses	1–100
Adenoviruses	10–20
BACTERIAL VIRUSES	
Coliphage T4	1
Coliphage T7	1.5–4
PLANT VIRUSES	
Tobacco mosaic virus	5×10^4–10^6

trate soluble components, obtained from breakdown of the virions or produced during intracellular viral synthesis.

The **ratio of the number of viral particles** (determined by electron microscopy) **to the number of infectious units** measures the **efficiency of infection,** which varies widely among different viruses, and even for the same virus assayed in different hosts. As is shown in Table 44–4, for most viruses the ratio is larger than unity. This result is due in part to the presence of noninfectious particles and in part to the failure of potentially infectious particles to reproduce. However, even with the highest ratio of particles to infectivity, infection is initiated by a single virion, since the dose–response curve remains linear. The ratio of total viral particles to hemagglutinating units is very high: about $10^{6.3}$ for influenza virus and 10^5 for polyoma virus.

Appendix

QUANTITATIVE ASPECTS OF INFECTION
Distribution of Viral Particles per Cell: Poisson Distribution

In a cell suspension mixed with a viral sample, individual cells are infected by different numbers of viral particles, and it is often important to know the distribution, i.e., the proportions of cells infected by zero, one, two, etc., viral particles.

These proportions depend on the **average number of viral particles per cell,** known as the **multiplicity of infection** (m). The relevant viral particles are those that initiate infection of a cell; inactive particles or particles that, for whatever reason, never enter a cell are neglected. Hence, m is related to the total number of viral particles (N) and of cells (C) by the relation $m = aN/C$, where a is the proportion of viral particles that initiates infection.

The proportion P(k) of cells infected by k viral particles is given by the *Poisson distribution*, assuming that the cells are all identical in their ability to be infected. In fact, cells vary in size, surface properties, and so forth, but usually the deviations are small enough to be negligible, at least as a first approximation.

According to the Poisson distribution:

$$P(k) = \frac{e^{-m}m^k}{k!} \qquad (1)$$

The value of m can be derived from the known values of N and C if a can be determined; otherwise m can be calculated from the experimentally determinable proportion of uninfected cells, P(0). By making $k = 0$ in equation 1,

$$P(0) = e^{-m}, \text{ and} \qquad (2)$$

$$m = -\ln P(0) \qquad (3)$$

where ln stands for the natural logarithm.

The use of equations 1, 2, and 3 will now be illustrated with reference to two practical problems.

PROBLEM 1. 10^7 cells are exposed to virus. At the end of the adsorption period there are 10^5 infected cells. What is the multiplicity of infection?

$$P(0) = 0.99, m = -\ln(0.99) = 0.01$$

This problem emphasizes the point that the multiplicity of infection can assume any value from 0 to ∞. Values smaller than unity indicate that a small fraction of the cells is infected, mostly by single viral particles.

PROBLEM 2. What is the multiplicity of infection required for infecting 95% of the cells of a population?

$$P(0) = 5\% = 0.05, m = -\ln(0.05) = 3$$

The point of this problem is that even at very high multiplicities a certain proportion of the cell remains uninfected. The multiplicity of infection required to reduce the proportion of uninfected cells to a certain value can be calculated from equation 3.

Classes of Cells in an Infected Population

It is usually important to determine the proportion of only three classes of cells: **uninfected cells** (k = 0); **cells**

with a **single infection** (k = 1); and **cells with a multiple infection** (k > 1). The proportions are:

Uninfected cells: $P(0) = e^{-m}$
Cells with single infection: $P(1) = me^{-m}$
Cells with multiple infection: $P(>1) = 1 - e^{-m}(m + 1)$*

PROBLEM 3. How do we determine the various classes of infected cells if the multiplicity of infection is 10?

$$P(0) = e^{-10} = 4.5 \times 10^{-5}$$

$$P(1) = 10 \times 4.5 \times 10^{-5} = 4.5 \times 10^{-4}$$

$$P(>1) = 1 - (4.5 \times 10^{-5})(10 + 1) = 1 - (4.95 \times 10^{-4}) = 99.95\%$$

With a population of 10^7 cells there are $4.5 \times 10^{-5} \times 10^7 = 450$ uninfected cells and 4500 cells with single infection; all the others have multiple infection.

PROBLEM 4. How do we determine the composition of the population of infected cells if the multiplicity of infection is 10^{-3}, or 0.001?

$$P(0) = e^{0.001} = 0.9990 = 9.99 \times 10^{-1} = 99.9\%$$

$$P(1) = 0.001 \times e^{0.001} = 10^{-3} \times 9.99 \times 10^{-1} = 9.99 \times 10^{-4} = 0.0999\%$$

$$P(>1) = 1 - 0.9990(0.001 + 1) = 0.000001 = 10^{-6}$$

With a population of 10^7 cells there are $9.99 \times 10^{-4} \times 10^7 = 9900$ cells with single infection and $10^{-6} \times 10^7 = 10$ cells with multiple infection; most of the cells are uninfected.

MEASUREMENT OF THE INFECTIOUS TITER OF A VIRAL SAMPLE

To measure infectious titer, a viral sample containing an unknown number (N) of infectious viral particles is mixed with a known number (C) of cells. N is then calculated from the proportion of cells that remains uninfected according to equation 3: $m = -\ln P(0)$; since, as defined above, $m = a(N/C)$, $N = mC/a = -C \ln P(0)/a$, or

$$aN = -C \ln P(0) \qquad (4)$$

Usually the factor a is not determinable, and therefore the number (N) of infectious viral particles present in the sample to be assayed cannot be calculated. In its place one obtains the product aN, the number of **infectious units.**

This is the basis for all measurements of the infectious viral titer. Its **application** is different in the plaque method and in the endpoint method.

---

* This value is obtained by subtracting from unity (the sum of all probabilities for any value of k) the probabilities P(0) and P(1).

Plaque Method

In the plaque method the number of plaques equals the number of infectious units. The actual number of cells employed in the assay is irrelevant, provided that it is in large excess over the number of infectious viral particles, so that m is very small; uncertainties connected with the counting of the cells are therefore eliminated.

THE DOSE–RESPONSE CURVE OF THE PLAQUE ASSAY. As stated above, the number of plaques that develop on a series of cell cultures infected with different dilutions of the same viral sample is proportional to the concentration of the virus. We shall now show that this linearity proves that a single infectious viral particle is sufficient to infect a cell (**single-hit kinetics**).

Let us assume that more than one particle, say two particles, is required. There would then be two types of uninfected cells: those with no infectious viral particles and those with just one such particle. According to the Poisson distribution the proportions of cells in these two classes are e^{-m} and me^{-m}, respectively. Thus, under the foregoing assumption, $P(0) = e^{-m}(1 + m)$, which, for very small values of m, approximates to $P(0) = 1 - 1/2m^2$. Therefore, $P(i) = 1/2m^2$, and the dose–response curve would be parabolic rather than linear (see Fig. 44–17). If more than two particles were required to infect a cell, the curvature of the dose–response curve would be even more pronounced.

Endpoint Method

In the endpoint method the virus to be assayed is added to a number of test units (e.g., cultures or animals), each consisting of a large number of cells. A test unit is now equivalent to a single cell of the plaque assay. Therefore, m is the multiplicity of infection of a test unit, rather than of a cell.

The virus titer can be calculated from the proportion of noninfected units, $P(0)$, at the endpoint dilution, according to equation 3: $m = -\ln P(0)$. If at the endpoint $m = 1$ (i.e., there is one infectious unit per test unit), then on the average $P(0) = 0.37$.

Another approach, especially useful in quantitating virulence or host resistance, is the **method of Reed and Muench,** which is applicable to an assay involving a series of progressive dilutions of a virus. A constant volume of each dilution is inoculated in each animal of a group. An empirical pooling of the results obtained at all dilutions gives the dose at which 50% of the animals are infected (ID_{50}) or killed (LD_{50}).

In the example of Table 44–5, none of the dilutions gives a 50% endpoint; this lies between the third and the fourth dilution. The LD_{50} is calculated from the cumulated values, assuming that the proportion of the animals affected varies linearly with $\log_{10}$ dilution. The interpolated value is given by

$$h \frac{\text{\% animals affected at dilution next above 50\%} - 50\%}{\text{\% animals affected at dilution next above 50\%} - \text{\% animals affected at dilution below 50\%}}$$

In this formula h is the log of a dilution step. The interpolated value is then added arithmetically (i.e., with the proper sign) to the log of the total dilution at the step just above 50% affected animals. In the example, interpolated

$$\text{value} = h \frac{71 - 50}{71 - 13} = h \frac{21}{58} = 0.36h \text{ (approximated to 0.4h)}.$$

If $h = 1/10$, $\log h = -1$; total dilution at the third step is $(1/10)^3 = 10^{-3}$, and log dilution $= -3$. Interpolated value then is $-1 \times 0.4 = -0.4$. The log $LD^{50} = -3 + (-0.4) = -3.4$. The LD_{50} titer is expressed as $= 10^{-3.4}$; i.e., the virus sample contains $10^{3.4}$ LD_{50} doses. If, instead, $h = 1/2$, $\log h = -0.3$, total dilution at the third step $= (1/2)^3 = 1/8$, and log dilution $= -0.9$. Interpolated value $= -0.3 \times 0.4 = -0.12$; log $LD^{50} = -0.9 + (-0.12) = -1.02$, and LD_{50} titer $= 10^{-1.2}$.

TABLE 44–5. Example of Endpoint Titration

Dilution Step	Mortality Ratio	Died	Survived	Total Dead*	Total Survived*	Mortality Ratio	Mortality %
1	6/6	6	0	17	0	17/17	100
2	6/6	6	0	11	0	11/11	100
3	4/6	4	2	5	2	5/7	71
4	1/6	1	5	1	7	1/8	13
5	0/6	0	6	0	13	0/13	0

* Cumulated values for the total number of animals that died or survived are obtained by adding in the directions indicated by the arrows.
 (Modified from Lennette EH: General principles underlying laboratory diagnosis of virus and rickettsia infections. In Lennette EH, Schmidt NH [eds]: Diagnostic Procedures of Virus and Rickettsia Disease, p 45. New York, American Public Health Association, 1964)

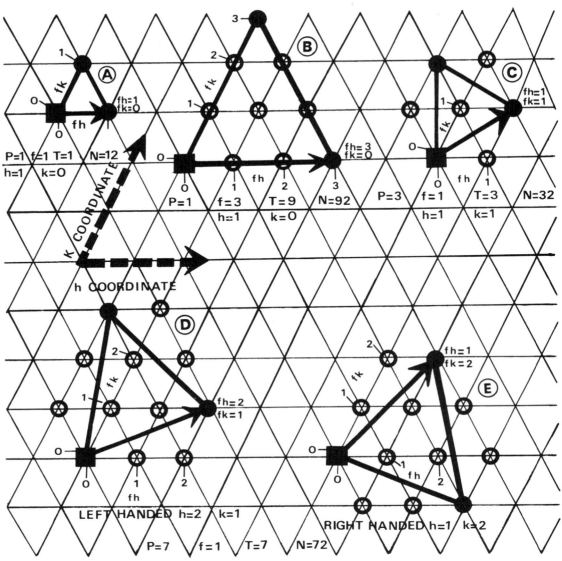

Figure 44–18. This figure shows how the triangular faces of the basic icosahedron are subdivided into smaller triangles, giving rise to the derived faces of complex icosahedrons. The triangles formed by thin lines outline the derived faces, whereas thick lines outline the faces of the basic icosahedron. The pentons (*closed circles*) are located at the corners of the basic faces, the hexons (*open circles*) at the corners of the derived faces. (*A*) The face of the basic icosahedron (T = 1) in which the derived and basic faces are identical. (*B*) The basic face is subdivided into nine derived faces: T = 9. The derived faces are all enclosed in a single basic face, using the same edges. (*C*) The basic face is subdivided into three derived faces: T = 3. The edges of basic and derived faces do not coincide: the basic face in fact contains six derived half-faces (see also Fig. 44–4). (*D* and *E*) T = 7. As in *C*, the edges do not coincide; moreover, the basic face can be subdivided into seven derived faces in two ways, generating a **left-handed** or a **right-handed** icosahedron. The relationship between derived and basic faces is in all cases given by the equation T = (fh)² + (fh)(fk) + (fk)² (see text), where fh and fk are the coordinates along the triangulation lines for a given value of T, measured from the full square. The parameter *P* (from T = Pf²) determines the relationship of the derived to the basic edges, which overlap for P = 1 (*A* and *B*). Such capsids are **symmetric,** as are capsids for which h = k (*C*); with other values of *P* capsids are **asymmetric** (*D*, *E*).

The multiplicity corresponding to one LD_{50} (50% survival) is calculated from the relation $e^{-LD50} = 0.50$. Therefore, $LD_{50} = -\ln(0.50) = 0.70$. One LD_{50} corresponds to 0.70 infectious units.

PRECISION OF VARIOUS ASSAY PROCEDURES
Plaque Method

The statistical precision is measured by the **standard deviation** (σ) of the Poisson distribution, which is equal to the square root of the number of plaques counted. If the number of plaques counted is not too small, 95% of all observations made should fall within two standard deviations from the mean in either direction (i.e., $\pm 2\sigma$). Thus, 4σ is the expected range of variability of the assay. If n replicate assays are made, $\sigma = \sqrt{\bar{x}}/n$, where $\bar{x}$ is the mean value of plaque numbers in the replicate assays. The standard deviation relative to the mean (**coefficient of variation**) serves as a relative measure of precision: $\sigma/\sqrt{\bar{x}}$. This is $\sqrt{\bar{x}}/\bar{x} = 1/\sqrt{\bar{x}}$ for a single assay, and $1/\sqrt{n\bar{x}}$ for n replicate assays. The smaller the coefficient of variation, the higher the precision, which therefore increases as the square root of the number of plaques.

Example: If a total of 100 plaques are counted, the standard deviation is 10. If the same assay is repeated many times, its results will fall between 80 and 120 plaques in 95% of the cases; the coefficient of variation is 1/10. If 400 plaques are counted, the coefficient of variation is 1/20.

Reed and Muench Method

An approximate value, empirically derived, of the standard deviation of the titer determined by this method is $\sigma = \sqrt{0.79hR/U}$, where h is the log of the dilution factor employed at each step of the serial dilution of the virus, U is the number of test units used at each dilution, and R is the interquartile range, namely, the difference between the log of the dilution at which $P(i)$ is 0.25 and 0.75, respectively. In this calculation σ is expressed in logarithmic units. For the data of Table 44–5, with six assay units (animals) at each dilution, $h = 1.0$ and $R = 1.0$ (both in $\log_{10}$ units); $\sigma = \sqrt{0.79/6} = 0.36$ (in $\log_{10}$ units). The range of variation of the LD_{50} is therefore ± 0.72 in $\log_{10}$ units, and the highest expected value (within the 95% confidence limits) is 28 times (antilog of 1.44) the lowest value.

NUMBER OF CAPSOMERS IN ICOSAHEDRAL CAPSIDS

The number of capsomers that can exist in an icosahedral capsid is $10T + 2$. In fact, of the 60T subunits, 60 are in 12 pentons, $60(T - 1)$ in $10(T - 1)$ hexons, giving a total of $12 + 10(T - 1) = 10T + 2$. For describing all possible

TABLE 44–6. Value of Capsid Parameters and Numbers of Capsomers Found in Icosahedral Viruses

p*	f*	T*	No. of Capsomers	No. of Hexons
1	1	1	12	0
	3	9	92	80
	4	16	162	150
	5	25	252	240
3	1	3	32	20
	7	147	1472	1460
7	1	7	72	60

* For explanation of p, f, and T, see text.

arrangements of capsomers it is useful to represent the surface of the icosahedrons in a sheet covered with a grid of triangles, establishing two coordinates, h and k (Fig. 44–18). Each triangle is equivalent to one of the triangles inscribed in each face of the basic icosahedrons, according to the triangulation number. It is possible to define the relationship of the inscribed triangle to the face by outlining the face of the icosahedrons along the grid. In *A* there is only one triangle per face; that is, we deal with the basic icosahedron, $T = 1$. In *B* the side of the face reaches the third intersection, so that nine triangles are inscribed in the face; hence, $T = 9$. In these cases the sides of the face follow the grid, and the icosahedron is said to be **symmetric** with respect to the coordinates. In other cases (*C, D, E*) the sides of the face do not follow the grid: the icosahedron is **asymmetric** and can occur in either a left-handed (*D*) or right-handed (*E*) form.

With this system of coordinates, $T = Pf^2$, where f can be any integer, and P is $h^2 + hk + k^2$, where h and k are any two integers without common factors. The product Pf^2 measures the surface of the icosahedral face in units equal to the surface of the triangles of the grid. Identically, $T = (fh)^2 + (fh)(fk) + (fk)^2$, so that fh and fk can be used as coordinates, as done in Figure 44–18. When either h or k equals zero, the icosahedron is symmetric, with $T = n^2$, n being the number of intervals between capsomers along one side of the icosahedral face. For instance, it is easy to see that for the capsid of Figure 44–7, $T = 25$.

Only some of the permissible numbers of capsomers are found in icosahedral viruses; some are listed in Table 44–6.

Selected Reading

Abad-Zapatero C, Abdel-Meguid SS, Johnson JE et al: Structure of southern bean mosaic virus at 2.8 Å resolution. Nature 286:33, 1980

Baroudy BM, Venkatesan S, Moss B: Incompletely base-paired flip-flop terminal loops link the two DNA strands of the vaccinia virus genome into one uninterrupted polynucleotide chain. Cell 28:315, 1982

Burnett RM: The structure of the adenovirus capsid. J Mol Biol 185:125, 1985

Carp RI, Merz PA, Kascsak RI et al: Nature of the scrapie agent: Current status. J Gen Virol 66:1357, 1985

Caspar DLD: Design principles in virus particle construction. In Horsfall F, Tamm I (eds): Viral and Rickettsial Infections in Man, p 51. Philadelphia, JB Lippincott, 1965

Francki RIB: Plant virus satellites. Annu Rev Microbiol 39:151, 1985

Hogle JM, Chow M, Filman DJ: Three-dimensional structure of poliovirus at 2.9 resolution. Science 229:1358, 1985

Kirkegaard K, Baltimore D: The mechanism of RNA recombination in poliovirus. Cell 47:433, 1986

Klug A: Architectural design of spherical viruses. Nature 303:378, 1983

Matthews REF: Viral taxonomy for the nonvirologist. Annu Rev Microbiol 39:451, 1985

Newcomb WW, Boring JW, Brown JC: Ion etching of human adenovirus 2: Structure of the core. J Virol 51:52, 1984

Riesner D, Gross HJ: Viroids. Annu Rev Biochem 54:531, 1985

Robinson IK, Harrison SC: Structure of the expanded state of tomato bushy stunt virus. Nature 297:563, 1982

Stanley J: The molecular biology of geminiviruses. Adv Virus Res 30:139, 1985

Summers J, Mason WS: Replication of the genome of a hepatitis B–like virus by reverse transcription of an RNA intermediate. Cell 29:403, 1982

Wang K-S, Choo Q-L, Weiner AJ et al: Structure, sequence and expression of the hepatitis delta (δ) viral genome. Nature 323:508, 1986

45

Renato Dulbecco

Multiplication and Genetics of Bacteriophages

Model Systems

In spite of marked differences in structure and in genetic complexity, all viruses are similar in many basic aspects of multiplication. For many decades animal and plant viruses could be studied only in the intact host, and so the interaction of viruses with cells was first worked out with bacteriophages, especially the **"T-even" phages of Escherichia coli** (T2, T4, and T6). Though they were thought to be the simplest possible organisms, they turned out, in fact, to be among the most complex of all viruses, but their complexity was instrumental to many discoveries.

With respect to their effects of the host cells, bacteriophages are divided into two classes: **virulent phages,** which multiply without integration and often kill the host bacteria, and **temperate phages,** which integrate their genome in the host DNA, giving rise to the phenomenon of **lysogeny.** We will consider first the virulent phages, using as a model the T-even coliphages.

Multiplication

STRUCTURE

The virions of even-numbered T phages are made up of a head and a tail (Fig. 45–1). The **head** has the shape of two halves of an icosahedron connected by a short hexagonal prism, and it contains the DNA in association with polyamines, several internal proteins, and small peptides. The pentons and hexons of the head (see Chap. 44) are

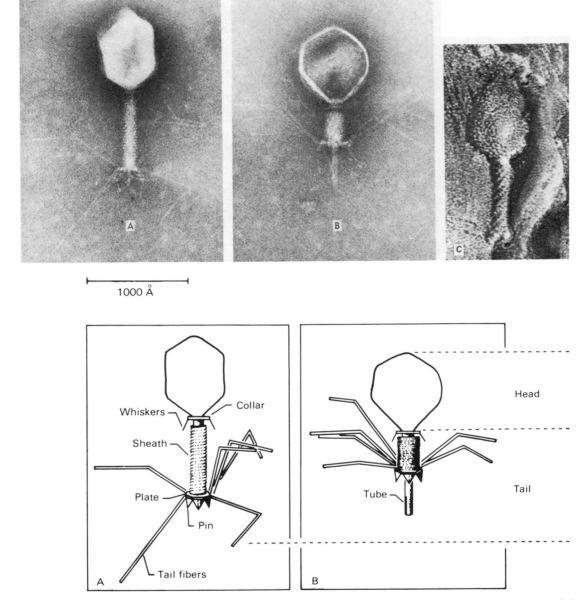

Figure 45-1. Electron micrographs of bacteriophage T2. (*A*) Phage before injection, with a full head and an extended sheath. (*B*) Phage after injection, with an empty head and a contracted sheath. (*C*) Helical structure of the sheath. *A* and *B*, negative staining. *C*, obtained by freeze-etching: the phage was embedded in ice, which was then fractured; the fracture was covered with a thin layer of evaporated metal, which was then photographed. (*A* and *B*, courtesy of E. Boy de la Tour; *C*, courtesy of M. E. Bayer.)

made up of different monomers. In other phages, such as *Salmonella* phages P1 and P2 and coliphage λ, the head is strictly icosahedral (see Fig. 44–3*F*). The T-even phage **tail** consists of a central helical **tube** (through which the viral DNA passes during cell infection), surrounded by a helical sheath capable of contraction. The **sheath** is con-

nected to the head through a thin disc or **collar** and to a **base plate** at the tip end. The plate is the organ of attachment to the wall of the host cell. It is hexagonal and of complex structure; it has a **pin** at every corner and is connected to six long, thin **tail fibers.** Each part of the virion is made up of several kinds of protein molecules.

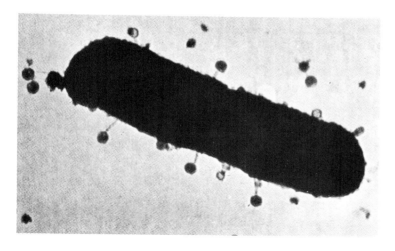

Figure 45–2. Electron micrograph of particles of phage T5 adsorbed to an *Escherichia coli* cell. The virions attach by the tips of their tails. Note also that the heads of some particles are clear (electron-transparent), having injected their DNA into the cells; others are dark (electron-opaque) and still contain their DNA. (Anderson TF: Cold Spring Harbor Symp Quant Biol 18:197, 1953)

About twenty are needed for the construction of a T4 head (see Fig. 45–14). Most other bacteriophages have tails, which vary greatly in dimensions and structure. Some small icosahedral phages, such as φX174, have no tail.

INFECTION OF HOST CELLS

The first step in infection is a highly specific interaction of the phage's **adsorption** organelle, such as the tail, with **receptors** on the surface of the host cell; then the DNA is **released** from the capsid and enters the cell.

Adsorption

All virions have a specialized structure for adsorption. In the T-even coliphages it is the base plate with its appendages. Electron microscopy shows that the tips of the **fibers** attach first and reversibly to the host cell receptors and are followed by the **tail pins,** which attach irreversibly. With all the tailed phages the adsorbed vi-

rion acquires a characteristic position with the tail perpendicular to the cell wall (Fig. 45–2).

The host cell receptors are proteins or lipopolysaccharide sugars located on the outer membrane where it is in contact with the inner membrane. Isolated receptors can bind to the phage tail, blocking adsorption of the phage to bacteria. This interaction of receptors with tail fibers is highly specific: mutations causing small changes in the receptors make the cells **resistant** to the phage. This principle is applied to *Salmonella* typing with the use of phages that adsorb to various forms of the O Ag. In T-even phages, resistance can be overcome by phage **host-range mutations** affecting the tail fibers. These coordinated changes illustrate the connection between the evolution of viruses and that of the host cells.

Separation of Nucleic Acid From Coat

In one of the most significant experiments of modern biology, Hershey and Chase demonstrated in 1952 that at infection the viral nucleic acid separates from the capsids (Fig. 45–3). They labeled the protein of T2 with ^{35}S or

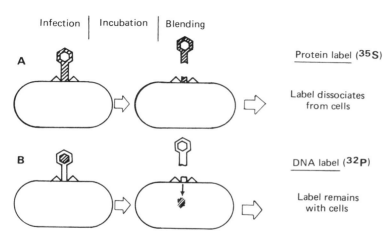

Figure 45–3. The Hershey and Chase experiment, showing the separation of viral DNA and protein at infection. (*A*) The phage protein was labeled by propagation in a medium containing ^{35}SO$_4^{2-}$. (*B*) Phage DNA was labeled by propagation in a medium containing ^{32}PO$_4^{2-}$. Phage was adsorbed to host bacteria, and after 10 minutes at 37°c, the culture was blended. Most of the ^{32}P remained associated with the cells, whereas most of the ^{35}S came off. Labeled components are shaded.

the DNA with ^{32}P, allowed the virus to infect bacteria, and by violent agitation in a blender, sheared the tails of the adsorbed virions. The experiment yielded two results that, at the time, seemed astonishing: (1) 80% of the ^{35}S label came off, whereas essentially all the ^{32}P label remained with the cells, and, since it was DNase-resistant, it was *within* the cells. (2) The blended bacteria produced progeny phage. These results strongly suggested that **phage DNA carries the genetic information of the phage into the cell.** This result provided the first evidence that the crucial event in viral infection is the penetration of the viral genome into the cell, and that the protein coat has only the function of guiding it there.

Mechanism of Penetration of the Nucleic Acid

The T-even phages have a highly specialized mechanism for releasing their DNA. After the tip of the tail has become anchored to the cell surface, the **contraction of the tail sheath** pulls the collar and the phage head toward the plate (see Fig. 45–1), pushing the tube through the cell wall locally digested by an enzyme contained in the tail. Because of this action, as well as its shape, the virion has been likened to a hypodermic syringe, and the release of the nucleic acid is called **injection.**

Contraction of the tail is the result of a **chain of conformational changes** initiated by the attachment of fibers and pins to the cell. The hexagonal base plate becomes starlike and separates from the tube, then the sheath shortens and thickens. Contraction is an irreversible reciprocal shift of the monomers, driven by the release of potential energy in the bonds between them. Many other tailed phages (such as T5 or λ) lack a contractile sheath and an injection mechanism. They simply release their DNA upon irreversible binding to the bacterial receptors. The released DNA, together with some proteins, penetrates through pores in the membranes, with which it remains associated. There it is exposed to nucleases but is protected by associated proteins and by DNA modifications (see DNA Modifications, below) or by other mechanisms.

TRANSFECTION. Bacteria can be infected by purified phage DNA after pretreatment with Ca^{2+} or conversion to spheroplasts; this property is important in DNA cloning (see Chap. 46). The efficiency of this transfection, however, is very low, because the DNA is likely to be degraded by exonucleases. The efficiency is higher (about 10^{-4}) for DNAs that are resistant to the enzymes because they either rapidly cyclize in the cells (e.g., λ DNA) or have a protein covalently bound at the ends (*Bacillus subtilis* phage φ29).

Effect of Phage Attachment on Cellular Metabolism

The attachment of the T-even and other phages *per se*, without expression of viral genes (e.g., in the presence of chloramphenicol to prevent new protein synthesis), causes a **disorganization of the cell plasma membrane.** It causes massive flows of ions in either direction, depolarizing the membrane; even larger molecules, such as nucleotides, leak to the medium. These alterations cause profound **metabolic disturbances,** including an almost immediate cessation of cellular protein synthesis. The effects are only transient, however, and are rapidly **reversed** by the incorporation of several phage-specified proteins in the membrane. These proteins also make the cell "immune" to superinfection by phage of the same type. DNA-less phages (**ghosts**), lacking the needed genes, do not reverse the changes, and cause cell lysis.

MULTIPLICATION CYCLE

The process of viral multiplication, after penetration of the parental nucleic acid, involves many sequential steps, which end in the release of newly synthesized **progeny virions.** Analysis of this **multiplication cycle** requires synchrony of cell infection, which is achieved, as in the classic work of Delbrück, by allowing virus adsorption for only a brief time (**one-step conditions**). The remaining virus is then made ineffective either by diluting the culture or by adding phage-specific antiserum. If the **multiplicity of infection**—that is, the average number of viral particles that infect a cell—is high (e.g. >3), essentially all cells are infected; if it is low (e.g. <1), a large proportion of cells are not infected (see Appendix to Chap. 44). The cells that are infected and release progeny virus (**infectious centers**) can be enumerated because they produce plaques in the regular assay used for the virus (see Assay of Infectivity, Chap. 44).

One-Step Multiplication Curve

The multiplication curve, such as that of Figure 45–4, describes the production of progeny phage as a function of the time after infection. If the cells are disrupted immediately after the DNA is injected, they do not produce plaques. This temporary disappearance of infectivity, called **eclipse,** is due to the inability of the naked viral DNA to infect bacteria under ordinary conditions. After the eclipse period, infectious phage starts appearing in the cells, where it accumulates until it is released by the **lysis (burst)** of the cells. In a bacterial culture, lysis is detected by a drop in turbidity. The time interval between infection and the beginning of release is the **latent period,** which varies with the type of phage and the culture conditions. The average number of infectious units of virus per cell at the end of replication represents the **viral yield.**

With the T-even phages, lysis is delayed by more than an hour if an infected culture is heavily reinoculated before the time of normal lysis (**lysis inhibition**). The resulting increase in viral yield is useful for the purification of virions. **Rapid-lysis (r) mutants,** which are defective in a membrane protein, do not delay lysis, and their

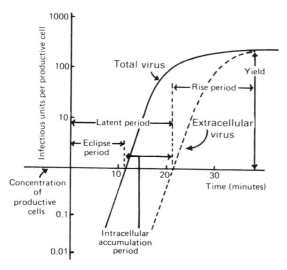

Figure 45–4. Diagram of the multiplication curve of bacteriophage T2. Bacteria and phage were mixed, and adsorption was allowed for 2 minutes; antiphage serum was then added to neutralize unadsorbed phage. The bacteria were recovered by centrifugation and were resuspended in a large volume of medium at 37°C in order to minimize readsorption of progeny phage to the bacteria. A sample was immediately plated to determine the concentration of productive cells (i.e., those able to produce phage). Other samples were taken from time to time and divided into two aliquots: one was shaken with chloroform to disrupt the bacteria and was then assayed **(total virus);** the other was freed of bacteria by centrifugation, and the supernatant was assayed **(extracellular virus).** The titers are compared with the concentration of productive cells as 1.0.

difference in plaque morphology from r+ (wild type; Fig. 45–5) is a valuable marker for genetic studies.

SYNTHESIS OF VIRAL MACROMOLECULES

A fundamental observation with T-even phages is that infection of bacteria causes a **profound rearrangement of all macromolecular syntheses.** Within a few minutes the synthesis of all DNA, RNA, and protein directed by the **cellular** genome ceases. The effect is unrelated to the transient inhibition produced by membrane damage during adsorption; it requires the expression of viral genes. Soon the cellular syntheses are entirely replaced by viral syntheses. This shift represents **the basis of viral parasitism:** the substitution of viral genes for cellular genes in directing the synthesizing machinery of the cell. With other phages the degree to which cellular functions are replaced by viral functions varies greatly, being minimal for some small filamentous phages the replication of which does not impede the multiplication of the host cells.

In cells infected by T-even and other large phages, the metabolic shift is determined by many new, viral proteins and enzymes that are synthesized after infection. Some of them turn off cellular syntheses, others carry out new viral syntheses.

Cessation of Synthesis of Host Macromolecules

Host macromolecular syntheses are stopped by three kinds of **turn-off proteins:** (1) Some phage proteins affect the host transcriptase, making it unable to recognize host promoters and consequently **shutting off host RNA syntheses.** (2) Some cause the **bacterial nucleoid to unfold;** the host DNA attaches to the cellular membrane where it is degraded. (3) Some cause **inhibition of host protein synthesis:** T-even phages cause the cleavage of a host tRNA, while T7 induces a translational repressor.

REGULATION OF TRANSCRIPTION OF PHAGE GENES

To achieve a smooth transition between cellular and viral syntheses, expression of genes is strictly regulated. An

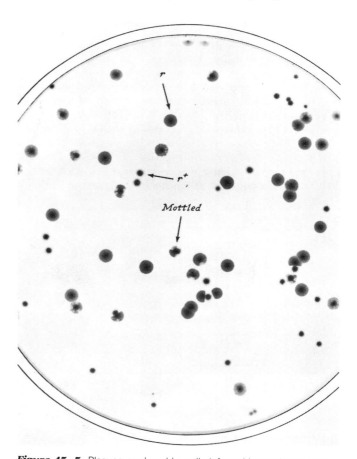

Figure 45–5. Plaques produced by cells infected by a mixture of T2r and T2r+ phage. Plaques of r type are large and without a halo; those of r+ type (wild type) are small and surrounded by a halo. The halo is produced by infected cells, with lysis inhibition caused by r+ phage. Cells infected by both r and r+ phage produce mottled plaques with a sectored halo (dark sectors, r phage; clear sectors, r+ phage). (Molecular Biology of Bacterial Viruses by Gunther S. Stent. W. H. Freeman and Company. Copyright © 1963)

important aspect of regulation is the **orderly temporal sequence** of expression of the genes. The regular succession of phage functions is determined mainly at the transcription level. Transcription of phage DNA generates viral mRNA, which, like bacterial mRNAs, are usually polycistronic.

The basic regulatory mechanism is the **successive appearance of new transcriptases** that recognize different sets of promoters or terminators. Some of the new transcriptases are specified entirely by viral genes, others are generated by changing the host transcriptase either by enzymatic modification or by association of viral proteins. For instance, coliphage N4 injects a transcriptase present in the virions; coliphages T3 and T7 cause the synthesis of new transcriptases after infection; T-even phages change the initiation specificity of the host transcriptase; and λ phage changes its termination specificity. **Changes of the phage DNA** also participate in this regulation: nicks and gaps may appear at some stage, as with T-even phages, or the DNA may enter the cells stepwise, as with phage T5, so that only a group of genes can be transcribed initially. The transcription changes occurring at any one stage are brought about by products or phage genes expressed at a previous stage.

We will now consider the **succession of transcription modes.** With the large phages, such as coliphage T4 or *B. subtilis* phage SPO1, several classes of genes are transcribed before replication of the phage DNA begins. **Immediate early and delayed genes** (Fig. 45–6), which are transcribed by the unaltered host RNA polymerase before replication of the phage DNA begins, encode

products that shut off cellular macromolecule syntheses and participate in the replication of the phage DNA. **Quasi-late** or **middle genes** also make products required for DNA replication and recombination but, in contrast to early genes, continue to be transcribed throughout infection. Their transcription begins at special promoters and requires other phage-specified proteins that interact with the host RNA polymerase. **Late genes** are transcribed by the host transcriptase in association with other phage-specified proteins after DNA replication has begun. Their products are capsid proteins and enzymes for lysing the cells.

With phage T4, late genes are transcribed from one DNA strand, called the "right" strand, whereas all the others are transcribed from the "left" strand, accentuating the differences between the two classes. Some genes are transcribed at different times and belong to more than one class. Coliphage N4 uses the injected transcriptases for early transcription after the DNA strands are separated by DNA binding proteins; for middle transcription it uses a new viral transcriptase made as a result of early transcription; for late transcription it uses the host transcriptase.

ANTITERMINATION. In the large phages, certain transcriptions units (Fig. 45–7) are subject first to immediate transcription, restricted to the promoter-proximal part by the host terminator *rho* factor. Later, delayed early transcription extends to the promotor-distal part, after proteins specified by immediate early genes prevent termination. Antitermination is especially important in the regulation of λ transcription (see Chap. 46).

POSTTRANSCRIPTIONAL REGULATION

Although most regulation occurs at transcription, there are important examples of posttranscriptional regulation. The **stability of the messengers** is highly variable. Thus the messenger for the T4 helix-destabilizing protein, gene product 32, is stabilized by the interaction of its 5' leader sequence with transacting factors present in the infected cells. Another regulation mechanism is **intron splicing.** Phage transcripts, unlike those of eukaryotic viruses, usually do not have introns. However, the gene for thymidylate synthase (TS) of phage T4 has a 1-Kb intron, with a sequence similar to those of eukaryotic class I introns, such as that of **Tetrahymena rRNA.** Like other class I introns, the sequence is removed from the transcript by self-splicing, carried out by the RNA itself. The intron has regulatory function, for early in infection only the first exon of the gene is translated, generating an enzyme involved in binding of tetrahydrofolate to dUMP; later, after the intron is spliced out, TS is made. Introns also exist in other T4 genes.

Further regulation of gene expression occurs at

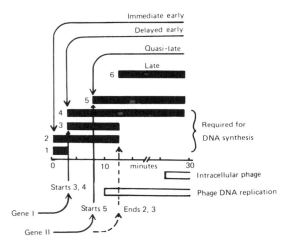

Figure 45–6. Program of transcription of *Bacillus subtilis* phage SPO1. Black bars (1,2,3 . . . 6) indicate the periods during which various classes of genes are transcribed. Phage genes I and II affect the beginning of synthesis of certain classes and the end of others, as indicated. Time is given in minutes from infection. (Modified from Gage P, Geiduschek EP: J Mol Biol 57:279, 1971. Copyright by Academic Press, Inc. [London], Ltd.)

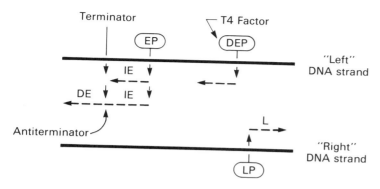

Figure 45–7. Transcription of different sets of genes on T4 DNA. Dashed arrows indicate newly synthesized RNA. *EP,* early promoters; *DEP,* delayed early promoters; *LP,* late promoters; *IE,* immediate early RNA; *DE,* delayed early RNA; *L,* late RNA. Delayed early messengers are transcribed either by interfering with termination of immediate early transcription (antitermination) or by initiating at new promoters. Early and late messengers are transcribed on different strands.

translation. For instance, the gene 32 protein inhibits its own production by binding to the messenger near the ribosome attachment site; that is, it is self-regulated. Many phages specify new tRNAs; they are essential to only some bacterial strains, which may be closer to those in which the phages have evolved.

VIRAL DNA REPLICATION

Many phages, such as T-even or T7 phages, have linear DNAs (see Chap. 44). These DNAs replicate in two phases: in the first one the amount of DNA increases; in the others, mature molecules are generated. In the **first phase** DNA replication **begins at a fixed internal origin** and proceeds **bidirectionally;** electron microscopy shows the origin of replication as a growing loop like that of Figure 45–8. The very long DNAs of T-even phages have several origins.

After several rounds of replication the mode changes to a **second phase,** which is characterized by giant molecules, called **concatemers,** generated by breakage and reunion, i.e., **recombination** (Fig. 45–9). The recombination intermediates characterized by single-strand nicks and gaps are the replication forks that from then on carry out most of the replication. The long concatemers generated by this mechanism have many branching points, and in electron micrographs appear as complex entanglements of filaments. With T-even phages, recombination continues to occur at a high frequency after concatemers are formed, leading to the dispersion of the parental DNA into many progeny molecules, each of which contains a small parental segment covalently linked to newly synthesized DNA.

Recombination is an essential step in the multiplication of T4 and other phages with linear DNA: mutants in recombination genes stop replication in the middle of infection. Recombination is important because it allows complete replication of their genomes. In their bidirectional replication from an internal origin, both DNA strands remain incompletely replicated at opposite ends

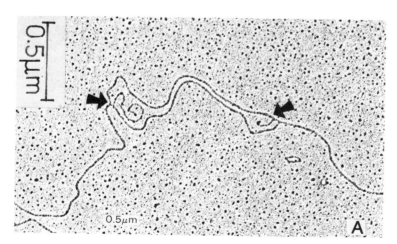

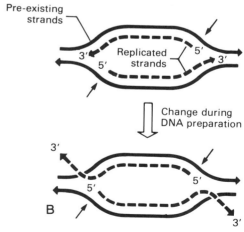

Figure 45–8. Evidence for bidirectional replication of T4 DNA. The electron micrograph of partially replicated DNA (*A*) shows a symmetric **growing loop,** with a "whisker" of single-stranded DNA (*arrows*). The sensitivity to specific nucleases shows that a whisker represents the 3' end of each growing strand. In all likelihood the whisker is formed as shown in *B,* because at each end of the loop the 3'-ended strand grows more rapidly than the 5' end (which is synthesized backward), and during the preparation of the DNA the two parental strands snap back to the point where replication is complete. (Modified from Delius et al: Proc Natl Acad Sci USA 68:3049, 1971)

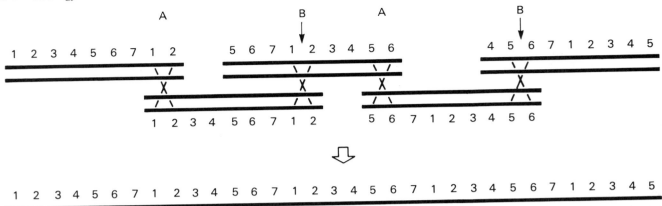

Figure 45–9. Concatemer formation by recombination. Crossovers of type B (*arrows*), between different parental molecules, cause genetic recombination, but those of type A, between identical molecules, do not. Numbers (arbitrary) indicate genes.

of a molecule if the synthesis of the last Okazaki segment (which grows backward) does not begin at the last nucleotide (Fig. 45–10). Joining the ends in concatemers permits their complete replication. At the end of replication the concatemers are cut in different ways to generate **mature molecules** with repetitious ends, which, in the case of T7 are completed by new synthesis (Fig. 45–11).

Biochemistry of Replication

After release from the virions, phage DNA, together with phage-specified proteins, becomes associated with the **cell plasma membrane,** where replication begins. Precursors derive from the medium and, for T-even phages, also from host DNA breakdown. For elaborating these precursors and synthesizing them into phage DNA, large phages (such as the T-even) specify many enzymes (Fig. 45–12), whereas the smallest phages depend almost entirely on enzymes of the host. With all phages a specific protein complex (**primosome**) is required during the first phase for **initiating phage DNA replication** at spe-

cific initiation sequences. As with bacterial DNA, phage primosomes cause the synthesis of short primer RNA segments, which are then elongated by DNA polymerase. With T-even phages, nicked molecules, which are intermediates in DNA recombination, act as primers in the second phase of replication.

B. subtilis phage φ29 DNA has a special method for initiating replication. The viral protein p3 is **covalently linked to both 5′ ends,** and it interacts with dATP to form a covalent complex, p3dAMP, the 3-OH group of which provides the replication primer. Synthesis then proceeds directly to the end of each strand. The linear, 18-Kb-long helical DNA can be replicated *in vitro* through the use of only p3 and a viral DNA polymerase.

The elongation of DNA chains is carried out by a complex of enzymes (at least 12 with T4)—some viral, some cellular—which includes DNA polymerase and accessory proteins, DNA unfolding protein, RNA polymerase, endonucleases and topoisomerases, a viral equivalent of *rec* A protein (see Recombination, Chap. 8) for the sec-

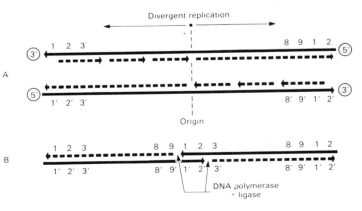

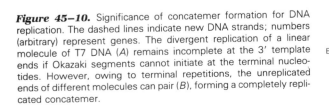

Figure 45–10. Significance of concatemer formation for DNA replication. The dashed lines indicate new DNA strands; numbers (arbitrary) represent genes. The divergent replication of a linear molecule of T7 DNA (*A*) remains incomplete at the 3′ template ends if Okazaki segments cannot initiate at the terminal nucleotides. However, owing to terminal repetitions, the unreplicated ends of different molecules can pair (*B*), forming a completely replicated concatemer.

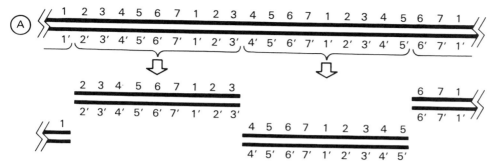

T4-type permuted molecules with repetitious ends

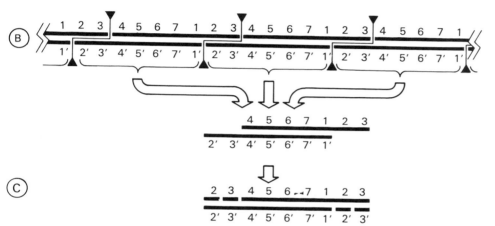

T7-type molecules with repetitious ends

Figure 45–11. Production of mature viral DNA molecules from concatemers. (*A*) Mechanisms generating permuted molecules with repetitious ends, as with phage T4. Constant lengths are cut off the concatemer, irrespective of sequences. (*B, C*) Mechanisms generating nonpermuted molecules with cohesive or repetitious ends. (*B*) Nucleases make staggered cuts on the two strands, recognizing specific sequences, and generate molecules with cohesive ends similar to those of mature phage λ DNA (see Chap. 46). (*C*) If the short strand at each end is continued by DNA polymerase as a complement of the longer strand, T7-type molecules with repetitious ends are generated. Black arrowheads indicate endonucleolytic nicking; numbers indicate sequences, primed numbers complementary sequences; dashed segments are replicated after cutting.

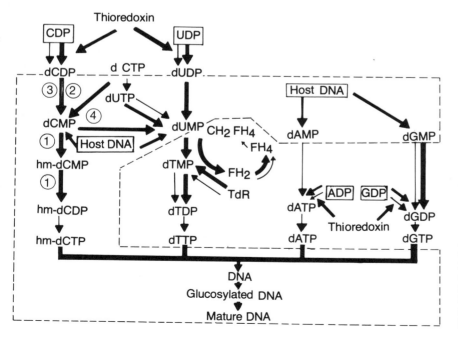

Figure 45–12. Enzymatic reactions involved in the synthesis of T4 DNA. Heavy arrows indicate enzymes specified by phage genes (whether or not there is a similar cellular enzyme with the same function). Thin arrows indicate cellular enzymes. The dashed line surrounds the machinery exclusive to T4 DNA reproduction. It includes enzymes for utilizing host DNA as a source of precursors, for replacing cytidine with hydroxymethylcytosine, and for synthesizing mature DNA. Boxed compounds are the normal source of precursors for DNA replication. *Circled numbers:* see text.

ond phase of phage T4, and enzymes producing nucleoside triphosphates. This arrangement favors speed of synthesis.

Phages with special bases in their DNAs (see Chap. 44) specify enzymes not only for synthesizing the special nucleotide (see Fig. 45–12) but also for preventing incorporation of the usual one. Thus, T-even phages, which have **5-hydroxymethylcytosine** in their DNA, specify enzymes such as those indicated as 1, 3, and 4 in Fig. 45–12, that dephosphorylate dCTP, dCDP, and dCMP. This is an essential step because cytosine-containing phage DNA is unsuitable as a template for late transcription and is broken down by the nucleases of these phages that degrade the host DNA.

MATURATION AND RELEASE

Maturation and release are the last two of the series of events caused by the expression of viral genes in the infected cells. In **maturation** the various components become assembled to form complete or **mature** infectious virions; in **release** the virions leave the infected cells.

Assembly of the Capsid

The assembly of T-even bacteriophage is interesting as a model for the formation of biological structures containing many different component proteins and for their stabilization. The assembly was elucidated by using **conditionally lethal mutations** of phage genes, which under nonpermissive conditions each block the morphogenetic process at a specific step. Upon lysis the cells yield partly

and often erroneously assembled structures recognizable by electron microscopy (Fig. 45–13). In addition, certain pairs of defective lysates give rise to **complementation** *in vitro;* i.e., the accumulated incomplete structures can assemble spontaneously, when mixed, to form infectious virus. Such complementation studies have shown that the T4 capsid is assembled through three **independent subassembly lines,** which produce the phage head, the fiberless tails, and the tail fibers, respectively (Fig. 45–14), using mostly proteins belonging to the late class. Assembly of the head tail is initiated on the inner layer of the cell's plasma membrane, in connection with cellular proteins. The three structures, when completed, spontaneously assemble into capsids.

METHOD. In contrast to the simple icosahedral and helical capsids discussed in Chapter 44, the T4 head is not generated by self-assembly of the main capsid protein. By itself, this protein aggregates randomly into "lumps" or long cylinders (see Fig. 45–13). In fact, the head is assembled through a series of **sequential steps,** in which each protein molecule (except the first) undergoes a conformational change as it is assembled, revealing the binding site that is recognized by the next molecule.

The assembly of the head capsid is preceded by the formation of a **core** by self-assembly of a **scaffolding protein** and other proteins. Capsid proteins, especially the major capsid protein p23, assemble around the core, yielding first the rounded **procapsid I** (see Fig. 45–14). Formation of the cornered and expanded **procapsid II** occurs after the molecules of the main capsid protein p23 are shortened at their amino terminus by a protease,

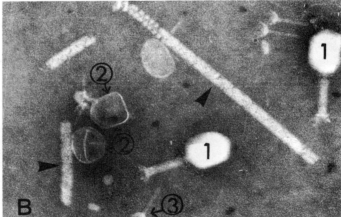

Figure 45–13. Electron micrographs of **polyheads** (A) and **polysheaths** (B, arrow) present in lysates produced by mutants of phage T4. In B one also sees "empty" head membranes (2) and tubes attached to base plates (3). Both micrographs also contain some normal virions (1) because the cells were simultaneously infected with wild-type phage. The polyheads contain hexagonal capsomers that are no longer recognizable in regular phage, owing to further assembly steps; the polysheath has the diameter of a contracted regular sheath. (A, courtesy of E. Boy de la Tour; B, Boy de la Tour E: J Ultrastruct Res 11:545, 1964)

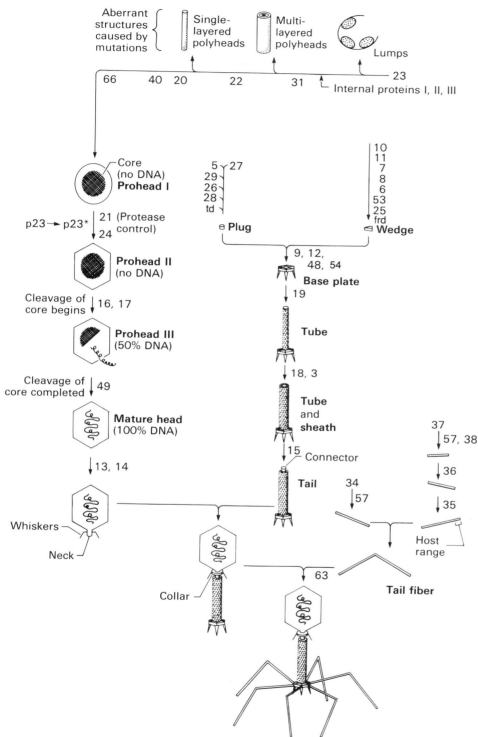

Figure 45–14. Assembly of T4 virions. Assembly occurs in three major subassemblies: head, tail, and tail fibers. Numbers indicate the T4 genes participating in a given step. *p23*, protein specified by gene 23; *p23**, product of cleavage of p23. The aberrant structures in head assembly accumulate when a mutation prevents the function of the next gene in the assembly line. "Lumps" are disorganized masses of p23 at the plasma membrane. (Data from Matthews CK, Kutter EN, Mosig J, Berget PB: Bacteriophage T4. Washington, American Society of Microbiology, 1983)

present in the core, to stabilize the structure. The core proteins are split by the protease into small fragments, some of which remain in the head (**internal peptides**). In *Salmonella* phage P22 the scaffolding protein is released intact from the maturing head and is reutilized for assembly of new proheads.

These mechanisms allow the orderly assembly of numerous components, synthesized at the same time in the same cell, to produce a capsid of a size adequate for holding the viral genome and to provide methods for capturing and holding the DNA (see below). Such mechanisms must be the result of a long evolution, during which the ability of the main components to form an icosahedral capsid by self-assembly was lost in order to allow the addition of other structures.

ASSOCIATION OF DNA AND CAPSID. Prohead I does not contain DNA. Cleavage of p23 causes extensive rearrangements, which expand the head and change its shape. These rearrangements also reveal chemical groups the binding of which to DNA probably provides the energy for "sucking" concatemeric DNA into the empty head, producing a "full" head. This process can be produced *in vitro* through the use of a mutant phage that produces heads but not DNA. DNA folding is facilitated by a **packaging enzyme** (the fragment cleaved off the p23 protein), polyamines, and basic peptides.

There are several DNA-packaging mechanisms. After the T4 head is filled with concatemeric DNA, the excess is cut off by an **endonuclease that works only when the head is complete.** Because this enzyme does not recognize DNA sequences, a "**headful**" of DNA is packaged; its length is precisely determined by the size of the capsid. With phages λ and T7 the nuclease recognizes special sequences, the positions of which precisely determine the length of the packaged DNA. Phages T1 and P22 are assembled by an intermediate mechanism in which packaging of a concatemer is initiated by cleavage at a specific **pac site** and is then continued by the headful mechanism. The encapsidated DNA need not be that of the phage: with some phages random fragments of cellular DNA can be encapsidated, giving rise to **generalized transduction** (see Chap. 46).

In the assembly of the hollow tail, which takes place separately, the precise length appears to be determined by the length of an internal tape protein. The assembled tail is then joined to the full head to produce complete (mature) virions.

Release

With T-even phages several gene products alter the plasma membrane, and then the **phage lysozyme** (from gene *e*) crosses the altered membrane and attacks the cell wall, causing lysis. With the very small phage φX174, lysis does not involve lysozyme but a host autolysin. Fila-

mentous phages are released by an entirely different mechanism, without lysis (see below).

DNA MODIFICATIONS: HOST-INDUCED RESTRICTION AND MODIFICATION

The mechanisms that allow the injected phage DNA to escape the action of membrane nucleases were unveiled by studying puzzling quasi-hereditary changes of the phage caused by the host. These studies led to the discovery of the **restriction endonucleases,** which now play such a central role in DNA cloning and sequencing (Fig. 45–15).

Methylation

Restriction and modification dependent on methylation were discovered by studying the behavior of phage λ infecting *E. coli* K12 cells lysogenic for phage P1 (designated K12 [P1]) instead of the nonlysogenic cells, which are the regular λ host. K12 (P1) cells are resistant to λ; rare infected cells, however, yield progeny phage, which can then grow regularly in K12 (P1) cells. It was soon recognized that the event **does not represent the selection of mutants,** for when the progeny phage is grown through a single cycle in cells not lysogenic for P1, the new progeny phage is again incapable of growing in K12 (P1). The explanation is that K12 (P1) cells contain a restriction endonuclease, specified by P1, which breaks down the unmethylated λ DNA entering the cells (see Fig. 45–15A). However, P1 also specifies the **modifying enzyme** that methylates cytosine in the DNA targets for the restriction endonuclease, protecting the DNA. This is a requirement for the survival of P1 itself and its host cell. When unmethylated λ DNA enters K12 (P1) cells, some molecules are methylated by the modifying enzyme before they are broken down. Made resistant, they multiply, generating DNA molecules that are immediately methylated. A single passage in nonlysogenic K12, which lacks the modifying enzyme, again yields unmethylated molecules, sensitive to the P1 restriction endonuclease. T3 and T7, although possessing an unmethylated DNA, can grow in K12 (P1) cells, because a viral enzyme methylates the DNA soon after its entry into the cells.

Glucosylation (see Fig. 45–15B)

E. coli B, the usual host for T-even phages, has a restriction endonuclease in the plasma membrane that breaks down unglucosylated phage DNA. In the T-even phages the DNA is glucosylated and therefore protected when entering the cells, and the newly formed DNA is rapidly glucosylated by phage enzymes (se Fig. 45–12). In bacteria lacking uridine diphosphate glucose (UDPG) the progeny viral DNA remains unglucosylated. The resulting virions, designated T*, are unable to grow; i.e., they are **restricted,** in any *E. coli* B strains, because their unglucosylated DNA is broken down. T* phage can, however,

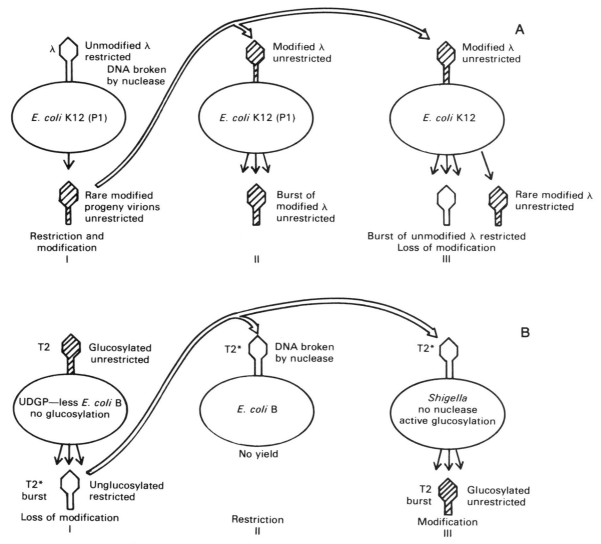

Figure 45–15. Two examples of host-induced restriction and modification. Shaded outlines indicate modified phage, which is unrestricted in the bacterial strains employed. (*A*) Restriction and modification in the same host. The regular λ phage is restricted and at the same time modified (by methylation) in *E. coli* K12 (P1) (*I*); the modified phage then grows regularly in the restricting host (*II*). After one growth cycle in *E. coli* K12 the modification is lost, except for rare virions that inherit a DNA strand from the parent (*III*). (*B*) Restriction and modification occur in different hosts. The DNA present in wild-type T2 is glucosylated (i.e., modified). Unglucosylated phage T2* is produced after a growth cycle in a nonglucosylating host (*I*). T2* is restricted in *E. coli* B (*II*); it grows in *Shigella,* in which it is modified (*III*) to yield regular T2 again.

multiply in *Shigella,* which lacks the restriction enzymes. Moreover, since *Shigella* makes UDPG, the progeny DNA is glucosylated and is again unrestricted in *E. coli* B. In a parallel situation, a phage mutant that is unable to glucosylate (gt⁻) cannot grow in *E. coli;* it grows in *Shigella,* but it is not modified.

Significance

It is likely that restriction endonucleases, recognizing unmethylated targets, developed in bacterial evolution as a **defense against foreign DNAs including phages.** At the same time, modifying enzymes had to appear to protect the cell's own DNA. Phages, in turn, have developed defenses against the nucleases. It is likely that the replacement of base C with HMC in T-even phages is a defense against a restriction endonuclease, which cannot attack the modified HMC-containing DNA. This change, however, makes the phages susceptible to other nuclease systems of probably later development, against which glucosylation is the defense.

Phage Genetics

MUTATIONS

Genetics studies with phages have been instrumental in the elucidation of the nature of the gene, the molecular mechanism of mutation, and recombination. Because phages are haploid, many mutations prevent propagation, i.e., are **lethal.** Among the nonlethal mutations, **conditionally lethal mutants** that are temperature or suppressor sensitive can be isolated in most genes. Also useful are **plaque-type mutants,** such as the **rapid lysis (r) mutants** (see Fig. 45–5) and the **host-range (h) mutants** (Fig. 45–16), which have already been described (see Infection of Host Cells, above). The r_{II} mutants are also conditionally lethal because they multiply in *E. coli* B but not in *E. coli* K12(λ).

The study of wild-type **recombinants** in genetic

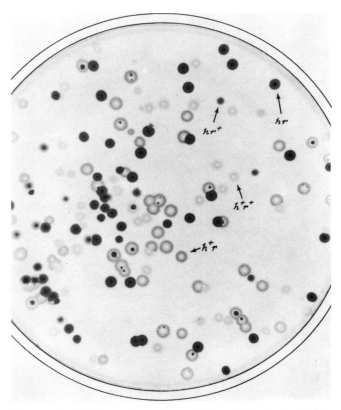

Figure 45–16. Plaques formed by a mixture of T2 phages carrying mutations at the h (host range) and the r locus, placed on a mixture of *E. coli* B and *E. coli* B/2 (i.e., resistant to T2h⁺ but sensitive to T2h). Phages with the h and those with h⁺ allele produce, respectively, clear plaques (dark areas in the photograph) and turbid plaques (gray areas in the photograph); phages with the r allele are larger than those with the r⁺ allele. Thus, all four possible combinations can be distinguished: T2h⁺r⁺ (wild-type), T2hr, T2h⁺r, and T2hr⁺. (From Molecular Biology of Bacterial Viruses by Gunther S. Stent. W. H. Freeman and Company. Copyright © 1963).

crosses between two mutants, and of **complementation,** has established a fairly complete **phage map** (see below) and defined the limits of many genes. In complementation two mutants in different genes, each unable to multiply alone under restrictive conditions, multiply together when they infect the same cell, in which each supplies in trans the function missing in the other.

HOMOLOGOUS RECOMBINATION

The study of recombination is based on the proportion of recombinants (**recombination frequencies**) in the lysate of a culture infected by two mutant strains. This study, especially with the larger bacteriophages, is complicated because recombination takes place during DNA multiplication and involves many DNA molecules, which repeatedly recombine within the same cell. The observed recombination frequencies must therefore be subjected to a suitable mathematical analysis, which then yields the **distances between markers.** This is the basis for establishing the genetic map. The relationship of recombination frequencies and **physical distances** between markers is rather uniform for any given phage, although deviations are observed in some regions of the genome: for example, hot spots for recombination at the replication origins. The relationship varies for different phages. For instance, a map unit (corresponding to 1% recombination) is equivalent to about 100 nucleotide pairs in T4 but about 2000 in λ. The difference can be attributed to the much brisker breaking and rejoining activity of replicating T4 DNA.

A Physical Map

A physical map is derived through the use of restriction endonucleases for producing characteristic DNA fragments. The many available enzymes, which recognize different target sequences, yield many well-characterized fragments from a given viral DNA in a wide range of sizes, easily separable by gel electrophoresis. The overlaps of the fragments with larger pieces obtained by incomplete digestion reveal the **sequential order** of the fragments, thus allowing a physical mapping of the restriction sites.

Genes can be located on restriction maps by determining which fragments overlap a given marker. One method is **marker rescue by fragments:** transfection of a purified wild-type DNA fragment into a cell infected by a mutant phage can generate wild-type progeny by recombination if the fragment overlaps the mutation. Another method involves **transfection of a partial heteroduplex** containing a mutant complete strand and a wild-type fragment strand (Fig. 45–17). Wild-type progeny obtained either by DNA synthesis completing the fragment strand or by error correction in the heterozygous region localizes the gene as overlapping the mutation.

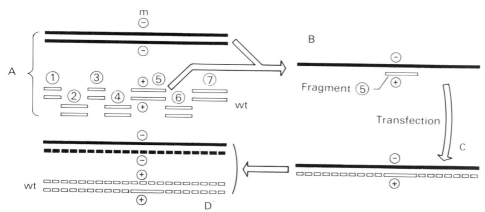

Figure 45–17. Marker rescue from synthetic partial heteroduplex. (*A*) Wild-type (*wt*) viral DNA is cut by restriction endonucleases, yielding characteristic fragments. Mutant DNA (mutation at −) and purified fragment 5 (which contains the corresponding wt allele +) are denatured and hybridized together. One of the products is the partial heteroduplex (*B*), which is introduced into suitable cells by transfection. (*C*) The heteroduplex is completed by DNA synthesis (*dashed lines*) and after replication yields a wt molecule (*D*). The mutant strand of the heteroduplex can also be converted to wt by error correction.

With the advent of DNA cloning and rapid sequencing technology, the genomes of many phages have been completely sequenced. Genes with known functions can be identified on such sequences by hybridization with DNA complementary to the messengers (cDNAs). This approach has led to a detailed knowledge of many genes not only with respect to their structural sequences but also to the regulatory sequences such as promoters, ribosome-binding sites, and termination signals.

Organization of the Genome

The arrangement of the genome of a phage has important implications for its function, as will be illustrated here with phage T4 and T7. The organization of the λ genome will be considered in Chapter 46.

Concerning the organization of the T4 genome, it should first be remarked that although the DNA in the virions is linear, the map (Fig. 45–18) is circular: if, starting at any marker, the map is completely covered by a series of crosses between pairs of markers at relatively close distance, the terminal marker is closely linked to the starting marker. The map is circular because the DNA molecules in the virions are **circularly permuted:** they are generated from a periodic concatemer by cutting off segments longer than the periodic unit. A segment can initiate at any gene and has long repetitious ends. Hence, any two genes that are adjacent to each other in the concatemers are also adjacent in most of the mature DNA molecules. In contrast, the map of phage T7 is linear (Fig. 45–19), because mature molecules are isolated by cutting concatemers at specific sequences.

The T4 map contains about 140 genes, and the functions of most are known. The map shows a high degree of organization: genes tend to be **clustered** according to their functions. Most of the genes for the proteins of the virions are together in one third of the genome, whereas the rest contains mostly genes for the various viral functions involved in DNA replication, transcription, and lysis. Genes specifying an organelle (head, tail, tail fibers) or a metabolic function (especially DNA replication) are often contiguous. A similar arrangement is found in the much simpler T7 genome. The significance of clustering is probably **functional:** contiguous genes can be efficiently regulated together, and if they are transcribed together on a polycistronic messenger, their products are generated in close proximity and timing. This pattern favors their assembly into multiprotein complexes, both in the virions and in multi-enzyme complexes in which substrates can flow rapidly from one enzyme to another.

Of the T4 genes, some are **essential,** their mutations blocking phage multiplication; others are not, their mutations only reducing the multiplication efficiency. Essential are almost all genes for the virion proteins but only half of those for metabolic functions. The essential genes include those specifying key enzymes for DNA replication or transcription, such as DNA polymerase, recombination enzymes, DNA-unfolding protein, subunits of the transcriptase, enzymes that generate dHMC or eliminate dC, and those that glucosylate the viral DNA (see Fig. 45–12). Of the **nonessential** genes, some perform functions that are not strictly necessary. Others duplicate cellular genes but work better because their products interact more efficiently with other viral proteins or with recognition signals on the viral DNA or RNA.

Among the three T-even phages (T2, T4, T6), most genes are highly conserved, implying that they are functionally important. Evidence for evolutionary divergence exists in the genes for the part of the tail fibers that recognizes the bacterial receptors. This evolution was probably promoted by the occurrence of phage-resistant bacterial mutants.

GENETIC REACTIVATION OF ULTRAVIOLET-INACTIVATED PHAGE

Phages have contributed much to our understanding of the biology of radiations, because effects on individual genes can be measured accurately. Thus in cells infected with ultraviolet-irradiated ("UV'd") phage, the **functional survival** of a phage gene can be determined directly

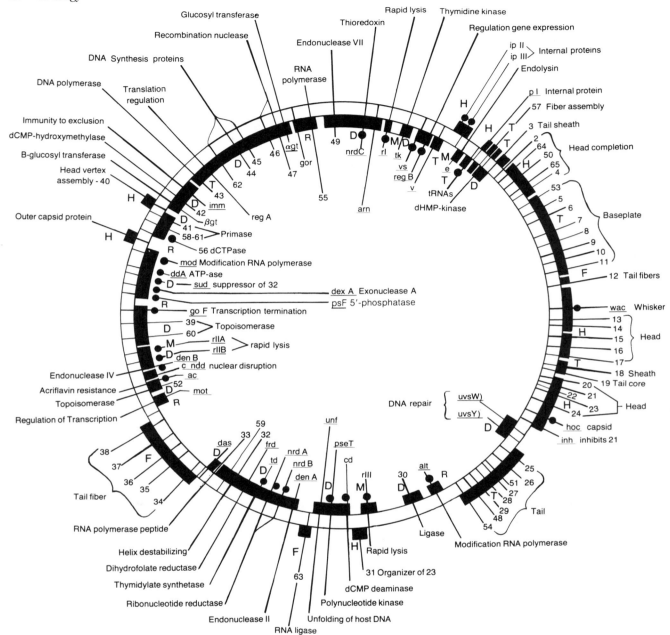

Figure 45—18. The circular genetic map of phage T4. Genes are indicated by numbers or acronyms. Blocks of genes with related functions are shown filled in black. Those on the inside of the circle specify functional proteins needed for DNA replication (*D*), mRNA synthesis (*R*), and translation (*T*) and proteins that participate in membrane functions (*M*). Genes on the outside specify structural proteins of the head (*H*), tail (*T*), or fibers (*F*). Genes identified by dots are dispensable. (Data from Matthews CK, Kutter EN, Mosig J, Berget PB: Bacteriophage T4. Washington, American Society of Microbiology, 1983)

from the amount of a gene product synthesized, or indirectly from its ability to complement a mutant gene infecting the same cells (Fig. 45—20). The functional survival depends on the size of the gene, its distance from the promoter (since UV damage interrupts transcription),

and the efficiency of **damage repair.** Large phages, such as T4, employ several genes in repairing damage to their DNA. Small phages (e.g., ϕX174 or λ) rely in large part or completely on repair mechanisms of the host, which is then said to carry out **host-cell reactivation.**

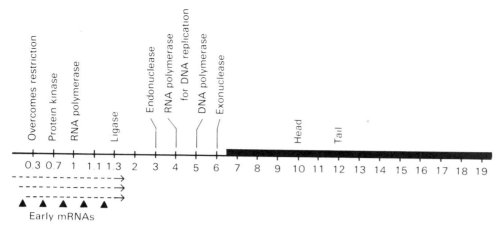

Figure 45–19. The linear genetic map of phage T7. The black bar indicates the late genes. Arrowheads are processing points of the early mRNAs that initiate on several closely spaced but distinct promoters.

In addition to the usual forms of repair of UV damage, T-even phages display strong reactivation based on recombination. One mechanism is **multiplicity reactivation.** When a cell is infected by irradiated phage, the probability of yielding infectious virus increases disproportionately to the multiplicity of infection (Fig. 45–21). The explanation is that different DNA molecules will have their UV lesions in different locations, and replicas of the undamaged segments can recombine to form concatemers containing the complete information of an intact molecule (Fig. 45–22). If recombination is prevented

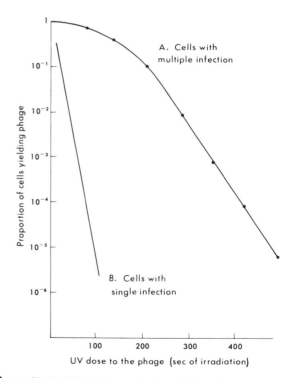

Figure 45–21. Multiplicity reactivation of UV-irradiated T2. In curve *A*, bacteria were each infected with an average of four T2 phages; the curve shows the fraction of the cells able to yield infectious phage for different UV doses given to the phage. Curve *B* shows the results obtained when the cells were infected at low multiplicity (i.e., mostly single infection). (Modified from Dulbecco R: J Bacteriol 63:199, 1952)

Figure 45–20. Kinetics of ultraviolet (*UV*) damage to a gene and to a genome. Functional survival of the $r_{II}B$ gene is compared with survival of the entire genome of T4 bacteriophage particles irradiated with UV light. Survival of the genome was measured from the fraction of K12(λ) cells yielding phage after single infection by irradiated $T4r_{II}B^+$ particles (*dashed line*). Survival of the $r_{II}B$ gene was measured by simultaneously infecting the same cells with unirradiated $r_{II}B$ mutant phage, which cannot multiply in this host unless complemented with an undamaged $r_{II}B^+$ allele. (Data from Krieg D: Virology 8:80, 1959)

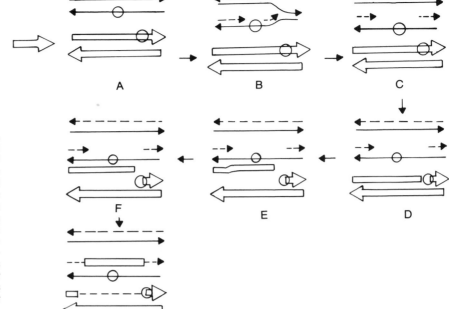

Figure 45–22. Reconstitution of an intact DNA strand (*G, black arrow*) from two UV-damaged strands (*A, open arrows*) by recombination. UV damages are indicated by crosses. *A* shows the homologous section of two DNA molecules. (*B*) The upper molecules replicate, but the new strand is interrupted at the UV lesion. (*C*) The gap is widened by the nuclease specified by genes 46 and 47. (*D*) The unreplicated damaged strand of the lower molecule is cut adjacent to the damage. (*E*) The cut lower strand pairs with the upper molecule at the gap. (*F*) The crossover is resolved by an endonuclease. (*G*) After repair synthesis, an undamaged strand (*arrow*) is formed. (Modified from Matthews CK, Kutter EN, Mosig J, Berget PB: Bacteriophage T4. Washington, American Society of Microbiology, 1983)

by mutations, multiplicity reactivation does not occur. This type of reactivation is much less evident with other phages in which recombination is less active.

A similar type of reactivation is **marker rescue** in a cell simultaneously infected by a lethally irradiated virion and by one or more undamaged virions. Recombination can incorporate a mutated gene from the irradiated genome into the undamaged DNA and hence into the progeny. Effective recombination must involve a partial replica of the UV'd phage containing the marker allele. Since this replica will initiate at a replication origin and terminate at the nearest unrepaired UV damage on each side, the probability of marker rescue decreases with the number of lesions in the irradiated DNA and with distance from the origin of replication.

Multiplication of Bacteriophages With Cyclic Single-Stranded DNA

The class of bacteriophages with cyclic single-stranded DNA includes two groups of phages with very different virions: **icosahedral** (φX174 and G4) and **filamentous** (f1 and M13). Surprisingly, the methods of infection and replication are similar.

These phages have only nine or ten genes, several of them expressing similar functions in the two groups. The genes encode structural proteins of the virions, proteins needed in DNA replication, and proteins for maturation.

INFECTION

Infection relies strongly on a **pilot protein,** which in φX174 is located in spikes at the 12 corners of the capsid, and in M13 at one end of the filamentous virion. This protein performs several important and seemingly unrelated functions: (1) it **causes the adsorption** of the virions to the cell receptors; (2) it **carries the viral DNA into the cells;** (3) it **initiates the replication** of the viral DNA, probably by linking it to the replicating machinery of the host at the cell plasma membrane; and (4) it is important for phage **morphogenesis.** The versatility of this protein is an example of the **genetic economy** of these phages. This economy finds another striking example in extensive gene overlaps in both φX174 and G4.

Most filamentous phages adsorb to the tip of pili specified by the F plasmid (see Plasmids, Chap. 7). As the DNA penetrates into the cell, the capsid protein becomes incorporated into the cell plasma membrane and is later reutilized during virus release (see below).

DNA REPLICATION

Replication of the phage DNA depends mostly on cellular enzymes. It takes place in three phases (Fig. 45–23).

1. Synthesis, by host enzymes, of a **complementary (minus) strand** on the infecting viral (plus) strand to form the **parental replicative form (RF).** This parental

RF, after being made superhelical by cellular gyrase, is transcribed into mRNAs with positive polarity, which are templates for the viral proteins.

2. **Replication of the parental RF by the rolling-circle model** (phase 1), generating ten to 20 **progeny RFs** per cell for φX174 and 100 to 200 for f1. This replication is initiated by a **multifunctional protein,** specified by a viral gene (protein A with φX74; see Fig. 45–23). As an endonuclease it nicks the positive strand (coming from the infecting virion) of the supercoiled parental RF at the origin of replication and becomes covalently bound, through a tyrosyl–dAMP phosphodiester bond, to the 5' end it generated. The bound protein moves along the negative strand, separating the parental strands from each other ahead of replication while a new progeny positive strand is made by the host DNA polymerase. The two strands are kept separate by single-strand binding protein. As soon as replication is completed, the multifunctional protein cuts off the positive progeny strand, which becomes free. Then, acting as ligase, it causes cyclization of the released strand; the energy of the protein–DNA bond is utilized to seal the nick. The cyclic progeny strand can then be used to make a new RF by building a negative strand (as above) or for virion formation in phase 2.

3. In phase 2, **asymmetric synthesis of positive progeny strands** on progeny RFs occurs as in phase 1, except that a new viral protein (C with φX174) binds at the origin, causing the association of the growing progeny viral strand with a procapsid, forming a virion. The virions are then released by cell lysis. Some mutations will block both single-stranded synthesis and virion formation. Presumably they affect proteins that control the interaction of the procapsid with the nascent DNA strand.

Throughout replication the negative strand of the parental RF remains unnicked and acts as the template for all the positive strands that end up in progeny virions. Therefore, replication of these phages follows a **stamping-machine model.**

MORPHOGENESIS

Major differences exist between the two kinds of phage in the formation of mature virions and their release. With the **icosahedral φX174** the progeny positive strands become associated with virion proteins as they are synthesized.

In contrast, with **filamentous** phages the progeny viral strand, after becoming associated with a phage-specified DNA-binding protein, attaches to the plasma membrane at points of adhesion with the outer membrane. The inner membrane contains the main capsid protein and accessory proteins in transmembrane position, including both newly synthesized molecules and those ini-

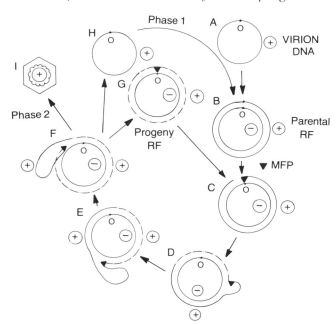

Figure 45–23. Replication of DNA of single-stranded phages. The virion DNA (the + strand; *A*) enters the cell, where it is converted to the ds parental RF (*B*). The multifunctional protein (*MFP, black triangle*) cleaves the + strand at the origin and binds to it (*C*). It causes the synthesis of a new + strand, displacing the old one (*D, E, F*), which it finally ligates into a closed circle (*G*); the protein remains bound to the replicated RF, which then commences a new cycle. The single + strand (*H*) is converted into a new RF in the first phase of replication; in the second phase it associates with a procapsid (*I*) and becomes incorporated in a progeny virion. (Data from Eisenberg S, Griffith J, Kornberg A: Proc Natl Acad Sci USA 74:3198–3202, 1977; Aoyama A, Hayashi M: Cell 47:99–106, 1986)

tially imported by the infecting virions. As the viral DNA is extended through the membrane, it picks up protein monomers of either origin and releases the DNA-binding protein, which remains in the cytoplasm and is then reutilized for further single-strand synthesis. In this way virions cross the inner membrane without damaging it and leave the cells through gaps in the outer membrane. The **length of the virions depends on the length of the viral DNA,** a characteristic exploited for DNA cloning (see Vectors for DNA Cloning, Chap. 46).

The interaction between the capsid protein of filamentous phages and the cell membrane is **intimate and balanced.** Cell growth is retarded only a little, although about 1000 virions are excreted at each cell generation. However, phage mutations affecting virion proteins can kill the cells. This steady-state virus–cell interaction resembles that occurring with some animal viruses (see Chap. 51). Although there is no cell lysis, the slower growth of the infected cells causes the formation of **turbid plaques,** that is, with a lower bacterial density than in the surrounding lawn.

RNA Phage

The small RNA phages have an icosahedral capsid (T = 3) with the addition of one or two molecules of a protein called A protein, similar in function to the pilot protein of phage ϕX. RNA phages, like filamentous DNA phages, attach to F pili of male bacteria; i.e., they are male specific. Their RNAs show **extensive self-complementarity** and are therefore able to form complex secondary and tertiary structures. These phages are divided into several groups differing in serology but with considerable homology of RNA sequences: f2, MS2, and R17 are in one group, Qβ in another. The genomes of these phages have similar, exceptionally simple organization: they contain only four genes, of which two partially overlap.

RNA REPLICATION

The viral RNA acts as both genome and messenger. Its replication, which is similar to that of some animal viruses (see Chap. 48), involves special intermediates (Fig. 45–24). In cells infected with RNA-labeled virions the label is found in two forms: entirely double-stranded molecules, completely resistant to RNase, called **replicative form (RF);** and molecules partially RNase resistant, called **replicative intermediate (RI),** which have a double-stranded backbone with one or two single-stranded tails. The pattern of labeling after brief pulses of a radioactive precursor shows that the RF is produced by building a **complementary** negative strand on the infecting **viral** positive strand; the subsequent synthesis of a third, positive progeny strand on the double-stranded RF converts it into the RI, from which successive progeny strands are then released, as Figure 45–24 shows. The synthesis of the positive strands can be **semiconservative or conservative** in different RIs; only in the former is the label parental RNA accessible to RNase degradation.

RNA Replicase

The replicases of various phages are highly specific for phages of the same group; they do not replicate cellular RNAs. They recognize two CCC sequences placed in the proper steric arrangement by the secondary structure of the RNA, one of which is present at the 5′ end of all phage RNAs.

The replicase of phage Qβ is made up of **four subunits,** of which only one (subunit II) is phage specified; **the others are cellular proteins involved in protein synthesis.** Subunit I is the ribosomal protein S1; subunits III and IV correspond to two elongation factors of protein synthesis: EF-Tu and EF-Ts. The complex of subunits II, III, and IV can replicate the Qβ **negative** strand; replication of the **positive** (viral) strand requires subunit I and an additional host ribosomal protein. The phage-specified **subunit II is the true polymerase,** for it can carry out chain elongation alone. The interaction of the phage subunit with the protein synthesis factors perhaps indicates an evolutionary relationship between phage RNA and cellular messengers, as already suggested for some plant viruses (see Chap. 44).

The RNA replicase makes mistakes at a much higher frequency than DNA polymerase, probably because it lacks the editing function of the latter enzyme. As a result, **each viable phage differs at one or more bases** from the average population.

REGULATION OF TRANSLATION

In spite of the great simplicity of the genome, these viruses have a fine regulation of gene expression, perfectly attuned to the needs of multiplication. This regulation takes place at the level of translation. It takes advantage of changes of the **secondary structure of the RNA** and of its **interaction with proteins.** Thus the ribosome-binding region of the **A gene** is normally buried in the

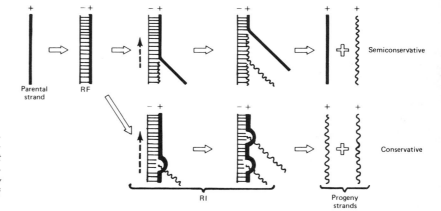

Figure 45–24. Intermediates in RNA replication. They can be distinguished because in the semiconservative type of replication the parental strand (*straight heavy line*) is exposed to RNase attack during replication, whereas in the conservative model it is not. *Wavy lines,* progeny strands; *dashed arrows,* direction of replication.

RNA folds; it may be accessible only once in the life of an RNA molecule (i.e., during synthesis), for as soon as the RNA folds up, it is permanently covered. The result is that very few A molecules are made. The **replicase gene** is translated early in infection, and the **coat protein** starts later, when new replicase molecules are no longer needed. Then the coat protein represses the translation of the gene of replicase by binding strongly to its ribosome-binding site. In this way the synthetic activity is ultimately concentrated on the synthesis of coat protein to encapsidate the RNA, and on a viral protein that triggers lysis to release the virions. Another approach to regulation is **gene overlap.** The coat protein gene overlaps with the **lysis gene,** which specifies a lysis protein made in minute amounts in infected cells. The lysis gene is out of phase and is read only occasionally by ribosomes that translate the coat gene.

Selected Reading

Alberts BM: The DNA enzymology of protein machines: Cold Spring Harbor Symp Quant Biol 49:1, 1984

Barrell BG, Air GM, Hutchinson CA III: Overlapping genes in bacteriophage φX174. Nature 264:34, 1976

Celis JE, Smith JD, Brenner J: Correlation between genetic and translational maps of gene 23 in bacteriophage T4. Nature 241:130, 1973

Eisenberg S, Griffith J, Kornberg A: φX174 cistron A protein is a multifunctional enzyme in DNA replication. Proc Natl Acad Sci USA 74:3198, 1977

Gage LP, Geiduschek EP: RNA synthesis during bacteriophage SPO1 development: Six classes of SPO1 RNA. J Mol Biol 57:279, 1971

Hsiao CL, Black LW: DNA packaging and the pathway of bacteriophage T4 head assembly. Proc Natl Acad Sci USA 74:3652, 1977

Kikuchi Y, King J: Genetic control of bacteriophage T4 baseplate morphogenesis. I. Sequential assembly of the major precursor *in vivo* and *in vitro*. J Mol Biol 99:645, 1975

Landers TA, Blumenthal T, Weber K: Function and structure in ribonucleic acid phage Qβ ribonucleic acid replicase. J Biol Chem 249:5801, 1974

Matthews CK, Kutter EM, Mosig G, Berget PB: Bacteriophage T4. Washington, American Society for Microbiology, 1983

Rashed I, Oberer E: Ff coliphages: Structural and functional relationship. Microbiol Rev 50:401, 1986

Sanger F, Air GM, Barrell BG et al: Nucleotide sequence of bacteriophage φX174 DNA. Nature 265:687, 1977

Schmidt FJ: RNA splicing in prokaryotes: Bacteriophage T4 leads the way. Cell 41:339, 1985

Zinder ND, Horiuchi K: Multiregulation element of filamentous bacteriophages. Microbiol Rev 49:101, 1985

46

Renato Dulbecco

Lysogeny and Transducing Bacteriophages

Lysogeny

Most of the bacteriophages described in the preceding chapter are **virulent;** i.e., they multiply vegetatively and kill the cells at the end of the growth cycle. The **temperate** phages, in contrast, besides multiplying vegetatively, can also produce the phenomenon of **lysogeny,** recognized in the early 1920s: the indefinite persistence of the phage DNA in their host cells, without phage production. Occasionally, however, the viral DNA in a **lysogenic cell** will initiate vegetative multiplication, generating mature virions. Lysogeny favors the persistence and spreading of a virus in a more subtle way than virulence: as Burnet has pointed out in connection with viral diseases in higher organisms, the best-adapted parasites are those that do not rapidly kill their hosts and thus deprive themselves of the opportunity to spread.

Temperate phages have provocative implications for many biological problems: they throw light on the origin of viruses and the evolution of bacteria, they provide an important mechanism for gene transfer between bacteria **(transduction),** and they supply a model for viral oncogenesis (see Chaps. 64 and 65), and for some forms of animal virus latency.

NATURE OF LYSOGENY

Lysogeny characterizes many bacterial strains freshly isolated from their natural environment. Such lysogenic cultures contain a low concentration of bacteriophage, which can be recognized because it lyses certain other related bacterial strains, known as sensitive or **indicator**

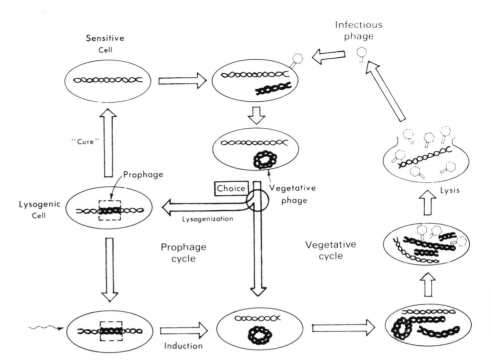

Figure 46–1. Development of a temperate bacteriophage.

strains. When a sensitive bacterial strain is infected by a temperate bacteriophage, one of two responses is seen (Fig. 46–1): some cells are lysed by phage multiplication, and others are lysogenized. Lysogenic strains thus produced are designated by the name of the sensitive strain followed by that of the lysogenizing phage in parenthesis, e.g., *Escherichia coli* K12(λ). Because temperate phages lyse only a fraction of the sensitive cells that they infect, they produce **turbid plaques.**

A bacterial strain can easily be recognized as lysogenic by streaking it on a solid medium across a strain sensitive to the phage released; a narrow zone of lysis is seen along the border of the lysogenic strain (Fig. 46–2). Since lysogeny cannot be recognized unless such a sensitive strain is available, many bacterial strains—perhaps most of those known—may be unrecognized as lysogens. Furthermore, many strains are lysogenic for several different phages.

The systems used most in experimental work on lysogeny are λ and related phages, active on *E. coli* K12; Mu, also active on *E. coli;* and P1 and P2, active on *Shigella dysenteriae* or on several strains of *E. coli.*

Relation to the Vegetative Growth Cycle

Lysogenic strains are not simply phage-contaminated bacterial cultures, since the ability to produce phage could not be eliminated by repeated cloning of the bacteria or by growth in the presence of phage-specific antiserum to prevent cell infection by virions present in the medium. In 1925 Borden recognized that it was a hereditary property of the cells. Moreover, since disruption of the lysogenic cells does not yield infectious phage, the phage must be present in the cells in a noninfectious form. However, the ability of lysogenic cultures to produce virus without obvious lysis remained puzzling until Lwoff, in 1950, patiently observing the behavior of single cells in microdroplets, showed that phage is produced by **a small proportion of the cells;** these lyse and release phage in a burst (**induction**) just like cells infected by phage T4. The other cells of the culture do not give rise to a productive infection and are said to be **immune*** to the released phage. The phage adsorbs to the immune cells and injects its DNA, but **the DNA does not multiply** and does not cause cell lysis. **Immunity, therefore, is different from resistance,** which, as noted in the preceding chapter, prevents adsorption and injection (see Infection of Host Cells, Chap. 45).

These experiments made it clear that lysogeny involves a special, stably inherited, noninfectious form of the virus, called **prophage,** associated with immunity. The prophage occasionally shifts abruptly to the vegetative form and then reproduces just like a virulent phage. Lwoff further showed that the shift from the prophage cycle to the lytic cycle, normally a rare event, could be **induced** in all the cells of a culture by moderate ultraviolet irradiation (see Fig. 46–2).

---

* This term is totally unrelated to immunity as studied in immunology.

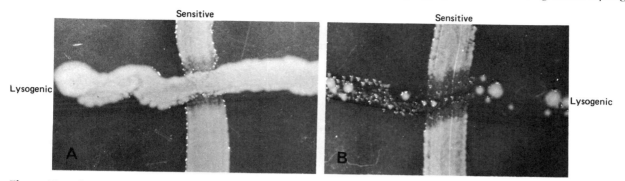

Figure 46–2. Cross-streaking of lysogenic and sensitive strains of *Escherichia coli* on nutrient agar. (*A*) Untreated. (*B*) Exposed to a small dose of ultraviolet (UV) light, after streaking, to induce the lysogenic cells. In *A* note the narrow bands of lysis of the sensitive strain (vertical streak) flanking the lysogenic strain (horizontal streak). In *B* note that the inducing treatment, by causing cell lysis, markedly reduces the colony density of the lysogenic streak, and the accompanying release of infectious phage causes pronounced lysis of the sensitive strain in the area of crossing.

THE VEGETATIVE CYCLE

The vegetative cycle of temperate phages is similar to that of virulent phages (see Chap. 45) but with some modifications. Virions of λ contain a double-stranded linear DNA, 48.5 Kb long, with a complementary single-stranded segment 12 nucleotides long at each 5′ end (Fig. 46–3). Under annealing conditions *in vitro*, these **"cohesive" ends** pair, generating a cyclic molecule with two staggered nicks. Infection with labeled phage reveals a similar cyclization *in vivo*, with ligase closing the two nicks. The evolution of the cohesive ends can be explained by the requirement for both a linear DNA in the virions, to allow encapsidation, and a cyclic form intracellularly, during replication and lysogenization (see The Prophage Cycle, below).

The Genetics of λ

As with other phages, the genes of λ are organized in three functional blocks independently regulated (Fig. 46–4): a left block specifying many functions needed for vegetative growth or lysogenization (left operon), a right block for DNA replication and capsid formation (right operon), and a central control block (immunity operon).

In Figure 46–4 the genome is represented as circular because within the cells the viral DNA is circular while the genes are expressed, i.e., during transcription. The **vegetative genetic map** (i.e., that based on recombination during vegetative growth) is linear, with the ends coinciding with those of the virion DNA (see Genetic Recombination, Chap. 45). This recombination is called **generalized** (to distinguish it from the specialized form, discussed below) and can be carried out either by the bacterial *rec BC* and *rec A* system (see Recombination, Chap. 8) or by a phage system (gene *red*, which encodes the recombination enzyme, and *gam*, which inhibits the host *rec BC* system; see Replication, below, for the role of *red* and *gam* genes).

Vegetative Transcription

The extraordinarily detailed study of λ transcription has led to a fairly complete understanding of the switching between vegetative and prophage cycles, which is an important model for the switching of genes in differentiation. Aspects important for the vegetative cycle will be reviewed now, and those specifically related to lysogenization will be reviewed later in this chapter.

The general pattern of vegetative λ transcription (see

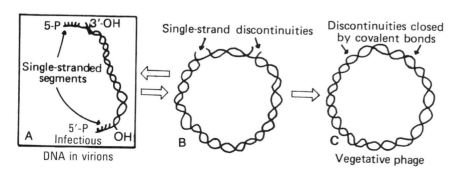

Figure 46–3. Different forms of phage λ DNA. (*A*) In the virions the DNA exists as a linear double-stranded molecule with complementary ends. (*B*) Under conditions of nucleic acid hybridization, the linear molecule can reversibly close into a ring by base pairing of the single-stranded ends. (*C*) Within the cell, the DNA forms completely covalently closed rings.

Fig. 46–4) is similar to that of T4 (see Chap. 45). We distinguish **immediate early** messengers (which appear in the presence of chloramphenicol), **delayed early,** and **late messengers**—all synthesized by the host transcriptase. As with T4, the transition from immediate to delayed early is produced by **interference with termination.** Thus, *in vitro*, in the presence of the bacterial termination factor rho, two short immediate early RNAs are formed. One is transcribed from the "left" DNA strand and extends leftward from promoter PL through gene N to the tL terminator; the other, transcribed from the "right" DNA strand, extends rightward from promoter PR through gene *cro* to the tR1 terminator, and some of this transcription continues further, through genes O and P

to tR2. The products of the genes expressed in these transcripts have important functions in viral development and lysogeny, which will be reviewed below.

The sequence of subsequent transcriptions is regulated as follows. After gene N is transcribed and translated, its product (indicated as pN), an antiterminator, allows both the leftward and the rightward transcriptions to proceed further. The mechanism of this antitermination is known in detail: pN binds to a characteristic *nut* (N-utilization) site downstream of each promoter, where it associates with cellular proteins (*nus*—for N-utilization substance—A, B, and E), forming a termination-resistant complex with the RNA polymerase.

The extended rightward transcription reaches as far

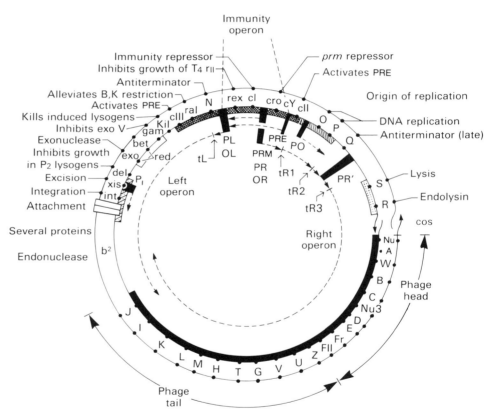

Figure 46–4. Genetic and functional organization of phage λ DNA. Capital letters indicate genes identified by nonsense mutations, which include most of those required for the vegetative cycle. The genes are arranged in functional groups, identified by the following bands (*left to right*): *dashed and dotted,* specialized recombination; *white,* generalized recombination; *crosshatched,* regulation; *dashed,* DNA replication; *dotted,* lysis; *black,* capsid. The DNA is represented in the cyclic configuration in which transcription occurs. In the virions the DNA is linear, with ends at *cos*. The map is linear; that is, markers close to *cos* but at opposite sides have the highest recombination frequencies. Transcription is indicated by dashed lines and proceeds in the directions indicated by the arrows. There are three main transcription segments designed as operons: leftward, rightward, and immunity operons. *PL, OL,* leftward promoter and operator; *PR, OR,* rightward promoter and operator; *PRE,* controller for repressor establishment; *PRM,* promoter for repressor maintenance; *PO,* promoter for replication-related transcription; *P₁,* promoter for *int* transcription; *tL, tR1, tR2,* terminators neutralized by the N gene product; *PR',* promoter for late transcription; *tR3,* terminator neutralized by the Q gene product.

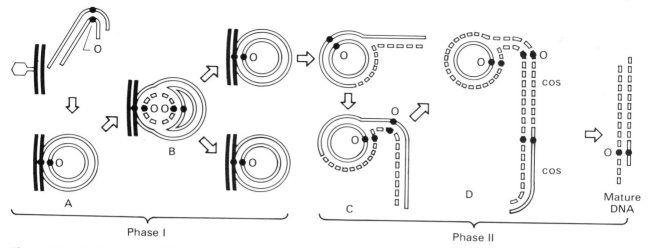

Figure 46–5. Replication of λ DNA. In phase I the DNA injected by the infecting virion first cyclizes (*A*), then (*B*) it replicates symmetrically in association with the cell plasma membrane (*double line*), initiating at a fixed origin (*O*). In phase II the DNA, free of the membrane, replicates asymmetrically (*C*), by the rolling circle model, generating linear concatemers. These are then cut at *cos* sites (*D*) to generate mature molecules with cohesive ends.

as gene Q, the product of which is another antiterminator. Together with protein *nus* A it binds to a *qut* (Q-utilization) site close to the promoter for late transcription (PR′) just to the right of Q. The binding allows this transcription, which also proceeds rightward, to avoid terminator tR3 and thus to continue through the long series of genes for the vegetative proteins all the way to gene J and into the b₂ region.

Late transcription is strongly increased by DNA replication but also occurs in its absence.

Replication

The replication of λ DNA utilizes exclusively exogenous precursors, because the bacterial DNA is not destroyed. Replication occurs in two stages (Fig. 46–5). First the cyclic parental DNA associates with the cell membrane and replicates symmetrically, generating cyclic molecules. Initiation, at a unique site within gene O, requires the functions of viral genes O and P, as well as host functions. Replication proceeds in opposite directions and terminates where the two forks meet. It is initiated by the binding of the λO protein to the origin, followed by the λP protein and by several host proteins, giving rise to a *replication complex*.

In the second, *late* stage the progeny DNA leaves the membrane and replicates according to the rolling-circle model, initiating at variable locations on the cyclic DNA and generating long concatemers. They are required for packaging the DNA in the capsid at maturation. The control for the transition from early to late replication depends on phage genes under N dependence.

The function of gene *gam* is important in replication because it inhibits the host's *rec* BC exonuclease V,

which would otherwise degrade the concatemers. *Red⁻ gam⁻* phage, which does not form concatemers, can, however, replicate in a *rec* BC⁺ host that is also *rec* A⁺. The combined action of the corresponding proteins carries out homologous recombination between circular molecules, if they contain the eight-base *chi* sequence, generating circular dimers and higher oligomers. *Red⁺ gam⁺* phage, however, cannot grow in *E. coli* lysogenized by phage P2 (*spi* phenotype; that is, susceptible to P2 interference), whereas *red⁻ gam⁻* phage can; this property is exploited in the construction of vectors for DNA cloning (see Bacteriophages as Vectors for DNA Cloning, below).

In the formation of **mature molecules with cohesive ends,** a **terminase,** specified by phage genes *NU1* and *A,* binds to concatemers or oligomers at the sites corresponding to cohesive termini (*cos;* see Fig. 46–5) and produces staggered single-strand cuts 12 nucleotides apart. The terminase also functions in **packaging** a mature DNA molecule into a prohead. Packaging is polarized, and the same terminase molecule can sequentially package adjacent units of concatemer. Further stages in head assembly and capsid maturation generally follow the T-even model. Virions are released when the product of gene S stops cellular metabolism and weakens the cellular membrane, allowing the lysozyme produced by gene R to lyse the cell wall.

THE LYSOGENIC STATE

When a sensitive bacterium is infected by a temperate phage, the entering DNA has a **"choice"** between vegetative multiplication and lysogenization (see Fig. 46–1). The proportion of infected cells that are lysogenized varies from a small percentage to nearly all, depending on the system and the conditions.

The lysogenic state is determined by the activity of the **regulatory region** of the λ genome, which not only causes integration of the phage genome in the cellular DNA but also generates immunity. Immunity emerged as the central feature of lysogeny when it was shown, by Jacob and Wollman, to be produced by repression of phage genes, much like the repression of bacterial operons.

Immunity and Repression

Lysogenic cells contain the **immunity repressor** but no vegetative proteins or mRNAs; in induced cells the opposite is true. The repressor is specified by the cI gene in the immunity region. Mutations that alter the repressor (cI) prevent lysogenization but allow vegetative growth, thus producing clear plaques.* **Immunity to exogenous infection** is also the result of repression: it is not produced by cI⁻ phage, and when normal lysogenic cells are induced, it breaks down simultaneously with repression. In the presence of repression, infecting DNA cyclizes but is not replicated; it survives for many cell generations as an **abortive prophage.**

The λ immunity repressor, isolated by Ptashne, is an acidic protein with a monomer molecular weight of 26,000. In vitro, this compound, in oligomeric form, binds strongly to the DNA of the immunity region at two **operators** (OR, OL; see Fig. 46–4) that regulate transcription starting at the **rightward and leftward promoters** (PR, PL). This binding normally prevents the expression of most viral functions outside the immunity operon.

In **virulent mutants** (*vir*), both OR and OL are defective and fail to bind the repressor. Like the cI⁻ mutants that fail to make repressor, *vir* mutants do not lysogenize and form **clear plaques.** However, there is an important difference: immune cells, containing cI repressor, can be lysed by *vir* but not by cI⁻ phages, because the latter are sensitive to the repressor.

SPECIFICITY OF IMMUNITY. Immunity is highly specific: even closely related phages form cI repressors of different specificities, and each phage recognizes exclusively its own. No point mutation in a repressor gene is known to change its specificity into that of a different phage, even a closely related one. **Heteroimmune** recombinants, which have most genetic properties of one phage but the immunity of another, have played an important part in unravelling the mechanisms of lysogeny.

* Clear plaques are also produced by mutations in genes cII, cIII, and cY (*PRE*), which, as discussed below, participate in lysogenization but not in repressor formation.

Regulation of cI Repressor Formation

The problem of how lysogenization is produced, maintained, and reversed has attracted a large number of investigators. The basic mechanism is the antagonism between the immunity (*cI*) repressor, which causes immunity, and the *cro* repressor, which prevents immunity. Their complex balance involves two promoters for *cI* (for establishment and then for maintenance), both negative and positive feedback, and host and phage factors that influence the rate of synthesis of the two repressors.

CHOICE BETWEEN LYSOGENIZATION AND VEGETATIVE MULTIPLICATION. In the early period after infection (about 20 minutes) the host transcriptase initiates transcription at the PL and PR promoters (Fig. 46–6A). Gene *cro*, dependent on PR, immediately starts producing the *cro* repressor (*croR*); gene N, dependent on PL, simultaneously produces the pN antiterminator; *pN* in turn permits transcription of genes cII and cIII, the combined products of which transiently activate the **promoter for repressor establishment (PRE),** thus initiating transcription of the immunity operon and synthesis of the immunity repressor (*cIR*).

During this early period, *cIR* and *croR* both accumulate, and their actions on OR, the righthand operator, are the key to the choice between lysogenization and vegetative growth. OR is made up of three sites: OR1, OR2, and OR3, the interactions of which with the two repressors are outlined in Figure 46–7. OR1 adjoins the PR promoter, and OR3 adjoins PRM, the **promoter for CI repressor maintenance,** where the permanent transcription of the immunity operon initiates during the lysogenic state. *cIR* binds mainly to OR1 and OR2, with two effects: it blocks transcription of the right operon, and therefore of gene *cro*, at PR; and it promotes transcription of gene cI at PRM. *CroR* binds mainly to OR3, repressing transcription of gene cI at PRM; weaker effects on OR1 and OR2 repress transcription of gene *cro* itself at PR. The different effects result from the different binding affinities of the two repressors with the three sites, which in turn depend on the base sequences and the tertiary structure of the DNA. The overall result is an antagonistic action of the two repressors on PRM, **cIR stimulating its own synthesis** by positive feedback (Fig. 46–6B); lesser effects, noticeable only at high repressor concentrations, cause **autoregulation of both repressors** by negative feedback. The balance favors lysogenization over vegetative growth.

Once *cIR* and *croR* have reached sufficient concentrations, all transcriptions (and soon all syntheses of viral products) come to a halt. The role of *PRE* is now ended. At this stage the choice is determined by the concentrations of the two repressors. If *cIR* predominates (see Fig.

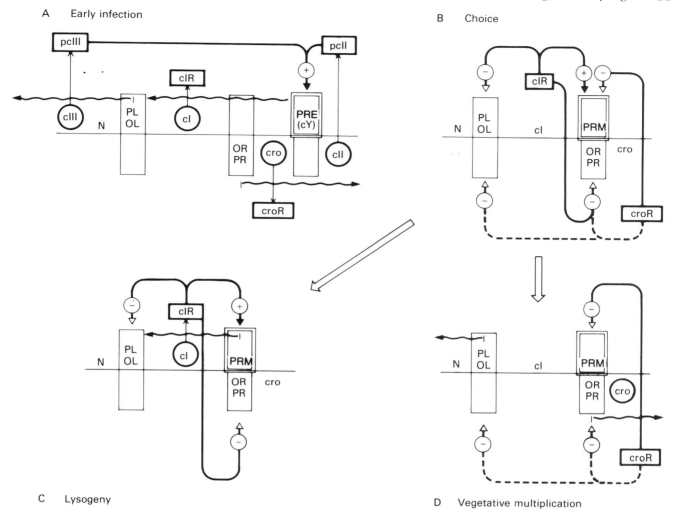

A   Early infection

B   Choice

C   Lysogeny

D   Vegetative multiplication

Figure 46–6. Regulation of repression in phage λ and determination of choice between vegetative replication and lysogeny. In the early period of infection (*A*) the left and right promoters (**PL, PR**) become active, allowing transcription (*wavy lines*) that, in the presence of pN, expresses **cII** and **cIII**. The products of these two genes (**pcII** and **pcIII**) activate the **controller for cI repressor establishment** (*PRE*), allowing transcription of the immunity operon (positive control, indicated by + and black arrowhead) and synthesis of immunity repressor, **cIR**. Cro repressor, **croR,** is also made. After about 20 minutes the various products have accumulated, and new synthesis temporarily ceases. (*B*) The choice is then determined by the competition between the negative control (−, *white arrowheads*) of the *cro* repressor (**croR**) and the positive control of the immunity repressor (**cIR**) on **PRM**. If **cIR** wins out (*C*), it activates **PRM,** keeps it permanently activated by positive feedback control, and blocks **PL** and **PR,** resulting in the lysogenic state; if **croR** wins out (*D*), **PRM** remains inactive, and transcription initiated at **PL** and **PR** causes vegetative multiplication. Active genes are circled. Heavy lines indicate the main actions of the gene products (*in rectangular boxes*); dashed lines indicate the weaker repression of the *cro* repressor on **PL** and **PR,** which forms a negative feedback control of vegetative multiplication. *OL, OR*, leftward and rightward operators.

46–6*C*), *PRM* is activated, *cIR* synthesis restarts, and the lysogenic state becomes established. If, on the contrary, *croR* predominates (see Fig. 46–6*D*), *PRM* is not activated, there is no *cIR* synthesis, and vegetative multiplication is established under the control of *croR* (see Plasmidial Prophages, below).

Once the choice is made, it is maintained by positive feedback. The choice is therefore determined during the first 20 minutes by the rates of synthesis of the two re-

pressors. Many factors influence these rates, especially the concentration of the cII product (an activator of *PRE*). Two *E. coli* genes, *hfl* (high-frequency lysogenization) A and B, specify a protease that breaks down the *cII* product, inhibiting lysogenization; their mutations promote lysogenization. Other *E. coli* mutations inhibit the synthesis of immunity repressor by decreasing the synthesis (or the action) of cyclic AMP (cAMP), which promotes lysogenization over the more expensive lytic multiplica-

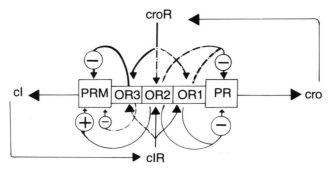

Figure 46–7. Interactions of cIR and croR with the three sites on **OR**. +, stimulation of transcription; −, repression; *solid lines* and *bold signs,* main effects; *dashed lines* and *thin signs,* secondary effects taking place only at high concentrations.

tion. This action parallels the effect of cAMP in bacterial metabolism, i.e., mediating the adaptation to poorer food sources, in response to hard times.

SPECIAL FEATURES OF THE CONTROL OF λ IMMUNITY.

The preceding brief review shows that the control of the immunity operon has many unusual features, which coordinate and fine-tune the action and synthesis of the cI repressor: (1) *cIR* represses the operators of the rightward and leftward operons, but it **promotes transcription** of the immunity operon by acting on *PRM.* (2) The independent control of the immunity operon at two sites, *PRE* and *PRM,* separates the **initial synthesis** of *cIR*—which takes place in all infected cells—from the **permanent synthesis,** which occurs only in lysogenic cells. (3) A single operator, OR, controls two different operons, responding to the same repressor with opposite consequences.

Plasmidial Prophages

Plasmidial prophages include those that are not integrated but persist as plasmids, with a constant number of copies per cell. Examples are P1 and λDV, a λ derivative. The mechanism by which λ is maintained throws light on the relationship between plasmids and temperate phages. It contains the λ origin of DNA replication, two genes for replication proteins, the promoters pR and pL, and gene *cro* (see Fig. 46–4). The *cro* gene is crucial

for the maintenance of the plasmidial states. The *cro* repressor keeps the number of copies of the plasmid at a steady level by binding to PR and repressing, in a concentration-dependent fashion, the genes for DNA replication: if the number of copies increases, the higher production of repressor increases repression of PR, causing replication to slow down.

The derivation of λDV from λ shows the relatedness of temperate phages to plasmids. Temperate phages can be considered plasmids that have acquired blocks of genes for a virion's proteins, lysis, and lysogenization.

THE PROPHAGE CYCLE

When lysogenization occurs, the vegetative λ DNA becomes inserted into the cellular DNA as a **prophage** and replicates with it. Most prophages are integrated at **fixed locations** on the bacterial chromosome. In *E. coli,* phage λ and related phages usually settle in a unique site; P2 can occupy at least nine distinct sites, but two preferentially; and μ, which is a special case, can integrate anywhere. Some prophages (e.g., P1) exist separate from the chromosome as plasmids. P1, however, must occasionally interact with the bacterial chromosome, because it gives rise to specialized transduction (see below).

Insertion

Genetic studies show that prophage λ is **linearly inserted** in the bacterial chromosome (Fig. 46–8), but the order of the genes is permuted from that determined during lytic multiplication: the order is int-cI-U instead of U-int-cI. Campbell explained this permutation by suggesting that in lysogenization the **viral DNA in cyclic form is inserted linearly into the bacterial chromosome by a single reciprocal crossover,** which opens the ring at a point different from that where the ends of the mature DNA meet (Fig. 46–9). In fact, insertion involves recombination between a **phage attachment site (att P),** 240 bases long, and a **bacterial attachment site (att B),** only 25 bases long; the two sites have 15 bases in common. The exchange generates two recombinant **prophage attachment sites** flanking the prophage, left and right (*att L* and *att R*).

This is a special form of **site-specific recombination,**

Figure 46–8. Evidence for linear insertion of the λ prophage in the bacterial chromosome. Prophage and bacterial genes were mapped by using deletions that entered the prophage from either side by taking advantage of the two *chl* (chlorate resistance) genes, A and D. To determine which prophage markers had also been deleted, the cells were induced and then superinfected with λ phage carrying distinguishable alleles of all the markers: the appearance (or nonappearance) of various prophage markers in the progeny phage indicated whether they were present in the partly deleted prophage. *Thin line,* phage DNA; *heavy lines,* bacterial DNA; *ara,* arabinose utilization; *gal,* galactose utilization; *blu,* stained blue by iodine; *bio,* biotin synthesis; *uvr,* ultraviolet light resistance. (Data from Adhya S et al: Proc Natl Acad Sci USA 61:956, 1968)

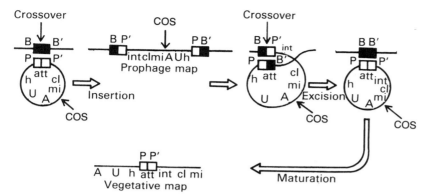

Figure 46–9. Campbell model for prophage integration explaining the permutation of the vegetative and prophage maps of λ DNA. Both the vegetative and the prophage maps can be derived from the same circle by opening it at different points: *cos* for vegetative multiplication and maturation and *att* for prophage insertion. The difference between the two maps is equivalent to shifting the block of markers (*int* to *mi*) from one end of the map to the other. For terminology, see Figure 46–4. *h*, host-range mutation; *mi*, minute plaque mutants; *heavy lines*, bacterial DNA; *thin lines*, phage DNA; *B* ■ *B'*, bacterial attachment site; *P* ☐ *P'*, phage attachment site; *B* ◪ *P'* and *P* ◪ *B'*, left and right prophage attachment sites, respectively, resulting from recombination between bacterial and phage sites.

similar to that carried out by some transposons (see Chap. 7), which requires little DNA homology. It starts with two single-strand cuts, staggered by seven base pairs within the homologous region. The cuts are carried out by the activity of a special **integrase,** specified by phage gene *int*. Recombination occurs in a complex containing the integrase, a host protein (integration host factor, IHF), and the two DNAs in **supercoiling** state: supercoiling favors the interaction of proteins with DNA.

Expression of *int* is regulated like that of *cI*, so the two main events of lysogenization—repression and integration—are coordinated. Thus, early in infection the cII-cIII products, which start *cI* repressor synthesis, also activate *int* transcription by acting at a **private promoter,** P$_I$ (see Fig. 46–4); in the lysogenic state, *int* continues to be expressed at a low level, because P$_I$ is insensitive to the cI repressor.

Excision

Excision of the prophage takes place when repression breaks down. It is the reversal of integration, i.e., a **reciprocal crossover between the attachment sites at the two ends** of the prophage, which yields a cyclic phage DNA molecule and an intact bacterial chromosome (see Fig. 46–9). Excision requires another viral function, that of gene *xis*, in addition to *int* and IHF (see Fig. 46–4). The *int* product is the basic recombination enzyme for both insertion and excision, and the *xis* product makes it conform to the different site configuration during excision (see Specificity of Attachment Sites, below).

Gene *xis*, in contrast to *int*, is transcribed from the PL promoter, under repressor control (see Fig. 46–4), and therefore is **not expressed in lysogenic cells.** The differential control of *int* and *xis* confers stability on the lysogenic state. If *xis* is expressed while repression persists, the prophage is detached but cannot multiply and is finally lost. IHF also performs a regulatory role because at high concentrations, such as those reached when the cells enter the resting phase, it inhibits excision, preventing phage multiplication under the unfavorable circumstances.

SPECIFICITY OF ATTACHMENT SITES. The differences in the sites involved in insertion and excision can be seen by considering the following equation of the prophage cycle:

$$\underset{\text{Site:}\quad\text{Phage}\quad\text{Bacterial}}{POP' + BOB'} \overset{int}{\underset{int\text{-}xis}{\rightleftharpoons}} \underset{\text{Left}\atop\text{prophage}}{BOP'} + \underset{\text{Right}\atop\text{prophage}}{POB'}$$

where P and P' represent phage half-sites, B and B' bacterial half-sites; O is the common central sequence, in which the crossover occurs. The sites existing after insertion (prophage sites) are recombinants between those of the phage and those of the bacterium. This difference explains the need for different enzymes in insertion and excision. Most important in determining the differences are the two parts of the prophage site, which is much longer than the bacterial site.

Induction of a Lysogenic Cell

The transition of prophage to vegetative phage represents a **breakdown of repression:** it can occur either spontaneously or, as already indicated, in response to an inducing stimulus. Then the disappearance of the repressor allows transcription to start at the PL and PR promoters. This transcription spreads to the whole genome (excluding the immunity region), as described above (see Vegetative Transcription), activating genes for prophage excision, DNA replication, recombination, virion proteins, and cell lysis. The prophage is excised as a covalently closed circle and begins to multiply as in the lytic cycle. Progeny virus is produced and leaves the cells as they burst.

Spontaneous induction may occur in rare cells as a result of the accidental activation of the induction machinery. Many prophages are induced by **ultraviolet (UV)**

light in doses too small to inactivate the phage. Prophages that are poorly inducible by UV light can be induced by thymine starvation, x-rays, alkylating agents, or some carcinogens. Some prophages cannot be induced at all; they do exhibit spontaneous induction but at a lower frequency than the inducible phages.

Induction of prophage λ has proved useful for the **identification of some carcinogenic chemicals**, which are strong inducers. Special lysogens are used, with cellular mutations for high permeability (since many carcinogens do not enter normal cells) and for elimination of excision repair (to enhance induction).

Mechanism of Induction

Induction is produced by conditions that remove the prophage from control by the repressor. Thus, when a lysogenic *Hfr* cell conjugates with a nonlysogenic F⁻ cell, induction occurs as soon as the prophage is introduced into the nonrepressive F⁻ cytoplasm (**zygotic induction**). A different mechanism is the infection of λ-lysogenic cells with a **virulent** λ mutant. Because the mutant is insensitive to the *cIR*, it produces in the cells all the gene products required for prophage excision and vegetative growth. The resident prophage is therefore induced despite the presence of its own repressor.

cIR inactivation is the most common mechanism of induction, generally as the consequence of changes initiated by UV light or other agents that produce, either directly or indirectly (i.e., during repair), **single-strand nicks or gaps** in DNA. The altered DNA causes activation of the cellular **rec A protease,** which cleaves the normal λ *cIR*. OR then synthesizes *croR*, which blocks *prm*.

This type of induction is part of a more general response of bacteria, whether or not lysogenic, to DNA damage. This "**SOS response**" (see Chap. 8), which tends to rescue the cell from genetic damage, includes activation of a new, error-prone DNA repair pathway and inhibition of bacterial division, with filamentous growth. The effect on phage allows it to abandon an irreparably damaged host.

EFFECT OF PROPHAGE ON HOST FUNCTIONS

Except for transducing phages (see below), which carry genes known to be derived from a recent bacterial host, most prophages exert no discernible effect on the bacterial phenotype other than immunity to superinfection. Certain prophages, however, change the cell's phenotype, either by expressing new functions (phage conversion) or by quantitatively modifying the expression of adjacent bacterial genes.

In **phage conversion** the new functions are viral, since they are abolished by phage mutations. Examples are changes of surface antigens in *Salmonella typhimu-*

rium and the formation of toxins by *Corynebacterium diphtheriae* and other bacteria. These effects are expressed by both vegetative phage and prophage, but other converting functions are expressed only by the prophage. For example, the resistance of *E. coli* K12(λ) to T-even phages with an r_{II} mutation (see Conditionally Lethal Mutants, Chap. 45) is caused by the *rex* gene, which is located in the immunity operon and hence expressed only by prophage.

PHAGE Mu
Transposonlike Properties, Bacterial Termini

Phage Mu is a temperate phage of considerable interest because it is a transposon (see Chap. 8) in phage form. It has important applications in studies of bacterial genetics owing to its ability to integrate into, and inactivate, any genes. The name *Mu* derives, in fact, from this ability to **induce mutations.**

The mature genome of Mu is linear double-stranded DNA, 37 Kb long. It has the remarkable feature of possessing **heterogeneous ends of bacterial origin.** In Mu DNA we must therefore distinguish the constant central **viral part,** which is the **Mu genome,** and the short, variable bacterial ends. In different Mu DNA molecules the bacterial ends are different, and in a large population of molecules they encompass essentially all host sequences. The mechanism that generates these ends will become apparent below.

Invertible Sequence

Another special feature is that a sequence (**G segment**), about 3 Kb long within the viral DNA, can appear in **either orientation** in different molecules. It contains the COOH end of the gene for the tail fibers, the remainder of which is in the adjacent noninvertible region. A different sequence is therefore transcribed in each orientation. By inverting the G segment the phage can switch its receptor-binding specificity from one group to another in the bacterial lipopolysaccharide, and therefore from one host to another. The frequent inversions occur by a site-specific recombination between two identical inverted repeat sequences, 34 base pairs long, which bracket the invertible region. The inversion is analogous in both mechanism and significance to the phase shift in *Salmonella* (see Chap. 8) and to a similar mechanism for changing host range in phage P1. The invertase specified by the Mu gene *Gin* is 60% to 70% identical in amino acid sequence to the invertase of the other two systems and is interchangeable with them. It is also related to the resolvases of some transposons (see Chap. 8).

Lysogenization and Its Consequences

The regulatory system that controls the choice between lytic and lysogenic development of bacteriophage Mu includes a repressor gene C, comparable to λ cI, and a gene *ner*, comparable to λ *cro*. The biology of this bacteriophage is dominated by the activity of the viral DNA component as a **transposon** (see Chap. 8). The Mu genome lacks the terminal repeats that are characteristic of transposons but has repeats near the ends, which perform the same function.

The events unfolding in the infected cells have been clarified by studying *in vitro* the interaction of two plasmids, one containing the two ends of the Mu genome, and the other used to imitate the host genome. An extract of induced Mu-infected cells provided the needed enzymes and other proteins. As Fig. 46–10 shows, the two plasmids, held by proteins in a stable **transposome,** undergo staggered cuts, and two of the four ends join to form a **transposition intermediate (TI).** This can evolve in two ways. (1) **Non-replicative transposition** (see Fig. 46–10A) transfers the ends of the Mu genome, **without**

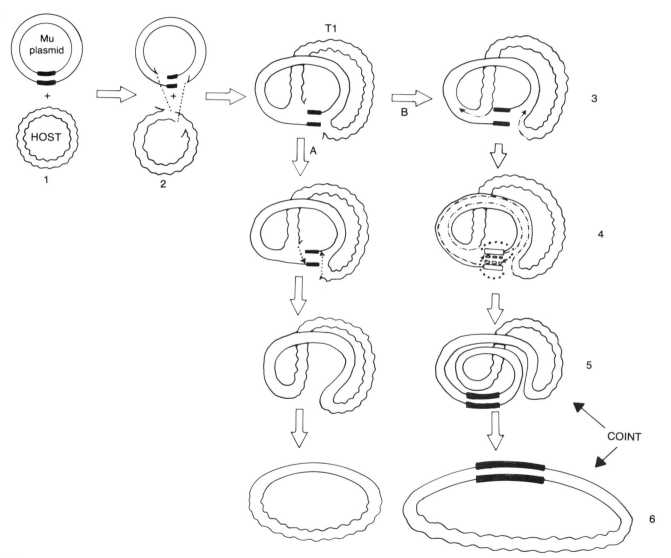

Figure 46–10. Transposition of the Mu genome *in vitro*. The donor is a plasmid (*thick lines*) containing Mu (*medium lines*) and the recipient a plasmid imitating the host genome (*thin lines; 1*). Cuts are produced at the two ends of Mu genome and in the host DNA (*2*); joining of the ends generates the transposition intermediate (*T1*). (*A*) Nonreplicative transposition by transfer of the genome without flanking sequences results in simple integration. (*B*) Replicative transposition. Elongation of the free ends of the host genome copies the Mu DNA (*3*), until the whole Mu genome with flanking plasmid sequences is replicated (*4*). An exchange at the two ends generates the cointegrate (*COINT; 5,6*). (Redrawn from Craigie R, Mizuuchi K: Cell 41:867–876, 1985)

the flanking plasmid sequences, to the host DNA, generating a simple insertion. (2) **Replicative transposition** (see Fig. 46–10*B*) extends the ends of the host genome to replicate the Mu-containing plasmid, **including the sequences flanking the Mu genome.** At the end of replication the free ends join, generating a **cointegrate,** which contains two copies of the Mu-containing plasmid integrated in the host genome. The choice between the two pathways is determined by the available enzymes, whether for DNA transfer or replication. Both simple insertion and replicative transposition require phage-specified protein A, the **transposase;** in addition, phage protein B increases the efficiency of transposition 100-fold.

When Mu infects a host bacterium, integration occurs by simple insertion of the genome, without the bacterial sequences, **at any place** in the bacterial DNA. Like transposons (see Chap. 8), the prophage is flanked by two 5-base-pair segments of duplicated host sequences. The ends of the prophage coincide with those of the Mu genome; therefore, the genetic maps of the phage and of the prophage are identical. The insertion of the prophage usually **inactivates** the bacterial gene in which it is inserted by interrupting its coding sequences, and by terminating transcription it can also inactivate distal genes in the same operon (**polarity).**

Induction does not cause excision, as with other prophages: instead, the prophage undergoes replicative transposition to a different site on the bacterial DNA. Replicative transposition goes on repeatedly, at each step doubling the number of Mu genomes, and generating many integrated copies of Mu DNA per cell.

A Mu DNA molecule leaves the host chromosome by a transposition that causes the detachment of a **cyclic hybrid molecule** containing a host segment adjacent to each end of the prophage. Because Mu DNA molecules may be integrated all over the host chromosome, the host component in such hybrids is highly variable. Encapsidation of Mu DNA contained in such hybrid rings occurs by the headful method, starting at a **packaging site** near one end of the phage genome and retaining only small parts of the host DNA from the cyclic hybrid. This method generates the linear molecules of Mu DNA with two variable bacterial ends, present in virions.

The hybrid rings may contain host markers or integrated episomes (**Mu-mediated mobilization of genes or episomes).** Conversely, phages or plasmids carrying Mu can be integrated into the chromosome by replicative transposition and are found flanked by two Mu genomes in identical orientation (**Mu-mediated integration of phage or episomes).**

SIGNIFICANCE OF LYSOGENY

Lysogeny indicates a close evolutionary relation between phages and transposing elements (see Chap. 8). All ly-sogeny depends on a site-specific recombination that is carried out by an enzyme with characteristics similar to those of the transposon resolvase. Moreover, transposable phage has all the characteristics of transposing elements.

Prophages give rise to **infectious heredity,** contributing new genetic characteristics to their host cells. Prophages that are not inducible and do not confer immunity cannot be distinguished from segments of cellular genetic material except by identifying their genes.

Phages as Transducing Agents

Transduction of bacterial genes from one cell to another by phages has been extensively used for mapping the bacterial chromosome (see Transduction, Chap. 7) and for studying gene regulation. Although now largely superseded in experimentation by DNA cloning and transfection, it still retains considerable interest as a natural process that interchanges bacterial and phage genes. The virologic aspects of the process will be considered here.

Two types can be distinguished: generalized transduction can transfer any bacterial genes, and restricted (specialized) transduction can transfer genes from only a very small region of the host chromosome adjacent to the prophage site.

GENERALIZED TRANSDUCTION

Generalized transduction is due to the encapsidation of cellular DNA in a phage coat by the headful mechanism (see Association of DNA and Capsid, Chap. 44). Apparently any marker of the donor bacterium can be transduced at a frequency of 10^{-5} to 10^{-8} per cell. The transducing DNA is not a replicon, and in the recipient cells a segment is incorporated into the host DNA by a double crossover, **in exchange for bacterial genes.** This homologous recombination is under the rec system of the host. Only closely linked markers can be **cotransduced** by the same phage particle because the piece of bacterial DNA carried by a phage corresponds to about 1% to 2% of the bacterial DNA.

Transduction is usually carried out with a high-titer phage preparation, which can be obtained from the donor strain either by lytic infection or by induction of lysogenic cells. Phage P1 is widely used in genetic studies of *Salmonella, E. coli,* and *Shigella.* Transduction occurs with different phages for many other bacterial genera.

Abortive Transduction

The introduced fragment of bacterial DNA (the **exogenote**) may persist without being integrated or replicated. Then it is transmitted **unilinearly,** i.e., from a cell to only one of its two daughters in which it expresses the functions of its genes. Such abortive transduction is easily recognized when the exogenote codes for an enzyme required for growth on a selective medium, for the restricted amount resulting from the unilinear inheritance yields **microcolonies** on minimal medium. The proportion of these to the large colonies generated by complete transduction reveals that **abortive transduction is several times more common;** hence, the probability of integration of the exogenote is rather small.

SPECIALIZED TRANSDUCTION

Phage λ can occasionally give rise to transduction in quite a different manner. It transfers only a **restricted group of genes** (the *gal* or *bio* regions) that are located near the prophage (see Fig. 46–11), and it is generated **only on induction of prophage,** but not (in contrast to generalized transduction) in lytic infection. The transducing genes are incorporated into the phage genome during abnormal excisions of the prophage.

The transducing λ virions have lost some λ DNA, to compensate in length for the incorporation of phage genes. As a result, most kinds of transducing particles are defective; i.e., they cannot multiply by themselves (e.g., λ*dgal*, λ*dbio*, where d stands for defective). However, they can replicate in mixed infection with regular λ as a **helper** to complement the missing functions. Some types of transducing particles (e.g., λ*gal*, λ*bio*) are infectious, because the missing phage genes are not essential for vegetative multiplication (Fig. 46–11).

Specialized transduction suggests an evolutionary relationship between bacteria and phages based on exchanges between their DNAs.

Bacteriophages as Vectors for DNA Cloning

BACTERIOPHAGE λ VECTORS

Phage λ is a suitable vector for DNA cloning because about one third of the genome (between genes J and N; see Fig. 46–4) is not needed for phage multiplication and can be replaced by foreign DNA. The total length of the hybrid DNA must be between 78% and 105% of that of the wild-type genome for efficient encapsidation. λ DNA contains multiple targets for restriction endonucleases, but variants with one or two targets, in the nonessential region, can be obtained by suitable selection or *in vitro* DNA recombination. Phage with a single target is suitable as **insertion vector:** its DNA is cut with the enzyme, and a fragment of foreign DNA is ligated in, reconstituting a

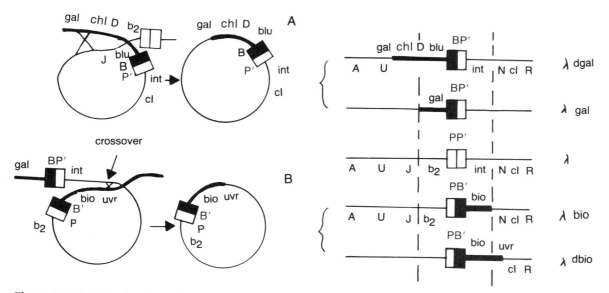

Figure 46–11. Production of transducing λ derivatives by crossovers outside the attachment sites. Crossovers on the left of the prophage (*A*) generate molecules containing *gal* and other bacterial genes (λ*dgal*) (where d stands for defective), which cannot replicate owing to loss of capsid genes. If the bacterium is deleted between *gal* and the prophage, *gal* genes can be incorporated in replacement of the nonessential b₂ region, and the phage is not defective (λ*gal*). Crossovers on the right (*B*) generate molecules containing *bio*, sometimes with other bacterial genes. Gene *bio* replaces recombination genes, and the phage can replicate (λ*bio*); if the replacement is longer (including N), the phage is defective (λ*dbio*). Phage λ*dgalbio*, which was useful for studying *int-xis*–promoted vegetative recombination, is obtained by recombination between λ*dgal* and λ*dbio*. The DNA between the two broken lines is not essential for replication. *Heavy lines,* bacterial DNA; *thin lines,* phage DNA.

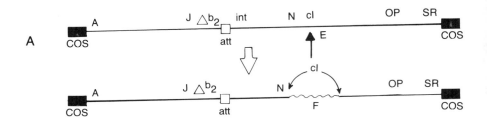

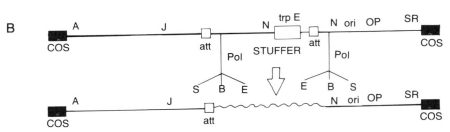

Figure 46–12. (*A*) Insertion vector λgt 10. This vector has a single site for restriction endonuclease EcoR1 (*E*) within the *cI* gene (*arrow*). The nonessential G₂ region has been partly deleted (Δb₂) to accommodate a longer insert. Insertion of a segment of foreign DNA (*wavy line*) inactivates the gene: the hybrid vector, *cI⁻*, makes clear plaques, whereas the intact vector makes turbid plaques. (*B*) Replacement vector EMBL3. This vector is the result of many manipulations, during which the *E. coli* tryptophan E gene (*trp E*) was inserted in the immunity region, and the region around N and *att* was duplicated. Two polylinkers (*Pol*), each carrying targets for three restriction endonucleases (*S*, Sal 1; *B*, Bam H1; *E*, EcoR1) were introduced near the two borders of the nonessential central region; their presence increases flexibility in the use of the vector. Using any of the three enzymes, the central "stuffer" segment is removed and is then replaced by the insertion of foreign DNA (*wavy line*). The resulting hybrid vectors are selected because, being red⁻ gamma⁻, they are Spi⁻, and therefore able to grow on a P2 lysogen, whereas the intact, Spi⁺, vector cannot. Lacking *int*, the hybrid vectors cannot lysogenize. For symbols, see λ map (Fig. 46–4).

complete genome with the addition. An example is λgt10, which has a single *EcoR1* target in the *cI* gene (Fig. 46–12*A*). Phage with an insertion of foreign DNA at that site becomes *cI⁻* and produces clear plaques, a useful marker for selecting phages with insertion. This vector can accommodate up to 7 Kb of DNA. Phage with two targets is suitable as a **replacement vector** (see Fig. 46–12*B*). The DNA between the two targets is removed and replaced with a foreign fragment. Such a vector can accommodate between about 5 Kb and 22 Kb.

A favorable feature of λ vectors is that **the hybrid DNA can be packaged into the phage capsid** *in vitro*, greatly simplifying the cloning process. *In vitro* packaging employs two phage strains. One has a mutation in the gene for the major capsid protein and is therefore unable to make heads. The other makes heads but has mutations that make it incapable of packaging the DNA. In infected bacteria each accumulates incomplete phage particles. When extracts of the two kinds of infected bacteria are mixed together with hybrid DNA of a proper size, the two mutations complement each other: the DNA is packaged, and complete phage is made. Upon infecting sensitive cells, this phage injects the hybrid DNA, which multiplies and can be recovered in pure form.

COSMIDS

λ Vectors cannot accommodate DNA fragments larger than 23 Kb and so are unsuitable for cloning large genes. This difficulty is overcome by using cosmid vectors. These are small plasmids with the sequence (*cos*) for making cohesive ends and for packaging λ DNA (Fig. 46–13). They also have the origin of DNA replication of the plasmid, a drug-resistance gene to aid in selection of recombinants, and several different restriction targets. They are packaged in λ capsids *in vitro* like regular λ DNA. After injection of bacteria, hybrid DNA multiplies as a plasmid. These vectors can accommodate fragments up to 45 Kb.

PHAGE M13 VECTORS

The filamentous phage M13, with cyclic, single-stranded DNA (see Chap. 45), has been transformed into a vector by first inserting the β-galatosidase gene in a nonessential sequence (Fig. 46–14). Within this gene a **polylinker** (i.e., a series of targets for various restriction endonucleases) has been placed in a way that does not prevent expression of the gene.

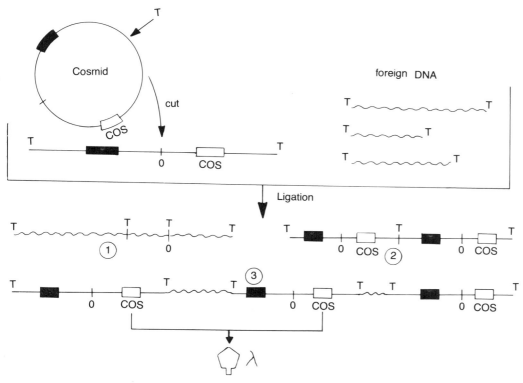

Figure 46–13. DNA cloning in a cosmid vector. The covalently closed vector contains a plasmid origin of replication (*0*), a gene for drug resistance (*black bar*), a target (*T*) for a restriction endonuclease, and the *cos* sequence. The vector and foreign DNA are cut with the restriction endonuclease and then ligated. Various products are formed: (*1*) a concatemer of foreign DNA fragments, (*2*) a concatemer of cosmids, and (*3*) a concatemer containing cosmids with foreign DNA insertions. Only the last ones are packaged in λ capsids because they possess *cos* sites at the proper distances. In an infected culture the vector-containing bacteria are readily selected because they produce drug-resistant colonies (Modified from Maniatis T, Fritsch EF, Sambrook T: Molecular Cloning. Cold Spring Harbor, NY, Cold Spring Harbor Laboratory, 1982)

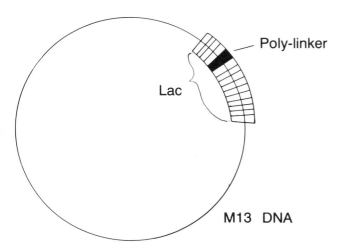

Figure 46–14. M13 cloning vector. A bacterial gene for β-galactosidase (*dashed*) is inserted in the nonessential region of the phage genome. The polylinker (*black*) within the gene by itself does not prevent β-galactosidase expression, which is abolished after foreign DNA is inserted at one of the targets. On plates containing a chromogenic β-galactoside, cells infected by the vector alone generate colored turbid plaques, while those with a foreign insertion make colorless plaques. (Modified from Maniatis T, Fritsch EF, Sambrook J: Molecular Cloning. Cold Spring Harbor, NY, Cold Spring Harbor Laboratory, 1982)

Through the use of one of the restriction sites of the polylinker, the foreign DNA is inserted into the double-stranded parental RF of the phage, which is then introduced into bacteria by transfection. The insertion destroys the function of the β-galactosidase gene. β-Galactosidase–negative turbid plaques produced by the transfected cells on indicator plates contain phage with the insert.

M13 vectors can accommodate widely different lengths of foreign DNA because during phage maturation the virions adjust their length to that of the DNA. The main use of this vector is for DNA sequencing: extraction of the progeny phage yields directly the single-stranded hybrid DNA that is needed for replication *in vitro*, in the presence of dideoxy nucleotides in the Sanger method.

Selected Reading

Buchari AI: Bacteriophage Mu as a transposition element. Annu Rev Genet 10:389, 1976

Craigie R, Mizuuchi K: Mechanism of transposition of bacteriophage Mu: Structure of a transposition intermediate. Cell 41:867, 1985

Friedman DI, Olson ER, Georgopulos C et al: Interaction of bacteriophage and host macromolecules in the growth of bacteriophage λ. Microbiol Rev 48:299, 1984

Friedman DI, Schauer AT, Olson ER et al: Proteins and nucleic acid sequences involved in regulation of gene expression by bacteriophage λ N transcription antitermination function. In Leive L (ed): Microbiology—1985, pp 271–176. Washington, American Society for Microbiology, 1985

Geider K: DNA cloning vectors utilizing functions of the filamentous phages of *Escherichia Coli*. J Gen Virol 67:2287, 1986

Glover DM: DNA Cloning: A Practical Approach. Washington, IRL Press, 1985

Ptashne M: A genetic switch; gene control of phage λ. Oxford, Cell Press and Blackwell Scientific Publications, 1986

Richet E, Abcarian P, Nash HA: The interaction of recombination proteins with supercoiled DNA: Defining the role of supercoiling in lambda integrative recombination. Cell 46:1011, 1986

Sadowski P: Site-specific recombinases: Changing partners and doing the twist. J Bacteriol 165:341, 1986

47

Renato Dulbecco

Animal Cells: Cultivation, Growth Regulation, Transformation

In animal cells growing in artificial media the effects of viruses can be detected by changes in the characteristics of the cells, and sometimes by cell death. The development of improved methods for cultivating the cells has been essential to the progress of animal virology. In particular, cell cultures provide quantitative techniques comparable to those used for bacteriophages, and with oncogenic viruses they provide methods for studying effects on the regulation of cell growth.

In this chapter we will consider the properties of cell cultures; the control of cell growth, including the roles of growth factors; and the genetic properties of cultured cells that are relevant for virology.

Characteristics of Cultures of Animal Cells

RELATION OF CELLS TO A SOLID SUPPORT. To prepare cell cultures containing separated animal cells, tissue fragments are first dissociated, usually with the aid of trypsin or collagenase. The cell suspension is then placed in a flat-bottomed **glass or plastic container** (a Petri dish, a flask, a bottle, or a test tube), together with a **liquid medium** (such as that devised by Eagle) containing required ions at isosmotic concentration, a number of amino acids and vitamins, and an animal serum in a proportion varying from a few percent to 50%. Bicarbonate is commonly used as a buffer, in equilibrium with

CO_2 (from 5% to 10%) in the air above the medium. After a variable lag the cells attach and spread on the bottom of the container and then start dividing mitotically, giving rise to a **primary culture.** Attachment to a rigid support is essential for the growth of normal cells (**anchorage dependence**) except those of the hemopoietic system.

Electron micrographs show that animal cells attach at a few points to the bottom of the vessel but elsewhere are separated by a layer of medium. The cells move actively, as shown by slow-motion pictures of living cell cultures under phase-contrast microscopy. The advancing part of a fibroblast is thin and rapidly moving (**ruffling**); the plasma membrane flows from the forward edge toward the nucleus, as seen from the motion of adhering particles. This flow is supported by the continuous arrival of membrane vesicles from the cytoplasm to the leading edge, where they fuse with the cell membrane, causing its expansion.

PRIMARY AND SECONDARY CULTURES. Primary cultures are maintained by changing the fluid two or three times a week. When the cultures become too crowded, the cells are detached from the vessel wall by either trypsin or the chelating agent EDTA, and portions are used to initiate new **secondary cultures (transfer).**

In both primary and secondary cultures the cells retain some of the characteristics of the tissue from which they were derived, and are mainly of two types: thin and elongated (**fibroblast like**), or polygonal and tending to form sheets (**epitheliumlike**). In addition, certain cells have a roundish outline and resemble epithelial cells but do not form sheets (**epithelioid cells**). The cells multiply to cover the bottom of the container with a continuous thin layer, often one cell thick (**monolayer**); if they are fibroblastic, they are **regularly oriented** parallel to each other. Primary cell cultures obtained from **cancerous tissues** usually differ from those of normal cells (see Cell Transformation, below).

CELL STRAINS AND CELL LINES. Cells from primary cultures can often be transferred serially a number of times. This process usually causes a selection of some cell type, which becomes predominant. The cells may then continue to multiply at a constant rate over many successive transfers, and the primary culture is said to have originated a **cell strain** (often called a **diploid cell strain**), the cells of which appear unaltered in morphologic and growth properties (Fig. 47–1). These cells must be transferred at a relatively high cell density to initiate a new culture, but eventually they undergo **culture senescence** and cannot be transferred any longer. For instance, with cultures of human cells the growth rate declines after about 50 duplications, and the life of the strain comes to an end.

During the multiplication of a cell strain some cells become **altered;** they acquire a different morphology, grow faster, and become able to start a culture from a small number of cells. The clone derived from one such cell is a **cell line;** in contrast to the cell strain in which it originated, it has unlimited life (**immortalization**). Cell lines derived from normal cells have a low **saturation density** under standard conditions (e.g., addition of 10%

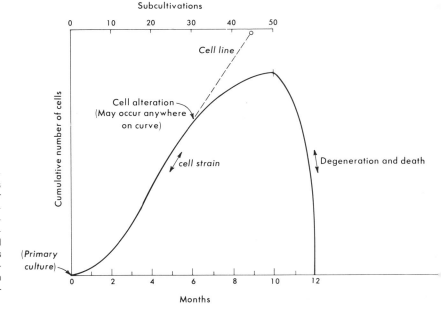

Figure 47–1. Multiplication of cultures of cells derived from normal tissues. The primary culture gives rise to a cell **strain,** the cells of which grow actively for many cell generations; then growth declines, and finally the culture stops growing and dies. During multiplication of the cell strain, altered cells may be produced, which continue to grow indefinitely and originate a cell **line.** The cumulative number of cells is calculated as if all cells derived from the original culture had been kept at every transfer. (Modified from Hayflick L: Analytic Cell Culture. National Cancer Institute Monograph, No. 7, 1962, p. 63)

serum and changes of medium two or three times weekly); they grow rapidly to form a monolayer and then slow down (**topoinhibition**).

MECHANISM OF CULTURE SENESCENCE. Senescence appears to depend directly on the **number of doublings** rather than on astronomical time. It does not depend on the special conditions of growth *in vitro*: **it also occurs *in vivo*,** as during serial transplantation from mouse to mouse of mouse mammary epithelium or of hemopoietic cells. Aging *in vitro* and *in vivo* are related: fibroblasts obtained from human donors of increasing ages or from patients with Werner's syndrome (premature aging), who have a shortened life span, can undergo fewer generations *in vitro*.

The mechanism of cellular senescence is obscure. A contributing factor may be the **accumulation of unrepaired damage** in cellular constituents. Accumulation of mutations in DNA may lower the growth ability of the cells and finally become lethal. Accidental errors in translation, causing changes in proteins that can in turn decrease the accuracy of information transfer (such as the aminoacyl-tRNA synthetases and DNA and RNA polymerases) may also cause progressive deterioration of cellular proteins, even independently of mutation. Indeed, in cultures of the mold *Neurospora* that are undergoing senescence, altered and partly inactive enzymes accumulate in the cells.

A second factor may be the **absence of efficient selection** against cumulative damage, because the culture of untransformed cells requires seeding at a relatively high cell density, and the cells also have their growth limited soon by **density-dependent growth inhibition.** In cell lines, in contrast, the cells initiate growth at low density, and selection against damaged cells is increased.

A third factor is the influence of **terminal differentiation** (see below), which limits life span *in vitro*. Thus, human keratinocytes on feeder layers of murine fibroblastlike 3T3 cells normally survive for about 50 cell generations (less if from old donors), then differentiate into squamous cells and die. If the cultures are grown in the presence of epidermal growth factor (EGF; see Growth Factors, below), their life span increases to about **150** generations, apparently because differentiation is delayed. Cellular changes preventing terminal differentiation may be responsible for lack of senescence in permanent cell lines.

LIQUID SUSPENSION CULTURES. Cell lines derived from cancers or from transformed cells (see below) sometimes produce cells with low adhesion to the container. These cells can be propagated in suspension by using a liquid medium poor in divalent ions and stirring constantly. Such suspension cultures (e.g., derivatives of murine L cells) are very useful for virological studies.

STORAGE. Because cell strains tend to exhaust their growth potential on repeated cultivation, lines tend to change continually; it is useful to keep cells of early passages **in the frozen state.** Large batches of cells are mixed with glycerol or dimethylsulfoxide and subdivided in a number of ampules, which are then sealed and frozen. The additives allow the cells to survive the freezing. The frozen cells can be maintained for years in liquid nitrogen **with unchanged characteristics;** when the ampules are thawed, most of the cells are viable and can initiate new cultures.

SPECIAL PROPERTIES OF CELLS IN CULTURES

COMMUNICATION BETWEEN CELLS. Though the cells of a culture are separate in many respects, **gap** junctions form where the plasma membranes of two cells come in contact, providing some continuity. Channels in the junctions are revealed by a low electrical resistance between the cells and by the passage of fluorescent substances (of molecular weight up to 2000), injected into a cell, to adjacent cells. Through the channels, metabolic intermediates are exchanged between cells in sufficient quantities to allow cells with a metabolic defect to grow in mixed culture with normal cells (**metabolic cooperation**).

Gap junctions are visible by freeze-fracture electron microscopy (Fig. 47–2) as areas of closely packed particles, 8 nm to 9 nm in diameter, spanning the membrane where the plasma membranes of two cells touch each other. These particles each have a central channel that admits water and solutes.

STATE OF DIFFERENTIATION. Animal cells in culture retain, at least in part, the state of differentiation they had in the animal, but it may not be easily recognizable. However, specific products can sometimes be formed, e.g., collagen from fibroblasts, cytokeratins from epithelia, casein from mammary gland cells, and specific hormones from pituitary and other hormone-producing cells. A differentiation potential may also be expressed *in vitro*: thus, skin keratinocytes differentiate into squamous cells, and 3T3 cells (skin "fibroblasts") into fat cells.

CLONING OF ANIMAL CELLS

Colonies can be obtained from most cell strains or lines by transferring cells to new cultures at a very high dilution. For cell lines the proportion of cells that give rise to colonies (**efficiency of plating**) approaches 100%, but for primary cultures or cell strains it is very small. The efficiency of plating can be greatly increased if the cells are

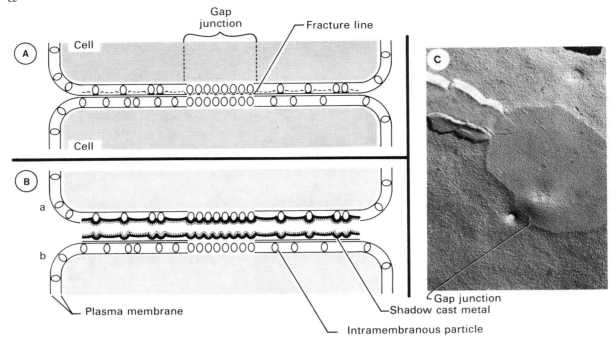

Figure 47–2. Intramembranous particles and gap junctions. A frozen cell preparation is fractured by hitting it with a knife. The fracture frequently goes through membranes, separating them into two layers. Where two cells are in contact (*A*) the fracture line (*broken line*) sometimes goes through a junction. The products of fracture (*B*) are coated with evaporated metal, and a replica of the surface is examined in the electron microscope. (*C*) Photograph of surface fracture *b* shows the gap junctions (*arrow*) and the intramembranous particles. (Courtesy of D. A. Goodenough)

mixed with a **feeder layer** of similar cells made incapable of multiplication by x-irradiation or mitomycin; these cells are still metabolically active and supply substances needed for growth. The efficiency of plating can also be increased by introducing individual cells into very small volumes of medium (as in sealed capillary tubes or in small drops of medium surrounded by paraffin oil),

which allow the cell products to reach an adequate concentration. In this way **clonal lines** are prepared.

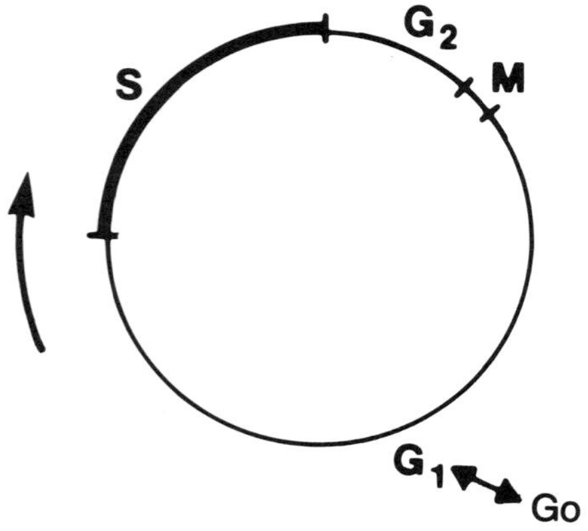

The Cell Growth Cycle

As with bacteria (see Chap. 3), the cells of a sparse culture in optimal medium multiply exponentially (i.e., with a fixed doubling time), although individual cells divide at random times. The cell growth cycle consists of four main phases (Fig. 47–3), each with a different biochemical and regulatory significance. DNA synthesis occupies only a fraction of the doubling time, the **synthetic (S) period,** which is separated from the **mitotic (M) period** by the **G2 period** (G for gap). After the mitotic phase and before the S period is the **G1 period**, which can vary enormously in length, depending on the cell type and the growth conditions.

The distribution of the cells of a culture in the various phases can be ascertained by **flow cytofluorometry,** as portrayed in Figure 47–4. From cells with a doubling time of 18 hours, typical lengths of the various periods are G1, 10 hours; S, 6 to 7 hours; G2, 1 hour; and M, about $\frac{1}{2}$ hour. Progress from one phase to another results from accumulation of specific substances within the cells, as

Figure 47–3. The cell growth cycle.

shown by the behavior of cell hybrids, obtained by the fusion of two cells. Thus, when the DNA of a cell in G1 is fused with a cell in S, it starts replicating, while it tends to condense into mitotic chromosomes when fused with a cell in mitosis.

SYNCHRONIZATION. The cell growth cycle is directly observable in synchronized cultures. The preferred method of synchronization is to start a culture with mitotic cells, which are weakly attached to the vessel and can be collected by shaking a randomly growing culture. However, after the first cycle, synchronization is rapidly lost because the G1 phases have different lengths in different cells.

GROWTH RESTRICTION IN G1. Cultures of untransformed cells stop growing after the depletion of serum or growth factors or, for certain cell types, of some amino acids or ions. The cultures then survive in a **quiescent state** known as **G0.** Growth resumes after addition of the depleted substance.

The growth of a quiescent culture can also be restored locally by removing a strip of cells (a **wound**) without replenishing the medium. Cells penetrating the wound from the edges initiate DNA synthesis within 12 hours and then divide. This phenomenon indicates that in such cultures growth stoppage also depends on local conditions (**topoinhibition**). A major source of topoinhibition is a boundary layer of the watery medium in contact with a continuous cell layer, through which medium factors can penetrate only by diffusion. Elimination of the boundary layer at the edges of a wound allows the factors to reach the cells much more efficiently by convection.

Cytofluorometry shows that **in quiescent untransformed cultures the cells are arrested in the G0 phase.** These cultures manifest a reduced uptake of glucose, phosphate, and other substances and a reduced synthesis of RNA and proteins; they display faster protein degradation and have most of the mRNAs free, rather than in polysomes.

The great variation in length of the G1 phase accounts for most of the variation in the total cycle time. Apparently, during the G1 phase the growth of untransformed cells must proceed through **restriction points** that are overcome by the availability of factors from the medium. A single factor may be sufficient for some cells, but several are needed for others. Serum contains many factors, but only one is usually limiting, causing the G1 lengths of different cells to follow a "single-event" distribution.

MECHANISM OF GROWTH CONTROL

We must distinguish between the **periodic doubling** of cell number and the **increase in cell mass.** Although

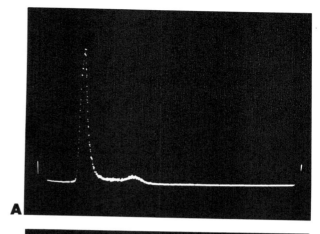

A

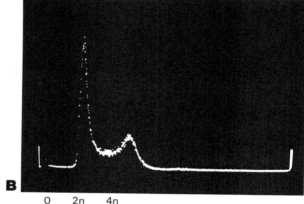

B

0   2n   4n

Figure 47–4. Flow cytofluorometry of fibroblastic cells. A single-cell suspension is exposed to an acridine dye, which, after intercalating in DNA, fluoresces green. The cells pass one at a time through a laser beam, and the excited fluorescence, which quantitates the amount of DNA, is measured. Each point in the graphs shows the number of cells that have the amount of DNA given in the abscissa. The tall peak at left measures the number of cells in G1 (with a diploid amount of DNA); the smaller peak at right measures the number of cells in G2 (with a tetraploid amount of DNA). Cells in S phase are between the two peaks. (*A*) Quiescent population, with most cells in G1 or G0. (*B*) Growing population, with a substantial proportion in S and G2. (Courtesy of R. E. Holley)

they go on at the same time in growing cultures, they are independently regulated. In early embryos, cells divide without increase in mass, and their size decreases at each division. Cultured cells can also be caused to initiate DNA replication without an increase in size by means of an alkaline shock, the rationale of which is discussed below. Studies with embryos show that the rapid periodic doubling involves only two phases, S and M.

The alternation of the two phases depends mainly on the alternative synthesis and destruction of a protein (**maturation-promoting factor,** MPF) that causes the cells to progress to the M phase. Influx or release of Ca^{2+}

from storage allows progress from M to S, apparently by causing MPF breakdown or inactivation. In somatic cells a protein, **cyclin,** perhaps similar to MPF, accumulates in the nucleus during S phase. DNA replication in S is controlled by the state of DNA or chromatin rather than by biosynthetic enzymes. Cycling is controlled by the cytoplasm: the enucleated cytoplasm of a fertilized oocyte undergoes periodic, cycle-connected changes, and plasmidial DNA injected into it replicates when the S phase is reached.

At later stages of embryonic development, and in adult animals, increase in cell mass occurs mainly during the G1 phase. In these cells multiple biochemical events take place after addition of a growth factor to quiescent cells (**pleiotypic response**). **Early events** consist in **ion fluxes,** such as Na^+-H^+ exchange activity, which increases the pH of the cytoplasm. This increase is required for growth: mutants lacking the ion exchange activity can grow only at alkaline pH. Ca^{2+} concentration increases by influx or by mobilization from membrane storage. The ionic changes activate energy-yielding and synthetic pathways. These early changes vary with cell type and growth factors, suggesting that different changes act as signals in different cells, perhaps depending on the specific blocks to be overcome in different cases.

The **late events** include many synthetic processes. The synthesis of polyamines increases; this is an essential step because its inhibition arrests cell growth. The polyamines may be needed for increasing transcription of DNA or for preparing it for replication. Later, the rate of protein synthesis increases by translation of mRNAs that had become dissociated from ribosomes in the period of quiescence; the synthesis includes cycle-specific proteins (e.g., receptors for other factors).

GROWTH FACTORS

Important for fibroblastic cells is a growth factor isolated by Ross from platelets (**platelet-derived growth factor, PDGF**). It is a small protein, made up of two chains, that is stored in the α granules of platelets and is released during blood clotting; hence, plasma has a much lower growth-promoting activity than serum. The **epidermal**

TABLE 47–1. Origins and Targets of Growth Factors

Growth Factor	Origin	Target(s)
Epidermal GF (EGF)	Submaxillary gland	Fibroblasts, epithelia
Tumor GF (TGFα) related to EGF	Cancer or embryonic	Fibroblasts, epithelia
TGFβ	Many cells	Many cell types, both stimulatory and inhibitory
Platelet-derived GF (PDGF)	Platelets	Fibroblasts, cells of mesodermal origin
Insulin	Pancreas	Fibroblasts
Insulinlike GF I	Liver	Fibroblasts, epithelia
Insulinlike GF II	Liver, placenta	Fibroblasts, epithelia
Interleukin-1 (IL-1)	Macrophages	Immune cells, astroglia; mediator of inflammation
Interleukin-2 (IL-2)	T-helper cells	T-helper cells, cytotoxic lymphocytes
Interleukin-3 (IL-3)	WEHI-3B leukemia	Hematopoietic stem cells; erythroid differentiation
Nerve GF (NGF)	Submaxillary gland	Sympathetic, sensory neurons
Endothelial GF or FGF (basic)	Bovine pituitary, brain, retina, adrenal, kidney	Endothelial cells, fibroblasts
Endothelial GF or FGF (acidic)	Bovine brain, retina	Endothelial cells, fibroblasts
Hemopoietin	Bladder, carcinoma line	Primitive multipotent hemopoietic cells
Erythropoietin	Liver	Erythroid precursor
Granulocyte-macrophage colony stimulating factor (GM-CSF)	Many tissues, especially lung	Hemopoietic stem cells, neutrophil activator
Granulocyte colony stimulating factor (G-CSF)	Id	Hemopoietic stem cells
Macrophage colony stimulating factor (M-CSF or CSF-1)	L cells	Macrophage progenitors; macrophage activator
B-cell stimulating factor 1 (BSF-1)	T cells	Resting B cell
Neurotransmitters		
Substance P	Neurons	Fibroblasts, vascular smooth muscle cells
Substance K	Neurons	Fibroblasts, vascular smooth muscle cells
Vasopressin	Neurons	Chondrocytes, bone marrow cells
Bombesin	Neurons	Bronchial epithelial cells
Transferrin	Liver	Lymphocytes
Thrombin	Plasma prothrombin	Fibroblasts
Phorbol esters*	Synthetic	Swiss 3T3 cells
Diglycerol analogs*	Synthetic	Swiss 3T3 cells
Ca^{2+} ionophore A23187*	Synthetic	T cells

* Drugs that simulate growth factor activity

growth factor (EGF), isolated by Cohen from the mouse submaxillary gland and also present in human urine, stimulates the growth of both epithelial and fibroblastic cells. Table 47–1 lists other growth factors, some of which also have other functions, such as proteolytic enzymes (thrombin), hormones (insulin), lectins (concanavalin A), ionophores, tumor promoters, and neurotransmitters. Cells produce factors required for their own growth, as shown by the enhancing effect of **conditioned medium** obtained from actively growing cultures on the growth of sparse cultures. **Tumor growth factors,** which cause normal cells to display characteristics of transformed cells, are produced by some normal and some transformed cells.

Growth factors overcome the blocks to replication that occur during the G0–G1 phases of the cell cycle. Best studied is the role of PDGF, which promotes the transition of fibroblastic cells from G0 to G1. Other growth factors, such as insulinlike growth factors, can then cooperate with PDGF in overcoming the subsequent G1 restriction points.

RECEPTORS FOR GROWTH FACTORS

As shown earlier for polypeptide hormones, growth factors act by binding to specific high-affinity receptors on the cell surface. These receptors are glycoproteins, with an **external,** a **transmembrane,** and a **cytoplasmic domain** (Fig. 47–5). Most receptors have a single polypeptide chain, but those for insulin and for insulinlike growth factor I have two chains, connected by disulfide bonds; the α-chain is external, while the β-chain has the transmembrane and the cytoplasmic domain. The binding of a factor to the external domain of a receptor has two effects. It changes the cytoplasmic domain, generating an internal signal that alters cellular characteristics, and it initiates a series of events that lead to disappearance of the receptor (**down regulation;** see below), a step in the regulation of the factor's activity.

CYTOPLASMIC SIGNALS

Two main mechanisms are known. In one, binding of the factor to the external domain activates a **protein kinase with specificity for tyrosine** in the cytoplasmic domain. Such receptors include those for EGF, PDGF, insulin, insulinlike growth factor I, and CSF-1. The responsible sequence of the cytoplasmic domain is similar in the various receptors. With ATP used as a phosphate donor, the kinase phosphorylates a number of different proteins along with itself. Which are significant for growth control is not known.

In the other mechanism, **phospholipase C splits phosphatidylinositol-4,5-bisphosphate (PIP2)** in the membrane (Fig. 47–6). This pathway, utilized by PDGF

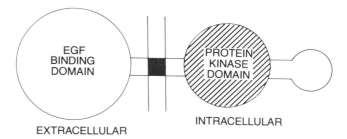

Figure 47–5. Arrangement of the EGF receptor showing the three domains: extracellular, transmembrane (*black*), and cytoplasmic, with two subdivisions, one of which is the protein kinase with tyrosine specificity. (Modified from Hunter T: Nature 311:414, 1984)

and many other receptors, is probably responsible for the early events in growth activation. Activation of the phospholipase may occur through **GTP-binding proteins** of the membranes, which, by hydrolyzing GTP to GDP, act as mediators between receptors and various kinds of regulatory proteins. For instance, adenyl cyclase is activated by such GTP-binding proteins in hormone-responsive cells.

The splitting of PIP2 generates two important activators of cellular functions: **inositol 1,4,5-triphosphate (IP3)** and **diacylglycerol (DG).** IP3 binds to receptors on the membranes of the endoplasmic reticulum, releasing bound Ca^{2+}; DG, in the presence of free Ca^{2+}, activates **protein kinase C,** a ubiquitous enzyme that phosphorylates serine and threonine (not tyrosine) in proteins. After binding DG in the presence of Ca^{2+}, the kinase sticks to the phospholipids of the membrane, where it phosphorylates various proteins. As shown below, these phosphorylations may have a regulatory role. A widely used growth and tumor protomer, TPA (tetradecanoyl phorbol acetate), acts like DG. The activation of protein kinase C increases the intracellular pH by about 0.15 units, probably by phosphorylating the Na^+-H^+ transporter, which is activated during growth.

IP3 and DG have a variety of effects, not all related to growth control. Both have very short half-lives (minutes), because IP3 is rapidly dephosphorylated, and DG is phosphorylated to phosphatidic acid. The short life of these compounds is incompatible with a complete mitogenic role. They lead, however, to a prolonged effect, because IP3 is replaced by its longer-lasting 1,3,4 isomer, and the active form of protein kinase C persists. These two compounds and the tyrosine protein kinase probably initiate a cascade of reactions that generate the late events.

RECEPTOR TURNOVER

FORMATION OF RECEPTORS. As the polypeptide chains of the receptor glycoproteins are synthesized, they pene-

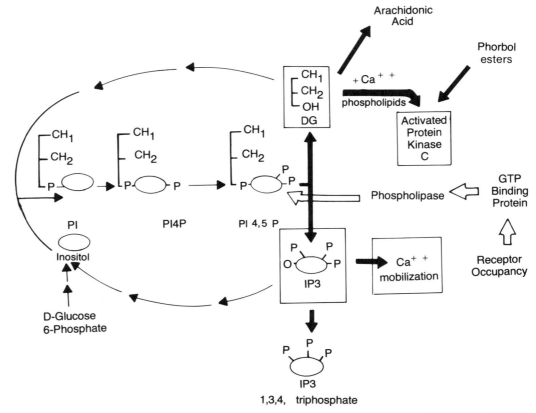

Figure 47–6. Inositol phospholipids in growth control. The thin lines indicate the turnover of inositol phospholipids. The boxes and heavy arrows indicate the compounds important in growth control; the open arrows indicate the activation pathway beginning at a receptor occupied by its ligand. *PI*, phosphatidylinositol; *PI4P*, phosphatidylinositol-4-phosphate; *PI4-5P*, phosphatidylinositol-4,5-diphosphate; *IP3*, inositol-1,4,5-triphosphate; *DG*, diacylglycerol. The significance of formation of arachidonic acid or inositol-1,3,4-triphosphate as well as inositol polyphosphates and cyclic phosphates (not shown) is unclear.

trate through the membrane of the rough endoplasmic reticulum (RER) by interaction of their **signal sequence** with a signal recognition ribonucleoprotein particle (SRP) and a membrane receptor (see Chap. 6). Within the tubules of the RER they remain connected to the RER membrane by an **anchor sequence** of hydrophobic amino acids. From the RER the proteins move in several steps, through the various sections of the Golgi stacks, where they undergo several modifications, such as addition of fatty acids and various kinds of sugars. Finally they reach the plasma membrane via transport vesicles.

INTERNALIZATION OF RECEPTORS. Receptors do not persist indefinitely at the cell surface because they are internalized (**endocytosis**) into the cells through special patches of membrane called **coated pits.** In electron micrographs of thin sections the pits, in the process of invagination, are seen to be coated at the cytoplasmic side, by a layer of **clathrin,** bound to the membrane protein (Fig. 47–7).

The invagination of the pits generates small vesicles that fuse together, forming **endosomes.** An ATP-driven proton pump makes the lumen of the endosome acidic (pH 4.5–5.5). At this pH some ligands (e.g., Fe^{3+} bound to transferrin) dissociate from their receptors, cross the endosomal membrane, and are released into the cytoplasm; others (e.g., hormones) dissociate but remain in the endosome; in all cases the receptors remain membrane bound. Many viruses follow a similar pathway (see Chap. 48).

Agents that raise the pH of the endosomes block these effects. They include weak bases such as ammonium chloride, the antimalarial agent chloroquine, the antiinfluenzal agent amantadine, and some ionophores (e.g., monensin).

Each endosome progressively becomes subdivided

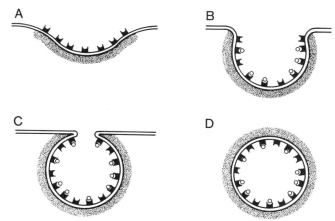

Figure 47–7. Endocytosis. A schematic representation from an electron micrograph. (*A*) Coated pit in a transverse section. The cell membrane (*double line*) is lined at the cytoplasmic side (*below*) by a layer of clathrin molecules. Receptors (*black*) protrude at the extracellular side. (*B*) As the ligand (*circles*) becomes bound to receptors, the pit deepens. (*C*) The pit closes into a vesicle still connected to the cell membrane. (*D*) The vesicle is free in the cytoplasm, as endosome, surrounded by the clathrin coating, with the ligand still attached to the receptors. See also Figure 48-2.

into two compartments: one fuses with a lysosome, and the other rejoins the cell surface. Most of the released ligands reach the lysosomes and are destroyed. Some kinds of receptors are recycled to the surface, but others (such as EGF receptors) go to the lysosomes and are destroyed, giving rise to the phenomenon of receptor **down regulation.** These different pathways satisfy dif-

ferent cell needs: recycling reutilizes transport molecules that carry necessary metabolic precursors into cells, whereas down regulation prevents hormones or growth factors from having an excessive action. Down regulation may also play a role in determining the late events following exposure of cells to a growth factor. In fact, similar events may be produced, in the absence of factor, by antibodies to the receptors, which cause clustering and internalization but no cytoplasmic signals.

Cell Transformation

Oncogenic viruses (see Chaps. 64 and 65) can cause mutationlike changes in cultured cells that affect their growth and other properties. Such **transformation** (Fig. 47–8) can also occur spontaneously during the serial growth of cell lines. Radiation and certain chemicals may cause similar transformation in cultures, as well as cancer *in vivo*, by inducing mutations (**somatic mutations).** Ames has shown that most oncogenic chemicals (or their metabolic derivatives) are also mutagenic.

Various transformed cells have different sets of the following properties, which are absent in resting untransformed cells:

CULTURE BEHAVIOR
  Increased culture thickness
  Random cell orientation
  Increased saturation density (decreased topoinhibition)
  Decreased serum or growth factor requirement

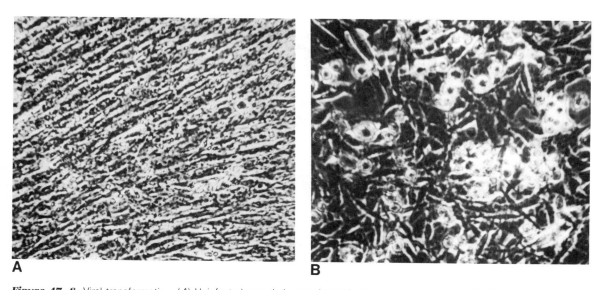

Figure 47–8. Viral transformation. (*A*) Uninfected crowded secondary culture of hamster embryo cells. Note that the cells are arranged in a thin (mostly single) layer with parallel orientation. (*B*) Similar culture transformed by polyoma virus (see Chap. 64). Note that cells lie on top of each other and are randomly oriented.

Increased efficiency of clone formation

Increased edge indentation and refractivity

Ability to form colonies on top of a layer of untransformed cells

Decreased anchorage dependence

Lack of G0 and arrest at various phases of the cell cycle, almost randomly, upon medium exhaustion

CELL SURFACE

Decreased fibronectin content (associated with disorganization of the attached cytoskeleton)

Shorter gangliosides

Reduced glycosyl transferase activity

Reduced adhesion to plastic

Increased receptor mobility (lectins)

Increased agglutination by lectins (Fig. 47–9)

Loss of receptors for hormones and toxins

Increased and unregulated transport activity (e.g., glucose, phosphate, nucleosides, K^+)

Increased glucose-binding protein

Presence of new antigens

METABOLIC CHARACTERISTICS

Increased protease production

Disaggregated microfilaments and reduction of actin-associated proteins

Increased aerobic glycolysis

Decreased intracellular concentration of cyclic AMP

Increased, abnormal collagen synthesis

Resurgent fetal functions

OTHER CHARACTERISTICS

Production of tumor by 10^6 or fewer cells inoculated

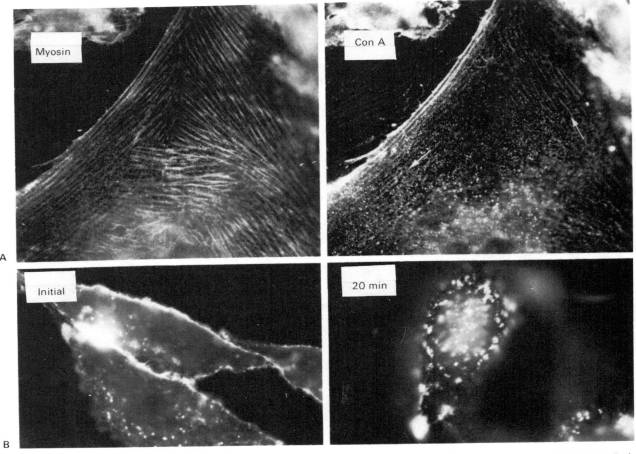

Figure 47–9. Surface and cytoskeletal alterations in transformed cells detected by the rearrangement of surface receptors induced by concanavalin A (conA, a plant lectin) and of myosin-containing cytoplasmic filaments. (*A*) At 20 minutes after exposure to conA, NRK (normal rat kidney) cells were fixed and treated with an anti-conA serum labeled with fluorescein (blue fluorescence, *right*), then treated with antimyosin serum labeled with rhodamine (red fluorescence, *left*). Through the use of appropriate light filters, the distributions of myosin (*left*) and of conA (*right*) were photographed in the same cell. The conA receptors, which were at first uniformly distributed over the cell surface, have formed very fine patches exactly aligned along the underlying myosin-containing stress fibers. (*B*; *right*) Large patches of conA receptors of Rous sarcoma virus–transformed NRK cells after 20 minutes of interaction with fluorescent conA. The initial uniform distribution is seen at left. (Ash JF, Singer SJ: Proc Natl Acad Sci USA 73:4575, 1976)

subcutaneously in immunologically accepting animals (syngeneic or athymic mice)
Chromosomal abnormalities
Absence of culture senescence

Some of the changes seen in transformed cells appear to derive from the persistent growth of transformed cells, since they are also found in growing untransformed cells. The genetic and epigenetic events controlling the various characteristics are not known; this is understandable given the limited methodology available for studying the complex genome in animal cells.

The spectrum of properties found in individual clones depends on the selection imposed during their isolation (e.g., whether by growth in agar, in low serum, or in dense cultures) and on the inducing agent (e.g., whether chemical or viral). Some kinds of transformed cells are malignant (i.e., they produce tumors in animals), others are not. The properties most closely associated with malignancy are loss of anchorage dependence, protease production, and reduction of fibronectin, but none is essential.

Genetic Studies With Cultured Cells: The Karyotype of Cultured Cells

The genetic characteristics of the cells are relevant for virological studies because many parameters of virus multiplication and its consequences depend on them. The analysis of the chromosomal constitution (**karyotype**) of tissue culture cells has acquired paramount importance in genetic studies since it became clear that karyotype anomalies of cultured cells are associated with certain human diseases. In addition, the karyotype gives an indication of the degree of abnormality that cells have attained during their cultivation *in vitro.*

Staining techniques allow not only the determination of the number of chromosomes in a cell but also their precise cytologic identification. Thus, characteristic **bands** are observed in mitotic chromosomes stained with the fluorescent dye quinacrine mustard and examined under ultraviolet light or stained after trypsin treatment.

In young cell strains, most cells tend to maintain the **diploid** (2n) chromosome number characteristic of the animal. The types of chromosomes are also usually normal, and the cells are said to be **euploid.** In contrast, the cells of older strains and cell lines (especially of transformed cells) deviate from the euploid pattern (**aneuploid** cells). The number of chromosomes may be different from diploid (**heteroploid**), either higher (usually between 3n and 4n, i.e., **hypertriploid**) or lower (**hypodiploid**). In **quasidiploid** cells the number of chromosomes is 2n, but their distribution is abnormal; for example, a chromosome of one pair may be missing and replaced by an extra chromosome of another pair. In addition, **chromosomal aberrations** (e.g., **translocations** and **deletions**) often involve highly characteristic morphologic abnormalities in individual chromosomes, which are useful as **markers** for cell identification.

Although in some cell lines most cells are diploid, the majority of cell lines are constituted of heteroploid, especially hypertriploid, cells. **Individual lines are heterogeneous,** with cells containing different numbers of chromosomes. The most frequent (**modal**) number remains constant if the cells are grown under a constant set of conditions, but a change in conditions often results in selection of a type with a different modal number. The variation encountered in heteroploid cultures reflects frequent unequal segregation of chromosomes at mitosis.

SOMATIC MUTATIONS. Most mutations occurring in tissue culture cells, like those observed in prokaryotes, result in **structural changes in a protein.** The mutations are revealed indirectly by physical changes of the protein (e.g., isoelectric point, electrophoretic mobility, heat stability) or directly by amino acid changes. In aneuploid cell cultures a mutational phenotype can also be brought about by a **change in chromosome balance,** due to the loss or the gain of certain chromosomes by mitotic segregation, without structural protein changes. Variants of this type occur and revert at a relatively high frequency, which is not enhanced by mutagenic agents.

The study of mutability suggests that in some cells of permanent lines, diploid genes tend to become **functionally haploid,** as if one of the two homologues were lost or inactivated. Thus in quasidiploid cell lines, many recessive mutations in diploid (autosomal) genes are phenotypically expressed almost as frequently as are mutations in haploid (X-linked) genes, though they would be expected to require two homologous mutations and hence to be much less common. Moreover, in mutant cells resistant to α-amanitin (which acts on RNA polymerase II), essentially all the polymerase molecules are resistant to the drug, although it is unlikely that both the genes specifying the enzyme were mutated. The molecular basis of these findings is not known.

Selected Reading

Ash JF, Singer SJ: Concanavalin A–induced transmembrane linkage of concanavalin A surface receptors to intracellular myosin-containing filaments. Proc Natl Acad Sci USA 73:4575, 1976
Beguinot L, Lyall RM, Willingham MC, Pastan I: Down-regulation of the epidermal growth factor receptor in KB cells is due to receptor internalization and subsequent degradation in lysosomes. Proc Natl Acad Sci USA 81:2384, 1984

Bell RM: Protein kinase C activation by diacylglycerol second messengers. Cell 45:631, 1986

Carpenter G: Properties of the receptor for epidermal growth factor. Cell 37:357, 1984

James R, Bradshaw RA: Polypeptide growth factors. Annu Rev Biochem 53:259, 1984

Newport JW, Kirschner MW: Regulation of the cell cycle during early xenopus development. Cell 37:731, 1984

Nishizuka Y: Studies and perspectives of protein kinase C. Science 233:305, 1986

Rheinwald JG, Green H: Epidermal growth factor and the multiplication of cultured human epidermal keratinocytes. Nature 265:421, 1977

Ross R, Raines EW, Bowen-Pope DF: The biology of platelet-derived growth factor. Cell 46:155, 1986

Vara F, Schneider JA, Rozengurt E: Ionic responses rapidly elicited by activation of protein kinase C in quiescent Swiss 3T3 cells. Proc Natl Acad Sci USA 82:2384, 1985

Yarden Y, Escobedo JA, Kuang W-J et al: Structure of the receptor for platelet-derived growth factor helps define a family of closely related growth factor receptors. Nature 323:226, 1986

48

Renato Dulbecco

Multiplication and Genetics of Animal Viruses

Multiplication

The multiplication of many animal viruses follows the pattern of bacteriophage multiplication described in preceding chapters, but with important differences. For instance, as discussed in Chapter 44, some RNA genomes are segmented; some DNA genomes have unusual termini (such as inverted, repeated nucleotide sequences or covalent crosslinking of linear strands); some virions contain a transcriptase and other enzymes; and some viruses express their genetic information in a unique manner, by reverse flow from RNA to DNA. Moreover, animal viruses differ from bacteriophages in their interactions with the surface of the host cells (which do not have rigid walls) and in the mechanisms of release of their nucleic acid in the cell.

Animal viruses can be differentiated into virulent and moderate. Moderate viruses resemble temperate bacteriophages in their ability to establish stable relations with the host cells, but the differentiation is less sharp than with bacteriophages, and it is sometimes difficult to decide whether an animal virus is virulent or moderate.

The main characteristics of animal virus families are given in Table 48–1. Further details are found in later chapters.

HOST CELLS FOR VIRAL MULTIPLICATION

The first hosts for experimental or diagnostic work with animal viruses were adult animals, then chick embryos. Chick embryos have also contributed in an important way to the development of virology by conveniently pro-

TABLE 48–1. Characteristics of Animal Virus Families

Type	Nucleic Acid Strandedness	Symmetry of Nucleocapsid	Naked (N) or Enveloped (E)	Diameter of Virus (nm)	Family	Examples (Specific Viruses Mentioned in This Chapter)
RNA	Single-stranded	Icosahedral	N	21–30	Picornaviruses (see Chap. 55)	Poliovirus, Coxsackie virus, Mengo virus
			E	45	Togaviruses (see Chap. 60)	Semliki Forest virus, western equine encephalomyelitis virus, yellow fever virus, dengue virus
		Helical	E	80–120	Orthomyxoviruses (see Chap. 56)	Influenza virus, fowl plague virus
				125–300	Paramyxoviruses (see Chap. 57)	Sendai virus, measles virus, mumps virus, respiratory syncytial virus
				70–80x 130–240	Rhabdoviruses (see Chap. 59)	Rabies virus, vesicular stomatitis virus
				80–160	Coronaviruses (see Chap. 58)	
		Unknown	E	110–130	Arenaviruses (see Chap. 60)	Lymphocytic choriomeningitis virus, Lassa virus
				90–100	Bunyaviruses (see Chap. 60)	California encephalitis viruses
				100	Retroviruses (see Chap. 65)	Rous sarcoma virus, human T-lymphotropic virus, human immunodeficiency virus
	Double-stranded	Icosahedral	N	75–80	Reoviruses (see Chap. 62)	Reovirus, rotavirus
DNA	Double-stranded	Icosahedral	N	70–90	Adenoviruses (see Chap. 52)	Adenovirus
				43–53	Papovaviruses (see Chap. 64)	Polyoma virus, SV40, JC virus, BK virus
			E	180–200	Herpesviruses (see Chap. 53)	Herpes simplex virus, varicellazoster virus, cytomegalovirus, Epstein-Barr (EB) virus
		Complex	E*	200–250 250–350	Poxviruses (see Chap. 54)	Vaccinia virus, variola (smallpox) virus
				45	Hepadnaviruses (see Chap. 63)	Hepatitis B virus
	Single-stranded	Icosahedral	N	18–22	Parvoviruses (see Chap. 52)	Adeno-associated virus

* Lipid in outer coat, but no distinct envelope

viding a variety of cell types, susceptible to many viruses. Various cell types can be reached by inoculating the embryo by different routes (Fig. 48–1). These hosts have now been replaced almost completely by cultures of animal cells for detailed studies of viral replication.

Cell Strains and Cell Lines

Every type of animal cell culture discussed in Chapter 47 has found application in virology. The choice of species, tissue of origin, and type of culture (primary, cell strain, or cell line) depends on the virus and the experimental objectives. The systems used for the individual viral families will be given in the appropriate chapters.

HOST SUSCEPTIBILITY. Each animal virus can replicate only in a certain range of cells. Among **nonsusceptible cells** some have a block at an early step (e.g., they lack receptors for viral attachment or a factor required for expression of viral genes), so that the expression of all viral functions is prevented (**resistant cells**). Other cells

lack a factor required for a later step, so that some, but not all, viral activities are expressed (**nonpermissive cells**). In either case a **heterokaryon,** formed by fusing a susceptible and a nonsusceptible cell, has the function and is usually susceptible.

PRODUCTIVE INFECTION
Role of Nucleic Acid—Transfection

That cells of higher organisms can be infected by naked viral nucleic acid, yielding normal virions, was first shown for tobacco mosaic virus RNA by Gierer and Schramm. The same was soon shown for the RNA or DNA of many animal viruses. In conjunction with the Hershey and Chase experiment with bacteriophage (see Separation of Nucleic Acid From Coat, Chap. 45), this result established the exclusive genetic role of the viral nucleic acid.

There are several important differences between infections by nucleic acid (transfection) and by virions.

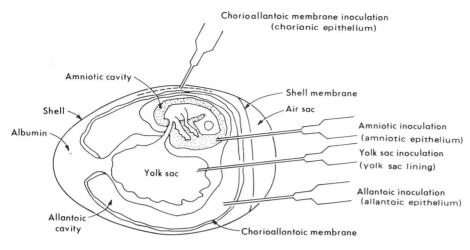

Figure 48–1. Chicken embryo (10–12 days old) and routes of inoculation to reach the various cell types (as indicated). For chorioallantoic membrane inoculation a hole is first drilled through the eggshell and shell membrane; the shell over the air sac is then perforated. Air enters between the shell membrane and the chorioallantoic membrane, creating an artificial air sac, where the sample is deposited, in contact with the chorionic epithelium. Yolk sac inoculation is usually carried out in younger (6-day-old) embryos, in which the yolk sac is larger.

1. **The efficiency of infection with nucleic acid is much lower,** by a factor of 10^{-6} to 10^{-8} in ordinary media, showing the important role of the viral coat in infectivity. The infectivity of nucleic acid is increased by several orders of magnitude in the presence of basic polymers (e.g., diethylaminoethyldextran) or by precipitation of the viral DNA onto cells with calcium phosphate. These additions appear to protect the nucleic acid against nucleases and to increase its uptake by the cells. Even under the most favorable conditions, however, the bare nucleic acid is no more than 1% as infectious as the corresponding virions. This limitation seems to arise from degradation of much of the nucleic acid within the cells. The efficiency is further increased by injecting the nucleic acid into the cells or by delivering it packaged in membranous vesicles (liposomes) that bind to the cell membrane and empty their contents into the cells.

2. **The host range is much wider with nucleic acids,** which can infect resistant cells. For instance, chicken cells, although resistant to poliovirus because they lack receptors for the virions, are susceptible to its RNA, but only a single cycle of viral multiplication takes place because the progeny are again virions and cannot spread to other cells.

3. **Infectious nucleic acid can be extracted even from heat-inactivated viruses** in which the protein of the capsid has been denatured; the nucleic acid can withstand much higher temperatures than the protein. The ability of nucleic acid infectivity to survive damage to the viral coat must be considered in the preparation of viral vaccines.

4. With some RNA viruses (e.g., poliovirus) **a DNA copy of the viral RNA is infectious.** This permits the preparation of viral genomes (such as those of vaccine strains) in large quantities by avoiding the high mutation rate in replication of RNA and its lability.

5. Finally, **the infectivity of nucleic acid is unaffected by virus-specific Abs,** which suggests that this form of a virus could be an effective infectious agent even in the presence of immunity. However, nucleases in body fluids probably greatly limit its role, because a single complete break in a molecule abolishes its infectivity. Indeed, naked viral nucleic acid plays a role in natural infection only as plant viroids (see Chap. 44), the tight secondary structure of which resists nucleases.

Of the animal viruses, papovaviruses, adenoviruses, some herpesviruses, togaviruses, and picornaviruses yield infectious nucleic acids. With retroviruses (see Chap. 65), infectious DNA can be extracted from infected cells or can be made by copying the viral RNA *in vitro*. Failure in other cases can be ascribed either to the difficulty of extracting a large DNA molecule intact or to **lack of virion enzymes** (e.g., a transcriptase) required for initiating the viral growth cycle.

Initial Steps of Viral Infection

Initially, as with bacteriophage infection, the viral nucleic acid must be made available for replication. Unlike bacteriophage, however, the entire animal virus nucleocapsid enters the cell, and the nucleic acid is then released. These early events can be investigated by studying the changes of viral infectivity or the fate of radioactively labeled viral components, or by electron microscopy. The following steps can be identified.

1. **Attachment or adsorption.** The virus becomes attached to the cells, and at this stage it can be **recovered in infectious form** without cell lysis by procedures that either destroy the receptors or weaken their bonds to the virions. Animal viruses have specialized attachment sites distributed over the surface of the virion: for instance, enveloped viruses such as orthomyxoviruses and paramyxoviruses attach through glycoprotein spikes (see The Envelope, Chap. 44), and adenoviruses attach through

the penton fibers (see Immunologic Characteristics, Chap. 50).

Adsorption occurs to **specific receptors,** which vary in number for different viruses from 5×10^2 to 5×10^5 per cell. Some receptors are glycoproteins (for myxoviruses or paramyxoviruses); others are phospholipids and glycolipids (for rhabdoviruses or some paramyxoviruses). They are usually macromolecules with specific physiological functions, such as complement receptors (for EBV), β-adrenergic receptor (for reovirus type 3), or Ia molecules of the major histocompatibility complex (for lactate dehydrogenase virus).

Whether or not receptors for a certain virus are present on a cell depends on the species and the tissue from which the cell derives and on its **physiologic state.** Cells lacking receptors for a certain virus are resistant to it, i.e., cannot be infected. Attachment is blocked by antibodies that bind to the viral or cellular sites involved. In epithelial cells the distribution of receptors may be polarized, i.e., confined to either the basolateral or the apical surface.

2. **Penetration** rapidly follows adsorption, and the virus can no longer be recovered from the intact cell. The most common mechanism of penetration is **receptor-mediated endocytosis,** the process by which many hormones and toxins enter cells: an area of cell membrane containing receptors with the attached virions—usually a clathrin-coated pit—invaginates to form a cytoplasmic vesicle. Virions can be recognized within the vesicles by electron microscopy (Fig. 48–2).

UNCOATING. The most common mechanism causing separation of the viral coat from the genome-containing core is an interaction of a coat component with the endosomal membrane. A key step is the **acidification** of the content of the endosome to a pH of about 5, owing to the activity of a proton pump present in the membrane. The low pH causes rearrangement of coat components, which then expose normally hidden hydrophobic sites. They bind to the lipid bilayer of the membrane, causing

extrusion of the viral core into the cytosol. For influenza virus (see Chap. 56) the acid-sensitive component is the HA_2 unit of the hemagglutinin; for adenoviruses it is the penton base. Ultimately the endosome fuses with a lysosome, where remnants of the viral coat as well as uncoated virions are destroyed. Paramyxoviruses are an exception: their fusion protein can carry out the fusion step at physiologic pH; hence, they penetrate and uncoat in a single step at the cell surface, rather than in endosomes.

Many weak bases, such as chloroquine or ammonium chloride, which accumulate in endosomes, prevent their acidification and block uncoating. Mutations in the proton pump have similar consequences. In both cases the cells become resistant to many viruses, as well as to hormones or toxins that use the same endocytic pathway. Amantadine, in contrast, affects only influenza type A and a few other viruses. Amantadine-resistant influenza mutants are altered in the hemagglutinin, which interacts with the endosomal membrane, and the connected matrix protein; these virions fuse with membranes in vitro at pH 7.0. This is a direct evidence for the role of membrane fusion in endosome for uncoating.

One-Step Multiplication Curves

As with bacteriophages, multiplication curves for animal viruses are obtained under one-step conditions (see Chap. 45) by means of cells from suspension cultures or from monolayer cultures dispersed by trypsin. For viruses the cell receptors of which can be easily destroyed (e.g., influenza viruses, polyomavirus) one-step conditions are possible even without dispersing the cell layers: the infected cultures are washed free of unadsorbed virus and then covered with a medium containing receptor-destroying enzyme, which prevents adsorption of released virus to the cells. For other viruses the same result is obtained by washing the monolayers with antiviral Ab after infection.

One-step multiplication curves of animal viruses (Fig. 48–3) show the same stages observed with bacterio-

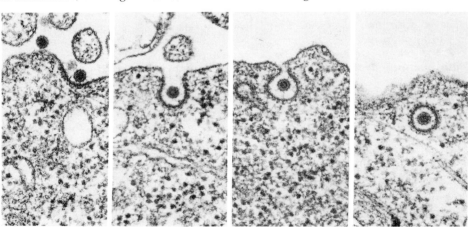

Figure 48–2. The first stages in infection of Semliki Forest virus, a togavirus. The electron micrographs show, from left to right, four stages in penetration. Of the two extracellular virions present in the first picture, one is attached to receptors in a coated pit; note the dark chathrin layer at the cytoplasmic site of the membrane. The virion-containing pit begins to deepen in the second picture and is almost closed in the third. In the fourth picture it has formed an endosome surrounded by chathrin and free in the cytoplasm. (Original magnification × 70,000; courtesy of J. Kartenbeck)

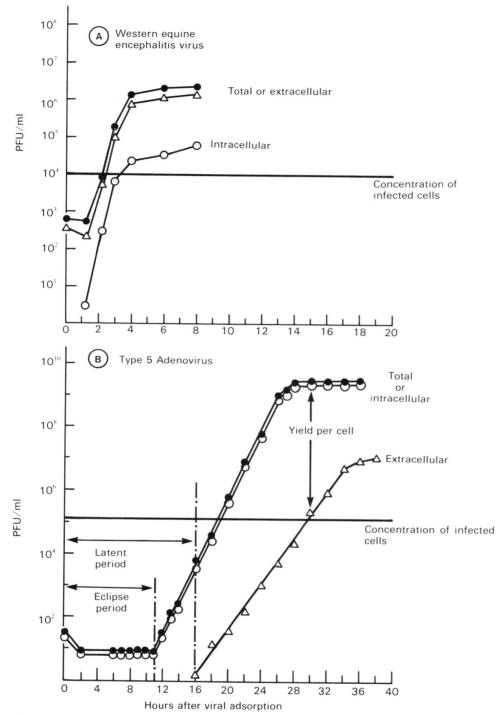

Figure 48–3. One-step multiplication curves of two viruses with intracellular accumulation periods of different lengths. Extracellular virus is measured in the medium surrounding the intact cells, intracellular virus after removal of the medium and disruption of the cells. (*A*) Western equine encephalitis virus multiplies in cultures of chick embryo cells with an extremely short accumulation period. (*B*) Type 5 adenovirus multiplies in cultures of KB cells with a long accumulation period. Intracellular (i.e., cell-associated) virus is measured by disrupting cells after they have been washed free of extracellular virus (i.e., virus already released into the medium). *PFU,* plaque-forming unit. (*A,* from data of Rubin H et al: J Exp Med 101:205, 1955; *B,* from data of H. S. Ginsberg and M. Dixon)

phages (see Chap. 45). The length of the **intracellular accumulation period** varies widely, however, with different viruses. It is very long with some, the progeny virions of which tend to remain within the cells, and is nonexistent with others, which mature and are released in the same act (by acquisition of an envelope at the cell surface). If the accumulation period is very short, the quantity of intracellular virus is at any time a small fraction of the total virus, as in Figure 48–3A.

Effect of Viral Infection on Host Macromolecular Synthesis

In these studies viral and cellular nucleic acids are separated and identified by their size, buoyant density, and configuration (e.g., cyclic), or their hybridization to the nucleic acid present in purified virions. Viral and cellular proteins can be distinguished immunologically or by their different rates of migration in acrylamide gel electrophoresis.

As with bacteriophages, **virulent viruses,** either DNA-containing (e.g., adenovirus, vaccinia virus, herpesvirus) or RNA-containing (e.g., poliovirus, Newcastle disease virus, reovirus), shut off cellular protein synthesis (Figs. 48–4 and 48–5) and disaggregate cellular polyribosomes, favoring a **shift to viral synthesis.** With most viruses the effect does not occur in the presence of inhibitors of protein synthesis (puromycin, cycloheximide), indicating that it is mediated by new virus-specified proteins. The replication and transcription of cellular DNA are

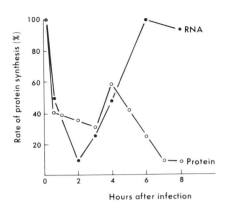

Figure 48–5. Inhibition of cellular RNA and protein synthesis in L cells infected with mengovirus (a picornavirus). The decline in incorporation of radioactive precursors begins immediately after infection. The resumption of synthesis at about 3 hours is due to synthesis of viral RNA and proteins. (Modified from Franklin RM, Baltimore D: Cold Spring Harbor Symp Quant Biol 27:175, 1962)

blocked by adenoviruses, herpesviruses, and poxviruses; chromosome breaks are often observed.

The **mechanisms of protein synthesis shut-off** vary even within the same viral family. Thus, among picornaviruses, poliovirus, using a viral protease, causes **cleavage of a 200-Kd cap-binding protein,** which is required for initiation of translation of capped cellular messengers. Viral messengers, being uncapped, are not affected. Viral mutants not carrying out the cleavage do not inhibit translation. In contrast, Mengo virus does not prevent initiation but causes an **elongation block of the initiation complexes** (ribosome, messenger, and Met-tRNA). With vesicular stomatitis virus the **leader sequence,** a transcript of a small part of the 3' end of the genome, also blocks the initiation step, probably in conjunction with viral or cellular proteins. Herpesvirus induces degradation of cellular mRNAs; and an adenovirus gene (E1b) prevents the transport of cellular mRNAs to the cytoplasm during the late part of infection.

In contrast to virulent viruses, **moderate viruses** (e.g., polyomavirus) may **stimulate** the synthesis of host DNA, mRNA, and protein. This phenomenon is of considerable interest for viral carcinogenesis (see Viral Multiplication in Cell Culture, Chap. 64).

SYNTHESIS OF DNA-CONTAINING VIRUSES

There are **three classes of DNA-containing viruses,** based on genome structure: I, **double-stranded linear;** II, **double-stranded circular;** and III, **single-stranded linear.** All the linear DNAs possess some form of inverted repetitions, either at the termini of the genomes (e.g., adenoviruses) or in their substructure (e.g., herpesviruses). In addition, within these classes distinctive

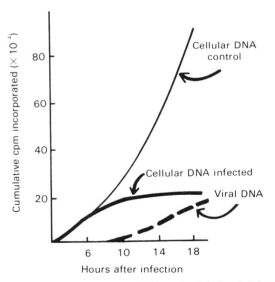

Figure 48–4. Inhibition of cellular DNA synthesis in L cells infected by equine abortion virus (a herpesvirus) in the presence of ^{3}H-thymidine. Viral DNA was separated from cellular DNA because of its higher buoyant density in CsCl. (Modified from O'Callaghan DJ et al: Virology 36:104, 1968)

structural variations also occur. Biosynthesis follows patterns similar to those described for bacteriophages, but the structural differences in the viral genomes appear to dictate different modes of transcription, methods of posttranscriptional processing, and forms of DNA replication. The general reactions will be described here; more specific details will be found in the chapters that follow.

Transcription

In the synthesis of DNA-containing animal viruses, as in eukaryotic cells and unlike the process in prokaryotes, transcription and translation are not coupled. Except for poxviruses, transcription occurs in the nucleus and translation in the cytoplasm. Generally, the **primary transcripts,** generated by RNA polymerase II, are larger than the mRNAs found on polyribosomes, and in some cases as much as 30% of the transcribed RNA remains untranslated in the nucleus (e.g., adenovirus transcripts; see Chap. 52). The viral messengers, however, like those of animal cells, are monocistronic.

Transcription has a **temporal organization.** With most DNA-containing viruses only a fraction of the genome is transcribed into **early** messengers before replication has begun, and after DNA synthesis the remainder is transcribed into **late** messengers. The complex viruses have **immediate early** genes, which are expressed in the presence of inhibitors of protein synthesis, and **delayed early** ones, which require protein synthesis for expression. Interference with DNA synthesis by temperature-sensitive mutations or by inhibitors shows that the switch to late transcription requires prior DNA replication.

Regulation is carried out by proteins present in the virions, or specified by viral or cellular genes, interacting with regulatory sequences at the 5' end of the genes. Fusion of such sequences to a marker gene with easily measurable effects (e.g., the herpes thymidine kinase gene) show that the few hundred nucleotides preceding the coding sequences are essential for control. They have properties related to those of the **enhancers** of tumor viruses (see Chap. 64): they respond in *trans* to substances produced by other genes and act in *cis* on the associated genes, which they either stimulate or inhibit. For instance, with herpes simplex virus, regulatory events acting mainly on transcription cause the sequential appearance of different classes of mRNAs: **immediate early (α), delayed early (β),** and **late (γ^1 and γ^2)** (Fig. 48–6). In addition, late genes, especially those of the γ^2 class, are activated after viral DNA replication begins (see Multiplication, Chap. 53). These regulatory properties are generally similar to those of large bacteriophages (see Chap. 45).

Different classes of genes may be transcribed from **different DNA strands** and therefore in opposite direc-

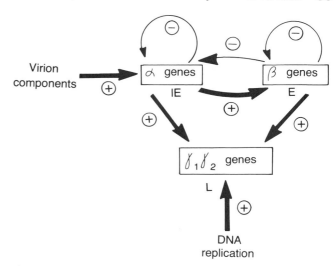

Figure 48–6. Regulation of transcription of the various classes of herpes simplex genes. +, *thick lines:* stimulation of transcription; −, *thin lines:* inhibition of transcription; *IE,* immediate early; *E,* early; *L,* late. α and β genes inhibit their own transcription through their products.

tions. Thus, with SV40, a papovavirus, unique regions on opposite strands code for early and late mRNAs (see Productive Infection, Chap. 64). In adenovirus (Fig. 48–7) **early transcription** occurs on both strands in five scattered regions of the genome. In contrast, synthesis of adenovirus **late messages** is mostly confined to the rightward-reading (r) strand: a unique large transcript is cleaved to generate five groups of messages (see Fig. 48–7) with identical 5' ends, each composed of a nontranslated cap and leader sequences, which are not contiguous to the message sequences in the genome. Each group has identical 3' ends. Similar is the arrangement of the two late mRNAs of both SV40 and polyomaviruses.

This complex method of producing mRNAs by **posttranscriptional processing** of primary transcripts serves to remove **intervening sequences** not destined to be translated. As with cellular mRNAs, their removal is essential for regulating the appearance of functional messages, and alternative splicings yield different messages and different proteins from the same segment of DNA. The complex regulation of transcription depends on proteins specified by both viral and cellular genes.

Like eukaryotic cell mRNAs, viral transcripts undergo other modifications: the addition of a **poly(A) chain** (100–200 adenine residues long) to the 3' end and a **methylated cap** at the 5' end. Except for poxviruses, these additions are made in the nucleus. As with cellular mRNA, the methylated capped 5' terminus appears essential for the stable attachment of viral mRNAs to the 40S ribosomal subunit and for effective translation.

The early transcription and processing of most DNA-containing viruses are carried out by host enzymes, us-

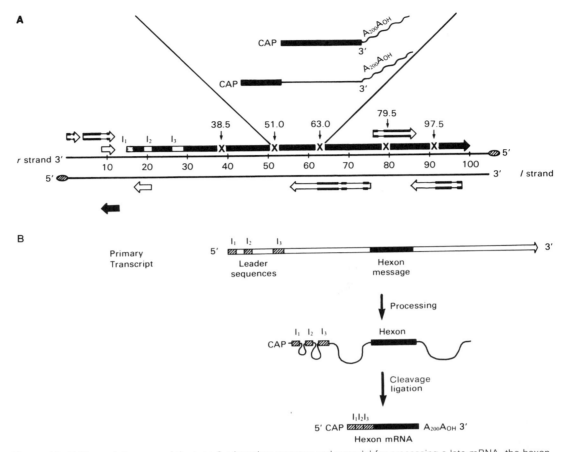

Figure 48–7. Transcription map of the type 2 adenovirus genome and a model for processing a late mRNA, the hexon message. (A) The encoded regions for the early and late mRNAs are indicated. *White arrows,* early transcripts; *black arrows,* late transcripts. The thickened lines on the early transcripts indicate the regions that are included in one of the major mRNAs processed from transcripts of each region. The possible sites for cleavage and polyadenylation of the late transcript to produce the 3' termini of the messages are designated by Xs or small arrows. The cleavage and splice sites (see *B*) are each identified by a unique nucleotide sequence. The enlarged segment between positions 51.0 and 63.0 represents two late messages processed from large transcripts. They have different 5' ends and identical 3' ends; the region of the message translated into protein is designated by the heavy segment. (B) Primary transcript for late mRNAs on the r strand, and suggested model for processing this transcript to form the hexon mRNA. Only one mRNA can be derived from each transcript.

ing RNA polymerase II as transcriptase. Poxviruses, however, are profoundly different: they use enzymes present in the core of the virions, and their transcripts, without intervening sequences, are not spliced.

Synthesis of Viral Proteins

Viral proteins are synthesized on cytoplasmic polysomes in a temporal sequence corresponding to that of the viral mRNAs (Fig. 48–8): early proteins participate in DNA replication and transcription, the late ones are predominantly structural proteins of the virions. With most viruses these proteins are transported to the nucleus, where assembly takes place. Posttranscriptional control may participate in determining the temporal sequence;

for instance, with adenovirus a small viral RNA, VA1, enhances translation of late protein by stabilizing the initiation factor 2.

DNA Replication

DNA replication utilizes precursors derived from the medium, since cellular DNA is not degraded. The smaller DNA viruses rely on the host cell DNA polymerase, but the more complex adenoviruses, herpesviruses, and poxviruses use virus-encoded polymerases. Synthesis begins toward the middle of the eclipse period (see Fig. 48–3B), after the early viral proteins are made: it is blocked by mutations in some early viral genes or by inhibition of protein synthesis shortly after infection.

The mode of replication is **semiconservative,** but the nature of the **replicative intermediates** depends on the devices used for achieving complete replication—a problem already noted in Chapter 45 (see Fig. 45–10). Several methods of replication can be recognized (Fig. 48–9).

1. **Adenoviruses** (see Fig. 48–9*A*) show asymmetric replication, which initiates at the 3' end of one of the strands. The growing strand uses as primer cytidylic acid bound to the precursor of the terminal protein (see Chap. 44). That protein, together with the viral DNA polymerase and other proteins, becomes associated with a terminal repetition at the 3' end of the template strand. The growing strand **displaces the preexisting strand** of the same polarity and, with the template strand, builds a complete duplex molecule. The displaced single strand in turn replicates in a similar way after generating a panhandle structure by pairing the inverted terminal repetitions.

2. Several viruses use **circular intermediates. Herpesvirus** (see Fig. 48–9*B*) has a linear genome terminated by direct repeats of 300 to 400 base pairs, often in multiple copies. After the infecting viral DNA reaches the cell's nucleus, the ends undergo limited exonucleolytic digestion (see Fig. 48–9*B, top*) and then pair to form circles. In the **first phase** they replicate as cyclic molecules (see Fig. 48–9*B, middle*); in the **second phase** they form tandem **concatemers** in head-to-tail connection, probably by rolling circle replication (see Fig. 48–9*B, bottom*). During maturation, unit-length molecules are cut from the concatemers.

Papovaviruses (see Fig. 48–9*C*) have cyclic DNA in their virions. Replication is **bidirectional** and **symmetric,** via cyclic intermediates.

3. Replication of **parvovirus DNA** (see Fig. 48–9*D*), a single-stranded linear molecule, is directed by the unusual structure of its ends, which have **terminal palindromes** capable of forming hairpins. The replicative intermediates of the defective **adeno-associated virus DNA** include special concatemers, predominantly of double length, in which **a plus and a minus strand are covalently linked in the same strand.** They are probably produced in two self-priming steps (see Fig. 48–9*D, top* and *second from top*). Progeny molecules are generated by strand displacement (see Fig. 48–9*D, second from bottom* and *bottom*) and from double-stranded molecules, either monomeric or concatemeric.

4. **Poxvirus DNA** (see Fig. 48–9*E*) has, in addition to its large size, another striking feature: the two complementary strands are joined. The links are at the end of inverted terminal repetitions, so the two ends form two equal palindromes. The replicative intermediates, present in the cytoplasm, are special concatemers containing pairs of genomes connected either head-to-head or tail-to-tail. A model generating these structures is

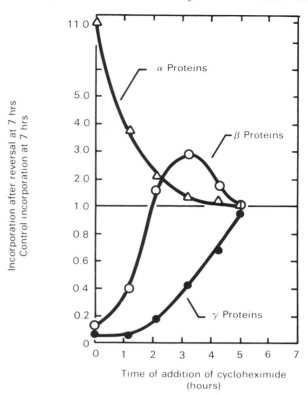

Figure 48–8. Sequential appearance of the three sets of type 1 herpes simplex virus proteins. These data show the rate of synthesis of representative α, β, and γ polypeptides following removal of cycloheximide 7 hours after infection. Cycloheximide was added to infected cultures at different times after infection (noted on the abscissa) and was removed from all cultures simultaneously. The relative rates of synthesis (plotted on the ordinate) were determined by measuring radioisotope incorporation into a specific polypeptide for 30 minutes after removal of cycloheximide as compared to the incorporation into the same viral polypeptide from infected cells to which inhibitor had not been added. (Modified from Honess RW, Roizman B: J Virol 14:8, 1974)

shown in Figure 48–9*E*, with head-to-head palindromes under the first part of the figure. Unit molecules are produced by staggered cuts and ligation.

5. **Hepatitis B** virus employs **reverse transcription** for multiplication (see Fig. 48–9*F*). The virions contain a partially double-stranded circular DNA with a complete negative strand (that is complementary to mRNA) in circular nicked form and an incomplete positive strand. Upon entering cells the positive strand is completed, generating a covalently closed, circular molecule, which is transcribed (see Fig. 48–9*F, top*). RNA transcripts are in turn reverse-transcribed into DNA by a viral enzyme in several steps, following very closely the model of retroviruses (see Chap. 65) including a jump of the nascent positive strand from one direct repeat (DR) to another (see Fig. 48–9*F, second and third from bottom*). The

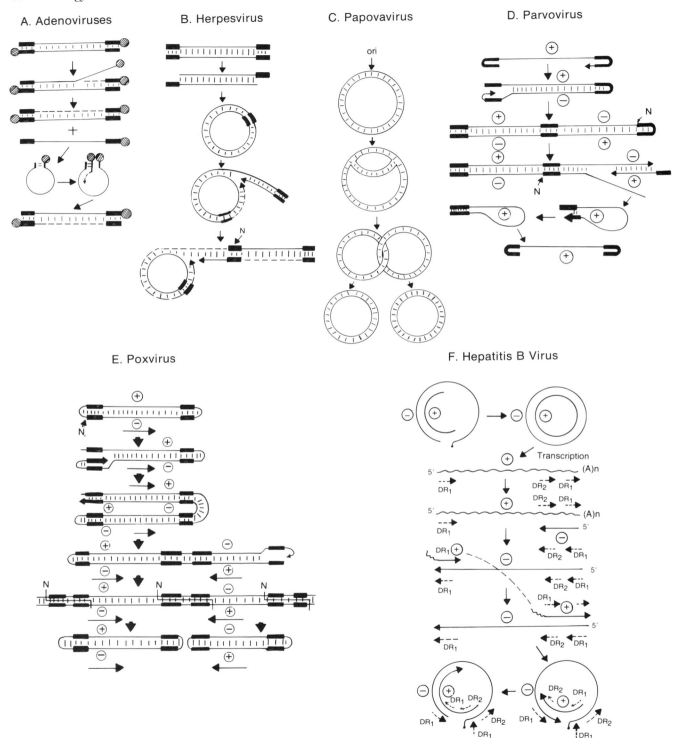

Figure 48–9. Models for the replication of viral DNAs. (*A*) *Dashed portions,* terminal protein. (*D*) +, −, strand polarity; *N,* endonuclease; *heavy lines,* palindromes; first and second from top are successive self-priming steps. (*E*) *Heavy arrows,* polarity of the helix; *N,* endonuclease. (*F*) *Wavy lines,* RNA; *thin lines,* DNA; *DR₁, DR₂,* direct repeats. (*D,* Senapathy P, Tratschin JD, Carter BJ: J Mol Biol 179:1, 1984; *E,* Moyer RW, Graves RL: Cell 27:391, 1981; *F,* Seeger C, Ganem D, Varmus HE: Science 232:477, 1986)

result is the uncompleted double-stranded molecule (see Fig. 48–9F, *bottom*).

With all viruses, as with bacteriophages, the newly synthesized viral DNA enters a pool from which it is subsequently removed to associate with virion structural proteins. Thus, if the infected cells are exposed to a short pulse of a radioactive DNA precursor at any time during the eclipse period, the label is distributed among virions finished at any subsequent time. In contrast to bacteriophage infection, viral DNA is made in excess, and much remains unused in the infected cells at the end of the multiplication cycle, often as a constituent of inclusion bodies.

Synthesis of RNA Viruses

The replication strategy of these genomes is dictated by the absence of multiple translation units within the same messenger, a characteristic of all animal cell messengers. To overcome this difficulty, three main strategies have developed: (1) With some viruses, the virion RNA, acting as messenger, is **translated monocistronically** into a **giant peptide,** which is then cleaved to generate distinct viral proteins. (2) In other viruses the virion RNA is **transcribed** to yield various **monocistronic mRNAs** by initiating transcription at different places. (3) Occasionally, the genome itself is a collection of **separate RNA** fragments that are transcribed into monocistronic mRNAs.

RNA-containing animal viruses can be placed in seven different classes, according to the nature of the viral RNA and its relation to the messenger, **which is taken to be of positive polarity** (Fig. 48–10).

In **classes I and II** the genomes have the same polarity as the messengers and are therefore defined as **positive-strand** viruses. In **class I** viruses (e.g., picornaviruses) the genome itself **acts as messenger,** specifying information for the synthesis of both structural and nonstructural proteins. The same RNA molecule must also initiate replication, because infection by a single viral particle occurs. Since RNA replication re-

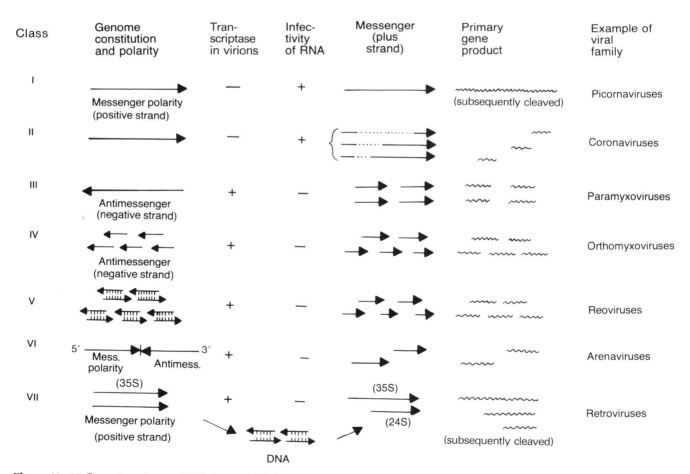

Figure 48–10. The various classes of RNA viruses and their primary modes of expression. The numbers of multiple genome pieces, messengers, and gene products are only diagrammatic representations and are not precise.

quires viral proteins (see below), the messenger function must be expressed first.

In **class II** viruses (e.g., coronaviruses) the genome generates first a negative-strand transcript, which is then transcribed into monocistronic mRNAs of different sizes. The way these are made is unique. Each begins with an identical short 5′ leader that is joined to transcripts beginning at the start of the various genes and then continuing to the 3′ end of the genome (**nested messengers**). These mRNAs are not produced by splicing a genomic-size transcript, because the virus, able to replicate in enucleated cells, does not require nuclear functions such as those required for splicing; instead, the leader probably remains attached to the polymerase and then acts as primer for the synthesis of the body of each messenger. Only one protein is made on each messenger, encoded in the 5′ end of its body.

In **class III and IV viruses** the genomes have the polarity opposite that of the messenger and are defined as **negative-strand genomes.** In **class III** (e.g., paramyxoviruses) a **virion transcriptase** transcribes the genomes into separate monocistronic messengers initiating at a single promoter. The transcriptase stops and restarts at each juncture between different genes, generating the various messengers. **Class IV** viruses (orthomyxoviruses) have the additional feature that the genome is in several distinct, nonoverlapping pieces of single-stranded RNA; each segment gives rise to its own messenger. Most genomic segments contain a single gene, but two segments contain two overlapping genes: one is expressed by a full-length messenger, the other by a shorter messenger obtained by the former by splicing. The replication of these viruses has a nuclear phase, during which splicing occurs by means of the same signals and enzymes as eukaryotic transcripts.

In the synthesis of their messengers, orthomyxoviruses follow the peculiar strategy of using as primers capped 5′-end fragments obtained by endonucleolytic cleavage of host messengers. The use of cellular 5′ ends might expose the viral messengers to a block of translation acting on cellular messengers in virus-infected cells (see Interferon, Chap. 49), but this effect is prevented by a viral protein.

Class V viruses (e.g., reoviruses) contain distinct, nonoverlapping segments of **double-stranded RNA** (ten in reoviruses); each is transcribed into an independent mRNA by a **virion transcriptase.** Most messengers are monocistronic, but one is bicistronic and expresses a second protein by initiating at an internal AUG in a different reading frame.

Class VI viruses (e.g., arenaviruses) have genomes that do not conform to any of the former classes. About half of the genome is of negative polarity and is transcribed into a messenger by a **virion transcriptase,** but the other half, of positive polarity, is transcribed twice: first a complete transcript of the genome is made, then the messenger is transcribed from this transcript. These viruses are said to have an **ambisense genome,** because the genetic information is inscribed in opposite directions in its two parts. This is unusual for RNA but not for dsDNA, in which information is often inscribed in opposite directions in a strand: some segments are transcribed from that strand, others from its complement. In arenaviruses this is a device for independently regulating genome replication and virion maturation (see Chap. 60).

Rhabdoviruses (negative-strand genome) have a related organization, because they make two leaders, one by transcribing the viral strand (+ leader), the other its complement (− leader). But they are not messengers; they remain untranslated: the positive-strand leader is probably involved in host shutdown, whereas the negative-strand leader binds to a specific protein of unknown function.

Class VII viruses (retroviruses) are unique because their genomes are transcribed into DNA, not RNA. They contain two identical single-stranded RNAs of positive polarity, with a poly(A) tail at the 3′-OH terminus and a cap at the 5′ end. In productive infection each is transcribed into DNA by a **reverse transcriptase** present in the virion; the functional mRNAs are then transcribed from this DNA (see Chap. 65).

Since RNA viruses of classes III to VII require a virion transcriptase for synthesizing a messenger, their purified viral RNAs are not infectious. Only those of viruses of classes I and II are infectious. Viruses of class VI behave aberrantly: the virion RNA is apparently of messenger, or positive, polarity but is not transcribed or translated.

With RNA virus there is no differentiation between early and late messengers in any of the preceding classes, with the possible exception of reoviruses (class V).

Viral Proteins

Viral proteins are synthesized in two different patterns that satisfy the monocistronic nature of the messengers.

1. When a virus has several messengers, each yields only one protein (two in exceptional cases; see above), and the number of viral proteins identifiable by acrylamide gel electrophoresis is equal or close to the number of messengers.

2. When, as with picornaviruses, the genome serves as a single messenger, a giant **polyprotein** is made, which is then cleaved (**processing**) to yield the viral proteins. The polyprotein is not normally recognized because it undergoes the first cleavage while being synthesized; inhibition of cleavage, by amino acid analogs incorporated into the precursor polyprotein or by high temperature, allows its detection. All viral proteins are generated in a series of successive cleavages (Fig. 48–11): the P1 region, at the amino terminus of the polyprotein, generates the

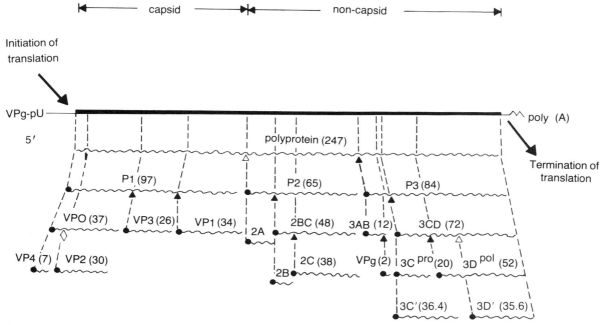

Figure 48–11. Posttranslational processing of the poliovirus polyprotein. *Solid line,* poliovirus genome; *wavy lines,* polypeptides. The polyprotein is rapidly cleaved into peptides P1, P2, and P3, which are then cleaved further. *Closed triangles,* cleavage at glutamine–glycine pairs by the virus-coded enzyme 3C; *open triangles,* cleavage at tyrosine–glycine pairs by viral enzyme 2A: *open diamonds,* cleavage at asparagine–serine pairs during capsid formation. 3C′ and 3D′ result from alternative cleavage. Numbers in parentheses indicate molecular weight in kilodaltons. (Modified from Emini EA, Schleif WA, Colanno RJ, Wimmer E: Virology 140:13, 1985)

structural proteins; the central P2 region generates proteins with unknown functions; and the P3 region at the carboxyl terminus generates noncapsid proteins, including the RNA polymerase and a cleavage enzyme. Cleavage is carried out by specific viral enzymes that are self-cleaved out of the polyprotein and recognize certain dipeptides, mainly glutamine-glycine, in the context of the secondary–tertiary structure of the protein.

REGULATION OF PRODUCTION. Regulation of class I viruses (picornaviruses and togaviruses) is governed by the posttranslation processing just described: proteins resulting from the first cleavage are available earlier in infection than those from later cleavages. With other viruses regulation is mainly transcriptional but is modulated by regulation of translation. This is evident when the synthesis of a gene product is shut off during the late period while the corresponding mRNA accumulates in the cytoplasm.

Replication of Single-Stranded RNA (Class I to V Viruses)

In all cases replication consists in building a **template strand** complementary to the viral strand and of the same length, which is then the template for **progeny viral strands.** These steps are carried out by a complex of enzymes of both viral and cellular origin, in association with the nucleocapsids of the infecting virions.

In many instances replication and transcription interfere with each other: with negative-strand viruses both template strands and transcripts are made from viral strands; and with positive-strand viruses a viral strand can be used as messenger or replication template. Initially in infection there is no interference: the messenger function is needed to provide proteins needed for replication. Later the supply of these proteins regulates the rate of replication. For instance, with vesicular stomatitis virus, a negative-strand virus, the amount of N protein regulates replication by binding to the progeny RNAs to form nucleocapsids. RNA replication can be abolished by preventing the binding of N-protein to the RNA, for instance by injecting into the cells a monoclonal antibody that combines with the free protein.

With the positive-strand poliovirus, replication occurs when the viral pVg protein becomes covalently linked at the 5′ ends of the RNA, apparently initiating the formation of a replication complex.

Messenger and progeny strands are sometimes differentiated structurally. For instance, with influenza virus (an orthomyxovirus), the messenger strands differ in two ways from progeny strands: they have a capped leader derived from cellular messenger, and they lack 17 to 22 nucleotides at the 3' end. Moreover, replication requires ongoing protein synthesis to provide the required proteins, whereas transcription does not.

REPLICATIVE INTERMEDIATES. In RNA replication the newly made template strand remains associated with the viral strand on which it is made, forming a double-stranded structure the length of the viral genome, known as RF (**replicating form**). Synthesis of new strands occurs by conservative asymmetric synthesis; a single viral strand is made, displacing the preexisting viral strand, which then becomes associated with capsid proteins to generate a new nucleocapsid (Fig. 48–12). An RF with a nascent viral strand is known as an **RI (replicative intermediate)**. RF molecules are fairly abundant during repli-

cation because after completion of a new strand the replicase appears to remain associated for some time with the template before reinitiating synthesis; RFs accumulate at the end of replication, when no more RIs are formed.

The mechanisms that form the double-stranded molecules are not well known. Poliovirus requires the association of the viral replicase with a protein known as "host factor," which can be replaced *in vitro* by oligo-U. The oligonucleotide acts as replication primer, suggesting that the host factor is a terminal uridyl transferase that adds a short chain of U's at the end of the poly(A). The chain of U's folding back along the poly(A) would form the replication primer.

SITE OF REPLICATION. The viral RNAs replicate in the cell's cytoplasm, except for orthomyxoviruses, which replicate in the nucleus. The **replicase,** present in infected cells, synthesizes new viral RNA strands of both polarities. Transcription occurs at the same site as replication. It is unclear whether replication and transcription are carried out by different enzymes or by the same enzyme variously modified by interaction with virion proteins, as is the case with RNA phages (see Chap. 45).

Replication of Double-Stranded RNA (Class VI Viruses)

Each segment of reovirus RNA is replicated independently. Replication is intimately tied to transcription of the genome by the **virions' transcriptase,** which generates mRNAs in the virion cores. A **replicase** uses the nascent messenger strand as template, making a negative strand, which then acts as template for a new positive strand. The two strands remain associated in a double-stranded molecule that ends up in a virion (Fig. 48–13). This replication is **asymmetric** and **conservative** because (1) the negative strand of the virion RNA

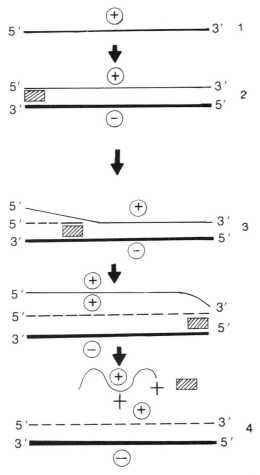

Figure 48–12. Replication of flavivirus RNA by strand displacement. The viral RNA (*1*) generates an RF (*2*). The replicase (*dashed*) builds a new + strand (*dashed*), displacing the RF + strand (thin line) (*3*). Finally, the preexisting + strand is released (*4*), while the RF and the replicase are ready to start a new cycle.

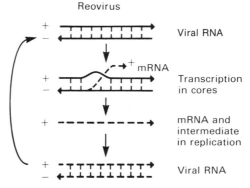

Figure 48–13. Transcription and replication of virion double-stranded RNA. In this conservative replication, information flows from **only one strand** (asymmetric), and both parental strands are conserved together.

serves as the initial template and (2) the parental RNA does not end up in the progeny (see Fig. 48–13).

Replication Through a DNA-Containing Replicative Intermediate (Class VII Viruses)

That the replication of retrovirus RNA occurs through a DNA intermediate was first shown by the blocking effect of inhibitors of DNA synthesis and transcription. To explain these observations Temin proposed that the viral replication proceeds through a double-stranded DNA RI, and that progeny viral RNA is obtained from it by regular transcription. Strong support for this theory was afforded by two discoveries: (1) a **reverse transcriptase,** present in virions, synthesizes DNA using the single-stranded viral RNA as template, and (2) viral DNA can be recognized in the infected cells, integrated into the cellular DNA as **provirus.** Transcription of the provirus by the cellular transcriptase yields the viral molecules that end up in virions. The complex process of reverse transcription is considered in detail in Chapter 65.

MATURATION AND RELEASE OF ANIMAL VIRUSES

Maturation proceeds differently for naked, enveloped, and complex viruses.

Naked Icosahedral Viruses

MATURATION. As with bacteriophages, maturation of naked viruses consists of two main processes: (1) assembly of the capsid and (2) its association with the nucleic acid. For **DNA viruses** the two steps are clearly separate, since DNA synthesis precedes the appearance of recognizable capsid components, sometimes by several hours. In contrast, with **naked icosahedral RNA viruses** the association of the capsid with the nucleic acid proceeds almost concurrently. The synthesis of viral RNA, measured either chemically or by infectious titer, is followed shortly by synthesis of mature virions; for poliovirus the time difference is 30 to 60 minutes. The fairly close temporal connection between synthesis and assembly may have evolved because of the inherent instability of the naked viral RNA in the cells.

ASSOCIATION OF THE CAPSID WITH THE NUCLEIC ACID.
With icosahedral animal viruses, as with bacteriophages, preassembled capsomers are joined to form **empty capsids (procapsids),** which are the precursors of virions. In fact, in pulse-chase experiments with labeled amino acids added to infected cells, the label is incorporated first into nascent polypeptide chains, then capsomers, later procapsids, and finally complete virions. With poliovirus the capsomers are pentameric; with adenovirus they are both pentamers and trimeric hexons (see Chap. 44). With both viruses the assembly of capsomers to form

the procapsid is accompanied by extensive reorganization, which is revealed by changes of serologic specificity or isoelectric point.

With a number of viruses (e.g., picornaviruses, adenoviruses), polypeptides are processed while the nucleic acid associates with the procapsid. For example, in poliovirus maturation, the final step is the cleavage of the precursor polypeptide VP_0 into $VP_2 + VP_4$ (see Fig. 48–12). Presumably the viral RNA penetrates between the capsomers from the outside, triggering the cleavage reaction. VP_4 may be responsible for locking the RNA into the capsid, because its loss at uncoating (see Initial Steps of Viral Infection, above) or during heat inactivation is accompanied by separation of RNA from the capsid. In adenovirus morphogenesis, the viral DNA, linked to the 55 Kd terminal protein, enters the procapsid together with core proteins; subsequently proteins of both procapsid and core undergo cleavage. For viruses that form concatemers in DNA replication, specific sequences act as signals for DNA cleavage into monomers and for packing into the procapsid.

With reoviruses, the genome of which is segmented (class VI), the various progeny RNA pieces appear to assemble within an inner capsid in which they are held together, forming the virion core. Undefined final steps in morphogenesis lead to association of this core with the outer capsid (see Chap. 62).

After they are assembled, icosahedral virions may become concentrated in large numbers at the site of maturation, forming the intracellular crystals (Fig. 48–14) frequently observed in thin sections of infected cells.

RELEASE. Naked icosahedral virions are released from infected cells in different ways, which depend on both the virus and the cell type. Poliovirus, for instance, is **rapidly released,** with death and lysis of HeLa cells: the study of single infected cells, contained in small drops of medium under paraffin oil (Fig. 48–15), shows a total yield of about 100 plaque-forming units released over a period of $\frac{1}{2}$ hour. In contrast, virions of DNA viruses that mature in the nucleus tend to **accumulate** within the infected cells over a long period; they are released when the cells undergo autolysis, but in some cases they are extruded without lysis.

Enveloped Viruses

MATURATION. Viral proteins are first associated with the nucleic acid to form the nucleocapsid, which is then surrounded by the envelope.

In **nucleocapsid formation** the proteins are all synthesized on cytoplasmic polysomes and are rapidly assembled into capsid components (recognizable by immunofluorescence or electron microscopy). With most RNA viruses the nucleocapsids accumulate in the cytoplasm (Fig. 48–16), but with orthomyxoviruses (e.g., influ-

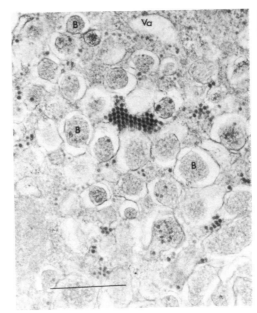
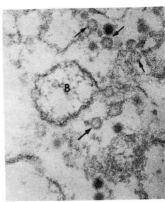

Figure 48–14. Electron micrograph of part of the cytoplasm of an HeLa cell infected by poliovirus (a picornavirus), showing a focus of viral reproduction. (*Left*) Many viral particles are present in the cytoplasmic matrix (some in small crystals) around or within membrane-bound bodies (*B*) and vacuoles (*Va*). (*Right*) Empty capsids (*arrows*). (Dales S et al: Virology 26:379, 1965)

enza and fowl plague virus) they are recognizable first in the nucleus (by 3 hours; Fig. 48–17) and later in the cytoplasm.

In **envelope assembly,** virus-specified envelope proteins go directly to the appropriate cell membrane (the plasma membrane, the endoplasmic reticulum, or the

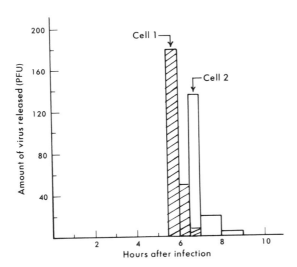

Figure 48–15. Kinetics of viral release from single monkey kidney cells infected by poliovirus. The cells were obtained by disrupting a monolayer of monkey kidney cells by trypsin; after being infected with poliovirus, each was introduced into a separate small drop of medium immersed in paraffin oil. Every half hour, the medium of each drop was removed, replaced with fresh medium, and assayed for infectivity by plaque assay. Note that with either cell the release was rapid (most virus came out in $\frac{1}{2}$ hour), and note also the difference in the latent periods. (Data from Lwoff A et al: Virology 1:128, 1955)

Golgi apparatus), displacing host proteins. In contrast, **the lipids and carbohydrates are those produced by the host cell,** as shown by their composition and by their specific activities in virus-producing cells labeled before infection with a radioactive precursor. The viral envelope has the lipid constitution of the membrane where its assembly takes place (e.g., the plasma membrane for orthomyxoviruses and paramyxoviruses, the nuclear membrane for herpesviruses). A given virus will differ in its lipids and carbohydrates when grown in different cells, with consequent differences in physical, biologic, and antigenic properties (see also The Envelope, Chap. 44). The viral proteins, however, to a certain extent also select the lipids with which they aggregate; thus, two types of influenza virus (A and B) grown in the same cells may differ in the proportions of individual phospholipids.

The **formation of the envelope glycoproteins** is best understood for viruses that bud at the plasma membrane (e.g., orthomyxoviruses). The glycoproteins are synthesized on polysomes bound to the endoplasmic reticulum; immediately the hydrophobic N-terminal **signal sequence** penetrates the membrane and is cleaved off as it emerges at the other side (see Chap. 6). A hydrophobic **anchor sequence** holds the polypeptide in the membrane, usually as a transmembrane protein. From the endoplasmic reticulum the polypeptide moves via transport vesicles to the Golgi apparatus, where it attains its full **glycosylation** and other modifications, such as **fatty acid acylation** (palmitic or myristic acid). Some proteins are **proteolytically cleaved,** generating two S–S bonded peptides, as in the fusion protein of paramyxoviruses. These changes are important for the function of the protein; thus, blocking N-linked glycosylation with tunicamycin or fatty acid acylation by mutation may prevent the appearance of the glycoprotein at the cell surface. After progressing through the various parts of the Golgi apparatus the glycoproteins, via other transport vesicles, reach the cell membrane, where they are exposed at the external surface of the cell.

The glycoproteins determine where virion maturation takes places: For instance the bunyavirus glycoproteins behave like intrinsic Golgi proteins, which reach the Golgi apparatus and are incapable of leaving it; as a result the virions mature and bud from the wall of Golgi vesicles into their lumen. From there they are transported to the surface by an unknown mechanism.

Matrix proteins that are present in viral envelopes are usually not glycosylated and stick to the cytoplasmic side of the plasma membrane through hydrophobic domains. Matrix proteins connect the cytoplasmic nucleocapsid with the cytoplasmic domains of the envelope glycoproteins and with the cell's cytoskeleton, and they gather the viral glycoprotein to form the virions (Fig. 48–18). The selection of viral glycoproteins is efficient but not exclusive: for instance, rhabdovirus virions contain 10% to 15% of nonviral glycoproteins. They may also contain glycoproteins specified by another virus infecting the same cell; such virions are known as **pseudotypes** (see Chap. 50).

FORMATION AND RELEASE OF VIRIONS. Envelopes are formed around the nucleocapsids by **budding of cellular membranes** (see Fig. 48–18). This budding is the result of an intimate adhesion of the nucleocapsid to the matrix (M) protein at the cytoplasmic side of the cell membrane where the viral glycoproteins are embedded;

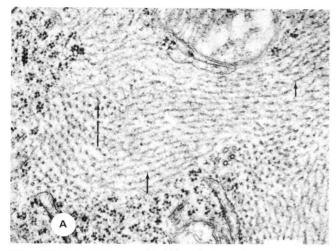
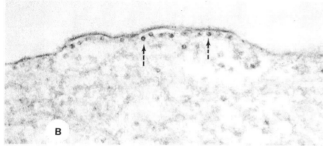

Figure 48–16. Maturation of an enveloped virus with a helical nucleocapsid (Sendai virus, a paramyxovirus) in infected chick embryo cells. (*A*) Accumulation of nucleocapsids in the cytoplasm, some cut transversely (*dashed arrow*), some longitudinally (*solid arrows*). (*B*) Transversely cut nucleocapsids under thickened areas of the plasma membrane, covered by spikes, preliminary to budding and virion formation. (Darlington RW et al: J Gen Virol 9:169, 1970)

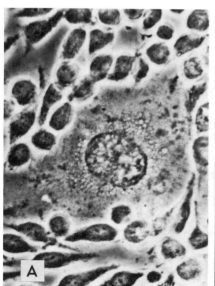
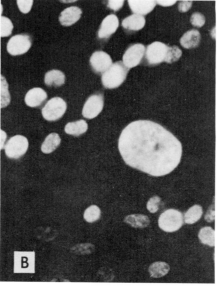

Figure 48–17. Localization of the nucleocapsid antigen of fowl plaque virus (an orthomyxovirus) in the nucleus 3 hours after infection of a culture of L cells. The cells were fixed and treated with fluorescent Ab to the viral Ag. (*A*) Phase contrast micrograph. (*B*) Micrograph of the same field in ultraviolet light, where only the Ag bound to the viral Ab is visible. The absence of fluorescence in the cytoplasm is especially evident in the giant cell. (Franklin R, Breitenfeld P: Virology 8:293, 1959)

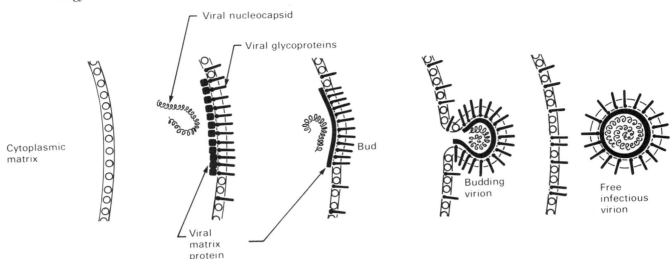

Figure 48–18. Budding of an enveloped virus (orthomyxovirus or paramyxovirus). *White circles,* host proteins of the plasma membrane (specified by cellular genes); *black spikes,* viral glycoproteins (the peplomers specified by viral genes), which become incorporated into the cell membrane, replacing host cell proteins, before budding of the viral particles begins. The viral **matrix (M) protein** attaches to the inner surface of the plasma membrane segment containing the viral glycoproteins and appears to serve as a recognition site for the nucleocapsid as well as a stabilizing structure.

the adhesion causes the membrane to curve into a protruding sphere surrounding the nucleocapsid. Interaction of matrix protein with actin-containing filaments seems to be important for budding.

The bud detaches from the membrane by a process that can be considered the reverse of penetration. If budding is from the surface membrane (e.g., paramyxoviruses), release occurs at the same time. If budding occurs in cytoplasmic vesicles (e.g., togaviruses), release requires subsequent fusion of the vesicle with the cell membrane. Either method of release is compatible with cell survival and can be very efficient, allowing a cell to release thousands of viral particles per hour for many hours. Herpesviruses bud out of the nuclear membrane into the cytoplasm and reach the outside through cytoplasmic channels and vesicles.

With some newly isolated influenza viruses (see Chap. 56) a deviation from the normal pattern of maturation leads to the formation of infectious cylindrical **filaments** with a diameter similar to that of the spherical particles. Their formation depends on genetic properties of the virus, type of host cell, and environment (e.g., in cell cultures it is greatly enhanced in the presence of vitamin A, alcohol, or surfactants).

A **polarized** viral budding is observed in cultures of epithelial cells in which the lipid and protein composition of the apical part of the plasma membrane differs from that of the basolateral part; the individuality of the two parts is maintained by the tight junctions between cells, which are barriers to diffusion within the plasma membrane. In MDCK cells (Madin-Darby canine kidney line), orthomyxoviruses and paramyxoviruses bud from the apical surface, whereas herpesviruses, rhabdoviruses, and retroviruses bud from the basolateral surface. The viral glycoproteins have distinguishing features that direct them to one or the other surface.

CELL SURFACE ALTERATIONS PRODUCED BY VIRAL MATURATION. With orthomyxoviruses and paramyxoviruses the viral glycoproteins incorporated in the membranes confer on the cell some properties of a giant virion. Thus, cells infected by these viruses may bind RBCs (**hemadsorption,** Fig. 48–19, the equivalent of hemagglutination) or viral antibodies, and paramyxovirus-infected cells may fuse with uninfected cells to form multinucleated syncytia, called **polykaryocytes** (Fig. 48–20), by fusion of their membranes. This fusion is equivalent to the fusion of the virion's envelope with the plasma membrane of the host cell at the onset of infection (see above, Uncoating). Polykaryocytes can also be formed by inactivated virions attached to the cell surface, provided that they have a functional fusion protein. This approach is used to fuse two different cell types, producing cell hybrids.

Complex Viruses

Maturation of the highly organized **poxviruses** takes place in cytoplasmic foci called "**factories.**" Membranes enclosing fibrillar material appear first; then viral DNA enters the particles when the membranes are almost complete, forming a dense, immature nucleoid (Fig. 48–21). As with other viruses, **peptide cleavage** appears to perform an important role. In contrast to simpler viruses, the poxvirus membrane contains **newly synthesized lipids** that differ in composition from cellular lip-

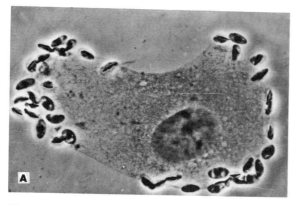
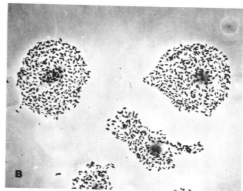

Figure 48–19. Hemadsorption of HeLa cells infected by Newcastle disease virus. Cells had been heavily irradiated with x-ray several days before infection; they stopped multiplying but increased in size and became giant cells, facilitating observations. The virus multiplies regularly in these cells. (*A*) Cell 5 hours after infection. The ability to adsorb chicken RBCs begins to appear at two opposite regions of the cell surface. At these regions new cell membrane appears to be laid down, allowing viral components to become incorporated together with the cellular components. (*B*) Cells at lower magnification, 9 hours after infection, showing that the entire cellular membrane has developed the capacity for hemadsorption. The RBCs are firmly attached to the cells and are not removed by repeated washing. (Marcus P: Cold Spring Harbor Symp Quant Biol 27:351, 1962)

ids. These viruses employ two pathways for releasing the progeny particles. Some become enveloped in Golgi-derived vesicles, which reach the cell surface, but the majority are released upon cell lysis.

The maturation of poxviruses after the precursors have been enclosed within the primitive membrane suggests that **poxviruses may be transitional forms toward a cellular organization.** Viral maturation, however, requires functions of the cellular nucleus, because it is blocked after the nucleus is removed or inactivated by UV light.

Assembly From Subunits in Precursor Pools

In a cell simultaneously infected by certain pairs of related viruses that differ in capsid Ags, such as poliovirus types 1 and 2, **phenotypic mixing** can occur because capsids made from building blocks of both viruses or either viruses may enclose a genome. Antiserum to either Ag may neutralize particles with mixed capsids (see Chap. 50). Hence, infection with a mixture of poliovirus of types 1 and 2 may yield six classes of virions (Fig. 48–22) with different combinations of genotype (RNA) and phenotype (protein). Similar mixing is seen with adenovirus, which yields capsids with fibers of different lengths and with random combinations of hexons, producing mixed antigenic types. Phenotypic mixing affecting envelope glycoproteins can occur even between unrelated viruses (such as rhabdoviruses and retroviruses), generating **pseudotypes.**

These observations show that the virions of animal viruses are assembled from building blocks more or less **randomly picked from pools.**

Genetics

Animal virus genetics has made impressive progress in recent years, owing to the extensive use of molecular methods. Yet it is not as well known as the genetics of bacteriophages, because the technical difficulties are

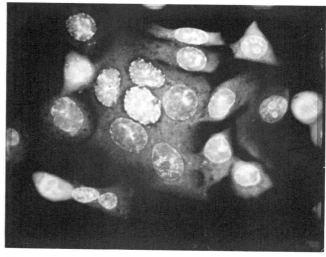

Figure 48–20. Formation of multinucleated syncytia (polykaryocytes) by Hep-2 cells infected by herpes simplex virus. Five cells have fused completely, and several others partly, into a central mass. Cells were stained with the fluorescent dye acridine orange and photographed in a dark field. (Roizman B: Cold Spring Harbor Symp Quant Biol 27:327, 1962)

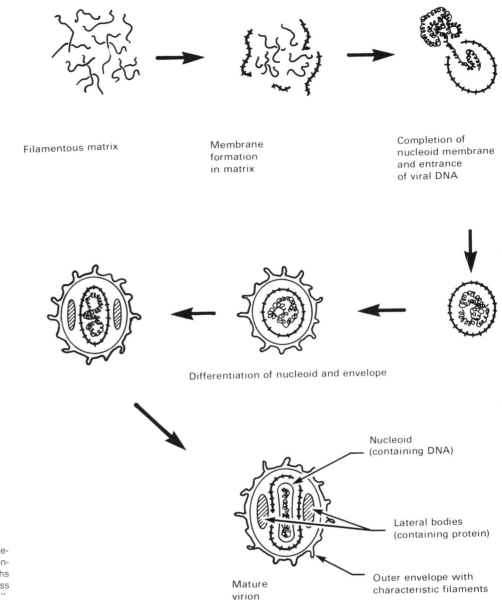

Filamentous matrix

Membrane
formation
in matrix

Completion of
nucleoid membrane
and entrance
of viral DNA

Differentiation of nucleoid and envelope

Nucleoid
(containing DNA)

Lateral bodies
(containing protein)

Outer envelope with
characteristic filaments

Mature
virion

Figure 48–21. Scheme of the development of vaccinia virions, reconstructed from electron micrographs (see Fig. 54–8). The entire process proceeds in "cytoplasmic factories."

MUTATIONS

Mutations of animal viruses occur **spontaneously** or can be **induced** by various chemicals, including nitrous acid, 5-bromodeoxyuridine (BUDR), hydroxylamine, and nitrosoguanidine (fluorouridine is also useful for RNA viruses). Mutations can also be engineered after the viral DNA (or the complementary DNA from RNA viruses) is

greater, and with some viruses recombination is infrequent.

cloned (see Chap. 8). With DNA viruses the spontaneous **mutation frequency** depends on the characteristics of the viral DNA polymerase, which can have either a mutator or an antimutator role (see Mutation, Chap. 8). The known **mutant phenotypes** are numerous and cover a larger range than bacteriophage mutations, because there are more ways for studying their properties (for instance, their various effects on animals).

The frequency of mutation, either spontaneous or induced, is higher with RNA than with DNA viruses because RNA polymerases and reverse transcriptases are

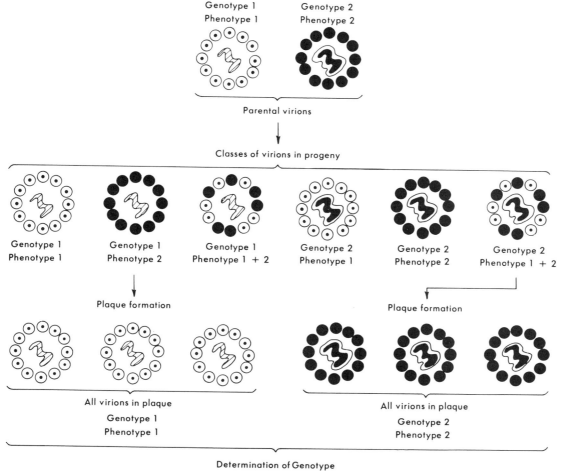

Genotype 1
Phenotype 1

Genotype 2
Phenotype 2

Parental virions

Classes of virions in progeny

Genotype 1
Phenotype 1

Genotype 1
Phenotype 2

Genotype 1
Phenotype 1 + 2

Genotype 2
Phenotype 1

Genotype 2
Phenotype 2

Genotype 2
Phenotype 1 + 2

Plaque formation

Plaque formation

All virions in plaque
Genotype 1
Phenotype 1

All virions in plaque
Genotype 2
Phenotype 2

Determination of Genotype

Genotype = the genetic information residing in the nucleic acid

Phenotype = the immunological characteristics residing in the capsid protein

Figure 48–22. Mechanism underlying phenotypic mixing of the antigenic specificity of poliovirus. Cells mixedly infected by types 1 and 2 produce virions of six genotype and phenotype combinations. Cloning of these by plaque formation yields unmixed virions, with phenotype determined by the genotype of the initiating virion, irrespective of the latter's phenotype.

much less accurate, by a factor of as much as 10^6. This high frequency causes **rapid drift** of the genome, as observed especially with othomyxoviruses and lentiviruses. As a result an RNA virus strain is always a heterogeneous collection of different genotypes, the composition of which depends on the selective conditions under which it is grown (see Viral Evolution, below).

Mutant Types

Only one class of **conditionally lethal mutations, temperature-sensitive (ts) mutations,** is useful in animal viruses. These mutations may affect the stability of the protein, its transport (by altering the membrane-binding domains), or its processing (such as cleavage or glycosylation). The ts mutations have been extremely valuable because they can be recovered in many (possibly all) genes. In spite of some defects, such as leakiness and a high reversion rate, they form the basis for most of our present knowledge of animal virus genetics. Some such strains of influenza virus are useful as vaccines because they are attenuated, i.e., less virulent.

Cold-sensitive mutants, which grow much better at 39°C than at 33°C, have occasionally been isolated. **Deletions** occur frequently in certain DNA genomes (such as parvoviruses); often the deleted sequences are located between direct repeats, 4 to 10 base pairs long. The mechanism seems to be slipped mispairing during replication. Frequent deletions in RNA genomes lead to formation of interfering particles and are discussed in Chapter 49.

Mutations affect many properties:

Host dependence (or host range). These mutants, occurring in many viruses, fail to multiply in certain nonpermissive cell types, which differ for the various viruses. Like the suppressor-sensitive mutants of phage, they can be propagated in permissive cells, are not leaky, and have a low reversion rate; however, they are limited to one or **a few genes** involved in overcoming the nonpermissiveness of the cells.

Plaque size or type (Fig. 48–23). These mutants are not as diverse as the corresponding phage mutants because animal virus plaques have less detail (they affect fewer and larger cells). Differences of plaque size may depend on differences either in features of the multiplication cycle or in the surface charges of the virions. Small charge differences affect plaque size because agar contains a sulfated polysaccharide that adsorbs the more highly charged virions, especially at certain pHs.

Cytopathic effect. Noncytopathic mutants produce plaques that are not lytic but are recognizable by other criteria (e.g., hemadsorption).

Pock type in poxviruses (see Chap. 54)

Surface properties, detected by physical methods or by appearance of extensive cell fusion (with herpesvirus)

Hemagglutination (in orthomyxoviruses)

Resistance to inactivation by a variety of agents

Resistance toward or dependence on inhibitory substances during multiplication

Pathogenic effect for animals

Functions of certain viral genes in the infected cells, such as production of thymidine kinase or induction of interferon

Physical changes, for instance, the length distribution of restriction endonuclease fragments with DNA, of oligonucleotides produced by RNAses with RNA, or the electrophoretic mobility of the proteins. Many mutations are **silent** and determine the extensive **polymorphism** observed in the base sequences of strains of the same virus.

Pleiotropism

Mutants selected for a given phenotypic alteration are often found to be changed in other properties as well. For instance, poliovirus mutants with altered chromatographic properties often (though not always) have decreased neurovirulence for monkeys. This pleiotropism reflects the effect of a single viral protein on several properties of the virus. Viral virulence is often affected because it depends on many viral functions. The connection between the properties is variable; for example, mutations that modify different charged groups of the viral capsid may have similar effects on chromatographic adsorption but different effects on the more specific adsorption to cells.

Pleiotropism has useful applications. Thus, by the application of characters that are detectable *in vitro* (such as temperature sensitivity of various functions or deficiency of cytopathic effect), it is possible to select **attenuated** (i.e., nonvirulent) strains for use in live virus vaccines. The more cumbersome animal testing is then reserved for final characterization. This approach has been used in the selection of live poliovirus vaccines.

Complementation

Complementation of ts mutants has been useful for determining the functional organization of the viral genome. When two ts mutants complement each other, the yield of a mixed infection, at the nonpermissive tem-

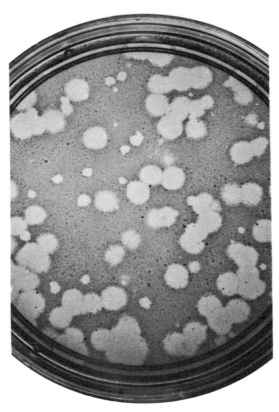

Figure 48–23. Two plaque-type mutants of fowl plague virus on a monolayer of chick embryo cells. The wild type produces large round plaques with fuzzy edges; a small-plaque mutant produces small plaques with irregularly indented outline and sharp edges. (Courtesy of H. R. Staiger)

perature, is greater than the sum of the separate yields at the same multiplicity of infection. Complementation is very efficient with the larger DNA viruses (adenoviruses, herpesviruses), for which the yield from mixed infection may approach 50% that of cells infected with wild-type virus. However, many animal viruses complement much less. Complementation is almost absent with picornaviruses, probably because their genome is translated into a larger polyprotein, and mutations that alter its tertiary structure prevent its cleavage into separate functional products. Probably for similar reasons, **asymmetric complementation** is observed with poliovirus: one mutant helps the other, but not vice versa.

In either DNA- or RNA-containing viruses, each complementation group usually corresponds to a different gene, as shown by different functional effects of different groups and by their effect on different proteins, recognizable by gel electrophoresis. Moreover, in viruses with a fragmented genome, in which each fragment is a gene, the number of the complementation groups is similar to that of the fragments.

GENETIC RECOMBINATION
DNA Viruses

In the large viruses with a linear DNA (adenoviruses, herpesviruses, poxviruses), recombination between pairs of ts mutants or between a ts and a plaque-type mutant occurs at frequencies comparable to those observed in some phages or bacteria. Generally, in recombination between mutants, the yield of mass cultures regularly contains the two reciprocal recombinant types, in comparable proportions. Recombination is also observed between viruses of different types within the same group (e.g., types 2 and 5 adenoviruses), but only at regions of high homology.

Little is known about the individual recombination events, except that they occur together with DNA replication, probably through the use of both viral and cellular enzymes. The proportion of recombinants between markers tested pairwise is additive. By a series of two- or three-factor crosses, the order of various mutations can be unambiguously established, in spite of the unusual structure or replication of these viral DNAs (see above). By this approach the herpesvirus map is 25 to 30 units long; hence, the frequency of recombination per unit length is much less than for phage T4, but of the same order as that of bacteriophage λ (see Chaps. 45 and 46). The maps are **linear;** i.e., distant markers are unlinked.

Maps have also been determined by molecular methods (Fig. 48–24), based on the fragments of the DNA generated by restriction endonucleases, the order of which is established from the overlaps of fragments produced by different enzymes. In **intertypic crosses,** differences in the fragments of the two parental DNAs reveal the crossover point in the DNA of a recombinant (see Fig. 48–24A). In **marker rescue** a fragment from wild-type virus is introduced into the cells by transfection together with viral DNA from a ts mutant; if the fragment overlaps the mutational site, wild-type virus can be generated by recombination (see Fig. 48–24B). With viruses having a cyclic DNA (e.g., polyomavirus), wild-type progeny virus can be obtained without recombination (see Fig. 48–24C): a strand of the wild-type fragment is annealed to the complementary complete strand of the mutated DNA, and the resulting hybrid is introduced into the cells. Extension of the fragment, by copying of the complementary strand, yields a wild-type strand if the fragment overlaps the mutation.

The maps obtained by molecular methods give the same gene order as those obtained by recombination, but the distances between genes do not agree well, suggesting that recombination occurs at different frequencies in different regions of the genome.

RNA Viruses—Reassortment

Burnet first recognized recombination between influenza virus strains in 1951. Some markers of this and other RNA viruses with **segmented genomes** (classes IV and VI) show high recombination frequencies, up to 50%. When the two parental viruses have corresponding segments that differ in size, these mutations can be assigned to different segments. In contrast, markers located to the same segment do not show recombination. Recombination among these viruses therefore results from random **reassortment** of segments. This is an important cause of variation: the major antigenic changes of influenza virus are primarily caused by reassortment between different species (see Evolution of Viral Antigens, Chap. 50).

RNA viruses with **continuous genomes** usually show little or no recombination; only picornaviruses and coronaviruses show measurable recombination, limited to some regions of the genome. This recombination probably arises from discontinuous replication by means of different templates, by a **copy choice** mechanism. An exceptionally high recombination frequency is found in the diploid retroviruses; recombination occurs in cells infected by **heterozygous virions,** in which the two RNA molecules carry different markers, probably by a copy choice mechanism.

GENE MAPS. For viruses **with a positive-strand continuous genome** that is translated into a single polypeptide chain (e.g., polioviruses), gene maps can be obtained **biochemically** (see Fig. 48–24D). A radioactive amino acid is added to the infected cell culture together with the antibiotic pactamycin, which inhibits initiation of protein synthesis. The label will be incorporated more by proteins corresponding to the distal (3′ end) part of the RNA, which has the highest chance of being still untrans-

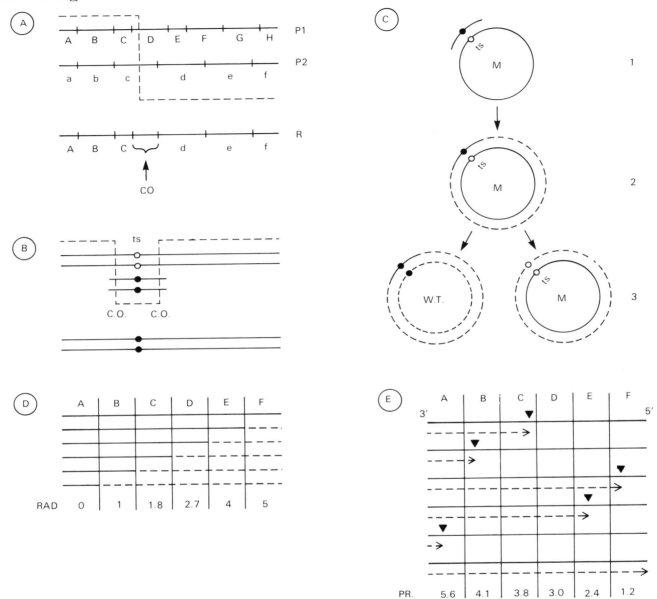

Figure 48–24. Molecular mapping. (A) In the intertypic cross of two viruses, **P1** and **P2,** the derivation of the DNA in the recombinant (R) can be ascertained from the restriction endonuclease fragments it contains; the position of the crossover point (CO) is delimited. (B) Mixed infection of cells with a viral DNA carrying a temperature-sensitive (ts) mutation and a wild-type (wt) fragment overlapping the mutation **(marker rescue)** produces a wt molecule by two crossovers (C.O.). (C) (1) A strand of a wt DNA fragment is annealed to a complete strand carrying a ts mutation (M); (2) after transfection (i.e., infection with pure DNA) the fragment is elongated; (3) replication yields a wt DNA molecule. (D) *Continuous lines,* a set of incomplete polyproteins at the time of addition of amino acid label + pactamycin; *dashed lines,* radioactive parts of the polyproteins when they are completed. The gradient of radioactivity (RAD) incorporated, after processing, establishes the order of genes (A to F). (E) *Continuous lines,* a set of viral genomes exposed to UV light; *arrowheads,* pyrimidine dimers; *dashed lines,* the transcribed segments; *numbers,* relative amounts of protein (PR.) synthesized for each gene (A to F). The gradient of synthesis establishes the order of genes.

lated when the label is added. The relative labeling of various proteins gives their location in the uncleaved polypeptide chain.

The gene order for several **negative-strand** viruses has also been determined by measuring the **effect of UV irradiation** of virions on the synthesis of the polypeptides specified by the various genes. The method is based on the blocking of the progress of transcription by the UV-induced pyrimidine dimers; therefore, a gene at the 5' end on the template strand has the highest sensitivity to the radiation because its transcription is blocked by any dimer along the whole genome, and the synthesis of the peptide it specifies undergoes the greatest reduction (Fig. 48–24E).

HETEROZYGOSITY. True heterozygosity is observed in retroviruses, which have two molecules of single-stranded RNA in the virion. **Multiploid heterozygosity** occurs in enveloped viruses that generate some particles with two nucleocapsids. This event is frequent in orthomyxoviruses and paramyxoviruses because two nucleocapsids can be readily accommodated in the floppy envelope; rhabdoviruses form occasional diploid particles of double length.

Reactivation of UV-Irradiated Viruses

CROSS-REACTIVATION; MULTIPLICITY REACTIVATION. With viruses that give rise to efficient recombination (large DNA viruses, influenza), markers from a UV-inactivated strain can be rescued by active virus simultaneously infecting the cells. The survival of a given marker, as a function of the UV dose, is much greater than the survival of the whole virus, as already seen with phages (see Genetic Reactivation, Chap. 45). Also, like bacteriophages, as multiplicity of infection with UV-inactivated virus is raised, the proportion of cells yielding infectious virus increases excessively. Both these reactivations are much more pronounced for RNA viruses with a segmented than with a continuous genome, in accord with the much higher frequency of recombination by reassortment of their genomic segments.

VIRAL EVOLUTION

The great ability to undergo variation is the background for the rapid evolution of viruses. The highest variability is expressed by RNA viruses, owing to the marked instability of RNA; further variability is contributed to some viruses by reassortment of genomic segments derived from different strains. Because of this tremendous variability, potentially no two RNA virions in the progeny of an infected cell should be identical. The extent of observed variability is, however, **controlled by selection** for effective interactions of virions with host cells and of viral proteins among themselves. Important in viral evo-

lution are viral antibodies and the type of host cells. Antibodies can explain the considerable evolution of influenza viruses in nature (see Chap. 50), and serial growth in different cell types causes considerable changes in a viral population.

The ability of viruses to evolve, however, seems to be subject to other types of control. This is especially seen in comparing different picornaviruses. Poliovirus exists in three serologic types, immunologically distinct but stable, whereas foot-and-mouth disease virus (FMDV) exists in seven main types with a larger number of subtypes, and new subtypes continuously emerge. The difference might be correlated with the ability of FMDV to infect a wide range of cloven-hoofed animals, whereas poliovirus infects only man. By shifting from one host to another, FMDV may have the opportunity of using more of its potential variability. Among picornaviruses, however, the rhinoviruses have a tremendous variety of strains, although they also infect only man. As discussed in Chapter 50, this variability may be based on the special selective conditions of viruses infecting mucosae. The limited evolution of poliovirus may be caused by internal restraints (e.g., in the processing step or in capsid assembly).

GENETIC CHANGES ASSOCIATED WITH CHANGES OF VIRULENCE. Many viruses cause severe diseases in animals or humans by attacking the cells of certain organs, such as the central nervous system or the liver. Such **virulent strains** can give rise to **attenuated strains,** which can multiply in the same host but fail to produce the disease (see Chap. 51). Some attenuated derivatives are used as live vaccines for immunizing animals or humans (see Chap. 50). They are usually obtained by growing a virulent strain serially through many generations in a different host, such as embryonated chicken eggs or cultures. For instance, attenuated poliovirus, lacking neurotropicity for humans and monkeys, is obtained by passaging the virus through monkey kidney cultures, and attenuated yellow fever virus and FMDV are obtained by passages through chick embryos. The passages probably select certain types of spontaneously occurring mutants. The approach is empiric, and the results vary from virus to virus; for viruses that can affect several hosts, attenuation in one host does not necessarily mean attenuation in another. Attenuation can also be obtained by selecting for variants with certain characteristics, such as cold-adapted for influenza or antibody-resistant for rhabdoviruses or bunyaviruses. In the reverse process, virulent strains can emerge spontaneously from attenuated ones and then remain stable.

Extensive genetic analyses have been carried out to try to identify the genes responsible for the changes of virulence. One approach is to compare the characteristics, and sometimes the base sequences, of closely related attenuated and virulent strains; the other is to determine the virulence of recombinants or reassortants between a virulent and an attenuated strain. These studies yielded different results in different systems. With some viruses many genetic changes contribute to determining virulence. For instance, the attenuated Sabin vaccine strains of poliovirus type 1 (see Chap. 55) differs from the original virulent strain in 55 bases and 21 amino acids, distributed all over the genome. Recombinants show that the 5' end of the genome is somewhat more important, but they afford no clear localization. With Sabin type 3 vaccine strain the situation is different: there are ten changes of bases and three of amino acids. Of these, a C → U change in the untranslated 5' end of the genome (N472) has predominant importance: its reversion to C, which occurs occasionally in vaccinated persons, restores the virulent phenotype. The revertant virus is strongly selected in humans but not in cultures. A similar precise localization of virulence-determining change restricted to a single amino acid was observed in the hemagglutinin of influenza virus and in the glycoprotein of rabies virus.

These differences in the changes involved in attenuation have important practical consequences. Spontaneous reversions to virulence occur easily when they can be brought about by a single change, but not when multiple changes are needed: whereas the type 3 Sabin vaccine strain is unstable, the type 1 vaccine is very stable.

USE OF ANIMAL VIRUSES AS VECTORS FOR IN VITRO DNA RECOMBINATION

Several animal viruses are useful vectors; some of them are oncogenic and will be considered in Chapter 65. Among the non-oncogenic ones, adenoviruses and herpesviruses are used. Here we will describe the use of vaccinia virus, which is suitable for the expression of recombinant genes in animal cells.

Construction of Recombinant Vaccinia Viruses

The principle is to introduce a foreign gene into a nonessential region of vaccinia virus DNA, such as the thymidine kinase gene (TK). For this purpose the selected gene is first introduced into the TK gene previously inserted into a plasmid vector. The TK gene is inactivated by the inserted gene, which becomes controlled by the TK gene promoter. The plasmid is then introduced by transfection into cells infected by wt vaccinia virus. Recombination between the resident and the transfected DNAs generates TK⁻ recombinant vaccinia DNA (**replication competent**), containing the inserted gene. During

growth in TK⁻ cells the recombinant virus is isolated by selection in medium containing 5-bromodeoxyuridine: for lack of the TK function, cells infected by the mutant do not incorporate the analog, and they escape its toxic effect. The cells express the foreign gene; if it specifies a membrane protein, the protein appears at the cell surface.

Selected Reading

Bishop DHL: Ambisense RNA genomes of arenaviruses and phleboviruses. Adv Virus Res 31:1, 1986

Boulan ER, Pendergast M: Polarized distribution of viral envelope proteins in the plasma membrane of infected epithelial cells. Cell 20:45, 1980

Evans DMA, Dunn G, Minor PD et al: Increased neurovirulence associated with a single nucleotide change in a noncoding region of the Sabin type 3 poliovaccine genome. Nature 314:548, 1985

Flint SJ: Regulation of adenovirus mRNA formation. Adv Virus Res 31:169, 1986

Futterer J, Winnaker EL: Adenovirus DNA replication. Current Topics Microbiol Immunol 111:41, 1984

Honess RW, Roizman B: Regulation of herpesvirus macromolecular synthesis. I. Cascade regulation of the synthesis of three groups of viral proteins. J Virol 14:8, 1974

Howard CR: The biology of hepadna viruses. J Gen Virol 67:1215, 1986

Kaariainen L, Ranki M: Inhibition of cell functions by RNA-virus infection. Annu Rev Microbiol 38:91, 1984

Kohn A: Membrane effects of cytopathogenic viruses. Prog Med Virol 31:109, 1985

Lodish HF, Porter M: Specific incorporation of host cell surface proteins into budding vesicular stomatitis virus particles. Cell 19:161, 1980

Mackett M, Smith GL: Vaccinia virus expression vectors. J Gen Virol 67:2067, 1986

Palese P, Young JF: Variation of influenza A, B, and C viruses. Science 215:1468, 1982

Scholtissek C, Koennecki I, Rott R: Host range recombinants of fowl plague (influenza A) virus. Virology 91:79, 1978

Seeger C, Ganem D, Varmus HE: Biochemical and genetic evidence for the hepatitis B virus replication strategy. Science 232:477, 1986

Senapathy P, Tratschin J-D, Carter BJ: Replication of adeno-associated virus DNA. J Mol Biol 178:179, 1984

Stanway G, Hughes PJ, Mountford RC et al: Comparison of the complete nucleotide sequences of the genomes of the neurovirulent poliovirus P3/Leon/37 and its attenuated Sabin vaccine derivative P3/Leon 12a₁b. Proc Natl Acad Sci USA 81:1539, 1984

Steinhauer DA, Holland JJ: Direct method for quantitation of extreme polymerase error frequencies at selected single base sites in viral RNA. J Virol 57:219, 1986

Stow NK, Subak-Sharpe JH, Wilkie NM: Physical mapping of herpes simplex virus type 1 mutations by marker rescue. J Virol 28:182, 1978

Wilson TMA: Nucleocapsid disassembly and early gene expression by positive-strand RNA viruses. J Gen Virol 66:1201, 1985

Renato Dulbecco

Interference With Viral Multiplication

Agents that interfere with viral multiplication are useful not only for therapy and prophylaxis but also for advancing our understanding of viral biology and of infection.

Control of Viral Diseases by Inhibition of Replication

Viral diseases result from a series of growth cycles that kill or alter cells (see Chap. 51). The maximal goal of antiviral treatment—to restore function to the infected cells—is usually unassailable because cellular macromolecules are damaged early in viral infections. Accordingly, the realistic goal is to stop viral replication and thus prevent spread to additional cells. But even this more limited goal presents considerable difficulties.

A major one is the problem of inhibiting the viruses without harming the cells. This selectivity is possible against bacteria because of their many metabolic, structural, and molecular differences from animal cells. Thus, sulfanilamide interferes with the function of *p*-aminobenzoic acid, which is a vitamin in bacterial but not in animal cells; penicillin interferes with the synthesis of the peptidoglycan, which is unique to bacteria; and streptomycin interacts with molecular features that are peculiar to bacterial ribosomes. The dependence of viral multiplication on cellular genes, in contrast, limits the points of differential attack.

Another limitation is that diseases become evident only after extensive viral multiplication and cellular alteration have occurred. Therefore, the most general approach to control is prophylaxis. Therapy is effective in localized viral diseases, such as herpetic keratoconjunctivitis (see Chap. 53), in which the killing of some uninfected cells can be tolerated if the damage is subsequently repaired. In addition, the special properties of herpesviruses permit the use of certain drugs for treating systemic herpes infections (see Agents Interfering With DNA Synthesis, below). With other viral diseases, therapy is limited to reducing their duration, the severity of symptoms, and the degree of viral shedding.

As with bacterial chemotherapy, a third important limitation of antiviral therapy is the emergence of **resistant mutants.** To avoid their selection, the principles valid for bacteria are equally applicable to viruses: adequate dosage, multidrug treatment, and avoidance of therapy unless clearly indicated. Fortunately, however, genetic resistance to two important antiviral agents—interferon and interferon inducers—does not seem to occur.

Viral Interference

When viruses of more than one type infect the same cell, each may multiply undisturbed by the presence of the others, except for possible recombination or phenotypic mixing. In certain combinations, however, the multiplication of one type of virus may be inhibited by the other. This inhibition is called **viral interference.**

The notion of interference developed first from observations with ring spot virus in tobacco plants. The initial lesions regress, but the virus persists, and if the plant is reinoculated with the same virus, no new lesions develop. Thus, the first infection interferes with the expression of the second infection. Subsequently it was found that in monkeys infection with a mild strain of yellow fever virus (a flavivirus) can prevent the usually lethal disease caused by a virulent strain or even by an antigenically unrelated flavivirus, showing that the protection is not due to Ab. Interference was later found with viruses in bacteria, thus opening the way for quantitative studies.

The study of interference with animal viruses took an important turn when Isaacs and Lindenmann, in 1957, discovered that influenza virus–infected cells produce a substance, which they called **interferon,** that accounts for many observed instances of viral interference.

DEMONSTRATION OF INTERFERENCE WITH ANIMAL VIRUSES

Interference was observed with many pairs of viruses in animals, but especially with influenza viruses in the allantoic epithelium of the chick embryo and recently with a variety of viruses in cell cultures. A typical experiment consists in inoculating the allantoic cavity with influenza A virus, as **interfering virus,** followed 24 hours later by influenza B virus, as **challenge virus:** the multiplication

of the second inoculum is partially or totally inhibited. The experiments can be simplified by using inactivated interfering virus; since it does not multiply, interference can be determined by measuring the yield of the challenge virus without the need to distinguish it from the interfering virus.

Interference depends on timing and on viral concentrations. Thus, if influenza A and B are inoculated **simultaneously** and at equal multiplicity in the allantoic cavity, they can multiply concurrently in the same cells, as shown by the production of phenotypically mixed progeny particles (see Assembly From Subunits in Precursor Pools, Chap. 48). Even viruses of different families can multiply in the same cells under proper circumstances.

We shall first consider in some detail the role of interferon in viral interference and shall then consider a heterogeneous group of other mechanisms.

INTERFERON

Interferon was discovered in the course of studying the effect of influenza virus inactivated by ultraviolet (UV) light on fragments of the chick chorioallantoic membrane maintained in an artificial medium. The supernatants, although devoid of viral particles, inhibited the multiplication of active influenza virus in fresh fragments. Subsequently, such "interferons" were shown to be produced by cells of many animal species infected by almost any animal virus, either DNA- or RNA-containing, and in tissue culture or in the animal.

Interferons (IFNs) are a family of small proteins first isolated from virus-infected or chemically activated fibroblasts and leukocytes (type I IFN); later, type II or immune IFN was isolated from T cells activated by mitogen or antigen. Type I IFNs are distinguished by serology into type α or β; type II is of one type, γ. IFN-α and IFN-β are stable at acidic pH (pH 2 in the cold), whereas IFN-γ is not. Cloning the genes showed the existence of a family of about twenty IFN-α genes, highly homologous to one another, including several inactive pseudogenes, and a single β gene, with 50% homology to α genes, all localized on human chromosome 9. These genes, which lack introns, encode proteins of 165 or 166 amino acids. Human IFN-α is not glycosylated, whereas IFN-β is. The single IFN-γ gene contains introns and has little homology to the other genes. It encodes a glycoprotein of 146 amino acids. At their 5′ ends the genes contain sequences with the characteristics of enhancers that control their expression.

IFNs are produced in large quantities by animal or human cells: IFN-α by leukocytes infected with Sendai virus or lymphoblastoid cell lines carrying the Epstein-Barr (EB) virus genome (see Chap. 64), and IFN-β by fibroblastic strains or lines (see Chap. 47) exposed to poly(I):poly(C) (see below).

A much more convenient source became available after IFN genes were cloned from cDNAs in bacterial vectors, which then express the IFN proteins in bacteria. Produced IFNs have all the effects of IFNs made in animal cells, although they are not glycosylated. With appropriate vectors, production is also possible in yeast or in murine cells. In all cases IFNs are purified by affinity chromatography or high performance liquid chromatography. These readily available and pure proteins made it possible to study the biological and antiviral activities of IFNs. Artificial recombinants between the various IFN genes obtained by DNA cloning express more IFNs with unusual properties.

IFN is usually **assayed** by determining its effect on the multiplication of a test virus, usually vesicular stomatitis virus (VSV, a rhabdovirus), which is very sensitive to IFN and infects cells of many vertebrate species. Serial IFN dilutions are added to the culture medium or to the agar overlay in a plaque assay, and the endpoint (a **unit**) is a 50% reduction in the viral yield or in the number of plaques. International units (IU) are expressed with reference to an international standard. A sensitive radioimmunoassay is also available, based on the use of monoclonal antibodies.

Effects

IFNs are very powerful drugs: a few thousand molecules are sufficient to confer resistance on a cell. Each IFN has **antiviral action, inhibits cell proliferation, and modulates the immune response,** especially by activating natural killer (NK) cells; IFN-γ powerfully activates macrophages. Some of the activities are enhanced by combining IFN-α or β with IFN-γ. The **effects are generally species specific;** for instance, purified IFN of chick origin is less than 0.1% as effective in mouse cells as in chick cells. However, IFN produced in monkey kidney cells is effective in human as well as in monkey cells, and human IFNs are active in cells of several mammals, including nonprimates. IFN produced by an artificial recombinant gene obtained from two human IFN-αs has an even broader range: it is quite active in murine and feline cells.

The described effects are part of a general regulatory action of IFNs on cellular functions, which results in **activation or inactivation of expression of many genes.** Among activated genes not related to the antiviral action are those for major histocompatibility complex (MHC) Ags of classes I and II, β_2-microglobulin, tubulin, certain "tumor-associated antigens," and at least 12 unidentified proteins. Inactivated are growth-related genes, such as those stimulated in fibroblastic cells by platelet-derived growth factor. MHC class I Ag genes are activated by all IFNs, class II genes by IFNγ. All the induced proteins are also present in cells not exposed to IFN but in much smaller amounts. The genes are activated at the level of transcription by the interaction of an unknown protein with a regulatory DNA sequence common to all of them.

Of special relevance among activated genes is the Mx gene. Cells of Mx$^+$ mice exposed to IFN-α or -β acquire an antiviral state—limited to orthomyxoviruses—whereas Mx$^-$ mice do not. In Mx$^+$ mice a 75-kd protein is induced in the nucleus; homologous proteins are formed in other mammals, including man. The findings suggest that the mechanism of resistance varies with the virus type.

Production After Viral Infection

IFNs are produced by cells infected with complete virions, either infectious or inactivated. Very few viral particles (e.g., with reovirus, a single physical particle) are sufficient to induce a cell. Under one-step conditions of viral multiplication the **synthesis of IFNs** begins after viral maturation is initiated; if not interrupted by an early block in the synthesis of host macromolecules, it continues at the same rate for 20 to 50 hours, then stops (Fig. 49–1). The IFNs are released extracellularly. If the cells survive longer, as after infection by UV-inactivated virus, they cannot produce IFN again in response to reinfection until after a **refractory period** of at least two cell divisions.

Usually, good IFN inducers are viruses that multiply slowly and do not block the synthesis of host protein early or markedly damage the cells. For example, an attenuated mutant of poliovirus is a much better IFN inducer than the wild type, which multiplies better; and the paramyxovirus of Newcastle disease multiplies well in chick embryo cells but causes little IFN production,

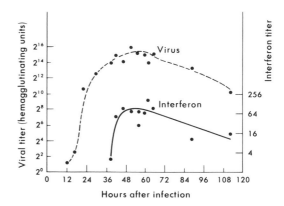

Figure 49–1. Time course of viral multiplication and interferon synthesis in the allantoic membrane of the chick embryo infected with influenza virus. Owing to the low multiplicity of infection, several cycles of viral growth were required before progeny virus could be detected; hence, the lag observed is much longer than the regular eclipse period of the virus. Note the considerable delay in the synthesis of IFN. (Modified from Smart KM, Kilbourne ED: J Exp Med 123:309, 1966)

while in human cells it causes a defective infection and induces abundant IFNs. However, many togaviruses induce large amounts of IFN, even though they multiply rapidly and have a pronounced cytopathic effect. Possible variables are effects on host macromolecule synthesis and amounts of inducer formed (see Chemical Induction, below).

Viral strains capable of high IFN production give rise to **autointerference** in endpoint assays: the dilutions containing the most virus may produce less virus because the IFN produced is sufficient to block further cycles of viral multiplication. Autointerference can also arise by a different mechanism (see Defective Interfering Particles, below).

Relation to Cells

Although animal cells of all types appear able to produce IFN-β, B-lymphocytes are the main producers of IFN-α. Thus, lethally x-irradiated mice grafted with rat bone marrow cells produce only rat-specific IFN. Moreover, antilymphocytic serum can inhibit IFN production in the animal. Bone marrow cells constitutively produce small

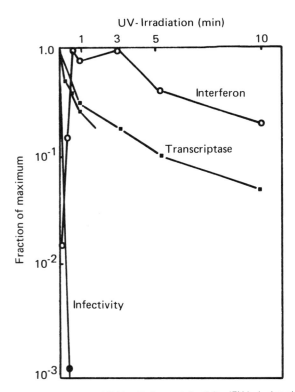

Figure 49–2. Effect of UV irradiation on infectivity, IFN-inducing ability, and transcriptase activity of Newcastle disease virus (a paramyxovirus). There is a marked enhancement of inducing ability after low UV doses, then it decays with a slope similar to that of transcriptase activity but much smaller than that of infectivity. (Modified from Clavell LA, Bratt MA: J Virol 8:500, 1971)

amounts of IFN, probably as regulator of cellular functions. Especially important is the regulatory role of IFN-γ. This IFN is produced by stimulated T-lymphocytes, and in vaccination against viral diseases it may activate macrophages, which then potentiate the action of Abs against cells (Ab-dependent cell cytotoxicity, see Chap. 50).

Chemical Induction

The nature of the stimulus to **interferon** production has been clarified by the Hilleman group's discovery that **double-stranded (ds)RNAs** such as reovirus RNA and certain synthetic polynucleotides can induce a large production of IFN in many animals and in tissue cultures. In human fibroblasts the effect of dsRNA is limited, however, to induction of IFN-β, whereas in the same cells viral infection induces both IFN-α and -β. The inducing activity resides in polyribonucleotides of a high molecular weight and resistant to enzymatic degradation, in which the 2′ position of the ribose is unsubstituted. One of the best inducers is the double-stranded synthetic polymer consisting of one chain of polyriboinoisinic and one of polyribocytidylic acid (poly[I : C]). Other inducers are bacteria, rickettsiae, bacterial endotoxin, and phytohemagglutinin. Some inducers are active only in certain cell types. In cells, poly(I : C) causes the formation of transacting factors that interact with the 5′ end of the IFN genes, allowing their transcription. Cells that are poor producers of the transactivators (such as the human HeLa cell line) produce little IFN upon induction.

ROLE OF dsRNA. It is likely that for most RNA viruses, dsRNA segments produced during replication (see Replication of Single-Stranded RNA, Chap. 48) mediate the induction of IFN. These viruses become much better inducers after mild UV irradiation, which inactivates viral genes required for initiation of progeny strand synthesis in the RIs, favoring accumulation of fully double-stranded RFs (Fig. 49–2). At high UV doses, however, the IFN-inducing activity decays in parallel to the activity of the transcriptase itself (and hence parallels the ability to form the RFs). Viruses containing dsRNA in the virions may induce without replication. For DNA viruses the inducer may be dsRNA resulting from symmetric transcription: double-stranded viral RNA extracted from cells infected with vaccinia (a DNA virus) induces IFN in tissue cultures.

CHARACTERISTICS OF PRODUCTION. In **virus-infected cells** the synthesis of IFN begins at about the time viral maturation initiates and then continues for many hours, unless the macromolecular syntheses of the host come to a halt. In cells exposed by poly(I : C), IFN production starts within about 2 hours; the rate of synthesis increases for several hours and then rapidly declines (**shut-off**). After shut-off, a new exposure of the cells to poly(I:C) does not restore IFN production for several hours (**hyporesponsiveness**; Fig. 49–3). Large concentrations of extracellular IFN inhibit the induction of IFN by either a virus or poly(I:C); in contrast, low concentrations enhance induction by an RNA virus (**priming**). The latter effect may be due to the demonstrable interference, by IFN, with the evolution of RFs into RIs.

Interferon is synthesized on membrane-bound polysomes and is then segregated into vesicles, which excrete it outside the cells. Until it is excreted, IFN does not act on the cell that produces it.

REGULATION OF SYNTHESIS.

After induction by poly(I:C), IFN mRNA makes its appearance, and this event—as well as IFN production—is prevented by actinomycin D.

At the time of shut-off, functional interferon mRNA disappears; however, if the cells are exposed at that time to actinomycin D, the mRNA persists, shut-off does not occur, and IFN production is increased. This increase is even more pronounced if the cells are exposed for several hours to a reversible inhibitor of protein synthesis before actinomycin D (Table 49–1). Such a **superinduction** is probably caused by a **posttranslational effect** that increases the life of IFN mRNA in the cells.

Because IFN is produced by cellular genes, it is clear why **viruses that block cellular mRNA or protein synthesis are poor inducers of IFN production.** Moreover, it is understandable that IFN synthesis fails in tissue culture cells simultaneously infected by a good inducer and a poor inducer: the poor inducer evidently inhibits the required cellular functions. Interactions of this type presumably also occur in animals and may influence the pathogenesis of viral infections.

MECHANISM OF INTERFERON ACTION

IFNs cause antiviral resistance not directly but by **activating cellular genes encoding antiviral proteins.** Interaction of IFNs with the cell surface induces the cell response. Thus, IFN bound to polysaccharide particles (Sepharose) is as active as free interferon. IFN-α and -β interact with the same receptors, which differ from those for IFN-γ. The receptors are glycoproteins. In human cells a gene specifying the receptors for IFN-α and -β is present on chromosome 21; that for IFN-γ is on chromosome 6, but it requires a gene present on chromosome 21 for activity. 21-Trisomic (Down syndrome) cells are especially sensitive to α and β IFNs.

Binding of IFN-β to its receptor is reported to cause a rapid increase of diacylglycerol in cells (see Chap. 47). This increase may be relevant for the gene activation

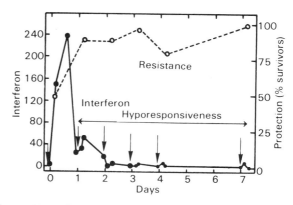

Figure 49–3. Refractory period following induction of interferon by double-stranded RNA injected into mice. After the first injection of RNA (*arrow at left*) there is a burst of interferon appearance in the serum, followed by the refractory period in which repeated injections (*other arrows*) do not cause further production. During this period, however, the mice are fully resistant to infection with encephalomyocarditis virus (a picornavirus), because the induced antiviral resistance persists. (Modified from Sharpe TJ et al: J Gen Virol 12:331, 1971)

caused by IFN. After binding IFN, the receptors are rapidly internalized through the endocytic pathway.

After the removal of extracellular IFN the inhibition of viral multiplication persists for a considerable period, the duration of which depends on the IFN concentration. The inhibition is overcome if cells are infected by virus at a high multiplicity of infection.

The **molecular mechanisms** of IFN-induced antiviral resistance are multiple and probably differ in different cell–virus systems. Effects on uncoating of the virus, on the stability or methylation of the viral RNA, and on transcription of the viral DNA have been reported. Retroviruses are affected at maturation, resulting in incorporation of nonviral proteins and altered processing of viral ones (see Chap. 65). However, *in vitro* studies with extracts of IFN-treated cells show that the **main target**

TABLE 49–1. *Effect of Cycloheximide and Actinomycin D on Interferon Production* *

Group	Cycloheximide (Hours of Treatment)	Actinomycin D at Hours 3–4	Interferon Yield
1	None	None	500
2	0–18	None	1,200
3	0–4	Yes	56,000
4	0–18	Yes	2,000

* Data show the effect that inhibition of RNA and protein synthesis has on interferon production in rabbit kidney cell cultures after poly(I : C) induction. The combination of the two inhibitors at suitable times and concentrations brings about dramatic enhancement of interferon production (compare groups 3 and 1).

(Data from Tan Yh et al: Virology 3:503, 1971)

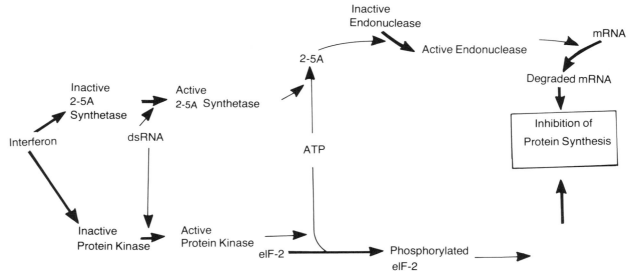

Figure 49–4. Effects of IFN and dsRNA on translation. *2-5A,* the adenine trinucleotide ''two-five A''; *eIF-2,* initiation factor.

of IFN action is translation, which is blocked by two mechanisms, involving a protein kinase and a nuclease (Fig. 49–4). In both the block requires the presence of minute amounts of **dsRNA,** which seems to signal to the cells the presence of a viral infection. As in the induction of interferon, the dsRNA may be that of a viral replicative intermediate, or it may result from symmetric transcription of the viral DNA; cellular dsRNA present in heterogeneous nuclear RNA is also adequate.

One of the **translation blocks** is caused by the activation of dsRNA-dependent **protein kinase,** which inactivates initiation factor eIF-2 by phosphorylating one of its subunits. The phorphorylation freezes the initiation complex formed by eIF-2, guanosine triphosphate (GTP), and Met-tRNAf with the small ribosomal subunit and mRNA. Because eIF-2 cannot be recycled, protein synthesis is inhibited or stopped. Some viruses can avoid this block: the VAI RNA of adenovirus prevents the activation of the kinase by dsRNA, reducing the effect of IFN.

The other translation block is due to **nucleolytic destruction of mRNA:** Kerr found that IFN induces the production of two **oligo A-synthetases** that, after activation by dsRNA, synthesize an **adenine oligonucleotide** containing three or more nucleotides of the form pppA2′p5′A2′p5′A with unusual 2′-5′ phosphodiester linkages—nicknamed ''**two-five A.**'' The oligonucleotide in turn activates a preexisting endoribonuclease, which cleaves mRNAs and ribosomal RNAs. Two-five A is important in antiviral resistance only in some virus–cell systems in which some analogs, which interfere with the RNase activity, inhibit the antiviral effect of IFN. In other systems, accumulation of two-five A in the cells has little effect on viral multiplication.

The blocks in translation may kill the infected cells,

but even so they halt the progress of infection. The two translation blocks occur in both infected and uninfected cells exposed to IFNs as part of the regulatory mechanism induced by this substance, but the effect is much more pronounced in virus-infected cells (containing dsRNA).

In mixed cultures of IFN-sensitive and IFN-resistant cells, antiviral resistance induced in the sensitive cells spreads to the resistant cells, presumably through channels of intercellular communication. Such a spread may favor the establishment of resistance throughout an organism.

Protective Role in Viral Infections

The protective role of endogenous IFN in viral infection of **cell cultures** is demonstrated by the establishment of **carrier cultures,** in which IFN produced in the cultures makes most of the cells resistant but cannot wipe out the infection; hence, only a small proportion are infected at any time (see Chap. 51).

A protective role of IFN **in animals** is suggested by many observations. (1) In mice recovering from influenza virus infection the titer of IFN is maximal at the time when the virus titer begins to decrease and before a rise in Abs can be detected. At this stage the IFN titer in the animals is sufficient to protect them against the lethal action of a togavirus. (2) Suckling mice, which are susceptible to coxsackievirus B1 (a picornavirus), produce little IFN in response to this virus, whereas adult mice, which are resistant, produce large amounts. Cortisone, which suppresses IFN production, makes adult animals susceptible. (3) Administration of a potent antiserum to IFN markedly increases the lethality of mouse hepatitis virus infection. (4) Mice naturally resistant to influenza virus or

VSV express a special protein the concentration of which is increased by IFN. This resistance-associated protein is maintained by the continued autocrine action of small quantities of IFN produced by the cells reacting with receptors of the same cells.

These studies suggest that IFN has a major protective role in at least some viral infections. Much depends on the **dynamics of the disease,** i.e., the relation between virus titer and IFN titer at various times. IFN is most effective when present before infection and when the dose of infecting virus is not too large (as at the beginning of most natural infections). The protection afforded may be especially useful because it develops more promptly than Ab production.

Clinically, effective **prophylaxis** was demonstrated against rhinovirus infection of human volunteers, with decreased incidence of infection and reduction of symptoms. Contacts of an infected patient can be protected by intranasal spray of large doses of IFN. Also reduced is cytomegalovirus reactivation in seropositive patients undergoing kidney transplant, with respect to both incidence of viremia and urinary shedding of the virus. In all cases the limiting factor is the extent of side-effects, which increase with dose in parallel with the beneficial effect.

POSSIBLE THERAPEUTIC USE. IFNs could theoretically be ideal antiviral agents, since they act on many different viruses and have high activity. However, their therapeutic value is limited by various factors: IFNs are effective only during relatively short periods and have no effect on viral synthesis that is already initiated in a cell. Moreover, at the high doses needed they have serious toxic effects on the host.

Attempts to use exogenous IFN for therapeutic purposes in human viral diseases have had some limited success. Thus, IFN-α had a prophylactic effect against influenza infection during epidemics, and local administration lessens the severity of respiratory diseases. In a large study, injection of large doses of recombinant IFN-α into genital warts (condyloma acuminatum) induced by papilloma virus (see Chap. 64) had a noticeable effect by causing their disappearance in one third of cases and reduced their surfaces in the other cases. Favorable effects have also been obtained after systemic treatment of warts of juvenile laryngeal papillomatosis, an aggressive disease also caused by a papilloma virus. A lessening of the pain of herpes zoster by IFN has been reported, as has a decreased spread of herpes simplex keratitis after local applications. However, the lesions produced by these viruses are not cured.

Effect on the Functions of Uninfected Cells

In vitro, IFNs inhibit cell replication and inhibit the activation of spleen lymphocytes by phytohemagglutinin. In human fibroblasts they increase the density of microfila-

ments at the cell membrane and the rigidity of the bilayer, while decreasing the mobility of surface receptors and endocytosis. *In vivo*, high doses inhibit liver regeneration, as well as the production of platelets and leukocytes, and enhance the expression of histocompatibility Ags on the lymphocyte surface.

IFN also profoundly affects the **immune response** in opposite ways: it reduces Ab production by stimulating T-suppressor cells, but it enhances the activity of cytolytic T cells; it also increases the cytotoxic activity of NK cells against virus-infected cells (see Chap. 50). These effects denote a **shift from humoral to cell-mediated immunity,** which has a defensive role in many viral infections. Some of these effects determine a pathogenetic role of IFN. Thus, administration of high doses to newborn mice induces a lethal liver degeneration and an autoimmune glomerulonephritis. Moreover, production of endogenous IFN contributes to the disease caused in mice by lymphocytic choriomeningitis virus (see Chap. 51), because it stimulates immune reactions. The combination of cell growth inhibition and enhancement of cell-mediated immunity accounts for the **antitumor** effect of IFN. Studies with highly purified IFN preparations show that all these effects are caused by the same molecule.

INTRINSIC INTERFERENCE (NOT MEDIATED BY INTERFERON)

An inability to detect IFN does not exclude its participation in an instance of viral interference, because the detection methods are relatively insensitive. However, if interference is established early in the infectious cycle, the participation of IFN can be considered unlikely because its production usually begins later.

Other mechanisms, grouped as intrinsic interference, have been studied with both bacteriophages and animal viruses.

Bacteriophage

Homologous or closely related phages (such as two mutants of the same strain, or T2 and T4) in the same cell must compete for the same precursors, cellular sites, and enzymes. The two strains can replicate more or less equally if they both infect the cell simultaneously and at low multiplicities. However, if they infect at different times or with different multiplicities, the phage with the advantage multiplies normally and interferes with (or even completely prevents) the multiplication of the other phage. In nonsimultaneous infection, interference appears to result from a change in the bacterial plasma membrane that prevents penetration of the DNA of the second phage.

With **unrelated** phages, one phage is excluded by a variety of mechanisms. Thus T-even bacteriophages probably exclude T1, T3, and λ by destroying the host cell DNA, the functions of which are needed for the multiplication of the excluded phages. Phage T4 excludes the RNA phage F2 both by inactivating a translation initiation factor that F2 requires and by rapidly degrading its RNA. If infection is not simultaneous, the phage that injects its DNA after another phage has taken over control of the cell machinery is always excluded.

Animal Viruses: Defective Interfering (DI) Particles

HOMOLOGOUS INTERFERENCE. As with bacteriophages, interference involving homologous animal viruses is most pronounced if one virus has an advantage, either in time or in multiplicity. This type of interference takes place with Newcastle disease virus (a paramyxovirus) at **adsorption,** through destruction of cellular receptors; with retroviruses at **penetration;** and with many viruses at **replication.**

Interference at replication is usually generated by **DI particles,** which accumulate after serial passages at a high multiplicity of infection. They are formed during infection with various kinds of RNA viruses, such as rhabdoviruses, togaviruses, orthomyxoviruses, and paramyxoviruses, coronaviruses, and of some DNA viruses (herpesviruses). With some viruses (e.g., VSV or Semliki Forest virus) the DI particles are smaller than regular particles and can therefore be obtained in pure form. They usually contain the normal virion proteins but have a shorter genome. DI particles are **replication defective:** they require as **helper** a regular virus coinfecting the same cells. In the early serial passages, DI particles rapidly increase in titer; then the yield of infectious virus, and finally the total particle yield, is progressively reduced (**autointerference**). In some cases these events lead to establishment of a **persistent infection** with a **carrier state** (see Chap. 51). Through autointerference the DI particles may limit the spread of viral infections in animals.

The genomes of DI particles are internally deleted but retain both ends, which are essential for the replication of RNA viruses. With DNA virus (e.g., herpesviruses) the origin of replication is always conserved and often repeated. These features show that to cause interference the DI genomes must replicate. They deprive the regular virus of its replicase by binding to it more effectively; moreover, they do not make a replicase of their own because they are always defective in the replicase gene.

The formation of DI genomes of RNA viruses is the consequence of the high variability of these genomes. With these viruses DI genomes are formed by a **copy choice** mechanism when the replicase, having replicated part of the template, skips to another part of the same or another template. With VSV and other negative-strand RNA viruses, the skipping generates defective genomes of four types: deletions, snapbacks, panhandles, and compounds (Fig. 49–5). In **deletions** the polymerase jumps to a site beyond, on the same template, skipping a segment. **Snapbacks** are formed when the

replicase, having transcribed part of the (+) strand, switches to using the just-made (−) strand as template; the resulting RNA, half of (−), half of (+) polarity, can produce a hairpin on annealing. A **panhandle** is formed by a similar mechanism, when the polymerase carrying a partial newly made (−) strand switches back to transcribing the extreme 5′ end of it; on annealing, the strand forms a panhandle. **Compound genomes** are made by a combination of deletions and snapbacks. Specific sequences may be involved in these irregularities. In fact, when a panhandle is formed in VSV the polymerase often jumps to a site 46 nucleotides from the 5′ end of the new (−) strand and then transcribes the remaining stretch; that site may be identified by its sequence. And the ends of deletions in Sendai virus DI particles have sequences related to transcription signals.

DI genomes may also have point mutations, not present in the regular viruses, and insertions, for instance of a cellular tRNA in the DI genomes of Sindbis virus.

The competition of DI genomes with competent genomes depends not only on the structure of the DI genome but also on that of the competent genome: different DI genomes may interfere to very different degrees with the same competent genome, and competent genomes may acquire mutations that make them resistant to the existing DI genomes. Subsequently, however, this resistance is overcome by new types of DI genome. During viral multiplication many types of DI genomes are continuously made; they are very heterogeneous. Both the DI genomes and the competent genome evolve continuously, increasing their ability to compete with each other.

HETEROLOGOUS INTERFERENCE. The mechanisms of heterologous interference vary in different systems, as shown by two examples. (1) Poliovirus arrests the multiplication of VSV and other RNA viruses, even when infecting later and at lower multiplicity. Apparently the mechanism is the poliovirus-induced alteration of the cap-binding protein (see Chap. 48): translation of the capped VSV or cellular messenger is prevented, but translation of uncapped poliovirus messengers is not. The interference may also involve other features of viral mRNAs because it does not occur with all RNA viruses. (2) Newcastle disease virus (NDV) RNA fails to replicate in cells previously infected by certain other RNA viruses, e.g., Sindbis virus (a togavirus). NDV RNA may form an inactive complex with the replicase induced by the interfering virus, because Sindbis mutants deficient in RNA replicase do not interfere.

SIGNIFICANCE OF VIRAL INTERFERENCE

Interference, both by IFN and by other mechanisms, is important in several aspects of viral infection. For in-

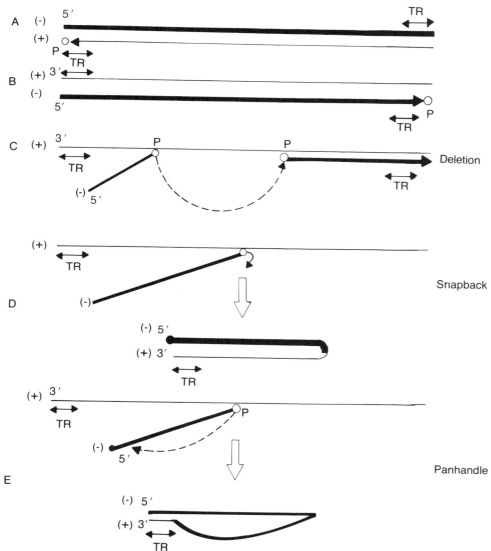

Figure 49–5. Models of production of DI particle genomes with VSV virus. (*A*) Normal transcription of the (−) genome to generate a full-length (+) transcript. (*B*) Transcription of the (+) genome with formation of competent progeny (−) genome. (*C*) Formation of a deleted DI genome. (*D*) Formation of a snapback DI genome. In the panhandle form (*E*) the 5′ end of the template is transcribed on a length of from 46 to several hundred nucleotides. The 46-nucleotide segment corresponds to the "leader," which appears to be especially important in replication. *Heavy lines*, genomic (−) RNA; *thin lines*, (+) RNA; *TR*, initiation of transcription.

stance, in human oral **vaccination** with attenuated poliomyelitis viruses (see Chap. 55) the three strains must be administered in a precise sequence or at specified concentration ratios to avoid interference of one strain with another. Similarly, the presence of various enteroviruses in the normal intestinal flora may hinder the establishment of infection by the vaccine strains. Viruses already present may also influence the response to a **naturally infecting virus.** For example, dengue virus infection in man is milder in the presence of an attenuated strain of yellow fever virus (both flaviviruses). IFN seems to play an important role in initiating recovery from some acute viral infections. The role of DI particles in animal infections is uncertain.

Chemical Inhibition of Viral Multiplication

SYNTHETIC AGENTS

The chemical structure of the more important synthetic agents is shown in Figure 49–6.

Selective Agents

Amantadine (adamantanamine), of peculiar structure, is active against influenza virus A and slightly against rubella and parainfluenza viruses. Being a base, it prevents the acidification of the endosomes and therefore the release of the viral genome to the cytoplasm (see

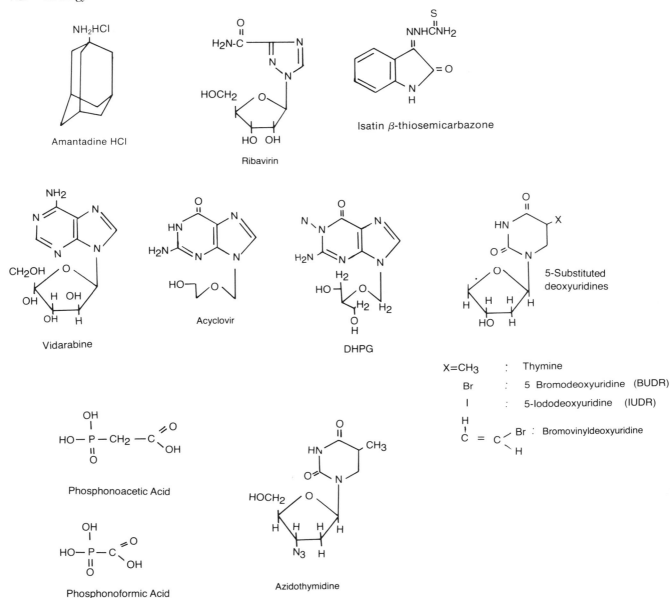

Figure 49–6. Chemical constitution of antiviral inhibitors.

Chap. 48). Drug-resistant mutants are changed in the hydrophobic part of the viral matrix protein, which is associated with the viral membrane and plays a crucial part in the release of the viral RNA from the endosome. Differences in the matrix protein between different viruses explain why the effect is limited to influenza A.

The compound, given orally, has prophylactic value against influenza A infections, attaining a 50% to 80% reduction of illness. Therapeutic effects are modest. The drug has limited side-effects, such as anxiety or insom-

nia, which rapidly dissipate on discontinuation of use. A methyl derivative, **rimantadine,** holds even better promise, because it is similarly effective but less toxic.

Ribavirin, a purine nucleoside analog, is phosphorylated in both virus-infected and uninfected cells. The phosphorylated derivative interferes with the synthesis of GMP, with resultant inhibition of both RNA and DNA synthesis and of mRNA capping. The toxic effect is stronger for virus-infected cells, which use more nucleic acid precursors than normal cells, owing to the synthesis of viral nucleic acids. *In vitro*, as well as in experimental animals, the drug has a broad antiviral spectrum that includes both DNA viruses (herpes simplex, vaccinia) and RNA viruses (influenza, parainfluenza, VSV).

In humans, ribavirin, especially by the IV route, is effective in the treatment of Lassa fever (see Arenaviruses, Chap. 62) if treatment is started within a week of onset of fever. Its administration by small-particle aerosol also appears to be somewhat effective in the treatment of influenza A and B within 24 hours of onset, and of respiratory syncytial virus infection in infants. Administered intravenously or orally, the drug causes hemolysis, which is rapidly reversible, but as an aerosol it has essentially no toxicity, probably because very little is absorbed.

Isatin-β-thiosemicarbazone and its N-methyl and N-ethyl derivatives inhibit the multiplication of several DNA and RNA viruses, especially poxviruses. They inhibit the synthesis of late poxvirus structural proteins, apparently by inactivating the late viral mRNAs. The result is the production of nearly spherical defective particles.

N-methylisatin-β-thiosemicarbazone (methisazone) had an impressive **prophylactic** success in preventing the spread of smallpox to contacts in an epidemic in Madras, India, in 1963. Among 1101 contacts vaccinated and treated with methisazone, three mild cases of smallpox occurred; in contrast, among 1126 vaccinated but not treated with the drug, 78 contracted smallpox, and 12 died. The much greater prophylactic effect of the drug can be attributed to its immediate effect on viral multiplication, contrasted with the delayed effect of the immune response to vaccination.

In contrast, the drug failed as a **therapeutic** agent in patients already suffering from smallpox. This failure is understandable, because the disease appears only after viral multiplication has reached a maximum.

Guanidine inhibits multiplication of some enteroviruses *in vitro* by interfering with the synthesis of viral single-stranded RNA without affecting cellular RNA synthesis. It does not show promise as a chemotherapeutic agent in animals, owing to the rapid emergence of **resistant viral mutants.** Viruses can also mutate to **dependence** on this drug.

Agents Interfering With DNA Synthesis

A considerable advance in antiviral chemotherapy has been the synthesis of analogs of purines and pyrimidines that inhibit the replication of certain viral DNAs much more than that of the cellular DNA. At the doses required for effective antiviral therapy they show little or moderate toxicity for the organism. Some of these compounds are the closest approximation to a selective antiviral therapy with limited cellular toxicity.

The most effective agents act against herpesviruses (see Chap. 53) and poxviruses (see Chap. 54), which, like the T-even bacteriophages (see Chap. 45), have large genomes that specify many enzymes involved in DNA synthesis. The viral enzymes that differ markedly from the corresponding cellular enzymes are used to phosphorylate the drugs or are selectively inhibited by the drugs. Viruses with smaller genomes, which depend much more on cellular enzymes, cannot be selectively inhibited.

With herpesviruses, enzymes important for chemotherapy are the viral thymidine kinase and DNA polymerase, both different from the cellular enzymes in substrate specificity. Some drugs are converted to monophosphates by the viral enzyme and then to triphosphates by cellular enzymes. Other drugs are phosphorylated entirely by cellular enzymes. The triphosphates then block viral replication either by inhibiting the viral polymerase or by becoming incorporated into viral DNA. These substances have little effect on uninfected cells.

Vidarabine (ara-A) (9-β-D arabinofuranosyladenine) is phosphorylated entirely by cellular enzymes and is therefore active against all human herpesviruses, including cytomegalovirus and EB virus, which do not specify a thymidine kinase. The drug is also effective against vaccinia virus. A similar analogue of cytidine, **cytarabine (ara-C, arabinosyl cytosine),** has a similar action but is less selective.

Vidarabine, administered intravenously, has beneficial effects on the treatment of severe varicella or herpes zoster infections in immunosuppressed patients and of the often fatal herpesvirus infection in newborns. The drug is very useful in the therapy of herpesvirus encephalitis in man, reducing its mortality from 70% to 30%. The treatment must be initiated very early in the disease, which requires brain biopsy for early diagnosis. Vidarabine-resistant mutants occur and usually have an altered DNA polymerase. A drawback of vidarabine is its rapid deamination in the body to the hypoxanthine derivative, which is much less effective; a carbocyclic derivative (in which a methylene group replaces the oxygen atom of the carbohydrate ring), **cyclaradine,** is resistant to deamination and retains antiviral activity.

Acyclovir, an analogue of guanosine lacking part of the sugar ring, is phosphorylated exclusively by the viral kinase and therefore has little effect against cytomegalovirus and EB virus. Acyclovir triphosphate inhibits viral DNA polymerase and is incorporated into the viral DNA. Administered intravenously in man, acyclovir is even more effective than vidarabine in the treatment of herpes simplex encephalitis and varicella zoster virus infection in immunocompromised patients. It is equally effective in the treatment of herpesvirus infection in newborns. Oral acyclovir is promising for the prophylaxis of genital herpesvirus infections and of herpesvirus infection in patients undergoing bone marrow transplantation or intensive cancer chemotherapy, because it can be administered for long periods with little toxicity. Acyclovir-resistant viral mutants of herpes simplex virus are altered in the thymidine kinase or polymerase genes; no such mutants have been observed with varicella zoster virus.

Very promising is a drug closely related to acyclovir and with a similar mode of action, **DHPG.** *In vitro* it is more effective against cytomegalovirus and EB virus than acyclovir itself, and it is beneficial against serious cytomegalovirus infection in immunodeficient persons. It is also effective against some acyclovir-resistant mutants of human herpesviruses. Its most frequent complication is neutropenia.

Other Pyrimidine Nucleoside Analogues

Halogenated derivatives of deoxyuridine, such as 5-bromo-deoxyuridine (BUdR) and **idoxuridine** (5-iodo-2-deoxyuridine, IUdR), were among the first antiviral compounds to be synthesized; however, they are less selective and more toxic than the preceding compounds. These analogues are taken up by cells and phosphorylated by the viral and cellular thymidine kinase; the phosphorylated derivatives are incorporated into DNA instead of thymidine in both virus-infected and uninfected cells. The DNA continues to replicate but causes the synthesis of altered proteins. Indeed, in cells infected by herpesvirus in the presence of idoxuridine, viral maturation fails owing to defects of the capsid protein. If the drug is later removed, virions are formed; they contain DNA replicated in the presence of the drug, in which iodouracil replaces thymine.

In spite of its toxicity, idoxuridine is valuable for the topical treatment of surface lesions (e.g., keratitis) produced by herpes simplex or vaccinia virus. Some herpesvirus mutants are resistant.

5-Halogenovinyl pyrimidine analogues are also promising, especially **bromovinyldeoxyuridine,** which is most effective against herpesvirus type 1 and varicella zoster viruses, owing to the specificity of their thymidine kinases. Other drugs effective *in vitro* or in animals, especially against cytomegalovirus, are various fluorosubstituted arabinosyl-pyrimidine derivatives.

Phosphonoacetic acid and phosphonoformic acid, which inhibit the herpesvirus-specified DNA polymerase, markedly reduce the replication of this virus in cultures, as well as the severity of experimental infection in animals. Drug-resistant mutants are altered in the viral DNA polymerase.

Agents Interfering With Reverse Transcription

Reverse transcription (RT) is a replication strategy employed by retroviruses, hepadnaviruses, and some eukaryotic transposons. Interference with RT is important in the chemotherapy of retroviruses that are highly pathogenic for humans, such as HIV (see Human Immunodeficiency Virus, Chap. 65); because RT plays no role in the multiplication or function of animal cells, a high selectivity can be achieved.

Several antiviral drugs inhibit RT *in vitro* with little effect on cellular DNA replication: **suramin** and **ribavirin** are among those, but they are too toxic for use in humans. More selective are **dideoxynucleosides,** such as **3′-azidothymidine** and **dideoxycytosine.** In either case the nucleotide obtained after phosphorylation by host enzymes is incorporated into DNA, where, for absence of a 3′-OH, it blocks further chain elongation. These drugs have beneficial although partial effect in the treatment of acquired immune deficiency syndrome (AIDS; see Human Immunodeficiency Virus, Chap. 65). Prolonged treatment, however, is toxic for the bone marrow.

General Considerations

The drugs active against herpesvirus are important in the treatment of this infection because in addition to alleviating the symptoms, they reduce viral shedding and the spread of the virus. They are effective only against the acute manifestations; they do not affect latent infections, in which there is little replication of the viral genome and the virus-specified enzymes involved in the antiviral effect may not be expressed. By inhibiting viral multiplication, the drugs also decrease the production of antibodies to the virus, with possible adverse effect on subsequent infection. As in the case of other viral infections, these drugs are most effective if used prophylactically; they can be therapeutically effective only during the very early stages of disease, when viral spread to the susceptible cells is limited. Once all susceptible cells are infected, the disease is no longer modifiable.

ANTIBIOTICS

Antiviral activity against poxviruses and retroviruses is displayed by derivatives of **rifamycin** (e.g., rifampin) and by the related antibiotics tolypomycin and streptovaricin. In bacteria rifampin is known to inhibit initiation by RNA polymerase; and with retroviruses these antibiotics inhibit the activity of the reverse transcriptase and cause the formation of defective, RNA-deficient virions. With poxviruses rifampin acts by **interfering with virus maturation:** viral membranes begin to form at the periphery of the filamentous matrix (see Complex Viruses, Chap. 48) but remain incomplete. If rifampin is then removed, the membranes close and form virions (Fig. 49–7).

The mechanism of action of rifampin on poxvirus maturation is unknown. It may conceivably alter the structure of the uncleaved precursor peptides, inactivate the cleaving enzyme, or block transcription of a few viral genes important in maturation. The activities of the vari-

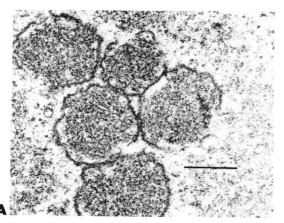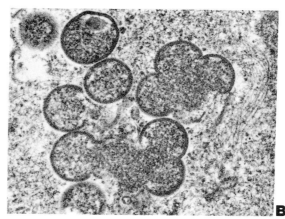

Figure 49–7. Effect of rifampin on the maturation of vaccinia virus in HeLa cells. (*A*) Electron micrograph from a thin section of infected cells treated for 8 hours with rifampin (100 μg/ml), showing the incomplete and disorganized viral membranes, each surrounding a matrix. Bar = 300 nm. (*B*) Thin section of similar cells 10 minutes after removal of rifampin, showing the rapid reorganization of the membranes to a morphology similar to that observed in normal maturation. (Courtesy of P. M. Grimley)

ous rifamycin derivatives toward retrovirus, poxvirus, or *Escherichia coli* transcriptase are uncorrelated, suggesting different modes of action. Poxvirus mutants resistant to rifampin and other antiviral rifamycin derivatives are readily isolated.

In vivo these compounds have some local effect; for instance, rifampin inhibits the development of vaccination lesions in man. They have little effect on systemic diseases in animals, perhaps because their high toxicity precludes the use of adequate doses.

Selected Reading

Balfour HH Jr: Acyclovir and other chemotherapy for herpes group viral infections. Annu Rev Med 35:279, 1984

Dolin R, Reichman RC, Madore HP et al: A controlled trial of amantadine and rimantadine in the prophylaxis of influenza A infection. N Engl J Med 307:580, 1982

Dolin R: Antiviral chemotherapy and chemoprophylaxis. Science 227:1296, 1985

Goeddel DV, Leung DW, Dull TJ et al: The structure of eight distinct cloned human leukocyte interferon cDNAs. Nature 290:20, 1981

Hall CB, McBride JT, Walsh EE et al: Aerosolized ribavirin treatment of infants with respiratory syncytial viral infection. N Engl J Med 308:1443, 1983

Hirsch MS, Schooley RT: Treatment of herpesvirus infections, parts 1 and 2. N Engl J Med 309:963, 1983

Lazzarini RA, Keene JD, Schubert M: The origins of defective interfering particles of the negative-strand RNA viruses. Cell 26:145, 1981

Lengyel P: Biochemistry of interferons and their actions. Annu Rev Biochem 51:251, 1982

McCormick JB et al: Lassa fever—effective therapy with ribavirin. N Engl J Med 314:20, 1986

O'Hara PJ, Nichol ST, Horodyski FM, Holland JJ: Vesicular stomatitis virus defective interfering particles can contain extensive genomic sequence rearrangements and base substitutions. Cell 36:915, 1984

Staeheli P, Haller O, Boll W et al: Mx protein: Constitutive expression in 3T3 cells transformed with cloned Mx cDNA confers selective resistance to influenza virus. Cell 44:147, 1986

Straus SE, Takiff HE, Seidlin M et al: Suppression of frequently recurring genital herpes. N Engl J Med 310:1545, 1984

Streissle G, Paessens A, Oediger H: New antiviral compounds. Adv Virus Res 30:83, 1985

Whitley RT et al: Vidarabine versus acyclovir therapy in herpes simplex encephalitis. N Engl J Med 314:144, 1986

Winship TR, Marcus PI: Interferon induction by viruses. VI. Reovirus: Virion genome dsRNA as the interferon inducer in aged chick embryo cells. J Interferon Res 1:4943, 1980

50

Renato Dulbecco

Viral Immunology

Induction of the Immune Response by Viruses

Viruses are usually strongly immunogenic and elicit two main types of response: a **humoral response,** caused by specific B cells with production of **antibodies** (Abs), and a **cellular response** caused by several varieties of T-lymphocyte clones, principally helper (T_h) and cytolytic (CTL) cells (see Chap. 15). B cells recognize free viral antigen (Ag), whereas T cells recognize viral Ags jointly with proteins of the major histocompatibility complex (MHC): CTLs with MHC-I and T_h with MHC-II proteins. Some specific cytolytic cells ($T_{h/c}$) recognize viral Ags together with MHC-II proteins. Participating in the organism's reactions are also "natural killer" (NK) cells, macrophages, various cells that are cytotoxic when activated by Abs (antibody-dependent cell cytotoxicity [ADCC]) and complement (complement-dependent cell cytotoxicity [CDCC]).

The repertoire of specificities of Abs and T cells relevant to an immune response is determined, as for other Ags, by rearrangements of the genes for immunoglobulins and for T-cell receptors (see Chaps. 16 and 17), as well as somatic mutations in the Ab genes. Abs and T cells formed in the same animal in response to a virus do not generally recognize the same antigenic groups (**epitopes**) of the virions. This is related to the different ways in which the antigen is presented to B or T cells. B cells see the free, unaltered proteins in their normal three-dimensional configuration on the viral surface; hence, Abs often recognize epitopes that depend upon the folded three-dimensional shape of the polypeptide chain. T cells usually see the Ag in denatured form or as fragments of the native antigen in conjunction with cellular MHC proteins. For instance, in the reaction to influenza hemagglutinin in mice, Abs almost regularly distinguish between different strains within the same subtype, whereas T cells show extensive cross-reactivities, sometimes even extending to different subtypes.

The characteristics of the immune reaction to the same virus may differ in different individuals, or in animals of non-inbred species, depending on their genetic constitutions; that is, depending on the available variable region genes for Abs or T-cell receptors and on the allele of the MHC genes expressed in different individuals.

Roles of the Immune Response

The humoral response and the cellular response have different consequences. Abs block the infectivity of virions (neutralization); those of the IgM or IgG class are especially relevant for defense against viral infections accompanied by viremia, whereas those of the IgA class are important in infections acquired through a mucosa (the nose, the intestine), because they can neutralize the virus at the entry portal. In contrast, the cellular response kills the virus-infected cells, which express at their surface viral proteins (such as the glycoproteins of enveloped viruses) and sometimes core proteins of these viruses (such as NP protein of influenza virus) or nonvirion proteins of nonenveloped viruses (such as the T Ag of papovaviruses).

Humoral Factors

Abs are elicited not only by surface components of intact virions and by internal components of disrupted virions, but also by viral products built into the surface of the infected cells or released by the cells (e.g., when they die). Although Abs provide the key to protection against many viral infections, they are sometimes pathogenic; e.g., deposits of Ab–Ag complexes in the kidneys of mice infected with lymphocytic choriomeningitis virus cause immune-complex disease. Abs are also useful in the laboratory for identifying, quantitating, and isolating virions (and also some of the unassembled components), for classifying viruses, and for serologic diagnosis of viral diseases. Especially useful for these purposes are **monoclonal Abs** obtained from **hybridomas** prepared with lymphocytes from mice immunized with viral material, or possibly from patients affected with viral diseases.

INTERACTIONS BETWEEN VIRIONS AND ANTIBODIES

Interactions of virions with Abs to different components of their coats have different consequences. Moreover, the many identical antigenic sites, regularly repeated on viral surfaces, interact with Abs in ways that would be impossible with isolated sites.

Reactions of viral Ags with their corresponding Abs are studied by both the usual immunologic tests and some special ones, such as hemagglutination inhibition and neutralization. Because neutralization is particularly characteristic of viruses and has widespread application, we shall consider it in detail.

NEUTRALIZATION

Viral neutralization consists of a decrease in the infectious titer of a viral preparation following its exposure to Abs. The loss of infectivity is brought about by interference by the bound Ab with any one of the steps leading to release of the viral genome into the host cells (see Chap. 48). Each one of these steps involves delicate molecular interactions between cellular and viral macromolecules in which the latter undergo characteristic conformational changes. The attachment of Abs to a virion macromolecule will perturb the step it carries out because the Ab binds with high affinity, preventing the physiologic conformational changes and often causing abnormal ones. In both cases viral replication is blocked. The **consequences** of the virion–Ab interaction therefore depend on many factors: (1) the structure of the virions; (2) the target of the Ab (e.g., Abs against the hemagglutinin but not against the neuraminidase neutralize influenza virus, and Abs against the hexon but not the penton neutralize adenovirus); (3) mutations affecting surface molecules that may alter susceptibility to certain Abs; (4) the type of Ab, especially its affinity for the components of virions; and (5) the number of Ab molecules attached to a virion. In addition, the interaction between viral and cellular macromolecules may be different in different cell types.

In neutralization we must distinguish readily reversible from stable virion–Ab complexes.

READILY REVERSIBLE VIRION–ANTIBODY COMPLEXES.
Readily reversible complexes are recognized by an experiment of the following kind. If influenza virus is mixed with Ab at 0°C and a sample of the mixture is added to an assay system (e.g., the allantoic cavity of chick embryos or cell cultures) $\frac{1}{2}$ hour later **without dilution,** a decrease of the viral titer in comparison with untreated virus may be seen; but if the neutralization mixture is **diluted** by a large factor before assay, the original titer is restored (Fig. 50–1). Hence, there is a freely reversible equilibrium between dissociation and re-formation of the Ab–virion complexes: decreasing the concentration of the reactants diminishes the rate of complex formation but not the rate of dissociation.

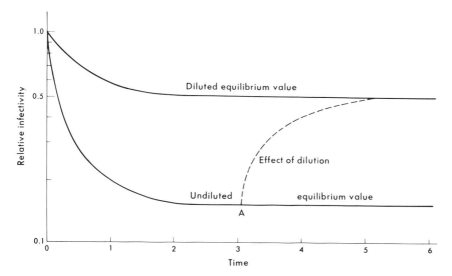

Figure 50–1. Time course of neutralization in a virus–Ab mixture with readily reversible Ab–Ag combinations. One mixture is undiluted; the other is diluted fivefold. Note the different equilibrium values reached. If at time *A* the undiluted mixture is diluted fivefold, the viral titer, corrected for dilution, increases, owing to dissociaton of virus–Ab complexes, and reaches the same equilibrium value as the originally diluted mixture.

VIRION–ANTIBODY COMPLEXES STABLE ON DILUTION. With time the virus–Ab complexes become more stable; if the assays are made **several hours after the virus has been mixed** with the Abs, neutralization persists after dilution (at neutral pH and physiologic ionic strength). Whereas reversible complexes form readily with little temperature dependence, the formation of irreversible complexes occurs very slowly at 0°C but rapidly at 37°C, suggesting the need for configurational changes.

Neither the virions nor the Abs are permanently changed in stable neutralization, for the unchanged components can be recovered. The neutralized virus can be reactivated by proteolytic cleavage of the bound Ab molecules into monovalent fragments and by other means, and intact Abs are recovered by dissociating the Ab–virus complexes at acid or alkaline pH (Fig. 50–2), by sonic vibration, or by extraction with a fluorocarbon.

PHYSICOCHEMICAL BASIS OF A STABLE ANTIBODY–VIRION ASSOCIATION. The firm association of Abs with virions originates from the multimeric nature of the viral coat, which allows a single Ab molecule to establish specific bonds with two sites on a single virion: electron micrographs show that an Ab molecule can bridge two subunits of the virion's surface (Fig. 50–3). The stability of the complex is high, because whenever one of the bonds dissociates, the other holds, and the dissociated one has time to become reestablished.

Mechanism

Reversible neutralization is probably due to interference with attachment of virions to the cellular receptors, because many sites for adsorption on a virion have each been covered by an Ab molecule (or a monovalent frag-

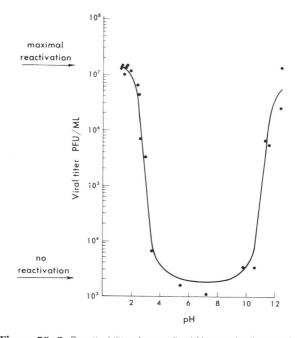

Figure 50–2. Reactivability of neutralized Newcastle disease virus at different pHs, showing the dissociability of the virus–Ab complexes at acid and alkaline pHs. Virus and Abs were incubated together until a relative infectivity of about 10^{-4} was obtained (assayed after dilution). Aliquots of the mixture were then diluted 1 : 100 in cold buffer at various pH values, and after 30 seconds the pH was returned to 7 by dilution in a neutral buffer. The samples were then assayed. (Granoff A: Virology 25:38, 1965)

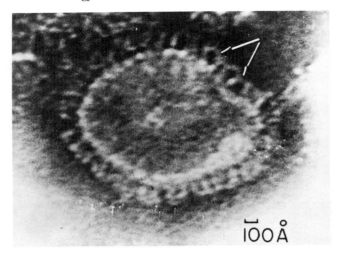

Figure 50–3. Electron micrograph of an influenza virion with Ab molecules attached to the spikes, some of them forming bridges between two spikes (*arrows*). (Lafferty KJ; Virology 21:91, 1963)

ment). In addition, aggregation by Abs may decrease the number of foci of infection.

Stable neutralization, in contrast, has a different mechanism: not only can it be shown that neutralized virions attach, but already attached virions can be neutralized. Moreover, **stable neutralization does not require saturation of the surface** of the virion with Ab molecules. Thus, phenotypically mixed virions with two types of surface monomers, produced by double infection (see Chap. 48; Fig. 50–4), can be neutralized by antisera specific for **either** monomer. In this type of neutralization the number of Ab molecules attached to a virion is much smaller than the number of protomers at the virion's surface. Fifty Ab molecules per virion are needed for 50% neutralization of influenza virus (1000 hemagglutinating units per virion) 3 or 4 molecules per virion for poliovirus (60 capsomers). Kinetic evidence shows that even a single Ab molecule can neutralize a virion (Fig. 50–5).

Stable neutralization is generally produced by Ab molecules that establish **contact with two antigenic sites** on different monomers of a virion, greatly increasing the stability of the complexes. That such double binding is obligatory for neutralization is shown by reactivation of poliovirus virions when the Ab is cleaved by papain and by their reneutralization if the Ab fragments are reconnected by an Ab to the neutralizing immunoglobulin.

The consequences of the double attachment are varied, as shown by some examples. (1) With poliovirus, neutralization often changes the isoelectric point (i.e., the balance of all + and − charges on the virion's surface) from pH 7 to 5.5 and also changes the accessibility of chemical groups on the virion's proteins to external

reagents. **The transition is all-or-none,** for there are no virions with intermediate isoelectric points. Apparently the binding of an Ab molecule to a capsomer **alters the capsomer's conformation;** the alteration then spreads to the whole capsid by a domino effect. (2) With bacteriophage M13, the DNA of which penetrates into the cells while the protomers of the capsid enter the plasma membrane (see Chap. 45), an Ab molecule blocks the latter process, presumably by holding two adjacent protomers together. Neutralization of some enveloped animal viruses may follow the M13 model. (3) With flavivirus the bound Ab **prevents the fusion** between the viral envelope and the membrane of the acidified endosome (see Chap. 48), blocking the release of the virion's core into the cytoplasm. The trapped virion is presumably handed over to a lysosome, where it is destroyed.

BOUND ANTIBODIES THAT DO NOT NEUTRALIZE. Since the kinetic data show that a single Ab can neutralize, while a higher number is actually bound when 50% of the virus is neutralized, the probability of neutralization by an attached Ab molecule is less than unity. One factor

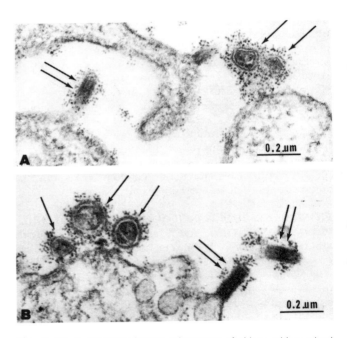

Figure 50–4. Immuno-electron microscopy of virions with a mixed envelope (pseudotypes) obtained by superinfecting a murine leukemia virus (MuLV)–producing cell with vesicular stomatitis virus (VSV). The virions are stained with ferritin-labeled anti-MuLV Abs: the ferritin molecules are recognizable as black spots. Single arrows point to uniformly labeled MuLV virions (spherical); double arrows to VSV (bullet-shaped) virions, some uniformly labeled (A; i.e., completely coated by MuLV-specific glycoproteins), others (B) with some glycoproteins VSV-specified and unlabeled and some MuLV-specified and labeled. (Chan JC et al: Virology 88:171, 1978)

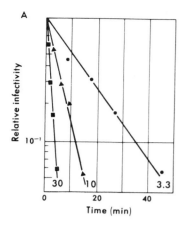
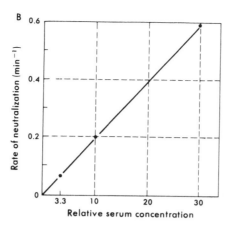

Figure 50–5. Kinetic curves of neutralization of poliovirus with stable Ab–virion complexes and Ab excess. Note the linearity of the curves in the semilog plot *A*, with different slopes corresponding to different relative concentrations of Abs (given by the numbers near each line). In *B* the slopes of the curves of *A* are plotted versus the concentration of the antiserum (in relative values), yielding a straight line. The linearity of the two types of curves implies that a single Ab molecule is sufficient to neutralize a virion. (Dulbecco R et al: Virology 2:162, 1956)

is heterogeneity of Ab molecules present in polyclonal sera; another is heterogeneity of the virions, even if clonal, owing to transient configurational changes. The virus–cell interaction is also important. Thus, in the endosomes, the lower pH favors partial dissociation of the Ag–Ab complexes, making some Abs incapable of interfering with the release of the viral genome. That the binding of Abs to the virions is not the sole factor in neutralization is dramatically demonstrated by effects of Abs on viruses (such as bunyviruses) that can grow in both mammalian and insect cells: the same Abs may neutralize the virus in one but not the other cell type.

SENSITIZATION. The binding of Abs to virions without neutralizing them gives rise to the phenomenon of sensitization: the virions can be neutralized by Abs to the bound immunoglobulin. Sensitizing Abs are probably weakly bound and are stabilized when they are crosslinked by the second Ab. A related phenomenon is the appearance of a **persistent fraction** at high Ab/virion ratios (Fig. 50–6). The persistent fraction is often neutralized at least in part by addition of Abs to the bound immunoglobulin. The persistent fraction is larger when virions are neutralized by individual monoclonal Abs rather than by a polyclonal serum. A mixture of several monoclonal Abs to different epitopes of the same virion monomer usually decreases the persistent fraction to the level observed with polyclonal sera, showing **cooperation between Abs of different specificities.** Thus, the binding of Abs may induce a configurational change in the monomer that increases its affinity for another epitope on the same monomer.

A bound non-neutralizing Ab may, in some cases, increase rather than decrease the infectious titer of a virus. For instance, in flavivirus infection of macrophages, the Fc domain of Ab bound at low concentrations interacts with Fc receptors on the cells, increasing virion attachment. At higher Ab concentrations the effect disappears because neutralization takes place.

VIRION SITES FOR NEUTRALIZATION. Only epitopes on molecules involved in the release of the viral genome into the cells are targets for neutralization. In influenza virions only the hemagglutinins are targets, not the neuraminidase molecules; and in the large herpesvirus virions the main targets are the C and D glycoproteins. But in poliovirus all antigenic sites recognizable on the capsid are such targets, because the whole capsid is a unit

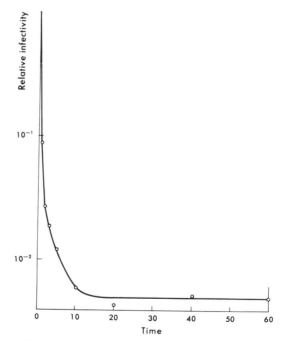

Figure 50–6. Kinetic curve of neutralization of western equine encephalitis virus (a togavirus), showing the rather abrupt change into a plateau as the time of incubation increases. This plateau is not justified by the A/V ratio employed, which would have allowed a far greater neutralization. The plateau corresponds to the persistent fraction. (Dulbecco R et al: Virology 2:162, 1956)

for releasing the nucleic acid. For adenovirus the main targets are the hexons, which are strongly interconnected and presumably work together for the release of DNA.

ANTIBODIES BOUND TO NON-NEUTRALIZING EPITOPES. Abs bound to non-neutralizing epitopes, together with those bound to neutralizing epitopes, are detected by nonbiological procedures such as complement fixation, immunoprecipitation, enzyme-linked sorption assay (ELISA), or radioimmunoassay (RAI). Some can also be detected by **neutralization in the presence of complement:** with enveloped virions the bound Ab activates the complement cascade (see Chap. 18), which then injects the C9 complexes into the virion envelope, causing its disruption. Complement can also inactivate virions by blanketing them. In primary infections of experimental animals, complement plays an important antiviral role initially, when Abs are of low affinity; thus, decomplementation by cobra venom factor (see Chap. 18) or genetic deficiency of component C5 increases the duration and severity of primary influenza.

PROTECTIVE ROLE OF NEUTRALIZING ANTIBODIES. The neutralizing power of a serum usually reflects the degree of protection in an infected animal. The correlation, however, is not always perfect. Discrepancies may be generated by differences in the neutralizability of a virus in the cells used for assay *in vitro* compared to those that the virus infects *in vivo*. For instance, the sera of mice protected from yellow fever did not neutralize the virus in Vero cells but did so in a mouse neuroblastoma cell line. Another reason for discrepancy is that an Ab that does not neutralize in cultures may act *in vivo* by activating host responses against the virus or the virus-infected cells (e.g., complement or macrophages). In addition, neutralizing Abs may fail to protect because rapid viral multiplication overcomes the neutralizing power. In the early period of immunization, low-affinity Abs act predominantly by activating complement and have low neutralizing power in cultures. The degree of protection is best estimated in cultures by carrying out **neutralization in the presence of complement.**

HEMAGGLUTINATION INHIBITION

With virions that agglutinate red blood cells (RBCs), adding the appropriate Abs to the virus before adding the RBCs decreases the hemagglutinating titer by hindering adsorption of the virions to the RBCs. In this **hemagglutination inhibition** (HI), unlike neutralization, Abs interfere with the adsorption of the virions, rather than with cell infection; and stable Ab–virion complexes do not appear to play a special role because univalent Ab frag-

ments are effective. The number of Ab molecules per virion may have to be high enough to cover all the sites of the virion involved in adsorption.

EVOLUTION OF VIRAL ANTIGENS

The great selectivity of neutralization can be understood on the basis of the structure, genetics, and evolution of viruses. Animal viruses that have evolved in the ecology of mammalian organisms have been opposed by the neutralizing Abs, which are able to block viral infection. Viral evolution must tend to select for mutations that change the antigenic determinants involved in neutralization. In contrast, other antigenic sites would tend to remain unchanged, because mutations affecting them would not be selected for and could even be detrimental. A virus would thus evolve from an original type to a variety of types, different in neutralization (and sometimes in HI) tests, but all retaining some of the original mosaic of antigenic determinants recognizable by complement fixation.

These evolutionary arguments are consistent with the observation that the clearest differentiation of types within a family is present in viruses of rather complex architecture, in which the Ags involved in the interaction with the cell vary more than the proteins of other virions. Thus, enveloped viruses have a strain-specific envelope but a cross-reactive internal capsid; adenoviruses have type-specific fibers and family-specific (but also type-specific) capsomers (see Chap. 52). Moreover, the C Ag of polioviruses, which appears only after heating, cross-reacts in all three viral types (see Chap. 55). The heating reveals antigenic sites that are normally hidden and hence are not affected by selective pressure.

The extent of antigenic variation differs widely among viruses. It is most extensive with lentiviruses (see Chap. 65) and influenza virus, an orthomyxovirus (see Chap. 56). In influenza virus the hemagglutinins are the main sites of neutralization (see The Envelope, Chap. 44). Epidemiologic studies show that this variation occurs by the appearance every decade or so of strains in which the hemagglutinins are genetically unrelated to those of the previous years (**antigenic shift**), followed by the progressively smaller changes in the new strain in the following years (**antigenic drift**; see Chap. 56). Antigenic shift is probably due to the appearance of reassortants, in which the hemagglutinin gene has been replaced by another from a virus present in an animal reservoir (e.g., avian or equine); antigenic drift, in contrast, is due to the progressive accumulation of mutations that partly overcome the immunity prevalent in the host population. The important role of selection pressure in drift is shown by the similar evolution of two strains with the same hemagglutinin subtype (H1) in the periods 1950–1957 and 1977–

present. The same two areas of the hemagglutinin evolved in both cases, although by means of different amino acid substitutions.

HABITAT AND SELECTION. Drift is probably favored by the habitat of influenza viruses, namely, cells lining the respiratory tract: here they are exposed only to IgA Abs, which tend to form reversible complexes with virions and hence are less likely to neutralize. The virus can then multiply (although at a reduced rate) even in an immune host, producing a large population as a source of mutants; these, being less sensitive to the Abs, can then be selected. In contrast, viruses that cause viremia are more effectively neutralized by IgG; hence, mutants with a somewhat decreased binding of neutralizing Abs may still be eliminated. Thus, viruses such as mumps and measles, which are structurally similar to influenza viruses, have persisted as single immunologic types with only minimal antigenic variation. Structural constraints may also limit variability, especially in icosahedral viruses, as suggested by the large number of rhinovirus and aftovirus serotypes compared with the restricted number in poliovirus (see Genetics, Chap. 48).

Diagnostic Use of Immunology

Immunologic tests can be used with standard antisera to identify and characterize a virus isolated from a patient and, with standard Ags, to detect antiviral Abs in the patient's serum. These two applications of viral immunology involve a number of general problems that will be dealt with before considering the diagnostic methodology (described in the appendix to this chapter).

TYPES OF VIRUS-SPECIFIC ANTIBODIES

Different kinds of viral preparations elicit the formation of different Abs. (1) Immunization with virions that cannot multiply in the host (e.g., with **killed virions**) produces Abs predominantly directed toward **surface components** of the virions; these Abs have neutralizing and HI activities against the virions, as well as complement-fixation (CF) and precipitation activities against Ags of the viral coat. (2) In contrast, **viruses that multiply in the host** and produce a cytopathic effect in some cells—as in natural infection or in vaccination with "live" vaccines—lead to the formation of Abs against **all the viral Ags,** including Abs for surface Ags, CF or precipitating Abs for both surface and internal Ags, and Abs for nonvirion Ags. (3) Immunization with **internal components of the virions** produces CF and precipitating Abs active only toward the Ags of these components. (4) Immunization with *peptides* reproducing segments of virion proteins elicits Abs, the properties of which depend both on the protein and the specific sequences reproduced (see Immunologic Prevention of Viral Diseases, below).

SPECIFICITY OF TEST METHODS

The Abs that react in the different tests may overlap, though they may not be altogether identical. Neutralization is primarily caused by Ab molecules specific for the sites of the virion that are involved in the release of nucleic acid into the cells, while CF usually involves additional surface or internal Ags. Neutralization probably requires molecules with a higher affinity for virions than do HI and CF. Only certain classes of immunoglobulins can participate in CF. After viral infection the titers of Abs to different components rise and fall with quite different time courses, as will be discussed in the chapters on specific viruses.

Because of their high specificity, immunologic methods can differentiate not only between viruses of different families but also between closely related viruses of the same family or subfamily. By these means, **family Ags** may be identified; each family or subfamily may be subdivided into types (**species**) on the basis of **type-specific Ags;** some types can even be subdivided further (**intratypic differentiation**). The levels in this classification are obviously somewhat arbitrary. Usually, Abs detected by neutralization tend to be less cross-reactive and thus are useful to define the immunologic type, whereas those detected by CF tend to be more cross-reactive and are useful to define the family. By proper procedures, however, such as immunization with purified Ags, highly specific CF Abs can be prepared.

The resolving power of Abs is maximized by using **monoclonal Abs;** each is produced from a single clone of B cells and is therefore endowed with a single specificity. Monoclonal Abs to influenza or rabies virus can reveal antigenic differences that are not picked up by sera of immune animals, which usually recognize many specificities. Whereas all the methods for measuring viral antigens are needed for classifying a new isolate, the method of choice for diagnostic purposes is ELISA (see Chap. 12), for its high sensitivity, ready availability of reagents, and low cost.

Cell-Mediated Immunity

IMMUNITY BASED ON CYTOTOXIC T-LYMPHOCYTES (CTLs)

As will be discussed in Chapter 51, this cell-mediated immunity appears to be very important not only in localizing viral infections and in ultimate recovery but also in the pathogenesis of viral diseases.

CTLs are usually found in the blood or the spleen and sometimes in abundance in the lymph nodes draining local infection sites (e.g., in herpes simplex infection of mice and rabbits). CTLs are also present in exudates within affected organs; in mice they are found in the lung after infection with influenza virus or in the cerebrospinal fluid after induction of meningitis by arenaviruses. CTLs kill virus-infected cells *in vitro* and probably *in vivo*.

In experimental animals primary CTLs reach maximal abundance about 6 days after a viral infection and then disappear as infection subsides. However, the organism subsequently contains, for a long time, virus-specific memory T cells, which can be recognized by culturing spleen cells together with virus-infected target cells: within a few days secondary CTLs appear in the culture, with much greater activity than in the initial response. A similar secondary response is probably produced in the body after a second infection.

Formation of CTLs is elicited by **cell-associated Ags** present at the cell surface, not only for enveloped viruses, the glycoproteins of which are incorporated into the plasma membrane (see Enveloped Viruses, Chap. 48), but also for other viruses when core or nonvirion proteins reach the cell surface. The specificity of the response depends on the types of viral proteins present at the cell surface. Thus, studies with reassortants between different influenza virus A types show that the hemagglutinin elicits both a type-specific and a cross-reactive immunity, and the matrix protein elicits cross-reactive immunity among several A types but not between A and B types. With vesicular stomatitis virus the response is directed mainly at the single-envelope glycoprotein. Larger viruses (herpesviruses, poxviruses) elicit both type- and group-specific responses because the viral glycoproteins contain determinants of both types.

Even noninfectious or inactivated viruses can elicit a cellular immune response, because their envelopes fuse with the cell plasma membrane in the initial stage of viral penetration. Moreover, the virions themselves may also be able to elicit the response after adsorbing to macrophages.

Both internal virion proteins and nonvirion proteins are often recognized by CTLs. An example is the nucleocapsid proteins of enveloped viruses, fragments of which reach the cell surface by an unknown route and are recognized very efficiently, giving rise mostly to cross-reactive CTLs. In animals infected by DNA-containing papovaviruses (see Chap. 64) the CTLs recognize the T Ag, a nonvirion Ag, exposed at the cell surface. Vaccinia recombinants expressing the protein of a foreign virus elicit a strong CTL response to it. Often, Abs to viral surface proteins do not block their interaction with CTLs, because the humoral and cellular responses recognize different epitopes: for instance, CTLs recognize Ags in association with MHC proteins rather than alone.

The spectrum of the initial and secondary cellular responses may differ. Thus, with influenza virus if the primary and the secondary infections are caused by the same viral type, the secondary response is type specific; but if the two types are different, it is cross-reactive. This mechanism can build a broad heterotypic cellular immunity, which tends to be highly cross-reactive.

ANTIBODY-DEPENDENT COMPLEMENT-INDEPENDENT CELL-MEDIATED CYTOTOXICITY

In the ADCC response the effector cells are **killer (K) cells,** which have monocyte or macrophage markers rather than T-cell markers and surface Igs. *In vitro* these cells kill virus-infected cells sensitized by IgG from immune donors, but not unsensitized targets. The cytotoxic cells are not themselves virus specific but acquire their specificity by reacting with the Fc region of the sensitizing Abs of suitable isotypes (see Chap. 14). The cytotoxic effect is inhibited by IgG F(ab')₂ fragments, which compete with the cell-bound IgG. This type of cytotoxic cell has been observed in humans and in animals infected by various enveloped viruses.

ADCC is very efficient *in vitro* against herpes simplex or varicella zoster–infected cells, preventing the usual spread of the virus from infected to neighboring uninfected cells; it may play a role in defense against human infections with these viruses.

NATURAL KILLER CELL CYTOTOXICITY

The so-called **natural killer (NK) cells,** which are found in the peripheral blood of most humans, are distinguishable from T- or B-lymphocytes, macrophages, and polymorphonuclear cells. Unlike K cells, they do not contain Fc receptors and are cytotoxic without requiring sensitizing Abs. They are not MHC restricted and attack a variety of isogeneic, allogeneic, or xenogeneic cell types, either infected by viruses or uninfected. NK cells are especially abundant in persons infected by many enveloped viruses or after vaccination with vaccinia virus. Their important role in the recovery from viral infection is suggested by an increased mortality of herpes simplex–infected mice after NK cells are depleted by means of the specific NK cells reagent anti-asialo GM1.

NK cells are not directly virus induced, because their presence is not related to a previous history of viral disease, but indirectly, by alpha **interferon** produced by virus-infected cells. In fact, interferon confers a viral specificity on NK cells by enhancing their cytotoxic action (as well as that of Ab-dependent K cells) on infected target cells while protecting uninfected cells from lysis. The NK action against virus-infected cells is blocked by monoclonal Abs specific for the viral proteins that they recognize as targets. There may be several classes of NK cells, recognizing different target molecules.

IMMUNOSUPPRESSION

The immune response is severely depressed in a number of viral infections by various mechanisms and with various consequences. Very important is immunodepression in humans infected by the lentivirus HIV (see Chap. 65), which infects and destroys T_h cells. As a result the patients become invaded by opportunistic parasites, including fungi, bacteria, and viruses. Measles virus blocks Ab synthesis by infecting lymphocytes. Immunodepression is observed in animals infected with oncogenic viruses (see Chaps. 64 and 65) containing either RNA (retroviruses) or DNA (Marek disease virus); it favors oncogenesis by helping the spread of the virus and reducing the rejection of transformed cells. This depression is manifested by an impaired responsiveness of lymphoid cells to T-cell mitogens (phytohemagglutinin or concanavalin A) and sometimes by a reduced response of B cells to immunization with other Ags. With herpesviruses, specific suppressor factors, produced by T suppressor cells, have been observed; this explains the earlier observation that thymectomy reduces the severity of the disease produced by a later infection. Cytomegalovirus induces suppression by infecting monocytes and inhibiting production of interleukin-1, which is essential for an immune response (see Chap. 15).

Immunologic Prevention of Viral Disease

Prevention is based on administration of vaccines, which generate humoral and cellular immunity against specific viruses. A vaccine may be made up of whole virions or of their components.

WHOLE VIRUS VACCINES

Whole virus vaccines contain nonpathogenic but immunogenic virions. Pathogenicity is eliminated either by chemical alterations that prevent expression and reproduction of the genome (**killed virus** vaccine) or by genetic changes that abolish pathogenicity but not the ability to reproduce (**attenuated live virus** vaccines). Both kinds of vaccine cause a polyclonal response, i.e., one directed at all or most of the epitopes of the surfaces of the virions; polyclonality is advantageous because neutralizing antibodies directed at different epitopes can collaborate (see Sensitization, above).

An example of killed vaccine is the Salk-type poliovirus vaccine, obtained by exposing the virus to formaldehyde. Attenuated vaccine strains are either isolated from a related host (e.g., vaccinia from cows to protect man against smallpox, Marek disease virus from turkeys to protect chickens) or selected for growth in a different cell type (e.g., poliovirus adapted to grow in monkey kidney cells to protect man against neurologic infection). Attenuated strains can also be obtained as mutants able to grow at low temperature (cold-adapted) or as recombinants or reassortants with defined genetic properties. A potential influenza live vaccine is based on a cold-adapted mutant, which can be extended to several serologic types by exchanging the RNA segment coding for the hemagglutinin (**cold-adapted reassortants**). A difficulty is the high reversion rate of the mutants. The genetic changes occurring during attenuation introduce some antigenic changes in the virus, but usually not sufficient to impair its immunizing activity.

Killed vaccines are administered by injection and elicit the formation of antibodies directed at surface epitopes of the virions, mostly IgG; they are less efficient in inducing an IgA response. The immunity they confer is of fairly short duration and must be maintained by repeated injections. They are unsuitable for viruses that are not easily neutralized (e.g., parainfluenza, respiratory syncytial virus) or are antigenically altered by the inactivating treatment (e.g., paramyxovirus F protein is damaged by formaldehyde). Abnormal responses, exacerbating rather than preventing subsequent disease, have been observed with formalin-killed vaccines to measles or respiratory syncytial virus. Attenuated live viruses, in contrast, can be administered orally or by inhalation; they initiate an inapparent infection that generates and maintains the immune response, both humoral and cellular, for many years. For viruses that enter the body at a mucosa site, they elicit production of IgA at that portal, a reaction that appears to be important for protection.

Drawbacks of attenuated live vaccines are their possible contamination with unknown infectious agents picked up from the organism or the cells in which they were prepared, genetic reversion to pathogenic forms (see Chap. 48), and sensitivity to interference (see Chap. 49) by other viruses present in the environment of the host. Another problem is the secondary spread of the vaccine strain, with possible enrichment of pathogenic revertants to which unvaccinated persons may be exposed.

COMPONENT VACCINES

Component vaccines have been made possible by the detailed knowledge of the viral genomes, of the structure of virions, and of how immunity works. Necessary for preparing component vaccines are (1) the identification of the components of the surfaces of the virions that elicit neutralizing Abs or other specific defenses; (2) the identification of neutralizing epitopes either by their alteration in **escape mutants** (i.e., mutants that escape neutralization by certain monoclonal Abs) or on theoretical grounds as special sequences of hydrophilic amino acids; and (3) the ability to produce virion proteins in large quantities by DNA cloning, or short peptides by artificial synthesis.

Component vaccines are of two types: subunit vaccines and peptide vaccines. **Subunit vaccines** employ whole viral proteins, which are injected either alone or in conjunction with adjuvants. An example is the hepatitis B vaccine, which was first made by means of molecules of the viral surface (HB_s) Ag present in the blood of chronic virus carriers. Through DNA cloning a similar vaccine can be made in yeast and in animal cells. For efficacy, the conformation of the protein in the vaccine must be similar to that in virions; for instance, the hepatitis B vaccine must contain dimers of the s Ag; monomers are not effective, perhaps because conformational changes alter the useful epitopes.

Peptide vaccines employ synthetically made peptides. The peptides are essentially haptens and elicit Ab formation only when they are conjugated to a protein, such as keyhole limpet hemocyanin. To boost the immunogenicity they are mixed with an adjuvant or are incorporated in membrane vesicles (liposomes). The conjugated peptides often elicit Abs of low affinity and not very cross-reactive with the intact protein; with some exceptions they are poor inducers of cellular immunity. Peptides derived from sequences that are known to bind neutralizing Abs sometimes do not elicit formation of such Abs but may **prime** the immune response: a subsequent exposure to complete virions will elicit a secondary response, much stronger than the primary response to the virions. Peptides can also cause a more intense cellular response in previously immunized animals.

ANTI-IDIOTYPIC ANTIBODIES (AB2) AS VACCINES. Abs (Ab2) raised against the variable region of the primary Ab (Ab1), which includes the binding sites, can replace Ag in inducing an immune response (Ab3) specific toward the Ag. Anti-idiotypic Abs (Ab2) capable of immunizing animals against polio or rabies viruses, among others, have in fact been prepared. An anti-idiotypic Ab can immunize by two different mechanisms. One is based on the fact that its binding site is the **internal image** of the Ag, because it is complementary to the complement of the Ag and can act like the Ag itself. A different mode of action is revealed by the ability of an Ab2 to immunize against poliovirus only mice of certain genetic constitutions, although all mice can be equally immunized by Ag. This Ab2 recognizes a special binding site of the primary Ab (Ab1) that is abundantly expressed in certain strains of mice. Ab2 activates the B cells displaying these Abs and acts by **idiotypic selection.** This type of anti-idiotypic Ab would not be suitable as vaccine in an outbred population because it is genetically restricted.

Anti-idiotypic Abs may also have an opposite effect: they may decrease the immune response to a virus, lowering the protection of the organism, perhaps through activation of specific T-suppressor cells.

VACCINES BASED ON VACCINIA VIRUS VECTORS. The surface proteins of many viruses, both DNA and RNA containing, have been introduced in vaccinia virus vectors (see Chap. 48). The proteins are regularly processed and are transported to the surface of the vector-infected cells in the proper orientation and configuration. Animals infected with such vectors have developed neutralizing Abs and T_cs specific for these proteins and have been protected from challenge with the corresponding virus. This is a promising approach for protecting animals and humans against viral disease.

LIMITATIONS OF VIRAL VACCINES

Vaccines are effective in protecting against diseases caused by viruses with genetically stable antigenicity, such as poliovirus. For viruses with a very large number of serologic subtypes, such as rhinoviruses or foot-and-mouth disease virus, vaccination is more problematic because a large number of vaccines would be required. Even greater difficulties are encountered with viruses that continuously vary, such as influenza viruses or the AIDS agents, HIVs. In the case of influenza, adequate protection can be obtained with killed vaccines made each year from the then-prevalent strain, because the emergence of variants is seasonal. If research identifies more invariant epitopes, then it might become possible to construct adequate peptide vaccines.

For practical use in humans, vaccines must satisfy stringent conditions of safety and efficacy. This caution delays the introduction of vaccines based on new principles, such as subunits, peptides, or anti-idiotypic vaccines. The use of vaccines based on vaccinia virus vectors entails infection of the organism with vaccinia virus, the consequences of which must be evaluated.

Appendix
MEASUREMENT OF NEUTRALIZING ANTIBODIES

The measurement of neutralizing antibodies rests on the points discussed in the body of this chapter. If the serum–virus mixtures are assayed **without dilution,** the relative concentration of Ab is derived from the proportion of neutralized virus, according to the **percentage law.** This method is simple to perform and measures both reversible and stable Ab–virion complexes; it is adequate and is widely used for diagnostic purposes, but it is unsuitable for accurate measurement.

The percentage law, recognized by Andrews and Elford in 1933, states that under conditions of Ab excess, the proportion (percentage) of virus neutralized by a

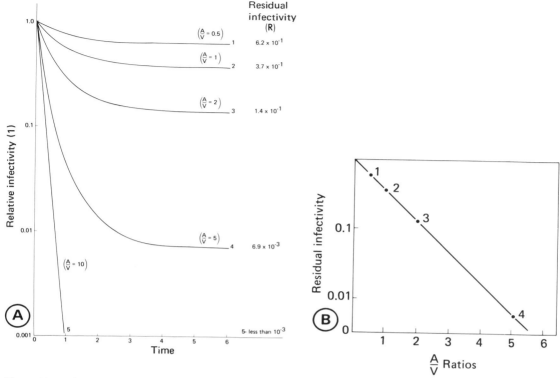

Figure 50–7. (*A*) Theoretical kinetic curves at different ratios of total Ab (*A*) total virus(*V*)–virion complexes. All the curves tend to plateau, the relative infectivity of which is a function of the ratio A/V (in arbitrary units). When the residual infectivity values are plotted semilogarithmically versus the A/V ratio, they generate a straight line (*B*). This result affords additional evidence that neutralization of a virion requires only the binding of a single Ab molecule.

given antiserum is constant, irrespective of the viral titer. This law can be deduced from the mass law.

$$V + A \underset{k_2}{\overset{k_1}{\rightleftharpoons}} \overline{VA}; \frac{k_1}{k_2} = \frac{(\overline{VA})}{(V)(A)}; k(A) = \frac{(\overline{VA})}{(V)} \quad (1)$$

where V indicates the free virus, A the free Ab (assumed to have a much higher concentration than bound Ab), VA the Ab–virion complexes, and k their dissociation constant.

The formation of stable Ab–virion complexes is more precisely measured by means of *kinetic curves in Ab excess.* The curves obtained closely approach straight lines passing through the origin; the slopes are proportional to the concentration of the Ab and to its affinity for the virions and are independent of the virions' concentration (see Fig. 50–5). These curves follow the equation*

---

* This equation is equivalent to that used in the appendix to Chapter 44; kt(A) is the average number of neutralizing Ab molecules per virion after time *t*, since molecules of Ab (which is in large excess) continue to attach to the virions at a constant rate. Thus, *I* is the fraction of virions that have **no neutralizing molecule.** As discussed in the appendix to Chapter 8, equation 1 generates single-hit curves and is valid if a single Ab molecule is sufficient to produce neutralization; otherwise, the curve would have an initial shoulder (i.e., it would be a multiple-hit curve).

$$I = e^{-kt(A)} \quad (2)$$

and, taking the logarithm of both sides:

$$\ln I = -kt(A)$$

$$k(A) = \frac{\ln I}{t} \quad (3)$$

where I is relative infectivity, t the time after mixing the virus and the Ab (in minutes), and k the affinity; the other symbols are as in equation 1.

In Ag excess the residual relative infectivity, determined after dilution of the virus–serum mixture (Fig. 50–7), theoretically obeys the relations

$$Res = e^{-c(A/V)} \quad (4)$$

where *Res* is residual infectivity, c is a constant, A denotes the total amount of Ab, and V denotes the total amount of virus.†

---

† This equation is derived similarly to equation 1 above, except that the average number of neutralizing Ab molecules per virion is c(A/V). Again, this equation requires that a single Ab molecule be sufficient for neutralization; otherwise the curve would have an initial shoulder.

METHODS FOR ANTIGENIC ANALYSIS OF VIRUSES

We shall consider here only those immunologic methods that are peculiar to viruses: neutralization, HI, and immuno-electron microscopy.

Neutralization Tests

Neutralization tests are used mostly in typing viral isolates and in characterizing related viruses.

DETERMINING NEUTRALIZING ANTIBODY TITERS BY A 50% ENDPOINT METHOD. The titration is usually carried out with **constant virus and varying serum.**

METHOD 1: ENDPOINT TEST. To carry out a constant virus–varying serum titration, the serum is inactivated at 56°C for 30 minutes to destroy labile substances that have antiviral activity or that affect neutralization. The serum is then diluted serially, usually in two steps, and each dilution is added to a constant amount of virus (usually between 30 ID_{50} and 100 ID_{50}; see Appendix, Chap. 44). The amount of virus must be adequate to infect all the assay units in the subsequent titration and at the same time small enough to detect a low concentration of Abs. The mixtures are incubated for a selected time (at either 37°C or 4°C, depending on the virus employed); then a constant volume of each is assayed for infectivity by inoculation into 5 to 10 test units of a suitable assay system, such as mice, chick embryos, or tissue culture tubes. The neutralization titer of the serum is the dilution at which 50% of the units are protected (50% endpoint), calculated by the method of Reed and Muench (see Appendix, Chap. 44).

The accuracy of the assay can be calculated by the same methods as used for viral assays (see Appendix, Chap. 44). (Since the serum dilutions are closely spaced, the assay is more precise than the corresponding viral assays, which usually employ more widely spaced dilutions.) The constant virus–varying serum titration relies on the constancy of the virus employed; this requirement, however, is not critical because, according to the percentage law, the proportion of virus neutralized is (at least largely) independent of the titer of the virus.

This method is statistically more accurate than the **constant serum–varying virus** method because in virus–serum mixtures a small change in Ab titer usually produces a much larger change in the infectious titer of the virus. The reason is the exponential relation between residual infectivity and Ab concentration of equation 3 above.

METHOD 2: PLAQUE-REDUCTION TEST. In the plaque-reduction test an inoculum of about 100 plaque-forming units is incubated with serial dilutions of the serum; each mixture is then added to a monolayer culture, which is overlaid with agar and incubated. The endpoint is an 80% reduction in the number of plaques. The precision of the method depends on the number of plaques at the endpoint (see Appendix, Chap. 44).

These two methods are used either qualitatively, to demonstrate the presence of virus-specific Ab, or quantitatively, to compare Ab titers in different sera.

METHOD 3: DETERMINING THE RATE OF NEUTRALIZATION BY THE PLAQUE METHOD. Determining the rate of neutralization by the plaque method measures the slope (κ) of the kinetic curves described above. The values obtained are extremely reproducible; differences of about 20% are usually significant. Values obtained with the same serum provide a sensitive basis for distinguishing viral strains, including laboratory mutants (e.g., in work with vaccine strains of poliovirus).

In this test a virus sample of known titer is mixed with the antiserum; samples are taken at intervals, diluted, and assayed for plaques. The logarithm of the ratio of the titer of the sample to the original titer (the relative infectivity, I) is plotted against the time of sampling, yielding a straight line through or near the origin. The slope of that line is κ, which is characteristic for the serum and the virus; it is determined from the relation $\kappa = -\ln I/t(A)$, from equation 2 above, where I is the relative infectivity determined after t minutes of incubation of the neutralizing mixture and A is the concentration of the serum.

Example: At a serum dilution of 10^{-3} the relative infectivity is 3×10^{-2} after a 10-minute incubation. Since $\ln 3 \times 10^{-2} = -3.5$, $\kappa = 3.5/(10 \times 10^{-3}) = 350$.

Hemagglutination Inhibition

For the HI test a serial twofold dilution of heat-inactivated serum is prepared in saline solution, and from each dilution 0.25 ml is mixed with 0.25 ml of a viral suspension containing 4 hemagglutinating units (defined in Chap. 44, see Hemagglutination Assay). (If nonspecific HI substances are known to be present in the serum or the viral preparation, they must be removed in advance.) To each mixture is added 0.125 ml of a 1% RBC suspension. The tubes are shaken and then incubated at the temperature and for the time required for optimal hemagglutination with the virus used. The agglutination pattern is read after incubation, as discussed in Chapter 44. The **HI titer** is the reciprocal of the highest serum dilution that completely prevents hemagglutination.

Example:

	Initial Serum Dilution								
	1:8	1:16	1:32	1:64	1:128	1:256	1:512	1:1024	HI Titer
A	0	+	+	+	+	+	+	+	+	8
B	0	0	0	0	+	+	+	+	+	64

Immuno-electron Microscopy

Special techniques are especially useful for certain viruses (such as human rotaviruses; see Chap. 62) that cannot be grown in cell cultures or in convenient animal hosts and do not produce hemagglutination but have a high titer in the blood or excretions of infected persons. When the virus-containing sample is mixed with appropriate Abs, electron microscopic examination reveals virions of characteristic shape with an Ab halo.

Complications of Immunologic Tests

Interfering substances occasionally obscure the significance of immunologic tests. Although they concern the specialist carrying out the test, some knowledge of their nature is useful for evaluating test results.

NEUTRALIZATION TEST. Human and animal sera contain **nonspecific viral inhibitors,** especially active against influenza, mumps, herpes simplex, and togaviruses. They result from interactions with properdin and the alternate complement pathway. They are heat labile and can be eliminated by incubation at 56°C for 30 minutes.

HEMAGGLUTINATION-INHIBITION TEST. Most sera contain inhibitors of hemagglutination that are not Abs; they must be removed. The inhibitors for influenza viruses are inactivated by receptor-destroying enzyme (RDE; see Mechanisms of Hemagglutination, Chap. 44), trypsin, or periodate; those for togaviruses and flaviviruses are inactivated by extraction with acetone-chloroform.

COMPLEMENT-FIXATION TEST. Performance of the CF test is frequently hampered by the presence of anticomplementary substances in crude tissue suspensions used as Ag. This is especially true of the infected brain suspension used with togaviruses; it can be freed of the anticomplementary factors by thorough extraction with acetone in cold or by extraction with a fluorocarbon.

Selected Reading

Ada GL, Jones PD: The immune response to influenza infection. Curr Top Microbiol Immunol 128:1, 1986

Air GM, Laver G: The molecular basis of antigenic variation in influenza virus. Adv Virus Res 31:53, 1986

Al-Ahdal MN, Nakamura I, Flanagan TD: Cytotoxic T-lymphocyte reactivity with individual Sendai virus glycoproteins. J Virol 54:53, 1985

Arnon R: Peptides as immunogens: Prospects for synthetic vaccines. Curr Top Microbiol Immunol 130:1, 1986

Brown F: Synthetic viral vaccines. Annu Rev Microbiol 38:221, 1984

Chow M, Yabrov R, Bittle J et al: Synthetic peptides from four separate regions of the poliovirus type 1 capsid protein VP1 induce neutralizing antibodies. Proc Natl Acad Sci USA 82:910, 1985

Emini EA, Jameson BA, Wimmer E: Priming for and induction of anti-poliovirus neutralizing antibodies by synthetic peptides. Nature 304:699, 1983

Emini EA, Ostapchuk P, Wimmer E: Bivalent attachment of antibody onto poliovirus leads to conformational alteration and neutralization. J Virol 48:547, 1983

Ertl HCJ, Finberg RW: Sendai virus–specific T-cell clones: Induction of cytolytic T cells by an anti-idiotypic antibody directed against a helper T-cell clone. Proc Natl Acad Sci USA 81:2850, 1984

Gerhard W, Webster RG: Antigenic drift in influenza A viruses. I. Selection and characterization of antigenic variants of A/PR/8/34(HON1) influenza virus with monoclonal antibody. J Exp Med 148:383, 1978

Gollins SW, Porterfield JS: A new mechanism for the neutralization of enveloped viruses by antiviral antibody. Nature 321:244, 1986

Hurwitz JL, Hackett CJ, McAndrew EC, Gerhard W: Murine T_H response to influenza virus: Recognition of hemagglutinin, neuraminidase, matrix, and nucleoproteins. J Immunol 134:1994, 1985

Kennedy RC, Eichberg JW, Lanford RE, Dreesman GR: Anti-idiotypic antibody vaccine for type B viral hepatitis in chimpanzees. Science 232:220, 1986

Laver WG, Air GM (eds): Immune recognition of protein antigens. Current Communications in Molecular Biology. Cold Spring Harbor, NY, Cold Spring Harbor Laboratory, 1985

Morrison LA, Lukacher AE, Braciale VL et al: Differences in antigen presentation to MHC class I– and Class II–restricted influenza virus–specific cytolytic T lymphocyte clones. J Exp Med 163:903, 1986

Raymond FL, Caton AJ, Cox NJ et al: The antigenicity and evolution of influenza H1 haemagglutinin, from 1950–1957 and 1977–1983: Two pathways from one gene. Virology 148:275, 1986

Reagan KJ, Wunner WH, Wiktor TJ, Koprowski H: Anti-idiotypic antibodies induce neutralizing antibodies to rabies virus glycoprotein. J Virol 48:660, 1983

Schwartz RH: The value of synthetic peptides or vaccines for eliciting T-cell immunity. Curr Top Microbiol Immunol 130:79, 1986

Stitz L, Althage A, Hengartner H, Zinkernagel R: Natural killer cells vs cytotoxic T cells in the peripheral blood of virus-infected mice. J Immunol 134:598, 1985

Uytdehaag HF, Bunschoten H, Weijer K, Osterhaus A: From Jenner to Jerne: Towards ideotype vaccines. Immunol Rev 90:93, 1986

Watson RJ, Enguist LW: Genetically engineered herpes simplex virus vaccines. Prog Med Virol 31:84, 1985

51

Harold S. Ginsberg

Pathogenesis of Viral Infections

The consequences of a viral infection depend on a number of viral and host factors that affect pathogenesis, including the number of infecting viral particles and their path to susceptible cells, the speed of viral multiplication and spread, the effect of the virus on cell functions, the intracellular state of the viral genome, the host's secondary responses to cellular injury (edema, inflammation), and the host's defenses, both immunologic and nonspecific. Viral infection was long thought to produce only acute clinical disease, but other host responses are being increasingly recognized. These include asymptomatic infections, induction of various cancers, chronic progressive neurologic disorders, and possible endocrine diseases.

This chapter deals with the general features of viral pathogenesis, and, where possible, it describes the mechanisms involved. Subsequent chapters discuss the particular characteristics and special effects of various viruses that primarily infect man.

Cellular and Viral Factors in Pathogenesis

The effects of viral infection on cells depend on both the characteristics of the virus and the susceptibility of the cells.

VIRAL CHARACTERISTICS

Viral virulence, like bacterial virulence, is under polygenic control and cannot be assigned to any single viral property; it is, however, frequently associated with several characteristics that promote viral multiplication and cell damage. Thus, virulent viruses multiply well at the elevated temperatures that arise during illness (i.e., above 39°C), induce interferon poorly and resist its inhibitory action, block the biosynthesis of host macromolecules, and damage cell lysosomes or alter the cell membranes of the infected cells, producing damage in the infected animal or cytopathic effects in cell cultures. **Attenuated mutants,** altered in various of these functions, including **conditionally lethal, temperature-sensitive (ts) mutants** (see Mutant Types, Chap. 48), produce less severe or no disease. However, viruses have genes that are nonessential for their replication but appear to be important for their virulence (e.g., the early region 3 of adenoviruses; see Chap. 52). Moreover, some viruses that are attenuated in their behavior in animals (e.g., poliovirus vaccine strains) cause the same cytopathic effects as wild-type virus in cultures, in which their multiplication is not restrained.

CELL SUSCEPTIBILITY

THE ROLE OF CELL RECEPTORS. The susceptibility of cells to viral infection is often determined by their early interactions with a virus: viral attachment or the release of its nucleic acid in the cells. With animal viruses, as was observed earlier with bacteriophages (see Chap. 45), resistance of the cells is often caused by failure of **viral adsorption;** hence, cells resistant to a virus may be susceptible to its extracted nucleic acid (see Productive Infection, Chap. 48). Indeed, differences in the adsorption of viruses to cells have been correlated with differences in organ susceptibility, and also with changes in host susceptibility with age (e.g., differences in types 1 and 3 reoviruses owing to their S1 genes; see Chap. 62).

Physiological and genetic factors affect the presence or activation of receptors for viral adsorption, as well as other cellular properties that influence susceptibility.

PHYSIOLOGICAL FACTORS. Cultivation may markedly alter the viral susceptibility of cells from that in the original organ. Hence, many viruses can be propagated in cells that are readily cultured, obviating the need for cells that are hard to culture, or for intact animals. For instance, polioviruses, which multiply in the nervous tissue but not in the kidney of a living monkey, multiply well in cultures derived from the kidneys, since receptors develop in the cultivated kidney cells.

Marked changes in susceptibility accompany the maturation of animals. Many viruses are much more virulent in newborn animals (e.g., coxsackieviruses, herpes simplex virus) and others in adults (e.g., polioviruses, lymphocytic choriomeningitis [LCM] virus). There are several mechanisms: with coxsackieviruses in mice the change in susceptibility is correlated with receptor activity, although it may also depend on changes in interferon,

endocrine function, and Ab production; with foot-and-mouth disease virus it involves the rate of viral multiplication; and with LCM virus it is due primarily to an immunologic mechanism (see below).

GENETIC FACTORS. Genetic differences in susceptibility have been demonstrated in mice (with togaviruses and influenza viruses) and in chickens (with oncogenic retroviruses). In crosses between resistant and susceptible animals the heterozygous first-generation (F_1) progeny are uniformly resistant or uniformly susceptible, depending on which allele is dominant. (Resistance is dominant with influenza and togaviruses, and susceptibility with retroviruses). Moreover, backcrosses of the F_1 person to the parent carrying the recessive allele yield 50% resistant animals, implying a difference in a single gene or a closely linked cluster. Some of these hereditary differences evidently involve the host cell–virus interactions, but others could reflect control of the immune response.

CELLULAR RESPONSES TO VIRAL INFECTIONS

Cells can respond to viral infection in four different ways: (1) no apparent change; (2) cytopathic effect and death; (3) hyperplasia, which may be followed by death (as in the pocks of poxviruses on the chorioallantoic membrane of the chick embryo; see Chap. 54); and (4) loss of growth control (topoinhibition or cell–cell contact inhibition), as in viral transformation of normal to cancer cells (see Chap. 64). The development of inclusion bodies and chromosomal aberrations may be special features of these cellular responses.

CYTOPATHIC EFFECTS. Virus-induced cell injury has been most extensively studied with cultured cells, since these are believed to reflect accurately the cell damage occurring *in vivo*, and their responses can be quantified. *In vitro* cell damage, termed the *cytopathic effect*, is recognized from various morphologic alterations, which are listed in Table 51–1; cell death usually follows.

The factors listed below appear to contribute to development of the various cytopathic effects.

EFFECTS ON SYNTHESIS OF CELLULAR MACROMOLECULES. Many virulent viruses cause an early depression of cellular syntheses. As noted in Chapter 48, DNA-containing viruses inhibit the synthesis of host cell DNA, but most do not affect host cell RNA and protein production until late in the multiplication cycle, whereas many RNA viruses inhibit host cell RNA and protein synthesis early in the multiplication cycle.

ALTERATION OF LYSOSOMES. Some viruses cause a reversible increase in lysosome permeability, without leakage

TABLE 51–1. Cellular Response to Viral Infection

Virus	Cell Type*	Cellular Response	Inclusion Body
Adenoviruses	HeLa	Cell rounding and clumping	Nuclear
	Rat embryo	Transformed	Nuclear
Herpesviruses (herpes simplex)	HeLa	Polykaryocytes (some strains); cell rounding	Nuclear
Poxviruses (variola)	HeLa	Slow rounding; hyperplastic foci	Cytoplasmic
Picornaviruses (polioviruses)	Monkey kidney	Cell lysis	None
Orthomyxoviruses (influenza viruses)	Monkey kidney	Slow rounding	None
Paramyxoviruses (parainfluenza virus)	Monkey kidney	Fusion of cell membranes; syncytial formation	Cytoplasmic
Coronaviruses	Human diploid	Minimal; syncytia rarely	None
Togaviruses (eastern equine encephalitis virus)	Mouse L	Cell lysis	None
Rubella virus	Human amnion	Slow enlargement and rounding	Cytoplasmic
Reoviruses	Monkey kidney	Enlargement and vacuolation	Cytoplasmic
Rabiesvirus	Hamster kidney	Usually none	Cytoplasmic

* With many viruses several cell types can be used; in such instances, a commonly used type is listed.

of enzymes from the organelles. This change, the cause of which is unknown, is shown by an increased binding of neutral red, the dye commonly used to stain the live cells in the plaque assay: the cells appear hyperstained and form "red plaques." Other viruses effect disruption of the organelles and discharge of their hydrolytic enzymes into the cytoplasm. The cells lose their ability to be stained with neutral red and form the usual "white plaques." This profound effect appears to be due to proteins synthesized late in the viral multiplication cycle, possibly capsid subunits.

ALTERATIONS OF THE CELL MEMBRANE. Many of the budding, enveloped viruses incorporate viral subunits, usually glycoproteins, into the infected cell membranes, as a prelude to the formation of the viral envelope (see Enveloped Viruses, Chap. 48). Even some viruses that do not bud from the cell surface, such as herpes simplex and vaccinia viruses, insert novel Ags into the plasma membrane. These changes may be recognized by reaction with virus-specific Abs, by **hemadsorption** (e.g., orthomyxoviruses and paramyxoviruses), or by absorption of increased quantities of plant lectins such as concanavalin A (e.g., poxviruses, paramyxoviruses). The inserted Ags make the cells targets for immunologic destruction by virus-specific Abs plus complement or by immune T-lymphocytes. In addition, effects on their membranes, as well as on the cytoskeleton, probably play a large role in altering the shape and function of cells. In a striking effect observed with paramyxoviruses, some herpesviruses, and the human immunodeficiency virus (HIV), the infected cells fuse with adjacent cells (i.e., establish a continuity between the plasma membranes, forming giant cells (**polykaryocytes**).

ABORTIVE INFECTION may also cause cytopathic effects although viral syntheses is incomplete: for example, in cultured HeLa cells influenza viruses synthesize Ags and damage the cells, though they do not form infectious virions.

Viral toxic effects produce cell damage in animals and in cell cultures, owing to the accumulation of virions or viral structural proteins. In mice, for example, the intravenous injection of a concentrated preparation of influenza, mumps, or vaccinia virus causes hemorrhages and cellular necrosis in various organs, resulting in death within 24 hours; a large intracerebral inoculum of influenza virus produces necrosis of brain cells. Addition of adenovirus fiber protein to KB cells inhibits cellular DNA synthesis, and the penton base effects cell rounding and clumping. All these effects are produced without synthesis of viral components or with synthesis of only incomplete particles.

DEVELOPMENT OF INCLUSION BODIES. Intracellular masses may arise as accumulations, either of virions or of unassembled viral components, in the nucleus (e.g., adenovirus), in the cytoplasm (e.g., rabiesvirus Negri bodies), or in both (e.g., measles). These inclusion bodies appear to disrupt the structure and function of the cells and to contribute to their death. Other inclusion bodies do not contain detectable virions or their components but are "scars" left by earlier viral multiplication (e.g., the eosinophilic, intranuclear inclusion bodies that eventually appear in cells infected by herpes simplex virus; see Chap. 53).

INDUCTION OF CHROMOSOMAL ABERRATIONS. In primary cultures, chromosomal aberrations such as breaks or constrictions are commonly seen after infections with measles and rubella viruses; with several adenoviruses and herpes, parainfluenza, mumps, polyoma, and Rous sarcoma viruses; and with simian virus 40 (SV40). During natural infections, measles virus produces similar chromosomal abnormalities in peripheral leukocytes. The alterations often appear to be an early expression of the cytopathic effect in cells that will die later. Some of these aberrations have characteristic features; for example, herpes simplex virus induces breaks only at certain sites of two specific chromosomes, and chromatid breaks may continue to occur during the multiplication of cells surviving infection by herpes simplex or polyoma virus, suggesting a persistent or latent infection of the cell clones.

CELL TRANSFORMATION. Certain viruses that produce tumors or leukemia (see Chaps. 64 and 65) may have several effects on cultured cells: (1) stimulation of the synthesis of cellular DNA (e.g., polyoma virus); (2) surface alterations recognizable by the incorporation or uncovering of new antigenic specificities distinct from those of virion subunits and by increased agglutinability by plant lectins; (3) chromosomal aberrations and sister chromatid exchanges; (4) disruption of the cytoskeleton system; and (5) alterations of the growth properties of the cells, resulting in cell hyperplasia because their division is no longer subject to topoinhibition (i.e., inhibition of growth in a dense culture). Moreover, growth is less dependent on serum in the culture medium and does not require anchorage to the surface of a culture vessel—colonies of transformed cells grow in soft agar or methylcellulose. This conversion of a normal cultured cell to one resembling a malignant cell has been termed **transformation** (see Chap. 64). DNA-containing viruses (adenoviruses, herpes simplex, polyoma, SV40) can transform only nonpermissive cells, but at least a portion of the viral genome persists and continues to function. In contrast, cells transformed by RNA viruses (e.g., avian leukosis, murine leukemia) are permissive and usually continue to produce virions.

Patterns of Disease

In a host, viruses cause three basic patterns of infection: localized, disseminated, and inapparent.

LOCALIZED

In localized infections, viral multiplication and cell damage remain localized near the site of entry (e.g., the skin or parts of the respiratory or gastrointestinal tract). When the virus spreads from the first infected cells to neighboring cells by diffusion across intercellular spaces and by cell contact, the result is a single lesion or a group of lesions, as with warts. In a less strictly localized pattern, when virus is transported by excretions or secretions within connected cavities, infection causes diffuse involvement of an organ, as with influenza, the common cold, or viral gastroenteritis. Virus may in time spread to distant sites, but this dissemination is not essential for production of the characteristic illness.

DISSEMINATED

Disseminated infections develop through several sequential steps, as illustrated by Fenner's classic investigation of ectromelia (mousepox), summarized in Figure 51–1. Mousepox virus enters through an abrasion of the skin and multiplies locally; from there it spreads rapidly to regional lymph nodes, where it also multiplies. The virus then enters the lymphatics and the bloodstream, and this primary viremia causes the dissemination of the virus to other susceptible organs, especially the liver and spleen. Viral multiplication results in necrotic lesions in these organs and a more intense secondary viremia, which disseminates the virus to the target organ, the skin. There the virus undergoes extensive multiplication, producing papules that eventually ulcerate. With the appearance of the papular rash the asymptomatic incubation period terminates, and clinical disease begins.

The temporal relation among viral multiplication in the various organs, development of lesions, and formation of Abs should be noted (see Fig. 51–1A). It is particularly striking that overt disease begins only after virus becomes widely disseminated in the body and has attained maximum titers in the blood and the spleen.

This model of dissemination is applicable not only to exanthematous diseases such as smallpox and measles but also to nonexanthematous diseases such as poliomyelitis and mumps. Thus, the target organ for poliovirus is the central nervous system, and for mumps virus the salivary and other glands. In some instances, such as poliovirus infections (see Chap. 55), primary and secondary viremias are not distinguishable.

The dissemination of neurotropic viruses to the nervous system may occur by transmission along nerves as well as by viremia. In mice, for instance, such centripetal transmission of herpes simplex virus after foot pad inoculation can be followed by assaying segments of nerves at various times after infection. The virus may conceivably travel either by axonal transport or by multiplication in endoneural cells (Schwann's cells and fibroblasts), in which viral Ags can be localized by immunofluorescence.

For many years, before refined structural studies of virions became possible, animal viruses were classified primarily in terms of their viscerotropism, neurotropism, or dermotropism. The grouping of viruses on the basis of their target organs is presented in Table 51–2. However, the target organ for a given virus (where the susceptible cells are damaged) and the type of disease produced bear no relation to the taxonomic position of the virus, as defined in Chapters 44 (see The Viral Particles) and 48 (see Table 48–1). In fact, such unrelated viruses as influenza and adenoviruses may produce diseases that cannot be clinically differentiated, and such related viruses as parainfluenza, mumps, and measles may produce completely different clinical syndromes.

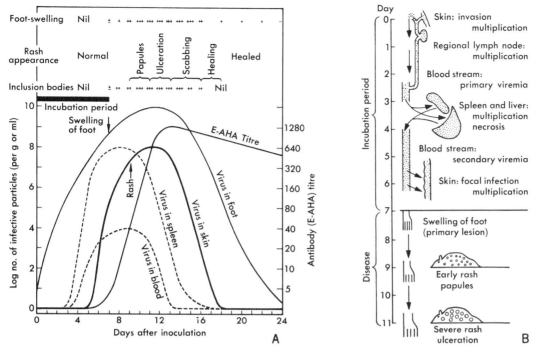

Figure 51—1. Sequential events in the pathogenesis of ectromelia (mousepox) in mice inoculated in the foot pad. (*A*) Relation among viral multiplication (in foot pad, spleen, blood, and skin), development of primary lesion and rash, and appearance of Abs (E-AHA). (*B*) Diagram of the dissemination of virus and the pathogenesis of the rash in mousepox. (Fenner F: Lancet 2:915, 1948)

INAPPARENT

Transient viral infections without overt disease (inapparent infections) are very common. Moreover, they have great epidemiologic importance, for they represent an often unrecognized source of dissemination of a virus, and they also confer immunity. For example, for every paralytic case of poliomyelitis in the United States before the days of widespread immunization, 100 to 200 inapparent infections could be detected serologically or by viral isolation.

Several factors are involved in the production of inapparent infections:

Moderate viruses or attenuated strains (as in live vaccines) usually cause inapparent infections.

When the host's defense mechanisms are effective, even viruses capable of causing acute disease may generate an inapparent infection. These defenses include the host's immunity, especially the ability to produce a prompt secondary response (Abs or cytotoxic T-lymphocytes), and the appearance of viral interference (see Chap. 49).

Failure of the virus to reach the target organ is also an expression of host defense, but of a more obscure

nature. Thus, as noted above, only about one of 200 nonimmune persons infected with poliovirus express symptoms of central nervous system (CNS) disease (see Chap. 55).

Effects of Viruses on Embryonic Development

The variation of susceptibility with age is especially striking for the embryonic period. Indeed, some viruses that produce mild disease in the adult produce extensive infection and severe malformations in the embryo. The role of viruses in the pathogenesis of congenital anomalies was not recognized until Gregg, in 1941, discovered that rubella virus may cause a variety of congenital anomalies if the mother is infected during the first 3 months of pregnancy (see Chap. 61). Of all the viral infections that may occur during the first trimester of pregnancy, rubella is the major cause of fetal death and congenital malformations. But a few other viruses also have teratogenic effects: **cytomegalovirus** induces a low incidence of microcephaly, motor disability, and chorioretinitis; **group B coxsackieviruses** are responsible for some congenital heart lesions; and **type 2 herpes simplex vi-**

TABLE 51–2. Grouping of Viruses by Pathogenic Characteristics in Man

Classification by Major Target Organs	Specific Virus(es)	Portal(s) of Entry	Other Affected Organs
Respiratory viruses	Influenza A, B, and C	Respiratory tract	
	Parainfluenza	Respiratory tract	
	Respiratory syncytial	Respiratory tract	
	Measles	Respiratory tract	Brain, skin, lung
	Mumps	Respiratory tract	CNS, testes, ovaries, pancreas
	Adenoviruses	Respiratory tract	
	Rhinoviruses	Respiratory tract	
	Coxsackieviruses (some)	Respiratory tract	CNS*
	Echoviruses (some)	Respiratory tract	CNS,* skin
	Reoviruses	Respiratory tract	?
	Lymphocytic choriomeningitis	Respiratory tract	CNS
	Coronavirus	Respiratory tract	
Enteric viruses	Polioviruses	Gastrointestinal tract	Muscles, CNS*
	Coxsackieviruses	Gastrointestinal tract	CNS,* skin
	Echoviruses	Gastrointestinal tract	CNS,* skin
	Rotaviruses (reovirus)	Gastrointestinal tract	
	Hepatitis A and B; non-A, non-B	Gastrointestinal tract, blood	Liver
Neurotropic viruses	Polioviruses	Gastrointestinal tract	Upper respiratory tract
	Coxsackieviruses	Gastrointestinal tract	Upper respiratory tract
	Echoviruses	Gastrointestinal tract	Upper respiratory tract
	Rabies	Skin and blood	
	Mumps	Respiratory tract	Testes, ovaries, pancreas
	Measles	Respiratory tract	Skin, lung
	Arboviruses	Blood	
	Herpes simplex virus	Respiratory tract, genitalia	Skin
	Virus B	Respiratory tract, blood	Respiratory tract
	Varicella zoster	Respiratory tract	Skin, cornea
	Lymphocytic choriomeningitis	Respiratory tract	Respiratory tract
	Kuru	Gastrointestinal tract	
	Creutzfeldt-Jakob	Unknown	
	BK and JC	Unknown	
Dermotropic viruses	Poxviruses	Respiratory tract	Respiratory tract, viscera, CNS
	Measles	Respiratory tract	Lung, brain
	Varicella zoster	Respiratory tract	CNS, cornea
	Coxsackieviruses	Gastrointestinal tract	Upper respiratory tract
	Echoviruses	Gastrointestinal tract	CNS, gastrointestinal tract
	Herpes simplex	Skin	Peripheral ganglia, cornea, genitalia
	Rubella	Respiratory tract	Respiratory tract
	Human wart (papilloma viruses)	Skin	
	Molluscum contagiosum	Skin	
Human T-cell tropic viruses	HTLV I and HTLV II	Blood	Spleen, lymph nodes
	Human immunodeficiency virus (HIV)	Blood	Organs containing lymphocytes, macrophages

* Major involvement in clinical disease
CNS, central nervous system

rus may cause microcephaly and other severe CNS malformations.

Passage of a virus across the placenta appears to be responsible for embryonic infection, and this probably occurs only when the mother is viremic. Multiplication of the virus in the placenta may favor transmission but is not strictly required, since the small coliphage ϕX174, which is unable to multiply in animals, is transmitted (though with very low efficiency).

Immunologic and Other Systemic Factors

CIRCULATING ANTIBODIES

PROTECTION. Abs in serum and extracellular fluids provide the main protection against primary viral infections; i.e., at the site of viral entry into the host. For those infections in which viremia is an essential link in the pathogenesis of the disease (i.e., measles, poliomyelitis, mumps, smallpox), the degree of protection is directly related to the level of neutralizing Abs in the blood when virus enters it. Furthermore, in experimental herpes simplex virus infections, the B-cell response limits the spread of virus to the CNS and reduces the establishment of latency in peripheral ganglia (see below). The mechanism by which Abs neutralize viruses has been considered in Chapter 50.

Protection of the respiratory and gastrointestinal tracts is associated with **IgA Abs** (see Chap. 14), which are secreted into the extracellular fluids. Hence, by inducing the secretion of IgA Abs, natural infections produce specific local as well as systemic immunity. Viral vaccines, particularly those containing live attenuated virus, also elicit the production of IgA Abs in the respiratory and gastrointestinal secretions. Although this feature theoretically affords a marked advantage to live viral vaccines, some vaccines produced with killed viruses and introduced parenterally, such as polioviruses and influenza viruses, have proved effective.

The protective role of Abs is also evident in the prophylactic effectiveness of **passive immunization.** Administration of immune serum or immune **γ-globulin** before infection or early in the incubation period can prevent or modify diseases with viremia and long incubation periods (greater than 12 days), such as measles, hepatitis A and B, poliomyelitis, and mumps. The striking protection of populations by some viral vaccines (Table 51–3) constitutes an additional demonstration of the prophylactic function of Abs. These specific vaccines will be discussed in the chapters that follow.

RECOVERY. Although humoral Abs generally develop during recovery from a viral disease, they appear to play a less prominent role in this process than in protection. Thus, intracellular virus may continue to increase, and pathologic lesions to evolve, even while Abs are being elaborated. Moreover, in most patients with agammaglobulinemia, recovery from viral diseases is usually normal, although some such affected children may have persistent and fatal echovirus infections. Furthermore, even patients with selective IgA deficiency do not develop more prolonged or more severe respiratory or enteric viral infections. These findings lend additional evidence that factors other than humoral Abs, such as natural killer cells, complement, and cellular immunity act to limit the course of and effect recovery from viral diseases.

The limited effect of humoral Abs on recovery is not surprising, since they are ineffective against intracellular viral precursors and virions. Furthermore, many viruses can spread directly to contiguous, uninfected cells, thus remaining inaccessible to Abs. However, Abs do serve an important function in restricting the dissemination of

TABLE 51–3. *Viral Diseases in Which Immunization Has Been Effective*

	Vaccine	
Disease	*Attenuated Virus*	*Inactivated Virus*
Smallpox	+	
Yellow fever	+	
Poliomyelitis	+	+
Measles	+	
Influenza	+ *	+
Mumps	+	+
Rabies	+ †	+
Adenovirus infection‡	+	+
Rubella	+	

* Experimental
† For veterinary use
‡ Caused by types 3, 4, 7, and 21

some viruses (e.g., polioviruses and togaviruses), the pathogenesis of which depends on a viremic stage.

PERSISTENCE OF ANTIBODIES. The time course and the persistence of Ab production and immunity vary with (1) the virus, (2) the nature of antigenic stimulus, and (3) the type of Ab. For example, (1) neutralizing Abs fall from their maximal level more rapidly, and to a lower titer, following influenza than following poliomyelitis infection; (2) immunity to measles persists for life following infection, but lasts only a few months following immunization with formalin-inactivated virus; (3) after infection, complement-fixing Abs generally appear earlier but decrease much sooner than neutralizing Abs.

Long-lasting immunity, with persistence of circulating Abs, follows infection with a number of viruses, especially those causing viremia. Thus, second attacks are extremely rare with measles, smallpox, yellow fever, or poliomyelitis, to mention only a few examples. In contrast, **second infections are common with most acute localized infections without viremia, particularly respiratory diseases** (probably because adequate quantities of neutralizing Abs do not persist in respiratory secretions, although sufficient circulating Abs are present). Adenovirus infections (see Chap. 52) are notable exceptions, perhaps because they frequently terminate in latent infections of lymphoid tissue in the respiratory and gastrointestinal tracts.

Latent infection with persistent synthesis of critical viral Ags offers the most reasonable explanation for the long-lasting immunity that follows many viral infections. **Repeated infections** by a prevalent virus or **secondary Ab response** in a disease with a long incubation period could also provide an explanation, but these possibilities seem much less plausible. Thus, neither of the latter mechanisms can account for the prolonged persistence of circulating Abs against smallpox or yellow fever in previously infected residents of the United States, where these diseases rarely, if ever, occur.

Panum's observations on a measles epidemic in the Faroe Islands offers a classic example of prolonged immunity in the absence of reexposure to the specific agent. Those persons who had been alive during the preceding epidemic, 67 years earlier, were immune, whereas the younger islanders were highly susceptible. Hence, not only did immunity persist in the absence of overt clinical reinfection, but the immune persons failed to infect their nonimmune contacts during all these years. This absence of transmission could be explained by incomplete viral multiplication, by the continued neutralization of the virus produced in the immune persons, or by persistence of memory cells rather than the existence of latent infection.

CELL-MEDIATED IMMUNITY

Dysgammaglobulinemias and drug-mediated immune suppression in humans have provided the strongest evidence that humoral Abs do not play the determinant role in recovery from many viral infections. Patients who lack immunoglobulins but develop cell-mediated immunity (CMI), which consists of cytolytic T-lymphocytes and Ab-dependent cell-mediated cytotoxicity (ADCC), ordinarily recover from viral diseases without difficulty, whereas patients with defective CMI but normal Abs recover poorly from certain viral infections. For example, in persons with defective CMI, smallpox immunization frequently leads to spreading of the virus, either with severe **generalized vaccinia** or extensive necrosis of the skin and muscle of the affected extremity (**vaccinia gangrenosa;** see Chap. 54). Moreover, these complications are unaffected by the administration of specific neutralizing Abs, but they are arrested by local injection of lymphoid cells from recently immunized donors, and development of a delayed hypersensitivity reaction to heat-inactivated vaccinia virus accompanies this recovery. Finally, in experimental animals, depression of CMI by antilymphocytic serum (which contains Abs to T cells) increases the severity or the duration of infection with a number of viruses, particularly those possessing envelopes (e.g., herpesviruses, poxviruses, and paramyxoviruses).

Cellular immunity also appears to play a critical role in **maintaining the latent viral infections:** activation of such infections has become common in patients with organ transplants or with malignancies whose CMI is suppressed by therapy. These latent infections include herpesviruses (varicella zoster, cytomegalovirus, and Epstein-Barr [EB] virus), adenoviruses, measles virus, hu-

man wart virus, and JC and BK viruses (papovaviruses similar to SV40). Herpesvirus infections are also often activated in patients with extensive burns (herpes simplex virus and cytomegalovirus) and in the aged (varicella zoster virus) owing to diminished CMI. Activation of these viruses is also a common and life-threatening event in patients with **acquired immune deficiency syndrome (AIDS)** owing to marked suppression of their T4-helper cells (see Chap. 65).

Pseudotolerance

Some viruses that infect the embryo or the newborn without damaging host cells give rise to apparent immunologic tolerance. Thus, chicks infected by avian leukosis virus produce virus throughout life without detectably producing neutralizing Abs. However, this phenomenon is not due to true tolerance, since Abs are formed, but they are complexed with the large amount of virus produced. Similarly, humans infected *in utero* with rubella virus or cytomegalovirus, and fetal mice infected with influenza virus, produce virus-specific Abs for long periods after birth. And although mice with **persistent LCM** or **murine leukemia virus** infection give birth to offspring who are viremic and lack detectable virus-specific circulating Abs, the infants synthesize Abs that complex with virus (see below). Such persistently infected mice, however, are deficient in cytotoxic T-cell production. Moreover, injection of specific cytotoxic T cells can reduce or even eliminate the persistent infection.

The virus produced in large amounts throughout the life of such **pseudotolerant** animals, usually with viremia, is **disseminated vertically** to their offspring through the ovum, placenta, or milk and **horizontally** to contacts through excretions and secretions. In such animals infection is asymptomatic for most of their life, but late in life a chronic disease may develop.

DISEASE BASED ON VIRUS-INDUCED IMMUNOLOGIC RESPONSE

The immune response, despite its protective and ameliorative effects, can also contribute to the production of disease, particularly with viruses that **antigenically alter the cell's surface membranes.** For instance, a severe, **hemorrhagic dengue,** often associated with a shock syndrome, occurs in those who have had prior infection with a different serotype, and children **immunized with inactivated measles** or **respiratory syncytial virus** develop unusually severe disease if subsequently infected by the same virus. These examples of **enhanced viral injury** could be due to one or a mixture of the following immunologic reactions with virus-infected cells: increased secondary response of cytotoxic T cells (see Chap. 50); specific Ab-dependent, cell-mediated lysis; Ab-mediated, complement(classic or alternate pathway)-dependent lysis; or enhanced binding of unneutralized vi-rus–Ab complexes to cell surface Fc receptors, thus increasing the number of cells infected. In another mechanism, circulating virus-specific **Ag–Ab complexes** may lodge in organs, such as the brain or kidney, induc-ing inflammation and disease.

The central role of the immune response in the devel-opment of some viral diseases is dramatically demon-strated in mice infected with **LCM** virus,* an arenavirus (see Chap. 60). In adult mice severe, often fatal disease follows about a week after intracerebral inoculation, but if the immune response has been suppressed (by neona-tal thymectomy, chemicals, x-irradiation, or antilympho-cytic serum), disease fails to develop although viral mul-tiplication and spread are unrestrained. Moreover, after infection *in utero* or at birth, specific CMI is not detect-able; the mice appear normal for 9 to 12 months in spite of widespread viral multiplication that produces persis-tent viremia and viruria, with viral Ags in most organs. Tissue injury can be initiated in such mice by transfer of spleen cells from a syngeneic immune donor but not by immune spleen cells treated with anti-θ serum (from ani-mals immunized with Ag from T-lymphocytes) or by large amounts of immune serum.

In adult mice, viral replication and spread are rela-tively restricted in both neural and extraneural tissues, but the cell-mediated antiviral immune response is quick to develop and to elicit lethal disease by attacking the membranes of a critical number of involved cells. In con-trast, in the fetus, neonate, or immunosuppressed adult mouse, with limited immunologic capabilities, infection proceeds unimpeded to eventual involvement of all tis-sues. The constant high level of virus that develops may conceivably depress the clonal expansion of virus-spe-cific T-lymphocytes, but Abs are produced, only to be complexed with excess circulating viral Ags and comple-ment. Filtration of these aggregates by the renal glo-meruli initiates an inflammatory response, culminating in glomerulonephritis. In addition, necrotic lesions ap-pear in the liver, brain, spleen, and other organs, appar-ently resulting from the reaction between virus-sensi-tized killer T-lymphocytes and viral Ags present on the surface of many cells. After intracerebral infection of an adult animal, the brain damage is generated by a similar T-cell–dependent immunologic mechanism. In contrast to the devastating effects of natural LCM infection, the immunization of mice with inactivated LCM produces humoral Abs that, upon subsequent infection, restrict viral spread and prevent disease.

These observations lead to the conclusion that in the absence of effective cellular immunity, LCM virus multi-plies harmlessly for a long time in mice, producing an inapparent infection similar to the persistent infection that it causes in cell cultures (see below). Thus, every aspect of the disease, in both acute and chronic infec-tion, can be shown to be immunologically mediated, and in this single animal model, LCM dramatically demon-strates both the benefits and the disadvantages of the immunologic response. **Anti-idiotypic Abs,** those di-rected against the Ab variable region (the antigen-com-bining site), usually develop during the immune re-sponse (see Chap. 14). Such Abs mimic viral Ag and, like viruses, may bind to receptors on the cell surface. This interaction of anti-idiotypic Abs and viral cell receptors could theoretically produce cell injury, although the pre-cise role of these Abs has not yet been elucidated in the pathogenesis of viral diseases. Induction of this type of complication following a viral infection seems possible when it is realized that the viral receptors have not evolved for the benefit of viruses, but rather to accommo-date substances required physiologically by the cell; for example, the human immunodeficiency (AIDS) virus (HIV) attaches to the T4-lymphocyte receptor, type 3 reovirus utilizes the mammalian β-adrenergic receptor, and EB virus combines with the complement C3d recep-tor on human B-lymphocytes.

NONSPECIFIC SYSTEMIC FACTORS

Nonspecific factors that influence resistance to viral in-fection include various **hormones, temperature, inhib-itors** other than Abs, **natural killer (NK) cells,** and **phagocytes.** Nutrition may also affect the course of viral infections; e.g., measles has devastating consequences in malnourished children of West Africa. However, malnu-trition influences so many aspects of the host defenses that specific analysis of cause and effect is difficult.

Infected cells may form and release an especially im-portant factor, **interferon,** which inhibits viral multipli-cation by preventing the synthesis of viral proteins, and thus interferes with the infection of other cells by many viruses. Interferon also activates NK cells. Accordingly, interferon not only prevents the infection of cells but also limits viral spread and assists in recovery. This agent has been discussed in Chapter 49.

Phagocytosis does not appear to be as important a defense mechanism in viral as in bacterial infections. On the contrary, some viruses impair the antibacterial activ-ity of **polymorphonuclear leukocytes** by producing leukopenia (e.g., measles virus) or by reducing phago-cytic function (e.g., influenza virus). **Macrophages,** how-ever, do appear to be important in viral infections; they rapidly take up certain viruses, and there is a correlation between host and macrophage susceptibility to viral in-fections (see Genetic Factors, above). Virus-infected mac-rophages can act as a source of infection for other cells, but with viruses that are unable to multiply in them,

---

* This virus can also infect humans, but it usually produces a mild respiratory infection, only occasionally followed by severe meningitis.

macrophages appear to play their usual role as scavengers.

Hormones have a potential effect on viral infections, as can be illustrated by several examples. Pregnancy increases the severity of several viral diseases: paralytic poliomyelitis is more frequent and more extensive; smallpox has a more severe course, and abortion is common; the complications of influenza, particularly pneumonia, are increased. Cortisone enhances the susceptibility of many animals to viral infection and commonly potentiates the severity of the disease; in humans it causes enlargement and perforation of herpetic corneal ulcers and induces extensive visceral spread of varicella virus that often terminates in severe pneumonia. These deleterious effects appear to result from the suppressive effects of cortisone on inflammatory reactions, CMI, and interferon production rather than from its action on the Ab response.

Temperature increase in the host may reduce viral replication by suppressing a temperature-sensitive step, by accelerating the inactivation of many heat-labile viruses (enveloped viruses), and by increasing interferon production. Conversely, in mice held at 4°C, rather than 25°C, after infection with coxsackievirus B1, viral multiplication is excessive in many organs, little interferon is produced, and the mortality is strikingly increased. A rise in body temperature may therefore contribute to recovery from viral disease.

Latent Viral Infections

In **latent infections,** overt disease is not produced, but the virus is not eradicated. This equilibrium between host and parasite is achieved in various ways by different viruses and hosts. The virus exists in latent infections either as an infectious and continuously replicating agent, termed a **persistent viral infection,** or in a truly latent **noninfectious occult form,** possibly as an integrated genome or an episomal agent.

Numerous experimental cell culture and animal models exemplify these two types of latent infections and permit understanding of both types of latent infections in humans. It is noteworthy that in humans, persistent viral infections may lead to chronic diseases, which appear some years after the initial viral infections. These persistent infections have been termed **slow viral infections** (see below) and are best exemplified by **subacute sclerosing panencephalitis** caused by a persistent, inapparent measles virus infection (see Chap. 57). In contrast, occult viral infections may lead to reactivation of the viral genome, perhaps akin to lysogeny in bacteria, resulting in productive viral replication and induction of acute disease, such as **fever blisters** caused by activated

herpes simplex virus (see Chap. 53) or **shingles** similarly initiated by **herpes zoster virus** (see Chap. 53).

LATENT PERSISTENT INFECTIONS

Enveloped viruses such as **paramyxoviruses** (measles, parainfluenza, and mumps viruses), some herpesviruses (e.g., EB virus), **retroviruses,** and **arenaviruses** appear particularly suited to initiate persistent infections. Infection appears to persist because the virus does not disrupt the essential **housekeeping functions** of the cells (e.g., DNA, RNA, and protein synthesis), although they may affect **luxury functions.** For example, **LCM virus of mice** (see above) may turn off growth hormone production in the anterior lobe of the pituitary gland without altering the infected cells' vital processes, thus leading to retarded growth of the infected mice. Some persistently infected cells, as in measles subacute sclerosing panencephalitis (see below), may be assisted by the remarkable capacity of humoral Abs to **redistribute ("cap") viral Ags on the plasma membrane** (Fig. 51–2). This phenomenon promotes shedding of viral Ags from the cell surface, leaving the cell surface free of viral glycoproteins and the infected cell protected from either cell-mediated (CTL) or Ab-dependent cell-mediated (ADCC) immunologic destruction.

Defective viral particles, the genomes of which are partially deleted, produce effective interference with homotypic, nondefective virus, and they are therefore termed **defective interfering** (or DI) particles (see Chap. 49). DI particles appear to play a central role in steady-state infections in cell cultures and possibly in the establishment and maintenance of *in vivo* persistent infections.

LATENT OCCULT VIRAL INFECTIONS IN ANIMAL HOSTS

There is an increasing recognition that some viruses, both DNA- and RNA-containing viruses, may become undetectable following a primary infection only to reappear and produce acute disease. The mechanism by which this latency is established and the virus then reactivated must be accomplished in different ways, as the following examples illustrate.

1. **Herpes simplex virus** has a special pattern of **latency** and **recurrence.** This virus usually infects humans between 6 and 18 months of age, and the virus persists but cannot be found except during recurrent acute episodes, such as herpes labialis (fever blisters; see Chap. 53). The form in which the latent occult virus persists between recurrent episodes is uncertain. Virus cannot be isolated from tissue homogenates, but by co-cultivating cells of sensory ganglia with susceptible cells,

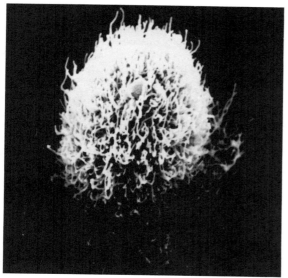

A

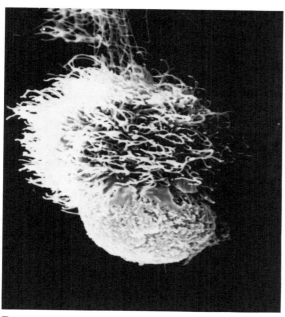

B

Figure 51–2. Distribution of measles virus Ags on the surface of infected HeLa cells before and after reaction with virus-specific Abs. (*A*) Scanning electron microscopy demonstrates the abundant fine microvilli randomly distributed when infected cells were not reacted with Abs. (In contrast, uninfected cells display shorter, thicker, and less abundant villi). (*B*) Marked redistribution ("capping") of viral Ags seen after infected cells were mixed with virus-specific Abs at 37°C. Serum without measles virus Abs or serum with Abs directed against Ags of other viruses did not cap the measles virus Ags. (Original magnification ×4000; Lampert PW et al: J Virol 15:1248, 1975)

virus has been detected in the human trigeminal (type 1 virus) and thoracic, lumbar, and sacral dorsal root ganglia (type 2 virus), as well as in sensory ganglia of experimentally infected mice, rabbits, and monkeys. DNA : DNA hybridization studies have detected the viral genome in normal brains as well as in peripheral ganglia. These data suggest that the DNA exists in a linear, unintegrated form, perhaps as episomes or in intact virions. It may be that, as in virus-carrier cultures, infection is confined to only a small proportion (about 0.01%–0.1%) of the ganglion cells by Abs, cellular immunity, viral interference (by interferon or DI particles), or metabolic factors. Because Abs are present, most of the extracellular virus is neutralized and goes undetected. Acute episodes, in which there is a burst of viral replication, probably depend on a transient change in the local level of immunity or changes in the susceptibility of the uninfected cells induced by a variety of physical and physiologic factors such as fever, intense sunlight, fatigue, or menstruation.

Experimentally, the role of cell susceptibility is shown by thymidine kinase–minus mutants, which cannot replicate in growth-arrested cells and do not establish latent infections of the trigeminal or cervical sensory ganglia following productive ocular infections. The role of shifts in the immune system is also illustrated by experimental infection of rabbits with herpes simplex virus. Herpes encephalitis can be reactivated 6 months after an acute encephalitis episode by inducing anaphylaxis with any Ag. Similarly, herpes keratitis can be provoked, after an acute corneal ulcer is healed, if the rabbit is made sensitive to horse serum and a corneal Arthus reaction is induced. **Nonspecific excitants** such as ultraviolet light, histamine or epinephrine injection, corticosteroids, or surgical manipulation of sensory ganglia and nerves can also **activate** experimental herpes simplex virus infection.

The other herpesviruses that infect humans also commonly produce latent infections: varicella zoster virus in sensory ganglia, cytomegalovirus in macrophages and lymphocytes, and EB virus in B-lymphocytes (see Chap. 53).

2. **Adenovirus infections** in humans are self-limited, but the virus frequently establishes a latent, persistent infection of tonsils and adenoids (see Pathogenesis, Chap. 52). Though these tissues fail to yield infectious virus when homogenized and tested in sensitive cell cultures, cultured fragments of about 85% of these "normal" tonsils and adenoids, after a variable time, show characteristic adenovirus-induced cytopathic changes and yield infectious virus. Viral DNA can be detected in tonsils and adenoids, as well as in peripheral lymphocytes.

Failure to recover infectious virus initially may be attributed to the paucity of virions, to their association with either Ab or receptor material, or to the absence of mature virions. The latent infection is probably not the result of lysogeny, since DNA in peripheral lymphocytes appears to be in a linear episomal form. The host's cytotoxic T cells probably fail to eradicate persistently infected cells because the carboxy-terminus of a 19,000 dalton glycoprotein (gp19kd), encoded in early region 3 (see Chap. 52), combines with one or more of the Class I major histocompatibility complex (MHC) Ags in the endoplasmic reticulum and prevents their transport to the infected cells' surfaces.

3. The **Shope rabbit papilloma virus** (see Chap. 64) illustrates latency due to the **replication of viral nucleic acid without viral maturation.** This virus produces warts equally well in the skin of domestic or of wild cottontail rabbits, but infectious virus and viral Ags can usually be detected only in the tumors of the wild animals,* in the keratinized cells of the outer epidermal layer but not in the growing basal layer. However, the basal layer of warts from either wild or domestic rabbits yields episomal, infectious, viral DNA when extracted with phenol.

In man the most dramatic example of **incomplete viral production** is the persistence and continued replication of **measles virus nucleocapsids** in lesions of **subacute sclerosing panencephalitis (SSPE),** which may develop years after acute measles infection (see Slow Viral Infections, below).

4. **Swine influenza virus,** an orthomyxovirus related to influenza A virus, illustrates a complex ecologic situation in which the virus is **latent in two intermediate hosts** and **requires assistance from a bacterium,** *Hemophilus influenzae suis,* to induce the acute respiratory disease in pigs (Fig. 51–3).

Shope showed that virus in the lung of a sick pig becomes associated, in occult form, with ova of the lungworm (a common parasite of most pigs), which is coughed up with pulmonary secretions, swallowed, and

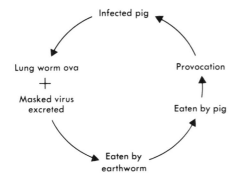

Figure 51–3. Natural history of swine influenza infection.

* Since the assay of this virus is extremely insensitive, it is not known whether failure to recover the virus signifies paucity or absence.

eventually passed in the pig's feces. The contaminated feces are then eaten by earthworms, in which the lungworm ova develop into larvae in which the virus can remain occult for at least 2 years. To complete the cycle the earthworm is eaten by a pig, and the lungworm larvae migrate to the pig's lungs, where they develop into mature lungworms. The virus remains noninfectious through this long odyssey, but when the parasitized pig is jolted by cold or by infection with *H. influenzae suis,* the occult virus is somehow induced to replicate; it may initiate an acute disease and induce viral spread. The nature of the viral occult form in the lungworm and earthworm and the exact role of the bacterium are unknown.

EPIDEMIOLOGIC SIGNIFICANCE OF LATENT INFECTIONS

As these examples indicate, latent viral infections affect the incidence and the pathogenesis of **acute viral diseases** in several ways. (1) In **herpes simplex** infection, **recurrent acute disease** is induced in the infected person when the balance that maintains the latent infection is intermittently disturbed (see Herpes Simplex Viruses, Pathogenesis, Chap. 53). (2) **Swine influenza** mimics the way in which a virus widely seeded among humans may initiate, in response to **environmental changes,** an **explosive epidemic by viral reactivation** in many areas at the same time, as noted in pandemics of influenza. Indeed, this intriguing animal model of latency may explain the manner in which influenza A virus "goes underground" only to **emerge,** often with **new antigenic characteristics,** many years later to produce widespread epidemics (see Epidemiology, Chap. 56). (3) A **reactivated latent (occult) virus may spread and initiate an epidemic** among susceptible contacts. Thus, a latent varicella virus, persisting in peripheral nerve ganglia after chickenpox, can be activated and produce the different clinical picture of herpes zoster (shingles), and the patient may then serve as a focus for initiating an epidemic of chickenpox (see Varicella–Herpes Zoster Virus, Epidemiology, Chap. 53). (4) Viral latency can also be seen in the **development of certain chronic diseases** dependent on an immunologic response, as for LCM, noted above, for SSPE, and possibly for some other so-called **slow viral infections.** (5) Some latent viral states **produce uncontrolled proliferation of cells,** i.e., tumors, as discussed in Chapters 64 and 65.

Slow Viral Infections

The term *slow virus* has become associated with those viruses that require prolonged periods of infection (often years) before disease appears. The term, however, is

TABLE 51–4. Examples of Slow Viral Infections

Disease	Host	Organ Primarily Affected	Virus
Kuru	Man	Brain	Unknown
Creutzfeldt-Jakob disease	Man	Brain	Unknown
SSPE	Man	Brain	Paramyxovirus (measles)
Progressive rubella panencephalitis	Man	Brain	Togavirus (rubella)
Progressive multifocal leukoencephalopathy	Man	Brain	Papovavirus (SV40, BK, and JC)
Scrapie	Sheep	Brain	Unknown—? viroid; ? prion
Mink encephalopathy	Mink	Brain	Unknown—? viroid
Visna	Sheep	Brain	Retrovirus
Maedi	Sheep	Lung	Retrovirus
Progressive pneumonia	Sheep	Lung	Retrovirus
LCM	Mouse	Kidney, brain, liver	Arenavirus
Canine demyelinating encephalomyelitis	Dog	Brain, spinal cord	Paramyxovirus (distemper)
Disseminated demyelinating encephalomyelitis	Mouse, rat	Brain, spinal cord	Coronavirus (mouse hepatitis virus)
Aleutian mink disease	Mink	Reticuloendothelial system	Parvovirus
Hard pad disease	Dog	Brain	Paramyxovirus (distemper)

somewhat misleading, for although the virus is patient and the disease process develops over a protracted period, viral multiplication may not be unusually slow. Moreover, diseases of this group may be caused not only by unusual or "unconventional" viruses but also by others that ordinarily cause acute diseases. Table 51–4 lists some chronic degenerative diseases, particularly of the CNS, that belong to this group. Five that occur in humans are discussed: SSPE, progressive encephalitis, kuru, Creutzfeldt-Jakob disease, and progressive multifocal leukoencephalopathy. Claims have been made that other chronic diseases may have similar origins, but this has not been proved (e.g., multiple sclerosis, systemic lupus erythematosus, diabetes mellitus). Viral diseases that have a long incubation period but an acute course (e.g., type B hepatitis, AIDS, and rabies) are not considered to be slow viral infections and are not discussed here.

SSPE is the best substantiated example showing that even some common human viruses (i.e., measles virus) may occasionally give rise to chronic degenerative diseases. This progressive degenerative neurologic disease of children and adolescents (causing mental and motor deterioration, myoclonic jerks, and electroencephalographic dysrhythmias) was unexpectedly discovered to be caused by measles virus. With this finding it became clear that a single virus could induce both an acute contagious disease and a chronic illness. The following data implicate measles virus in this severe chronic disease: (1) all patients have had measles several years (up to 13) prior to onset of SSPE. (2) All have unusually high titers of measles Abs, even in the spinal fluid, and the Ab levels (IgM as well as IgG) often increase as the disease progresses. (3) Affected brain cells have nuclear and cytoplasmic inclusions similar to those seen in measles infections (see Fig. 57–10); the inclusions consist of filamentous tubular structures indistinguishable from the

nucleocapsids seen in cells infected with measles virus (see Fig. 57–12). (4) Immunofluorescence study of the brain lesions reveals Ags that react with Abs to measles virus but not to distemper virus (a close relative of measles virus). (5) A virus very similar to measles virus has been isolated from brains of ferrets and newborn mice inoculated with brain material and by cocultivation of affected brain cells with cells that readily support measles virus multiplication (e.g., African green monkey kidney cells, HeLa cells); though virus has not been directly isolated from homogenates of brain cells, when the affected cells are cultured *in vitro*, they show typical inclusion bodies as well as viral Ags and viral nucleocapsids. The RNA of the SSPE virus contains all the nucleotide sequences of measles virus RNA, but about 10% consists of additional sequences, which may have been derived by recombination with the RNA of another virus.

The finding of measles Abs in the spinal fluid of patients with multiple sclerosis raised the possibility that this chronic neurologic disorder is also a complication of prior measles. How the persistently infected cells escape immunologic eradication in the presence of apparently undisturbed T-cell levels and function, normal amounts of all components of complement, and large quantities of measles-specific Abs still requires explanation. Examination of brain biopsies during early clinical stages of SSPE showed general decrease in expression and syntheses of the viral genome. During the final stages of the disease, replication of the viral RNA is decreased, but nucleocapsid (RNA genomes and NP protein) levels are increased. Moreover, matrix protein, which is required for budding and virion assembly, is markedly decreased or absent. Thus, the block in viral replication and failure to express viral glycoproteins on surfaces of infected cells account for the slow development of pathologic changes, the accumulation of cell-associated viral components, the inability to isolate infectious virus, and the failure of elimination of infected cells in the face of a normal immune response. Thus, the surfaces of infected cells do not show budding viral particles or even surface viral glycoproteins, which are present intracellularly, and the viral matrix (M) protein, which is required for virion assembly, is present in decreased amounts or is not detectable. Patients do not have anti-M Abs, although Abs to all other viral Ags are made. Indeed, humoral Abs could affect these findings, since Abs produce "capping" and shedding of viral Ags (see above), and it has been demonstrated that Abs suppress synthesis of M protein in persistently infected cultures.

Progressive rubella panencephalitis, similar to but more rapidly progressive than SSPE, develops in a rare child who previously had congenital or early childhood rubella. **Rubella virus** (see Chap. 61), rather than measles virus, is recovered by culturing the affected brain *in vitro* with or without cocultivation with cells susceptible to rubella virus multiplication. Owing to the rare occurrence of this disease, little is known of the characteristics of the infecting virus.

Progressive multifocal leukoencephalopathy is a rare **subacute demyelinating disease.** Two different species of **papovaviruses** (see Chap. 64) have been isolated from the brains of victims and are also seen in the intranuclear inclusion bodies of affected oligodendrocytes. All the viruses isolated are of the **SV40-polyoma subgroup;** two viruses are almost identical to SV40, but all the others, termed **JC virus,** are closely related and distinctly different from SV40 immunologically, chemically, and biologically. JC virus multiplies only in human cells from very few organs (primary fetal glial cells are most sensitive for viral isolation). JC virus replicates very slowly, so it is truly a "slow virus." This disease generally develops in patients with **immunologic defects due to disorders of the reticuloendothelial system** (such as Hodgkin's disease and leukemias) or to immunosuppressive therapy. Most JC virus infections occur during childhood, and about 75% of adults have circulating Abs. Thus, the emergent viruses appear to be **opportunists** liberated from a latent infection in an immunologically compromised host.

Kuru, another slow viral disease of man, was first observed in 1957 in the Fore tribe of cannibals living in Stone Age conditions in New Guinea (*kuru* = shivering or trembling in the Fore language). It is transmitted by consumption of the brains of deceased relatives, a tribal ceremonial ritual for children and young women. This degenerative disease of the cerebellum, manifested by ataxia, disturbed balance, clumsy gait, and tremor, progresses inexorably to death in less than a year after onset. A striking decrease in kuru was seen after the tribal chiefs prohibited this custom of cannibalism.

The pathologic findings do not include the customary inflammatory evidence of an infectious process but do resemble the findings in scrapie, a disease of sheep proved to be transmissible and caused by an unusual agent (i.e., difficult to inactivate and without detectable nucleic acid). This resemblance suggested a viral etiology for kuru, and Gajdusek and Gibbs, using brain material from kuru patients, transmitted the disease serially to chimpanzees. (Subsequently, it was also transmitted to New and Old World monkeys and even to mink and ferrets.) The degenerative process, which had the same clinical and pathologic characteristics as kuru in humans (subacute spongiform encephalopathy), appeared 18 to 30 months after the initial inoculation of chimpanzees and after 1 year in subsequent passages.

Creutzfeldt-Jakob disease, a fatal presenile dementia of midadult life that is not geographically restricted and hence not so exotic, like kuru shows a **spongiform encephalopathy** and appears to be a **chronic viral disease.** A similar disease has been serially transmitted from the brains of patients to chimpanzees, several species of monkeys, guinea pigs, mice, hamsters, ferrets, goats, and cats. Its epidemiology is not clear, but the disease has been accidently transmitted in humans by a corneal transplant from a person who subsequently developed the fatal disease, by growth hormone prepared from human pituitary glands, and by electroencephalographic electrodes sterilized only with 70% ethanol and formaldehyde vapor after the electrodes had been used on a patient with the disease.

THE VIRUSES

Owing in part to long incubation periods and cumbersome assays, the etiologic agents of kuru and Creutzfeldt-Jakob disease, presumably viruses, have not been well characterized. The properties described, however, have been so difficult to study that it has not been possible to place these agents in a category with any of the well-known viruses of man and other animals.

The unconventional viruses that are of necessity present in brain extracts are highly resistant to inactivation by the usual chemical and physical sterilizing agents: formaldehyde, β-propiolactone, proteases, nucleases (RNases and DNases), ultraviolet irradiation at 254 nm, and heat at 80°C (they are incompletely inactivated at 100°C). Moreover, electron microscopic examinations of infected brain tissues (with as much as $10^{12}LD_{50}/g$) and virus concentrated in CsCl and sucrose gradients (10^7–$10^8\ LD_{50}/ml$) did not reveal viral particles. Neither infectious nucleic acids nor viral Ags have been detected.

Neither immunosuppression (e.g., from x-ray or cyclophosphamide) nor immunopotentiation (e.g., with adjuvants) affects the pathogenesis of kuru or Creutzfeldt-Jakob disease in experimental animals, and B- and T-cell functions appear intact in the natural disease and in experimental infections. However, Abs directed against infectious extracts have not been detected in patients or infected animals. The characteristics of the agents of kuru and Creutzfeldt-Jakob disease are similar to those of scrapie and transmissible mink encephalopathy viruses. These indeed appear to be unique viruses, if they *are* viruses. Further characterization may show that the so-called slow or unconventional viruses are in fact a new type of infectious agent. Amyloid fibrils, associated with infectivity, have been purified from brains infected with unconventional viruses and termed **prions** or SAF (scrapie-associated fibrils). These fibrils are claimed to be free of nucleic acid. The amyloid present in the fibrils, which are found in abundance in the affected brains, differ only in quantity from amyloid molecules present in normal cells. It is possible that within such protein particles, a small nucleic acid molecule, such as a **viroid,** could be buried. A viroid consists of an infectious molecule of covalently closed circular single-stranded RNA (110,000–127,000 daltons) without associated protein (see Distinctive Properties, Chap. 44). Viroids depend entirely on host cell macromolecules for their replication. These agents have been described only as pathogens in plants. However, the single-stranded RNA of the **hepatitis delta virus,** which depends on the hepatitis B virus for its infectivity (see Chap. 63), has a high degree of nucleotide hormology with the potato spindle tuber viroid. Although the delta virus has not yet been proved to be a viroid, the history of earlier studies on the distribution of other novel microbes suggests that viroids will also be found in organisms other than plants, probably in animals.

The examples of slow viral infections of man are still few, and the evidence of causation is sparse. Nevertheless, suspicion of the role of viruses in chronic degenerative diseases is now high, and some conventional viruses (e.g., togaviruses, picornaviruses, paramyxoviruses) are being implicated in diseases such as multiple sclerosis and diabetes mellitus.

Selected Reading

Allison AC: Lysosomes in virus-infected cells. *Perspect Virol* 5:29, 1967

Blanden RV: Mechanisms of recovery from a generalized viral infection. II. Passive transfer of recovery mechanisms with immune lymphoid cells. J Exp Med 133:1074, 1971

Fulginiti VA, Kempe CH, Hathaway WE et al: Progressive vaccinia in immunologically deficient individuals. In Bergsivia D, Good RA (eds): Birth Defects: Immunologic Deficiency Diseases in Man, p 129. New York, National Foundation, 1968

Gajdusek DC: Unconventional viruses and the origin and disappearance of kuru. Science 197:943, 1977

Galloway DA, Fenoglio C, Shevchuk M, McDougall JK: Detection of herpes simplex RNA in human sensory ganglia. Virology 95:265, 1979

Haase AT, Gantz D, Eble B et al: Natural history of restricted synthesis and expression of measles virus genes in subacute sclerosing panencephalitis. Proc Natl Acad Sci USA 82:3020, 1985

Klein RJ: Initiation and maintenance of latent herpes simplex virus infections: The paradox of perpetual immobility and continuous movement. Rev Infect Dis 7:21, 1985

Notkins SL, Oldstone MBA (eds): Concepts in Viral Pathogenesis, I (1984) and II (1986). New York, Springer-Verlag

Ter Meulen V, Stephenson JR, Kreth HW: Subacute sclerosing panencephalitis. In Fraenkel-Conrat H, Wagner RR (eds): Comprehensive Virology, Vol 18, p 105. New York, Plenum Press, 1983

Walker DL, Padgett BL: Progressive multifocal leukoencephalopathy. In Fraenkel-Conrat H, Wagner RR (eds): Comprehensive Virology, Vol 18, p 161. New York, Plenum Press, 1983

52

Harold S. Ginsberg

Adenoviruses

Acute viral respiratory diseases continuously impose huge clinical and economic burdens. Thus, great efforts to isolate other major causative agents followed the isolation of influenza virus in 1933. The search was unsuccessful, however, until in 1953 two groups of investigators discovered **adenoviruses,** the first of several families now known to be etiologic agents of these acute infections. Rowe and colleagues, using cultures of human adenoids as a potentially favorable host for the elusive "common cold" virus, noted cytopathic changes in uninoculated cultures after prolonged incubation, as well as in cells inoculated with respiratory secretions. The pathologic alterations were shown to be due to the emergence of previously unidentified viruses from latent infections of adenoid tissues—hence the name *adenoviruses.* Hilleman and Werner, studying an epidemic of influenzalike disease in army recruits, isolated several similar cytopathic agents from respiratory secretions added to cultures of human upper respiratory tissues.

Adenoviruses cause acute respiratory and ocular infections. Although they are not the etiologic agents of the common cold, they are responsible for a small percentage of acute viral respiratory infections. It is evident, however, that certain specific types are etiologic agents of a broader range of infections such as hemorrhagic cystitis and infantile gastroenteritis.

General Characteristics

Several characteristics of adenoviruses (family **Adenoviridae**) are of particular interest: (1) Adenoviruses are simple DNA-containing viruses (i.e., composed of only DNA and protein) that multiply in the cell nucleus. (2) They induce **latent persistent** infections in tonsils, adenoids, and other lymphoid tissues of man, and they are readily activated. (3) Several adenoviruses are **onco-**

CHARACTERISTICS OF ADENOVIRUSES

Icosahedral symmetry
Diameter of 60–90 nm
Capsid contains 252 polygonal capsomers, 12 fibers, and four minor proteins.
Double-stranded DNA genome
Resistance to lipid solvents (absence of lipids)
Related by family cross-reacting soluble Ags (except for the chicken adenoviruses)
Multiplication in cell nuclei

genic for a number of newborn rodents (they were the first viruses of humans shown to have this property). (4) They serve as "helpers" for a group of small, defective DNA-containing viruses, the **adeno-associated viruses** (discussed at the end of this chapter), which cannot replicate in their absence. (Conversely, some adenoviruses cannot multiply efficiently in primary monkey cells unless the genetically unrelated simian virus 40 (SV40) is present as a helper [see Abortive Infections, Chap. 48].)

Adenoviruses are widespread in nature. The 93 accepted members of the adenovirus family have similar chemical and physical characteristics and a family cross-reactive Ag (see "Characteristics of Adenoviruses), but they are distinguished by Abs to their individual type-specific Ags: 41 are from humans and the rest from various other animals. Comparative studies of the viruses from humans permit their classification into several groups (Table 52–1).

Properties

STRUCTURE. Electron microscopy shows the virions to be 60 nm to 90 nm in diameter. In sections the viral particles have a dense central **core** and an outer coat, the capsid (Fig. 52–1). Negative staining reveals **icosahedral particles** with capsids composed of 252 **capsomers** (Fig. 52–2): 240 **hexons** make up the faces and edges of the equilateral triangles, and 12 **pentons** constitute the vertices. The hexons are truncated triangular or polygonal prisms with a central hole (Figs. 52–3 and 52–4). The pentons are more complex, consisting of a **polygonal base** with an attached **fiber,** the length of which varies with the viral type (Figs. 52–4 and 52–5). Four additional minor capsid proteins (IIIa, VI, VIII, and IX) are associated with the hexons or pentons in stoichiometric amounts (Fig. 52–6). These proteins confer stability on the capsid, form links with the core proteins, and function in virion assembly.

PHYSICAL AND CHEMICAL CHARACTERISTICS. Each virion contains one **linear, double-stranded DNA molecule** associated with **proteins** to form the **core.** The DNAs of different viral types vary in molecular weight and base composition (see Table 52–1): viruses within each subgroup share 70% to 95% of their nucleotide sequences (as shown by DNA–DNA, and DNA–mRNA hybridization), and the DNAs of viruses of different subgroups have only 5% to 20% homology.

The viral DNA has two novel features: (1) the terminal

TABLE 52-1. Physical, Chemical, Oncogenic, and Hemagglutinating Characteristics of Human Adenoviruses

| Subgroup (Subgenus) | Type | Oncogenic* Potential | Viral DNA | | | Agglutination of RBCs | |
			Percentage of Virion	Mol. Wt.	G + C (%)	Rhesus	Rat
A	12, 18, 31	High	11.6–12.5	20×10^6 (30 Kbp†)	47–49	0	Partial‡
B	3, 7, 11, 14, 16, 21, 34,§ 35	Weak	12.5–13.7	$23–25 \times 10^6$ (35–38 Kbp)	57–61	+	0
C	1, 2, 5, 6	None	12.5–13.7	23×10^6 (35 Kbp)	57–59	0	Partial
D	8–10, 13, 15, 17, 19, 20, 22–30, 32, 33, 36–39	None	12.5–13.7	$23–25 \times 10^6$ (35–38 Kbp)	57–60	+ or 0‖	+
E	4	None	12.5	23×10^6 (35 Kbp)	57	0	Partial
F–G#	40, 41						

* Highly oncogenic adenoviruses induce tumors in newborn hamsters within 2 months after inoculation; weakly oncogenic viruses induce tumors in fewer animals in 4 to 18 months. Even those viruses that are nononcogenic transform nonpermissive rodent cells *in vitro,* and the transformed cells produce tumors in syngeneic newborn animals and in nude mice.

† Kilobase pairs

‡ Complete hemagglutination occurs when heterologous Ab to a virus of the same subgroup is added to the reaction mixture, producing groups of fibers or aggregation of pentons into regular groups of 12.

§ Type 34 has hemagglutination characteristics of a subgroup D virus.

‖ Some types (9, 13, and 15) agglutinate rhesus RBCs but to a lower titer than rat RBCs.

These types have not yet been completely characterized. Hybridization shows a high degree of DNA homology, suggesting that they should be placed in the same subgenus, but their restriction endonuclease patterns are distinctly different.

RBCs, red blood cells

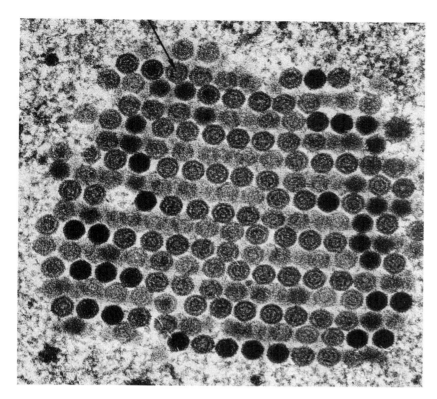

Figure 52–1. Thin section of a crystalline mass of adenovirus particles in a cell infected with type 4. Two types of particles can be seen: dense particles with no discernible internal structure, and less dense particles showing the central body of the viral particles, the core (*arrow*), and the capsid. The polygonal shape of adenoviruses is apparent in many particles. Differences in appearance of particles are probably due to the relation of the center of the virion to the plane of section. (Original magnification greater than ×110,000; courtesy of C. Morgan and H. M. Rose, Columbia University)

nucleotide sequences of each strand are **inverted repetitions** so that if the DNA is denatured, both strands form single-stranded circles through "panhandles" produced between the complementary ends, and (2) a small protein of about 55,000 daltons is covalently linked through the terminal deoxycytosine at the 5′ end of each strand (Fig. 52–7). The functions of these unique terminal structures of the viral genome are important in DNA replication.

The virion contains at least ten species of proteins associated with the capsid and the DNA–protein core (see Fig. 52–6). The capsid proteins have relatively strong noncovalent bonds between protomers in a capsomer, and weaker bonds between capsomers (see Chap. 44); hence, the capsid can be artificially disrupted into intact capsomers (see Figs. 52–3 and 52–4). Further dissociation of the hexons into their constituent polypeptide chains, in contrast, requires rigorous denaturing conditions (e.g., 6M guanidine hydrochloride and a sulfhydryl reagent to block formation of disulfide bonds). The penton base is the least stable of the capsomers. Noncovalent bonds associate the glycosylated fiber with the penton base.

The virion's core includes the covalently bonded **5′-terminal protein** and **two basic proteins** associated with the DNA to form a chromatinlike structure (see

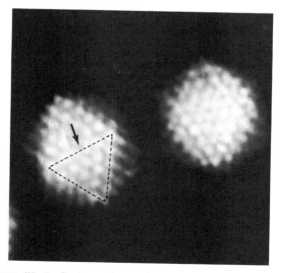

Figure 52–2. Electron micrograph of purified particles of type 5 adenovirus embedded in sodium phosphotungstate. The icosahedral symmetry (see Chap. 44) of the virion and subunit structure of the capsid are apparent. The arrow points to a virion's axis of twofold rotational symmetry. Capsomers at the apices of triangle are centers of the fivefold symmetry. The capsid parameters are P = 1, f = 5; hence, the capsid consists of 252 capsomers (see Number of Capsomers in Icosahedral Capsids, Appendix to Chap. 44). (Original magnification ×440,000, reduced)

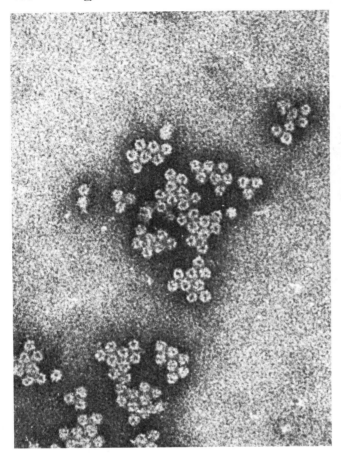

Figure 52–3. Electron micrograph of purified type 5 hexons embedded in sodium silicotungstate. The polygonal shape of the capsomer with the central hole is apparent. Groups of nine hexons are seen in different orientations so that both tops (those with large holes) and bottoms (small holes) are observed. The subunit structure can also be seen in many hexons. (Original magnification ×375,000; courtesy of M. V. Nermut, National Institute for Medical Research, London)

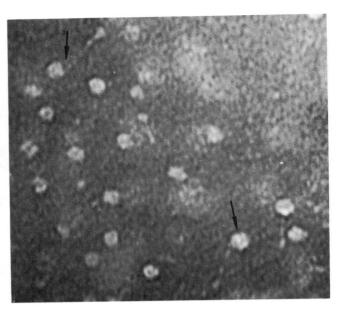

Figure 52–4. Electron micrograph of the capsid components of purified type 5 adenovirus particles disrupted at pH 10.5. Note the polygonal, hollow capsomers (i.e., the hexons, which are 7–8.5 nm in diameter with a central hole about 2.5 nm across) and the fibers (1–2.5 nm wide and 20 nm long), attached to polygonal bases (i.e., the pentons [*arrow*]). By actual count there are 12 pentons per virion. (Original magnification ×480,000, reduced)

Chap. 47). One of these (designated *protein VII*; see Fig. 52–6) is, like histones, rich in arginine (about 23 mol/dl).

STABILITY. Adenoviruses are relatively stable in homogenates of infected cells; they retain undiminished infectivity for several weeks at 4°C and for months at −25°C. Purified virions, however, are relatively unstable under all conditions of storage, owing primarily to the spontaneous release of pentons. Adenoviruses are resistant to lipid solvents.

HEMAGGLUTINATION. Human adenoviruses differ in their ability to agglutinate rhesus monkey or rat red blood cells (RBCs). Hemagglutination occurs when the tips of fibers on virions or on aggregated pentons (commonly arranged as dodecagons) bind to the RBC surface

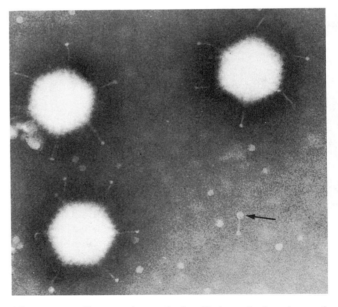

Figure 52–5. Electron micrograph of purified type 5 adenovirus particles embedded in sodium silicotungstate. Micrographs obtained in areas where the silicotungstate was thin revealed the fiber components of the penton projecting from corners of the virion. Free pentons (*arrows*) and hexons are also present. (Original magnification ×350,000; Valentine RC, Pereira HG: J Mol Biol 13:13, 1965)

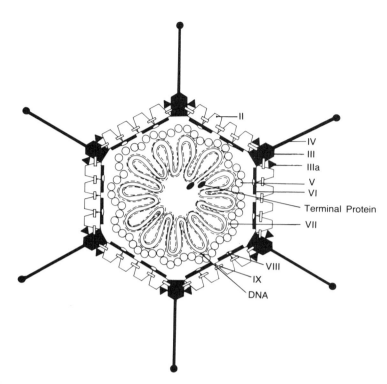

Figure 52–6. Model of an adenovirus particle, showing the apparent architectural interrelationships of the structural proteins (roman numerals) and the nucleoprotein core in the virion. The hexon (II), penton base (III), and fiber (IV) and the hexon-associated proteins (IIIa, VI, VIII, and IX) make up the capsid. Proteins V and VII are core proteins associated with the viral DNA; the TP (terminal protein) is covalently linked to the 5' end of the DNA.

and cause crosslinking. The nature of the fiber receptor on susceptible RBCs is not known. The combination of adenoviruses with cell receptors is stable, and spontaneous elution of the hemagglutinin (as seen with influenza viruses) does not occur.

There is a remarkable agreement among members of subgroups according to their hemagglutinin properties, viral oncogenicity, and DNA characteristics (see Table 52–1).

IMMUNOLOGIC CHARACTERISTICS. The major immunologic reactivities of adenoviruses are expressed by the hexon and penton proteins (Table 52–2). The **hexons** contain **family-reactive determinants,** which cross-react with a similar Ag in all except the avian adenoviruses. The hexons also possess a **type-specific reactive site,** which is the prevalent Ag exposed when hexons are assembled in virions (identified by neutralization titrations). Complement fixation (CF) and enzyme-linked immunosorbent assay (ELISA) titrations measure the family Ag on free hexons, but they are on the inner surface of the hexon in assembled virions.

The **pentons** provide minor Ags of the virions and a **family-reactive soluble Ag** found in infected cells. The **purified fibers** contain a **major type-specific Ag** as well as a **minor subgroup Ag.** Although the fiber is the organ of attachment to the host cell, Abs to the fiber or to the

intact penton only weakly reduce viral infectivity, probably by aggregating virions.

Neutralizing, hemagglutinin-inhibiting, and CF Abs appear about 7 days after the onset of illness and attain maximal titers after 2 to 3 weeks. Antibodies appear in

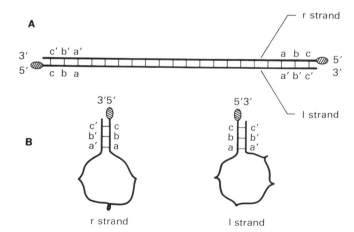

Figure 52–7. Diagram of the adenovirus genome, indicating the terminal inverted repetitions of nucleotide sequences and the protein covalently linked to a deoxycytosine at the 5' terminus of each strand. (A) The intact native linear viral DNA molecule. (B) The configuration assumed by each strand after denaturation owing to hybridization of the complementary 3' and 5' ends of the strands to each other, forming a "panhandle."

TABLE 52–2. *Characteristics of Major Type 5 Adenovirus Proteins*

| Property | Hexon Protein | Penton Proteins | | | Internal Proteins‖ |
		Complete†	Base†	Fiber	
Immunologic reactivity	Type-specific and family cross-reactive	Family cross-reactive	Family cross-reactive	Type specific	?
Biologic activity	None known	Cytopathic; attachment of virions to cells	Cytopathic	Blocks biosynthesis of macromolecules; inhibits viral multiplication	Probably aid in assembly of viral DNA, TP for DNA replication
Hemagglutination	0	Partial	0	Partial	0
Molecular weight of native protein	315,000‡	419,000	236,000§	183,000§	
Polypeptide chains	105,000		80,000	61,000	55,000 (TP) 48,000 (V) 19,000 (VII)

* Proteins of capsid ($\cong$58% of virion) consist of seven species; internal proteins ($\cong$30% of virion) consist of three species.

† A DNA endonuclease is closely associated with but physically separable from the penton base; its function during infection is unknown.

‡ Molecular weights of hexons and fibers vary with type (e.g., type 2 hexon is 360,000 daltons).

§ Estimated from sedimentation coefficient

‖ See Figure 52-6.

nasal secretions at the same time or within a week after the detection of serum Abs. The CF Abs begin to decline 2 to 3 months after infection but are usually still present 6 to 12 months after infection. Neutralizing and hemagglutinin-inhibiting Abs persist longer, decreasing in titer only twofold to threefold in 8 to 10 years. Minor rises in heterotypic neutralizing Abs may follow adenovirus infections, especially when Abs to several types are already present at the time of infection.

The same type of adenovirus rarely produces a second attack of disease. Such persistent type-specific immunity is unusual among viral respiratory diseases, resistance of relatively short duration being the rule (see Immunologic Characteristics in Chap. 56 and Immunologic Characteristics under Parainfluenza Viruses in Chap. 57). This prolonged immunity probably results from the common latent, persistent infections of lymphoid cells.

TOXIN PROPERTIES OF THE PENTON. In addition to their immunologic reactivities, the penton and its individual components possess striking biologic activities. Thus, the intact penton causes rounding and clumping of cultured cells and detaches them from their support. Therefore, the penton is also termed **toxin,** or **cell-detaching factor.** Hydrolysis of the penton's base by trypsin, leaving the fiber intact, destroys the cytopathic effect. It is perhaps this same property of the penton that permits the virion to penetrate into the cellular matrix (see Viral Multiplication). The purified fiber, which is present in infected cells as a soluble protein as well as in pentons, has a different toxic action: in cultured cells it blocks biosynthesis of DNA, RNA, and protein, stops cell division, and inhibits the capacity of cells to support the multiplication of related or unrelated viruses (see Table 52–2).

HOST RANGE. Adenoviruses from humans inoculated intranasally into cotton rats produce pulmonary disease that pathologically resembles that in man. Most adenoviruses of man, however, do not produce recognizable disease in common laboratory animals, but inapparent infections follow intravenous inoculation in rabbits or intranasal instillation in hamsters, piglets, guinea pigs, and dogs. In rabbits, type 5 virus persists for at least 6 months in the spleen, and it emerges when explants are cultured *in vitro* (similar to latent infection of human tonsils and adenoids; see Pathogenesis, below). Chick embryos are susceptible only to chicken adenoviruses. The members of subgroups A and B (see Table 52–1) produce tumors when inoculated in large amounts into newborn hamsters, rats, and mice (see Chap. 64). Rodent cells are nonpermissive for viral replication, but all adenoviruses can transform (immortalize) these cells. Only the E1A and E1B gene products are required to induce complete transformation of rodent cells (see Chap. 64).

A variety of **cultured mammalian cells** support the multiplication of adenoviruses to a high titer and evince characteristic cytopathic changes, including pathognomonic nuclear alterations (see Effect on Host Cells, below). Epithelium-like human cell lines (such as KB cells) and primary cultures of human embryonic kidney are most satisfactory for human adenoviruses; primary cultures of various other types of human and animal cells also support viral multiplication but give much lower yields. Human lymphocytes (T and B cells) also support viral replication; only low yields of virus are produced, and persistent infections can be maintained indefinitely in T cells.

MULTIPLICATION. The essential features of multiplication (Fig. 52–8) are similar for all adenovirus types. **Adsorption** to susceptible cultured cells is relatively slow, reaching a maximum after several hours. The viral particle then promptly penetrates the cell, primarily by a process analogous to phagocytosis; then, through the action of the penton base, when the pH of the phagocytic vacuole is reduced below pH 6 (Fig. 52–9), the endocytic vacuole's membrane is ruptured, permitting the virions to find their way into the cytoplasmic matrix for uncoating.

Uncoating of the viral DNA begins immediately after the virions have penetrated into the cytoplasm. It is detected biochemically when the nucleic acid becomes susceptible to DNase and by electron microscopy when the virion appears spherical rather than polygonal (see Fig. 52–9B). Initially, pentons and the immediate surrounding hexons are displaced, and this reduces the stability of the capsid. The other hexons and associated proteins then separate, and the naked viral core either enters the nucleus through nuclear pores or releases the viral DNA into a nuclear pocket (see Fig. 52–9C). The DNA thus gains access to the nucleus, where viral replication takes place (Fig. 52–10). Viral uncoating requires 1 to 2 hours.

Prior to and independent of viral DNA replication, **immediate early and early messenger RNAs** are transcribed from five separate regions of the genome (Fig. 52–11), corresponding to approximately 14% of the rightward(r)-reading strand and 27% of the leftward(l)-reading strand (designating the direction in which each strand is transcribed); these regions are identified by hybridization with restriction fragments of viral DNA and by electron microscopy of DNA–mRNA hybrids. Transcription from the early regions is not initiated simultaneously, but it is coordinately and sequentially regulated: E1A (see Fig. 52–11) is being transcribed within 1 hour after infection (termed **immediate early genes),** and its 13S mRNA gene product (a 289 amino acid protein) enhances transcription from the other early genes; E1B, E3, and E4 begin transcription about 2 hours after infection; and E2A and E2B are transcribed shortly thereafter. As early as 2 to 3 hours after infection, early messengers are present on polyribosmes and are translated into several early proteins, which are detected by SDS-polyacrylamide gel electrophoresis. Three of these proteins are required for replication of viral DNA: the E2A single-strand–specific DNA-binding protein (mol. wt. 60,000–72,000 daltons, depending on viral type); an E2B

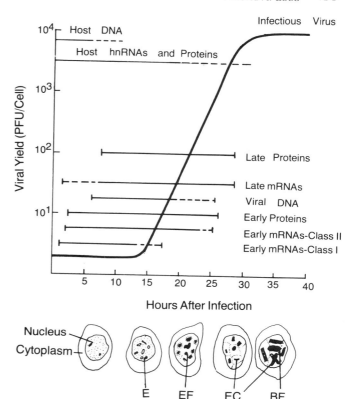

Figure 52–8. Diagram of the sequential events in the biosynthesis of type 5 adenovirus, its effect on synthesis of host macromolecules, and the concomitant development of nuclear alterations. *E,* eosinophilic masses; *EF,* eosinophilic masses with basophilic Feulgen-positive borders; *EC,* eosinophilic crystals; *BF,* basophilic Feulgen-positive masses; *PFU,* plaque-forming units. Types 1, 2, 5, and 6 have similar multiplication and nuclear changes.

DNA polymerase (about 140,000 daltons); and the DNA-terminal protein (55,000 daltons), which is required for initiation of DNA replication and is encoded in region E2B and synthesized as an 88,000-dalton precursor protein.

The activities of several enzymes involved in DNA synthesis may also increase prior to DNA replication, but these enzymes are not unique to virus-infected cells, and it is uncertain whether they are products of the viral or the host cell genome. Inhibition of the synthesis of early mRNA (e.g., by actinomycin C) or of early proteins (by cycloheximide or by amino acid analogues) prevents viral DNA synthesis.

Replication of viral DNA is semiconservative and is initiated at either end of the molecule with **the terminal deoxycytidine,** which is covalently linked to precursor terminal protein, serving as a primer. As elongation occurs, there is an asymmetric displacement of the other strand (see Fig. 52–11). Replication begins in the nucleus 6 to 8 hours after infection, attains its maximum rate by

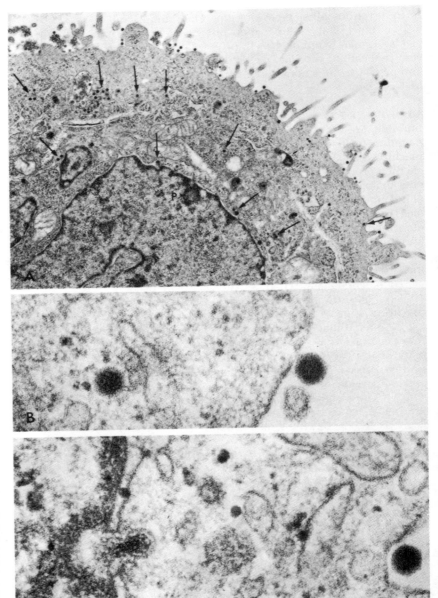

Figure 52–9. Adsorption, penetration, and uncoating of adenovirus in HeLa cells. (*A*) Numerous particles are adsorbed to the cell surface; others are present as free virions in the cytoplasm (*arrows*). Some particles in phagocytic vacuoles are also noted. A nuclear pocket (*P*) is also present. (Original magnification ×15,000) (*B*) Higher magnification of a polygonal virion adsorbed to the plasma membrane and a virion that has assumed a spherical form in the cytoplasm, probably after penetration from a phagocytic vacuole. (Original magnification ×150,000) (*C*) Partially uncoated viral particle releasing core material into a nuclear pocket. For comparison, an unaltered virion is shown on the cell surface prior to engulfment. (Original magnification ×150,000; Morgan C et al: J Virol 4:777, 1969)

18 to 20 hours, and practically ceases by 22 to 24 hours after infection.

Transcription of late mRNAs begins in abundance adequate to produce viral structural proteins shortly after the initiation of DNA replication, although a small number of transcripts encompassing the L1 and L2 regions (see Fig. 52–11) can be detected during the early phase of infection. Hybridization to separate DNA strands shows that late mRNAs are predominantly encoded in the *r* strand. Only one late mRNA appears to arise from the *l* strand (see Fig. 52–11). The mRNAs are generated by **processing of long primary transcripts and by splicing of noncontiguous leader sequences to sequences containing the message for a single protein** (see Chap. 48). Thus, at least 13 late messages are derived from primary transcripts stretching between the **major late promoter** (at 16.4 map units) and the end of the genome (100 map units). Processing probably regulates the relative proportions of late messengers, some of which are present in considerable abundance (e.g., those for the hexon and the 100K proteins). It follows that late proteins, which include the capsid proteins, are primarily products of the *r* strand; their formation depends on prior replication of the viral DNA. Although early mRNAs continue to be transcribed late, only about 30% of them are expressed (e.g., translation of the DNA-binding protein mRNA is meager during the period of late mRNA and protein synthesis).

Viral DNA and mRNAs are synthesized in the nucleus, and virions are assembled in the nucleus, but viral proteins, like host proteins, are synthesized on polyribosomes in the cytoplasm. **Translation of late viral mRNAs** is greatly facilitated by **VA I RNA** which is a small RNA transcribed from the *r* strand by the host RNA polymerase III (see Fig. 52–11). After their release from polyribosomes, the polypeptide chains are immediately transported into the nucleus, where they assemble into the multimeric viral capsid proteins (capsomers). **Assembly** of mature particles takes several steps that begin with formation of the procapsid about 2 to 4 hours after initiation of capsid protein production; this period is probably required to attain component pools of adequate size and terminates with entry of the viral DNA and final processing of the precursor structural proteins (pVI, pVII, and pVIII). The **eclipse period** lasts 13 to 17 hours.

EFFECTS ON HOST CELLS. Adenovirus infection has a profound effect on the physiology of host cells. Production of host DNA stops abruptly 8 to 10 hours after infection, and host biosynthesis of protein and RNA ceases 6 to 10 hours later (see Fig. 52–8). Accordingly, the division of infected cells also halts.

The hallmark of infection with adenoviruses is the development of characteristic **nuclear lesions** (see Fig. 52–8; Fig. 52–12), caused by the accumulation of unassembled viral components. In fact, the process of adenovirus assembly is quite inefficient; **only about 10% to 15% of the new viral DNA and proteins is incorporated into virions.** Some cellular changes may be produced by the E1B 55K protein and the E4 11K protein, which appears to collaborate to inhibit translation of host proteins, as well as the accumulation of fibers, which also blocks the synthesis of host macromolecules and of DNA endonuclease. The basophilic inclusion bodies (see Figs. 52–8 and 52–12) are composed of the excess viral DNA and structural proteins; the large basophilic crystals present in cells infected with type 3, 4, or 7 are made up of viral particles arranged in a crystalline lattice (see Figs. 52–1 and 52–12). In cells infected with subgroup C adenoviruses, prominent bar-shaped eosinophilic crystals are formed, mostly by the arginine-rich internal viral proteins.

Despite the extensive alterations, the infected cells remain intact, and the nuclei do not release the newly synthesized virions. Less than 1% of the total virus is in the culture fluid when the maximal viral titer is attained (as measured after cell disruption). The infected cells also remain metabolically active.

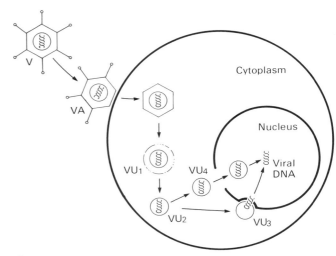

Figure 52–10. Diagram of uncoating of the adenovirus particle. *V*, intact virion; *VA*, attachment of the virion to the cell, followed by penetration of the virion into the cell; *VU_1*, the pentons are detached, leaving the virion somewhat spherical and the viral DNA susceptible to DNase; *VU_2*, the capsid disintegrates, leaving a viral core that migrates to the nucleus; *VU_3*, viral DNA is freed from the core into the nuclear pocket, and free DNA enters the nucleus; or, *VU_4*, the core enters the nucleus through a membrane pore, and DNA is then dissociated from proteins.

Pathogenesis

Progress in the study of pathogenesis has been impeded by the lack of satisfactory animal models. Accordingly, knowledge of human adenovirus infections is derived primarily from clinical observations and from experiments on volunteers. The recognized diseases (Table 52–3) predominantly involve the **respiratory tract,** the **eye,** and the **gastrointestinal tract.** The association of particular types with specific disease syndromes is striking. For example, **types 8, 19, and 37** are essentially the only adenoviruses associated with **epidemic keratoconjunctivitis,** the very **fastidious adenoviruses types 40 and 41** are the etiologic agents of **epidemic infantile gastroenteritis,** and type 11 adenovirus is an etiologic agent of **acute hemorrhage cystitis** in children (see Table 52–3). Adenoviruses usually cause either self-limited illnesses or inapparent infections, which are followed by complete recovery and persistent type-specific immunity.

Type 3 or 7 adenovirus has been isolated from a number of **fatal cases of nonbacterial pneumonia in infants,** and **type 7** from rare cases of **fatal pneumonia in military personnel.** The pulmonary lesions observed

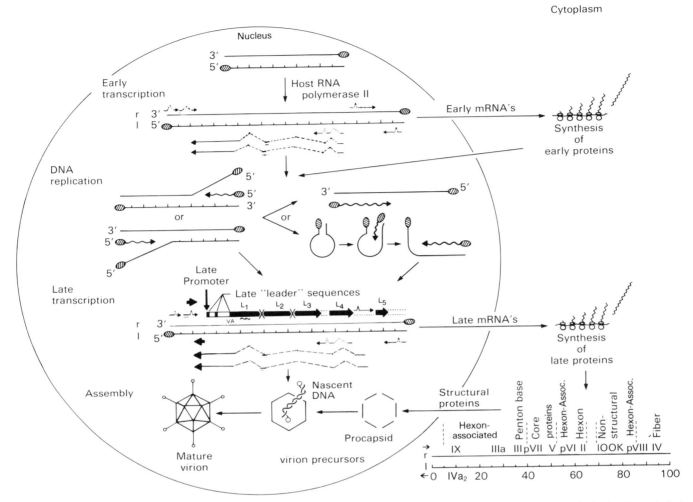

Figure 52–11. Diagram of the biosynthetic events in the multiplication of type 2 adenovirus (used as a model, since its transcription has been studied in greatest detail). Early mRNAs are transcribed from five separate regions of the genome; like late messages, they are processed from larger transcripts, and they have leader sequences transcribed from noncontiguous regions (the intervening absent sequences are indicated by a connecting caret [∧]). The semiconservative, asymmetric DNA replication is shown. A mechanism for replication of the displaced single strands, using the inverted terminal repetition to form a circlelike intermediate, is also suggested. The 80K 5′ precursor terminal protein with a covalently linked deoxycytidine acts as a primer in initiation of DNA replication. Late mRNAs, except that for protein IVa₂, have a single promoter at 16.4 map units on the r (rightward) strand. Note that each late "megatranscript" contains only one set of leader sequences, and therefore only a single mRNA can be derived from each transcript. Also indicated are the regions of the genome in which the late viral proteins are encoded, and their known functions. The left terminus of the viral DNA initiates entry of the viral genome into preformed empty capsids.

are those of nonbacterial bronchopneumonia, but in addition there are numerous bronchiolar epithelial cells containing central basophilic masses in the nuclei, closely resembling the **inclusion bodies** produced by the same adenovirus types in cell cultures (see Fig. 52–12). The acute, mononuclear pulmonary infiltration produced in cotton rats, which is pathologically similar to that in humans, results from early viral gene functions in virus-infected bronchial and bronchiolar epithelial cells calling forth an inflammatory response of monocyte/

macrophages and lymphocytes, including specific cytotoxic T cells.

Virus introduced by feeding, or swallowed in respiratory secretions, multiplies in cells of the gastrointestinal tract and is excreted in the feces (where it is still infectious, owing to its stability), but, except for **types 40 and 41,** it usually does not produce gastrointestinal disease. However, some other adenoviruses, most often type 1, 2, 5, or 6, have occasionally been isolated from cases of acute infectious diarrhea, severe mesenteric adenitis, and intussusception, though an etiologic relationship between virus and illness has not been clearly demonstrated. Viremia is not observed in human infections, and virus is not commonly transmitted to distant organs. However, in cases of immune deficiency, viruses have been isolated from peripheral blood cells, and cases of meningitis, hepatitis, and nephritis have been described.

Most persons are infected with one or more adenoviruses before the age of 15. As a corollary, 50% to 80% of tonsils and adenoids removed surgically yield an adenovirus when explants are cultured *in vitro*. The most frequent types are 1, 2, and 5, the types responsible for most infections in young children. Adenoviruses have also been isolated from cultured fragments of mesenteric lymph nodes, and viral DNA has been detected in rare (10^{-4} to 10^{-5}) peripheral lymphocytes. It thus appears that following an initial infection, the virus frequently becomes latent in the lymphoid tissues, where it may persist for long periods. Although recurrent illness has not been shown to arise from these latent infections under usual circumstances, activation of latent adenoviruses does occur in patients with immunosuppression, as in those with the acquired immune deficiency syndrome (AIDS) (see Chap. 65).

It appears likely that latent persistent infections are

TABLE 52–3. Clinical Syndromes Caused by Specific Types of Adenoviruses

Disease Syndrome	Adenovirus Type	
	Most Common	Less Common
Acute respiratory disease of recruits	4, 7	3, 11, 14, 21
Pharnygoconjunctival fever; pharyngitis	3	5, 7, 21
Conjunctivitis	3, 7	2, 5, 6, 9, 10, 11
Epidemic keratoconjunctivitis	8, 19	37
Nonbacterial pneumonia of infants*	7	
Acute infantile gastroenteritis	40, 41	
Acute hemorrhagic cystitis*	11	

* Least common of the diseases produced

readily established because the infected cells are not lysed, and viral particles or viral genomes remain protected within their nuclei. In nature, the occult virus is confined to relatively few cells. Culturing tissues *in vitro*, however, alters the cellular environment in ways (including dilution of Abs) that permit the virus to multiply

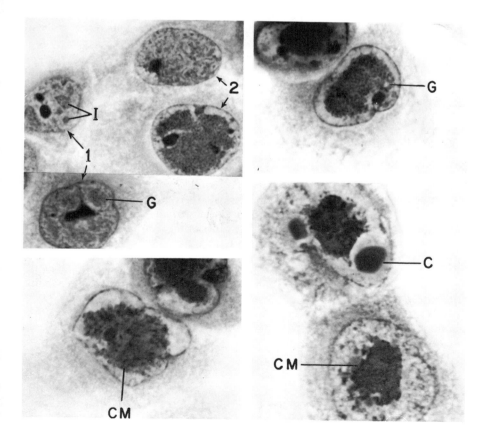

Figure 52–12. Sequential development of nuclear alterations in HeLa cells infected with type 7 adenovirus (types 3 and 4 produce similar effects). The earliest changes may be the formation of small eosinophilic inclusions (Feulgen-negative) (*I, 1*) and clusters of granules (*G*). Clusters of granules gradually become larger (*2*) and more prominent (Feulgen-positive) and form a large central mass (*CM*). In later stages the nucleus is enlarged, and crystalline masses (Feulgen-positive) are apparent (*C*). (Original magnification ×1050; Boyer GS et al: J Exp Med 110:327, 1959)

more rapidly, to spread to uninfected cells, and to produce detectable cytopathic changes.

As noted above, some adenoviruses produce tumors in experimental animals after inoculation of large amounts. Ardent search, however, has not uncovered evidence that these viruses cause cancer in humans.

Laboratory Diagnosis

Adenovirus infection can be diagnosed serologically and by isolation of the offending virus from respiratory and ocular secretions, urine, and feces. For isolation, infected material is inoculated into cultures of continuous lines of human cells (e.g., HeLa or KB) or into primary human embryo kidney cells. If virus is present, cytopathic changes (rounding and clumping of cells) develop after 2 to 14 days, the time depending on the quantity of virus in the infected materials. Types 40 and 41, the fastidious **enteroadenoviruses,** cannot be isolated in the commonly used primary cell cultures or most continuous cell lines owing to their poor replication; therefore, electron microscopic examination of stool extracts has primarily been used for their identification and immunologic typing.

The virus is identified as an adenovirus by indirect immunofluorescence or CF titration with a hyperimmune rabbit serum or a convalescent human serum. This procedure detects the cross-reactive hexon and penton family Ags. The specific type of adenovirus can be ascertained most conveniently through hemagglutination-inhibition titrations; the required number of titrations can be reduced considerably by first determining the hemagglutination subgroup (see Table 52–2). To establish a virus as a new serotype, neutralization titration is the method of choice. A rapid presumptive diagnosis of adenovirus infection, but not of the specific viral type, can be made by examination of cells in respiratory or ocular secretions or urine by means of immunofluorescence or ELISA with rabbit serum containing Abs to the family Ag commonly employed. However, since each adenovirus type has a unique nucleotide sequence, and hence unique distribution of targets for restriction endonucleases, the specific type can be precisely identified by the restriction fragment patterns of its DNA.

The **serologic diagnosis** of an adenovirus infection is accomplished most conveniently by **CF titration,** which detects cross-reactive family Ags. Unfortunately, this assay identifies fewer than 50% of new infections, because many people have a constant high level of Abs from prior infections. A more precise diagnosis requires neutralization titrations with acute- and convalescent-phase sera. **Hemagglutination-inhibition** assay* for Abs is practi-

cally as sensitive as neutralization, is simpler and less expensive, and is almost as accurate if nonspecific inhibitors are removed from the serum. However, owing to the type-specificity of the reaction, all the common adenovirus types must be used in the test if an adenovirus has not been isolated from the patient.

Epidemiology

Man provides the only known reservoir for strains of adenoviruses that infect humans. Person-to-person spread in respiratory and ocular secretions is the most common mode of viral transmission, though dissemination in swimming pools has also been implicated in epidemics of **pharyngoconjunctival fever** and **conjunctivitis.** The spread of **epidemic keratoconjunctivitis** caused by types 8, 19, and 37 adenoviruses appears to be associated with conjunctival trauma produced by dust and dirt in shipyards and factories, or with improperly sterilized optical instruments. Adenoviruses are commonly present in the feces of infected persons, even those producing respiratory and ocular infections, but only types 40 and 41, the agents of infantile gastroenteritis, appear to be transmitted by the fecal–oral route.

Despite the large number and the worldwide distribution of adenoviruses, their clinical importance is largely restricted to epidemics of **acute respiratory disease (ARD),** an influenzalike illness in military recruits, and to limited outbreaks among children (except for keratoconjunctivitis caused by types 8, 19, and 37). Infections are observed throughout the year, but the greatest incidence and largest epidemics occur in late fall and winter. Types 7, 4, and 3 (in order of decreasing importance) are the viruses most frequently responsible for epidemics of acute respiratory and ocular diseases (see Table 52–3); types 11, 14, 21, 40, and 41 have been increasingly implicated in epidemics. Peculiarly, type 4 adenovirus commonly causes ARD in military recruits but rarely produces infections in civilians. This epidemiologic behavior of type 4 is without parallel; its explanation is unknown.

A relatively high proportion of adults have Abs to one or more types of adenoviruses (particularly types 1–3, 5, and 7), indicating previous infections. However, epidemiologic studies indicate that adenoviruses annually cause at most 4% to 5% of viral respiratory illnesses in civilians.

Prevention and Control

Isolation of sick persons has little or no effect on the spread of adenoviruses, since many healthy persons are carriers. Immunization, however, offers an effective preventive measure, as would be expected from the lasting type-specific immunity produced by natural infections. A highly effective live virus vaccine is used primarily to

---

* Types 10 and 19 cross-react to such an extent that they cannot be distinguished by this procedure.

protect military recruits; a formalinized virus vaccine was experimentally successful in recruits, but its irregular antigenicity caused its abandonment. Successful immunization suppressed ARD caused by type 4 in recruits in the United States, but type 7 replaced it. Following continued immunization with types 4 and 7, other types have appeared (e.g., type 21, which had previously been responsible for epidemics of ARD in European defense forces).

Vaccines for military use should contain types 3, 4, 7, and 21 depending on the types prevalent. In closed populations, such as those in chronic-disease hospitals or homes for orphans, a vaccine containing types 1–7, 40, and 41 may be useful for infants and young children.

Parvoviruses

Small icosahedral viruses, 18 nm to 28 nm in diameter, were first discovered in association with adenoviruses and found to replicate in association with adenoviruses. These **adeno-associated viruses (AAVs)** led to the recognition of several other viruses of lower animals (latent rat viruses, minute mouse virus, and porcine virus) and another virus infecting humans (**human parvovirus**) that have physical and chemical properties similar to AAV. These viruses are grouped in the family *Parvoviridae* (its vernacular term is **parvovirus** (L. *parvus*, small)) and further divided into the subgroups (genera) composed of defective viruses, such as AAV (officially termed *Dependovirus*) and autonomous viruses (*Parvovirus*); a third genus of parvoviruses (*Densovirus*) infects only arthropods. **Defective parvovirus** depends totally on the multiplication of the unrelated adenoviruses or herpes simplex viruses (types 1 and 2); hence, it has been given the genus name *Dependovirus*. The viral genome consists of a linear molecule of single-stranded DNA with a molecular weight of only 1.5×10^6. The virions, however, can contain either a positive or a negative DNA strand that, when extracted from the viral particles, forms double-stranded molecules unless a reagent that blocks hydrogen bond formation (e.g., formalin) is present. The quantity of genetic information contained in this small DNA molecule is very limited and is utilized largely for specifying the three viral capsid proteins. The precise functions that the helper supplies are unknown, but studies with adenovirus mutants and microinjection of adenovirus early mRNAs indicate that AAV utilizes early adenovirus proteins for each step in AAV replication. Thus, E1A provides a function to initiate AAV transcription, E4 is utilized for DNA replication, and adenovirus VA I is utilized for AAV translation. It is striking that the multiplication of adenovirus is itself inhibited when it offers assistance to the defective AAV (and AAV also reduces adenovirus oncogenicity in hamsters). Although

production of infectious AAV in cells coinfected with an adenovirus resembles complementation by two genetically related defective mutants (see Chap. 48), cross-hybridization of DNAs extracted from AAVs and several types of adenoviruses failed to detect homologous regions.

Five distinct immunologic types of AAVs have been identified as contaminants of human and simian adenoviruses, and AAV Abs are frequently found in humans and monkeys. About 70% to 80% of humans acquire Abs to AAV types 1, 2, and 3 within the first decade of life, and approximately 85% of adults maintain detectable Abs. AAVs types 1–3 can occasionally be isolated from fecal, ocular, and respiratory specimens during acute adenovirus infections, but not during other illnesses. Type 5 AAV, however, was isolated from a penile, flat, condylomatous lesion, and its seroepidemiology differs from that of the other types: the highest Ab titers are found in those 15 to 20 years old, and only about 60% of adults possess Abs.

In cell cultures, AAV DNA integrates into the host DNA and replicates with it, only to be excised and induced to replicate when the latently infected cells are coinfected with an adenovirus. Thus, AAVs may have a similar virus–host cell relationship in man as in cell cultures. They have been suspected of playing a role in the pathogenesis of "slow" viral infections (see Chap. 51), but so far this role is unconfirmed, and they appear to be only silent, unobtrusive partners of adenovirus infection.

AUTONOMOUS PARVOVIRUSES

One virus isolated from human sera during screening for hepatitis B virus Ag (see Chap. 63) has been shown to be a human pathogen, termed **human parvovirus,** and a number of species from lower animals (latent rat viruses, feline panleukemia virus, minute mouse virus, porcine virus, and many others) have physical and chemical characteristics similar to AAV except that they replicate without assistance of a helper virus. It is striking that although they are not defective, the **autonomous parvoviruses require actively dividing cells for productive infection.**

Since the initial detections of human parvovirus (termed B19) were not associated with specific illnesses, intense interest was not aroused until this virus was identified with aplastic crisis in two children with sickle cell disease. The association of human parvovirus with such aplastic crises has been strengthened by isolation from other cases, and viral isolations have been extended to adults with polyarthralgia syndrome and with aplastic crisis in sickle cell disease, as well as with several other hereditary hemolytic anemias such as hereditary spherocytosis, β-thalassemia, and pyruvate kinase activity. The pathogenic effect of human parvovirus in humans broadened when viral isolation, as well as clinical, epidemiologic, and immunologic evidence, indicated that it was the long-searched-for etiologic agent of erythema infectiosum ("fifth" disease), an acute, febrile, self-limited childhood disease.* Fifth disease and transient aplastic crisis share similar epidemiologic features in that they both spread within families and to close contacts, and they appear to confer lasting immunity. Children 5 to 10 years of age are at greatest risk for both infections.

Human volunteer experiments in adults free of Abs have further confirmed the etiologic role of human parvovirus as well as its pathogenic potentials. One week after inoculation, the volunteers developed a viremia accompanied by mild illness consisting of fever, malaise, myalgia, pruritus, and excretion of virus in respiratory secretions. A week later, reticulocytopenia, reduced platelets, neutropenia and lymphopenia, and a slight drop in hemoglobin developed. Seventeen to 18 days after inoculation, rash and arthralgia—typical adult erythema infectiosum—appeared.

The human parvovirus has all the chemical and physical characteristics of well-characterized parvoviruses. Its DNA does not have homology with AAV, but it does hybridize with the DNAs of other autonomous parvoviruses, implying that human parvovirus is a member of the genus *Parvovirus*. Productive replication of human parvovirus has been demonstrated in nuclei of human progenitor erythroid cells in bone marrow cultures supplemented with erythropoietin. Replication of viral DNA is similar to that of other parvoviruses, and its multiplication profoundly inhibits cell growth. These data indicate that the human parvovirus (B19) replicates autonomously and requires actively dividing cells, like other parvoviruses of this genus. It is not certain, however, that

an appropriate helper virus might not enhance its replication.

Selected Reading

BOOKS AND REVIEW ARTICLES

Berns KI (ed): The Parvoviruses. New York, Plenum Press, 1984

Challberg MD, Kelley TJ: Eukaryotic DNA replication: Viral and plasmid model systems. Annu Rev Biochem 51:901, 1982

Ginsberg HS (ed): The Adenoviruses. New York, Plenum, 1984

Moran E, Mathews MB: Multiple functional domains in the adenovirus E1A gene (minireview). Cell 48:117, 1987

Young N, Mortimer P: Viruses and bone marrow failure. Blood 63:729, 1984

SPECIFIC ARTICLES

Anderson MJ, Higgins PG, Davis LR et al: Experimental parvoviral infection in humans. J Infect Dis 152:257, 1985

Berget SM, Moore C, Sharp PA: Spliced segments at the 5′ terminus of adenovirus 2 late mRNA. Proc Natl Acad Sci USA 74:171, 1977

Berk AJ, Sharp PA: Structure of the adenovirus 2 early mRNAs. Cell 14:695, 1978

Cassant YE, Cant B, Field AM, Widdows D: Parvovirus-like particles in human sera. Lancet 1:72, 1975

Catmore SF, Tattersall P: Characterization and molecular cloning of a human parvovirus genome. Science 226:1161, 1984

Chow LT, Gelinas RE, Broker TR, Roberts RJ: An amazing sequence arrangement at the 5′ ends of adenovirus 2 messenger RNA. Cell 12:1, 1977

Evans R, Fraser N, Ziff E et al: The initiation sites of RNA-transcription of AD2 DNA. Cell 12:733, 1977

Hilleman MR, Werner JR: Recovery of a new agent from patients with acute respiratory illness. Proc Soc Exp Biol Med 85:183, 1954

Horne RW, Brenner S, Waterson AP, Wildy P: The icosahedral form of an adenovirus. J Mol Biol I:84, 1959

Konarska MM, Grabowski PJ, Padgett RA, Sharp PA: Characterization of a branch site in lariat RNAs produced by splicing of mRNA precursors. Nature 313:552, 1985

Rowe WP, Huebner RJ, Gilmore LK et al: Isolation of a cytopathogenic agent from human adenoids undergoing spontaneous degeneration in tissue culture. Proc Soc Exp Biol Med 84:570, 1953

Trentin JJ, Yabe Y, Taylor G: The quest for human cancer viruses. Science 137:835, 1962

---

* Termed *fifth disease* because it was the fifth childhood rash disease recognized after scarlet fever, rubeola (measles), rubella (German measles), and epidemic pseudoscarlatina

53

Harold S. Ginsberg

Herpesviruses

General Characteristics

The herpesviruses, a family (Herpesviridae*) of structurally similar viruses, are named (Gr. *herpein*, to creep) for those members responsible for two common diseases of man: **herpes simplex** (fever blister) and **herpes zoster** (chickenpox [varicella] and shingles) (Gr. *zoster*, girdle). The skin lesions of the herpetic diseases illustrate the affinity of most herpesviruses for cells of ectodermal origin. Like the poxviruses, another group of DNA viruses, herpesviruses exhibit focal cytopathogenicity, producing vesicles or pocks in patients and in egg membranes. Another prominent characteristic is the production of **latent** and **recurrent** infections; although the mechanism is not clear, it is noteworthy that herpesviruses, like adenoviruses (which also initiate latent infections), are DNA viruses and replicate in the cell nucleus.

The unusual epidemiologic features of the common herpes infections long puzzled physicians until techniques became available for identifying and characterizing the responsible viruses. Thus, the multiple recurrence of fever blisters in certain persons was bewildering until about 1950, when Burnet in Australia and Buddingh in the United States showed that herpes simplex virus often becomes latent after initiating a primary infection, usually in children, and is then repeatedly activated by subsequent provocations (see Latent Infections, Chap. 51). A similar mechanism was suspected in **chickenpox** (varicella); i.e., that it **recurs as herpes zoster** (shingles), since an outbreak of chickenpox was often observed to follow the sporadic appearance of zoster in an adult. However, because the initial and subsequent syndromes are so different, the relation was not certain until Weller,

---

* There are three subfamilies: Alphaherpesvirinae (includes the herpes simplex types 1 and 2 viruses and the varicella zoster virus), Betaherpesvirinae (cytomegalovirus), and Gammaherpesvirinae (Epstein-Barr virus).

in 1954, isolated in tissue cultures the same virus from patients with the two diseases. The virus closely resembles that of herpes simplex.

Besides the herpes simplex and varicella zoster viruses, the major herpesviruses that infect man are cytomegalovirus (inclusion or salivary gland virus of man), and EB (Epstein-Barr) virus. These viruses are widely separated in evolution despite their structural similarities (see Characteristics of Members of the Herpesvirus Family). They differ strikingly in their DNA composition, although some similarities exist in the manner in which the general structures of the viral genomes have long and short unique sequences separated by repeated sequences (Fig. 53–1). Herpes simplex viruses are immunologically related to each other but show no relatedness to other family members; and the other three human herpesviruses antigenically cross-react only slightly.

Herpesviruses have also been found in every other eukaryotic species examined, from fungi to monkeys. Ex-

CHARACTERISTICS OF MEMBERS OF THE HERPESVIRUS FAMILY

Size: 180–200 nm
Symmetry of capsid: Icosahedral
Capsomers: 162; 9.5 × 12.5 nm
Lipid envelope: Present
Sensitivity to ether and chloroform: Inactivates infectivity
Nucleic acid: Double-stranded DNA*
Site of biosynthesis of viral DNA: Nucleus
Site of assembly of viral particles: Nucleus
Inclusion bodies: Intranuclear, eosinophilic†
Common family antigen: None

* Molecular weights (daltons): herpes simplex viruses approximately 96×10^6, varicella zoster virus $80-86 \times 10^6$, cytomegalovirus 145×10^6 (231 kilobase pairs), EB virus 114×10^6. Guanine-cytosine content (mol/dl): cytomegalovirus 58.8%, herpes simplex type 1 67%, herpes simplex type 2 69%, varicella zoster 46%, and EB virus 59%. DNA–DNA hybridizations indicate that 40% to 46% of the base sequences are homologous in types 1 and 2 herpes simplex viruses.

† Cytomegalovirus-infected cells may also contain basophilic cytoplasmic inclusion bodies.

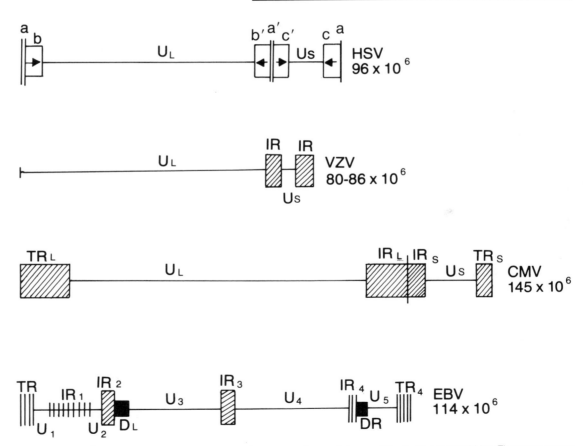

Figure 53–1. Diagrams of the physical structures of genomes of several viruses of the family Herpesviridae. The genomes are composed of unique (*U*) sequences with short (*S*) or a long (*L*) region, numbered where there are several (*EBV*). Terminal repeats (*TR*) bound the genome, and inverted repeats (*IR*) bound the unique regions. The rectangles contain reiterated sequences of more than 1000 base pairs. (Modified from maps presented by B. Roizman, L. D. Gelb, R. LaFemina, and G. Hayward, and Wathen and Stinski)

amples include **B virus of monkeys** (which may infect man accidentally), **pseudorabies virus** of pigs, **virus III** of rabbits, **cytomegaloviruses of animals** (inclusion virus of guinea pigs, inclusion virus of mice, and the agent of inclusion body rhinitis in pigs), oncogenic viruses (see Chap. 64) that produce **lymphoproliferative malignancies** in chickens (**Marek's disease virus**) and monkeys (**herpesvirus saimirii**), and several viruses that may be associated with renal carcinoma in frogs. These animal viruses appear to have only minor immunologic relatedness to human herpesviruses except for the marked cross-reactivity between herpes simplex virus and B virus. In this chapter only the viruses that infect man will be discussed.

Herpesvirus particles have a diameter from 180 nm (herpes simplex and varicella zoster viruses) to 200 nm (cytomegalovirus). Within a population of virions, many particles do not possess envelopes, and some are empty capsids (Fig. 53–2*B* and *D*). The virion components (Figs. 53–2 and 53–3) are arranged in (1) a DNA-containing **toroidal core** about 75 nm in diameter, (2) an **icosahedral**

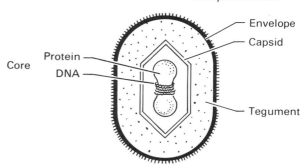

Figure 53–3. Diagram of a virion of herpes simplex virus. The DNA–protein core consists of DNA wrapped around an associated protein as on the spindle of a spool. The envelope contains a number of glycoproteins, the capsid is composed of at least four unique proteins, and the so-called tegument consists of about eight distinct polypeptides.

capsid 95 nm to 105 nm in diameter, (3) a surrounding **granular zone (tegument)** composed of globular proteins, and (4) an encompassing **envelope** possessing periodic short projections. The capsid is composed of 162 elongated hexagonal prisms (**capsomers**), each with a small, central hole.

The chemical composition of purified enveloped viral particles is consonant with their morphology. Herpes simplex virions contain 25 to 30 virus-specified proteins (70% of the virion); a large, linear, double-stranded DNA molecule (7% of the virion); envelope lipid (22% of the virion), which is chiefly host-specific phospholipid derived from the nuclear membrane; and small amounts of polyamines (spermine within the nucleocapsid and spermidine within the envelope). Other herpesviruses examined are similar.

Like other enveloped viruses, herpesviruses are relatively unstable at room temperature and are readily inactivated by lipid solvents.

Herpes Simplex Viruses

PROPERTIES

IMMUNOLOGIC CHARACTERISTICS. Types 1 and 2 herpes simplex viruses are the main immunologic variants that can be distinguished by neutralization titrations, although they cross-react strongly, and by their distinct clinical patterns. Some additional minor antigenic variations have been observed but do not warrant classification. Infected cells contain, along with virions, soluble Ags that elicit a delayed-type skin reaction in addition to the usual immunologic reactions. Infected cells are also antigenically altered by the insertion of four to five viral structural glycoproteins into their plasma membrane; the glycoprotein D (gp D) is particularly prominent in eliciting neutralizing Abs. These Ags make the infected

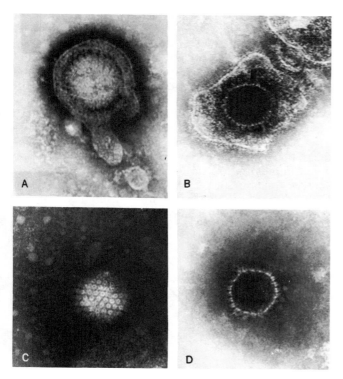

Figure 53–2. The four morphologic types of herpes simplex virus particles embedded in phosphotungstate. (*A*) Enveloped full particle showing the thick envelope surrounding the nucleocapsid. (*B*) Enveloped empty particle; the capsid does not contain viral DNA and therefore can be penetrated by the phosphotungstate. (*C*) Naked full particle; the structure of the capsomers is plainly visible. (*D*) Naked empty particle. (Original magnification ×200,000; Watson DH et al: Virology 19:250, 1963)

cells susceptible to damage and lysis by reaction with virus-specific Abs plus complement, or with specifically activated T-lymphocytes.

Specific Abs can be assayed by neutralization, enzyme-linked immunosorbent assay (ELISA) or complement fixation (CF) tests; they reach maximum titers about 14 days after infection. During the early stages after a primary infection, neutralizing Abs can be detected only in the presence of complement. Thereafter, complement is not required for neutralization, but its presence raises Ab titers fourfold to eightfold. Neutralizing Abs are primarily directed to the envelope glycoproteins; ELISA and CF Abs react with all the virion proteins.

Abs may drop to undetectable levels after the first infection, only to reappear with recurrent episodes. By adulthood, titers are generally high and persist indefinitely. Accordingly, an increase usually cannot be detected in recurrent adult disease, although infectious virus can be isolated readily from the lesion. Fetuses acquire maternal Abs via placental transfer, and they persist until about 4 months after birth; this persistence probably explains the common occurrence of primary infection in babies from 6 to 18 months of age.

Cell-mediated immunity also develops after primary infection and may be a major immunologic factor in maintaining a latent state. It appears that impairment of cellular immunity, as measured by diminished production of macrophage migration-inhibitory factor and diminished activity of sensitized T-lymphocytes, is correlated with episodic recurrences. Moreover, in immunosuppressed patients the virus is commonly activated and disseminated, leading to acute disease.

HOST RANGE. Humans are the natural host for herpes simplex viruses, but a relatively wide range of animals are also susceptible, including mice, guinea pigs, hamsters, and rabbits. The effects of infection depend on the route of inoculation. For example, inoculation of the cornea in the rabbit results in keratoconjunctivitis or keratitis, whereas intracerebral inoculation produces fatal encephalitis. The chick embryo has been a convenient host: the production of pocks on the chorioallantoic membrane affords a reproducible method for detection and assay, similar to that employed with poxviruses.

Many **cultured cell types** support multiplication of herpes simplex virus, undergo extensive cytopathic changes, and develop intranuclear inclusion bodies; chromosomal breaks and aberrations are also observed. The response of the cells varies with the strain of virus employed: some strains cause marked clumping of cells, and some produce typical plaques with suitable cells.

MULTIPLICATION. The **glycoproteins of the viral envelope** provide the normal **attachment** of the virions to susceptible cells. Following attachment, the viral envelope **glycoprotein B (gp B)** induces its **fusion** with the

cellular plasma membrane, permitting the nucleocapsid to enter directly into the cytoplasm (see Chap. 48). Intact virions may also enter via endocytosis, from which they are released into the cytoplasm by similar viral envelope–membrane fusion. In the cytoplasm the capsid migrates to a nuclear pore, where the viral DNA is released into the nucleus and initiates viral multiplication. The **eclipse period** is 5 to 6 hours in monolayer cell cultures, and virus increases exponentially until approximately 17 hours after infection (Fig. 53–4); each cell has then made 10^4 to 10^5 physical particles, of which about 100 are infectious. Virions are **released** by slow leakage from infected cells.

As with other DNA-containing viruses (see Chap. 48), the **biochemical events** are **sequentially regulated,** presenting a cascadelike effect. Thus, after the viral DNA enters the nucleoplasm, even in the absence of protein synthesis, the host cell RNA polymerase II transcribes noncontiguous, restricted regions of the viral genome to produce five **immediate early (α) mRNAs** (Fig. 53–5). If translation is blocked, only these α mRNAs accumulate in the cytoplasm, and the larger, unprocessed transcripts remain in the nucleus. If protein synthesis is permitted, α **proteins** are made, leading to transcription of other regions of the genome and production of **delayed early (β) mRNAs.** The β **proteins** block further synthesis of α proteins and lead to transcription of a third set of RNAs and their processing into **late γ mRNAs.** Thus, the synthesis and translation of the mRNAs are **coordinately regulated:** formation of the α proteins is necessary for synthesis of the β proteins, and both of these nonstructural and minor structural proteins are necessary for synthesis of the **late major structural γ proteins.** It is noteworthy that synthesis of all the γ proteins is not dependent on viral DNA replication: $\gamma 1$ proteins such as gp B and the major capsid protein VP5 are made in the absence of viral DNA synthesis, although they are synthesized in relatively low abundance; but $\gamma 2$ proteins

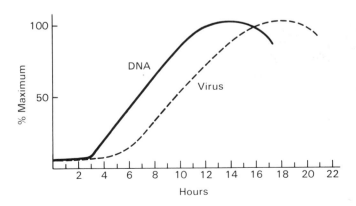

Figure 53–4. Temporal relationship between viral DNA replication and the production of infectious virus.

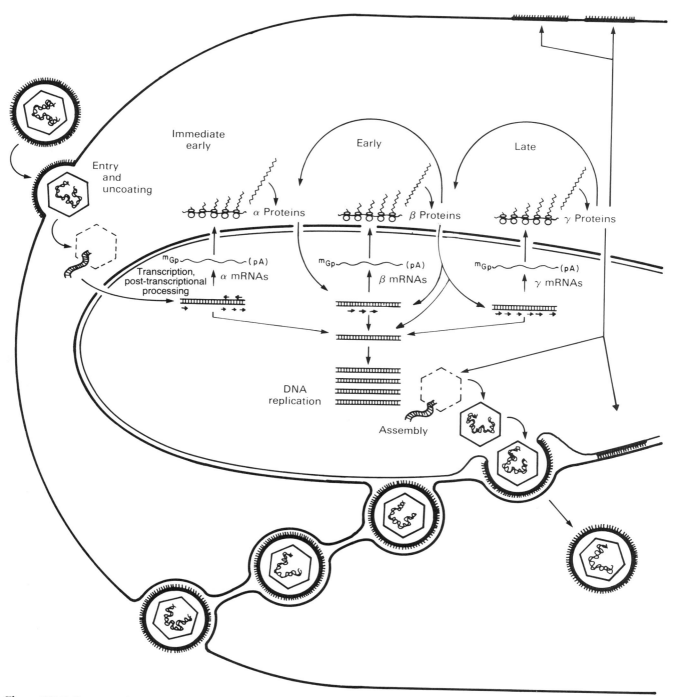

Figure 53—5. Sequence of events in the multiplication of herpes simplex virus from entry of the virus into the cell by fusion of the virion envelope with the membrane of the endocytic vacuole to assembly of virions and their exit from the cell through the endoplasmic reticulum. Also illustrated are transcription and coordinated sequential processing of mRNAs and synthesis of sets of proteins (α, β, γ) required for DNA replication and virion structures. (Modified from a diagram kindly supplied by B. Roizman, University of Chicago)

(e.g., glycoprotein C) strictly require amplification of viral DNA. The three sets of mRNAs produce an aggregate of about 50 virus-encoded proteins.

Viral DNA replication is carried out by both viral α and β proteins, which include a DNA polymerase and DNA-binding protein, and host cellular enzymes. The reactions are not yet precisely understood, but they appear to involve complex replicative intermediates, including "head-to-tail" concatemeric circular and linear–circular forms (see Chap. 48) generated by the reiterated nucleotide sequences (see Fig. 53–1). Three origins of DNA replication have been described: two in the terminal *c* reiterated sequences of the S segment (see Fig. 53–1) and one in the middle of the L segment near the genes encoding the DNA polymerase and DNA-binding protein. The concatemeric viral DNA is cleaved at a terminal reiterated *a* sequence, and it is packaged in preformed capsids (see Figs. 53–1 and 53–6). These particles are noninfectious and unstable until they acquire an envelope. Envelopment is initiated at sites on the inner nuclear membrane into which viral glycoproteins have been inserted. The nuclear membrane reduplicates (Fig. 53–7) to permit egress of viral particles into the cytoplasm, where it associates with the endoplasmic reticulum, and the viral particle appears either to complete envelopment or to become enveloped anew. Mature, infectious virus is slowly liberated from infected cells through the endoplasmic reticulum, but occasionally it also escapes by a process akin to reverse phagocytosis. Unlike the process with other enveloped viruses, envelopment and release of viral particles does not occur by budding from the plasma membrane. Rather, the plasma membrane is changed morphologically and contains virus-specific glycoproteins; this consequently makes the membrane a target for immunologic attack.

Only about 25% of the viral DNA and protein made in cell cultures is assembled into virions. The production of host DNA, RNA, and protein declines, concomitant with viral replication, and finally ceases within 3 to 5 hours after infection. Ultimately the productively infected cell dies.

The biosynthetic steps can be correlated with the development of nuclear inclusion bodies and the formation of viral particles. A basophilic, Feulgen-positive, granular mass of newly synthesized viral DNA develops centrally in the nucleus. The assembly of incomplete viral particles (see Fig. 53–6) begins within this material.

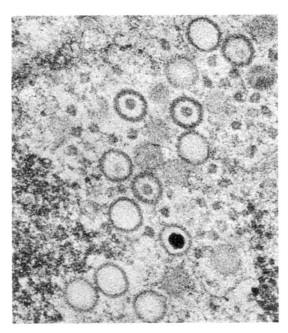

Figure 53–6. Formation of herpes simplex virus particles within the cell nucleus. Some capsids are in the process of assembling, while others are complete. Cores of varying density are forming within the capsids. Indistinct particles probably represent virus sectioned at one margin, with loss of density and overlapping structure. (Original magnification ×90,000; Nii S et al: J Virol 2:517, 1968)

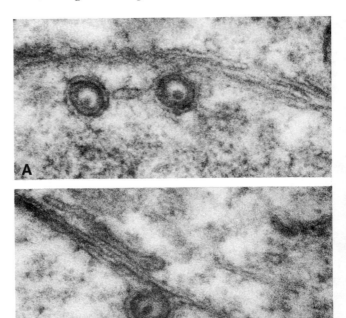

Figure 53–7. (*A* and *B*) Thin sections of mature particles of herpes simplex virus. The particles show the inner core and two or three membranes. In *B* the nuclear membrane partly surrounds a viral particle (apparently a step in the simultaneous assembly and egress of complete virions from the nucleus). (Original magnification ×87,000; Morgan C et al: J Exp Med 110:643, 1959)

The movement of viral particles from the nucleus into the cytoplasm is accompanied by the transport of soluble Ags into the cytoplasm; concomitantly the originally basophilic intranuclear inclusion body is converted into an eosinophilic, Feulgen-negative mass. Thus, the eosinophilic inclusion body that is usually observed in infected cells does not contain viral particles or specific viral Ags (detectable by immunofluorescence) but actually is the burnt-out remnant of a viral factory.

PATHOGENESIS

The most striking characteristic of herpes simplex virus infection is its propensity for persisting in a **quiescent** or **latent state** in humans, with recurrence of activity at irregular intervals. The initial infection occurs through a break in the mucous membranes (e.g., eye, mouth, throat, genitals) or skin, where local multiplication ensues. From this focus virus spreads to regional lymph nodes, where it multiplies further. On occasion virus disseminates into the blood and to distant organs. Viremia can occasionally be detected by isolation of virus from the cells of the blood buffy coat.

The initial infection ordinarily occurs in children 6 to 18 months of age; serologic surveys have demonstrated that it is most often inapparent. But 10% to 15% of those infected do develop a **primary disease,** usually **herpetic gingivostomatitis,** which is characterized by multiple vesicles in the oral mucous membranes and the mucocutaneous border. Similar lesions are less commonly seen in other regions, including the nostrils, the esophagus, the external genitalia or urethra, the cornea, and sites of trauma. More serious but rare complications include neonatal generalized infections, meningoencephalitis,* diffuse skin involvement in children with chronic eczema (Kaposi's varicelliform eruption, eczema herpeticum), and hepatitis.

During the initial infection, virus is taken up by adjacent nerve endings, and it migrates rapidly in axons of nerves to the ganglia. When the initial infection recedes, the virus persists, despite the presence of a high Ab titer, producing a latent infection in a sensory ganglion adjacent to the major site of the primary disease. The genome of the occult virus appears not to be integrated and probably persists in an episomal state; moreover, only α mRNAs are found. The balance may be readily upset, including viral replication in the nerve ganglion, thus provoking the second, **recurrent form** of herpes simplex disease. Because neutralizing Abs and cellular immunity are present, the new virus cannot disseminate, but it does migrate centrifugally from the ganglion along nerve axons to reach contiguous cells; recurrent herpes

---

* Encephalitis occurs more frequently in adults, either as a primary infection or as a flare-up of a latent infection.

simplex therefore usually remains localized. In a given person the clinical features are much the same with each episode. For example, if **gingivostomatitis** was the **primary disease,** the **recurrent form** is usually **herpes labialis** (fever blisters) in the same region of the lip; and if the primary disease was **herpetic keratitis,** the recurrent disease is also keratitis (which may lead to corneal scarring and is one of the major infectious causes of blindness).

Types 1 and 2 herpes simplex viruses differ significantly in their pathogenic potential. **Type 1 virus** is primarily associated with **oral** and **ocular lesions** and is transmitted in oral and respiratory secretions, whereas **type 2** is isolated primarily from **genital** and **anal lesions** and is passed through sexual contact. Changes in sexual mores, however, have somewhat altered this common pattern: occasionally, type 2 virus is isolated from oral lesions and type 1 virus from genital lesions. Mothers with **genital herpes** (a painful, persistent, recurrent infection) are the primary source of **neonatal infections** with type 2 virus, which are often severe and even fatal. It is also striking that in the United States more than 80% of women with cervical carcinoma have Abs to type 2 virus (see Chap. 64). However, a prospective study showed that this is not the result of a causal relation but probably only a reflection of lower socioeconomic status and, possibly, of a more promiscuous sexual activity of many who become victims of cervical carcinoma.

LATENCY. In experimental models in mice, labial strains of type 1 are characteristically found latent in neurons of cervical ganglia, and type 2 genital strains are found in sacral ganglia. Infectious virus is not detectable as such in ganglia, but it emerges when ganglion fragments are cultured *in vitro.* In humans the virus has similarly been detected in cultures of trigeminal ganglia (type 1) or sacral ganglia (type 2). Moreover, neurectomy of the facial branch of the trigeminal nerve (as therapy for persistent trigeminal neuralgia) characteristically activates latent herpes simplex virus, producing fever blisters. In experimental latent infections of mice and rabbits, trigeminal or sciatic nerve section likewise activates quiescent virus.

Viral multiplication and recurrent disease may also be induced in humans by naturally occurring factors such as heat, cold, or sunlight (ultraviolet light), and by immunologically unrelated hypersensitivity reactions, pituitary and adrenal hormones, and emotional disturbances. A few of these, such as epinephrine and hypersensitivity reactions, have evoked recurrences in experimental animals (see Latent Infections, Chap. 51).

LABORATORY DIAGNOSIS

The increasing occurrence of life-threatening herpesvirus infections in children with inherited T-cell deficiencies and in patients who are immunosuppressed because of infection (e.g., with acquired immune deficiency syndrome [AIDS] virus), transplantation, or cancer therapy has given great importance to rapid, accurate, and economic laboratory procedures for an etiologic diagnosis. The development of effective chemotherapeutic agents is making rapid diagnosis even more imperative. The following methods are available: (1) The quickest and most economical diagnostic test consists of demonstrating characteristic multinuclear giant cells containing intranuclear eosinophilic **inclusion bodies** in scrapings from the base of vesicles. This procedure cannot, however, distinguish a herpes simplex infection from one produced by varicella zoster virus or cytomegalovirus. (2) **Electron microscopic examination** of biopsied tissues reveals cells that contain viral particles in various stages of maturation. (3) **Specific Ags** may be detected in cells from the lesions by ELISA or by immunofluorescent techniques. (4) **Virus** can be **isolated** by inoculating material from lesions (especially vesicular fluid or scrapings) into susceptible tissue culture cells or newborn mice. (5) **A rise in Ab titer** is detected by neutralization, CF, indirect hemagglutination, or ELISA through the use of serum obtained early in the primary disease and again 14 to 21 days after onset (paired sera). Following primary infections there is often also a small rise in neutralizing Abs for varicella zoster virus. Patients with recurrent disease have a high initial titer of circulating Abs and often do not show a significant increase.

EPIDEMIOLOGY, CONTROL, AND TREATMENT

Close person-to-person contact is the most common mechanism of viral transmission. Man is the only known natural host and source of virus. Secretions from lesions about the mouth (herpes labialis and stomatitis) and genitalia are the most frequent sources. Virus is also transmissible on eating and drinking utensils and other fomites for a brief period after contamination. Occasionally virus is found in the saliva of healthy persons, particularly children after the primary infection; the shedding may even continue for weeks or months.

Rare cases of primary infection have been reported in adults. Approximately 80% of adults have relatively high titers of neutralizing and CF Abs, as well as cellular immunity, and a substantial fraction of this population is subject to recurrent herpes.

Control is not feasible because of the large numbers of persons with inapparent infections and minor recurrent lesions from which virus is shed. However, it is important, whenever possible, to prevent contact between infants and persons who have herpetic lesions. Vaccines to prevent genital herpes are being prepared and experimentally studied. Such type 2 virus vaccines are being developed by means of specially constructed, attenuated virus; synthetic polypeptides consisting of a portion of the specific protective viral glycoprotein Ag; or an expression vector producing such a specific Ag.

Chemical therapy of herpesvirus infection provides the best example of developing chemotherapeutic agents that act specifically on viral rather than on host macromolecular synthesis, taking advantage of the unique characteristics of the biochemical reactants (see Chap. 49). Thus, **acylovir** [9-(2-hydroxyethoxymethyl) guanine] has been used successfully to treat the mucocutaneous lesions of oral and genital herpes simplex as well as encephalitis (successful treatment of encephalitis requires that acylovir therapy begin immediately upon suspicion of the diagnosis, before results of brain biopsy are known). It is noteworthy that although treatment of patients with recurrent oral or genital lesions significantly reduces the recurrences during therapy, the recurrent pattern resumes shortly after therapy is discontinued. Moreover, after prolonged therapy, resistant viral mutants, whose target DNA polymerase is markedly less affected by the drug, appear.

Vidarabine (adenine arabinoside; 9-β-arabinofuranosyladenine), also an inhibitor of herpesvirus-encoded, DNA polymerase, as well as virus-specific ribonucleotide reductase, has been employed locally for keratitis and herpes labialis and intravenously in cases of encephalitis and disseminated herpes simplex infections. Although vidarabine has been shown to be effective, acyclovir is more easily administered, has proved to be more effective in treatment of encephalitis, and is probably the present drug of choice. A number of other nucleoside analogues have been synthesized and studied: 5-iodo-2'-deoxyuridine (idoxuridine, IUdR) is useful for local treatment of herpetitic keratitis, but recurrences often follow cessation of therapy, and drug-resistant herpesviruses may emerge; and trifluorothymidine has proved to be even more successful than IUdR in the local treatment of herpes keratitis.

B Virus (Herpesvirus Simiae)

In its natural host, the monkey, B virus produces latent infections like those of herpes simplex virus in humans. However, in humans B virus can produce acute, ascending myelitis and encephalomyelitis. The increased handling of monkeys and the widespread use of monkey kidney cells in the commercial production of poliovirus vaccines augmented the transmission of this virus to humans: at least 24 cases have occurred in the past 30 years. Recognition of the danger is critical, since the disease in humans is usually fatal.

B virus closely resembles herpes simplex virus: the two are antigenically related but not identical. Antiserum from rabbits immunized with B virus neutralizes herpes simplex virus; however, Abs to herpes simplex virus neutralize B virus only slightly.

Varicella–Herpes Zoster Virus

The viruses isolated from patients with varicella (chickenpox) and from those with herpes zoster (shingles) are physically and immunologically indistinguishable (Fig. 53–8). Their identity has been further established by the production of typical varicella in children following inoculation of herpes zoster vesicle fluid.

PROPERTIES

Varicella zoster virus is relatively unstable even at −40°C to −70°C; the infectivity of virus from tissue cultures cannot be maintained reliably for longer than 2 months. Vesicle fluid from a patient, however, remains infectious for many months at −70°C, perhaps because of the high titers of virus and the high concentrations of protein.

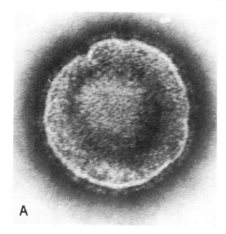

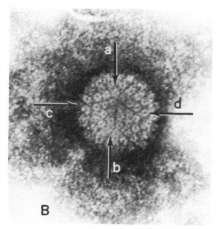

Figure 53–8. Morphology of varicella–herpes zoster virus particles embedded in sodium phosphotungstate. (*A*) Intact virion showing the envelope with surface projections and the centrally placed capsid. (*B*) Viral particle in which the envelope has ruptured, revealing the structure of the capsid more clearly. Arrows point to capsomers situated on axes of fivefold symmetry. (Original magnification ×200,000, reduced; Almeida JD et al: Virology 16:353, 1962)

IMMUNOLOGIC CHARACTERISTICS. The virions contain 30 to 33 polypeptides of which at least five are envelope-associated glycoproteins. A number of these Ags can be detected immunologically in vesicle fluids as well as in infected cells. In varicella, IgM Abs usually appear in the serum 2 to 3 days after onset of the exanthem; IgG Abs soon replace the IgM and continue to increase in titer for about 2 weeks. Titers of CF Abs decrease over a period of months, but neutralizing and immune adherence hemagglutinating Abs, as well as Abs to infected cell membrane Ags (detected by immunofluorescence), persist for many years after the primary infection. Epitopes inducing neutralizing Abs were identified on three of the glycoproteins; monoclonal Abs to two of the glycoproteins neutralized in the absence of complement, but one of the Ab sets required the addition of complement for viral neutralization. Cellular immunity, detected by peripheral blood and mucosal lymphocyte responsiveness to the viral membrane Ags, also persists and appears to play a protective role against reinfection, even in the absence of detectable circulating Abs. Most herpes zoster patients have a relatively high titer of viral IgG Abs at the onset of the disease, an indication that herpes zoster, like recurrent herpes simplex, occurs in previously infected, partially immune persons.

HOST RANGE. In contrast to herpes simplex virus, varicella virus does not cause reproducible disease in experimental animals or chick embryos, but it can be propagated in cultures of a variety of human and monkey cells. Little virus is found in the culture fluid, but stable virus can be obtained by sonic disruption of cells 24 to 36 hours after infection. Serial propagation is also accomplished by transfer of infected cells. Characteristic cytopathic effects and eosinophilic intranuclear inclusion bodies develop (see Fig. 53–9), and metaphase arrest and chromosomal aberrations have been observed.

MULTIPLICATION. Multiplication of varicella zoster virus is confined to the nucleus, and the developmental stages are similar to those of herpes simplex virus; the biochemical details have not been described. Viral DNA replication begins about 6 hours after infection, structural proteins can be detected 2 to 4 hours later, and a maximum quantity of virus is attained 24 to 36 hours after infection. Infectious virus after assembly does not emerge readily from infected cells in culture. Natural infections, however, are highly contagious, and the virus is present extracellularly in high titer in the vesicle fluid of lesions.

PATHOGENESIS

VARICELLA (CHICKENPOX). The primary disease produced in a host without immunity is usually a mild, self-limited illness of young children. The clinical picture strongly suggests that the virus is spread by respiratory secretions, enters the respiratory tract, multiplies locally and possibly in regional lymph nodes, produces viremia, and is disseminated by the blood to the skin and internal organs. The virus prefers ectodermal tissues, particularly in children.

After an incubation period of 14 to 16 days, fever occurs, followed within a day by a papular rash of the skin and mucous membranes. The papules rapidly become vesicular and are accompanied by itching. The lesions, which are painless (in contrast to herpes zoster), occur in successive crops, and all stages can be observed simultaneously.

In the infrequent adult cases the disease is more severe, often with a diffuse nodular pneumonia; the mortality may be as high as 20%. Varicella is also usually diffuse and intense in children receiving adrenocortical steroids and in persons with immune deficiencies.

The vesicles evolve from a ballooning and degeneration of the prickle cells of the skin, along with formation of giant cells with intranuclear eosinophilic inclusion bodies (Fig. 53–9). In disseminated fatal varicella, lesions containing similar giant cells appear in liver, lungs, and nervous tissue. Basically identical lesions may occur in dorsal root ganglia of patients with herpes zoster.

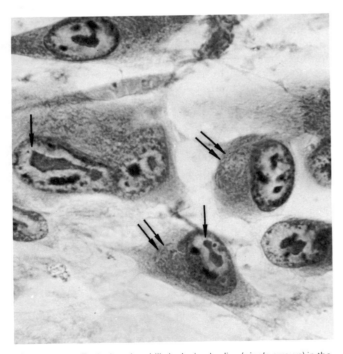

Figure 53–9. Typical eosinophilic inclusion bodies (*single arrows*) in the nuclei of human embryonic cells infected with varicella–herpes zoster virus. Poorly differentiated, pale eosinophilic bodies are also present in the paranuclear area of several cells (*double arrows*). (Original magnification ×1260; Weller TH et al: J Exp Med 108:843, 1958)

HERPES ZOSTER (SHINGLES). Herpes zoster (shingles) is the recurrent form of the disease, occurring predominantly in adults. It affects persons who were previously infected with the varicella zoster virus and who possess circulating Abs. The syndrome develops from an inflammatory involvement of sensory ganglia of spinal or cranial nerves, the virus having reaching the ganglia earlier during acute varicella by travelling along the nerves from the involved skin. The virus appears to remain latent in ganglionic nerve cells, and during activation it probably travels back along the nerve fibers to the skin.

Herpes zoster usually has a sudden onset of pain and tenderness along the distribution of the affected sensory nerve (frequently an intercostal nerve), accompanied by mild fever and malaise. A vesicular eruption, similar in pathology to varicella (except for its distribution), then occurs in crops along the distribution of the affected nerve; it is almost always unilateral. The vesicular eruption may last as long as 2 to 4 weeks; the pain may persist for additional weeks or months. Paralysis results if the inflammation spreads into the spinal cord or cranial nerves. Meningoencephalitis, which occurs rarely, is usually manifested as an acute illness with severe symptoms (headache, ataxia, coma, convulsions), but most patients recover completely.

Clinically, herpes zoster may be activated by trauma, by injection of certain drugs (arsenic, antimony), or by tuberculosis, cancer, or leukemia. Moreover, in persons with immunodeficiency states, particularly when these are induced in the therapy of lymphoproliferative diseases, the activated virus may disseminate and cause serious, often fatal, illness.

LABORATORY DIAGNOSIS

A clinical diagnosis of either varicella or herpes zoster seldom offers serious difficulties, but rarely the identification of chickenpox may require laboratory tests. This can be done most rapidly and easily by preparing a smear from the base of a vesicle and staining (by the Giemsa method) to detect typical varicella giant cells and cells with characteristic inclusion bodies. Immunofluorescence, ELISA, agar gel immunodiffusion, counterimmunoelectrophoresis, and electron microscopic examination of vesicular fluid provide rapid confirmation. For these rapid diagnostic techniques it is necessary to obtain clear viral fluid from early lesions—within the first 3 days for varicella and no more than 7 days with herpes zoster. In addition, virus may be isolated in cultured human or monkey cells and identified by serologic techniques. Antibody determinations on paired sera from patients may also be useful. Indirect immunofluorescence, immune adherence hemagglutination, and ELISA techniques are the most sensitive and rapid Ab assays. Antibody titrations are also useful to demonstrate susceptibility in seronegative adults or immunologically crippled children after exposure, as a basis for early prophylaxis with immune globulin.

EPIDEMIOLOGY AND CONTROL

Varicella virus is usually transmitted in respiratory secretions, producing a highly communicable disease with high clinical attack rate. Epidemics are common among children, especially in the winter and spring. Second attacks of chickenpox apparently do not occur. Herpes zoster, in contrast, is of low incidence, is not seasonal, is recurrent, and is predominantly confined to persons over 20 years of age. As has been pointed out, a case of herpes zoster may initiate an outbreak of chickenpox, and contact with chickenpox is said to provoke attacks of shingles in partially immune persons.

An attenuated viral vaccine has been developed and undergone intensive study in immunocompromised children, particularly those with leukemia receiving chemotherapy, as well in healthy, nonimmune young children. Mild rashes and fever follow immunization in as many as 35% to 40% of children undergoing therapy, and shingles may also subsequently appear. In leukemic children receiving maintenance chemotherapy, treatment is usually suspended from 1 week before to 1 week after immunization. Serum Abs appeared in approximately 90% of these children, the attack rate after exposure was reduced from about 90% to 20%, and all cases were extremely mild.

In view of the seriousness of varicella in adults, however, one might question the wisdom of attempts to prevent infection in healthy children unless the preventive procedure can offer as lasting protection as the natural disease. As in measles (see Measles, Prevention and Control, Chap. 57), chickenpox can be prevented or modified by administering high-Ab-titer IgG to contacts within 72 hours of exposure. Prophylaxis is of particular importance for susceptible adults and for children with impaired immunity. Adenine arabinoside (vidarabine) and acyclovir (see Chap. 49) have been used with apparent success to treat disseminated disease in the seriously ill (particularly in immunologically suppressed patients) as well as in some normal adults, but additional control studies are necessary.

Cytomegalovirus (Salivary Gland Virus) Group

Salivary gland virus disease of newborns is a severe, often fatal illness, usually affecting the salivary glands, brain, kidneys, liver, and lungs. M. G. Smith, in 1956, isolated the causative agent. The term *cytomegalovirus* (*CMV*) was applied to the group because of the large size of the infected cells and their huge intranuclear inclusion

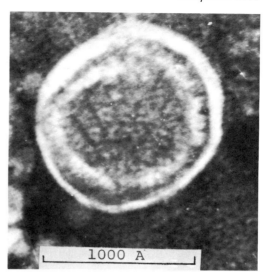

Figure 53–10. Enveloped full particle of human cytomegalovirus. (Original magnification ×405,000; Wright HT Jr et al: Virology 23:419, 1964)

bodies. Assignment to the herpesvirus family was based on the morphology of the viral particle (Fig. 53–10), the chemical composition of the virion (see Characteristics of Members of the Herpesvirus Family), and the characteristics of the intranuclear inclusion body present in infected cells (Figs. 53–11 and 53–12).

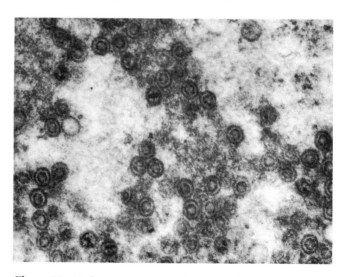

Figure 53–11. Electron micrograph of a portion of an intranuclear inclusion in a cell infected with human cytomegalovirus. The inclusion is made up of viral particles in various stages of development. Particles are composed of a central core about 40 nm in diameter, surrounded by a pale zone and, externally, by a thin membranous shell. Only a few particles have a dark central core, indicating the presence of nucleic acid. (Original magnification ×40,000; Becker P et al: Exp Mol Pathol 4:11, 1965)

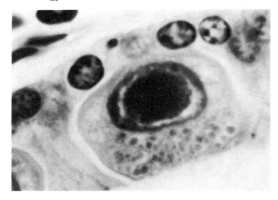

Figure 53–12. Epithelial duct cell from a human submaxillary gland, showing typical eosinophilic nuclear and basophilic cytoplasmic inclusions produced by infection with human cytomegalovirus. (Original magnification ×1500; Nelson JS, Wyatt JP: Medicine [Baltimore] 38:223, 1959)

Although primary disease is rare, infections are widespread: congenital and neonatal subclinical infections (acquired from the mother, before or at birth; see Pathogenesis) occur in 5% to 7% of live births. However, the incidence is much greater in those receiving blood transfusions, for example, in patients undergoing open heart surgery or organ transplants and receiving large quantities of blood. The virus commonly produces latent infections that may subsequently be activated by pregnancy, immunosuppression for organ transplantation or cancer chemotherapy, or viral infection (e.g., AIDS virus).

PROPERTIES

IMMUNOLOGIC CHARACTERISTICS. CMV-infected cells, like other herpesviruses, contain about 50 unique polypeptides of which 30 to 35 are virion structural proteins. The specific proteins that mediate the immune response, however, are only partially identified, but several have been shown to induce either humoral Ab or T-cell responses. Natural killer cell activity is also increased in acute CMV infections. Four membrane-associated glycoproteins have been characterized, of which three are disulfide linked and provide the antigenic epitopes for neutralizing Abs. As with herpes simplex virus, complement enhances the neutralizing capacity of Abs to some glycoprotein epitopes.

Human CMVs are not antigenically homogeneous, although their differences are not sufficient to warrant classification into distinct immunologic types. Moreover, DNA–DNA renaturation kinetics and restriction endonuclease maps indicate that the nucleotide sequences of different viral strains are largely homologous, although differences may be noted.

Humoral Abs develop relatively early during infection (IgM Abs are present at birth in the congenitally infected newborn) and persist at high levels during viral excretion. Cellular immunity, however, appears to play the major role in suppression of viral multiplication, leading to latent infection or, less commonly, to viral eradication.

HOST RANGE. Many species of animals are infected with their own specific CMVs, but no laboratory animal has proved susceptible to infection with the CMVs of humans. Virus has been isolated and propagated only in cultured human fibroblasts. *In vivo*, however, virus appears to multiply in a variety of cell types, including many of epithelial morphology. CMV also infects human lymphocytes and monocytes, but virus only undergoes an abortive replication cycle. In fact, in disseminated disease, virus has been detected in cells of essentially every organ.

MULTIPLICATION. Like herpes simplex viruses, viral replication is controlled by a coordinately regulated transcriptional program divided into three periods: **immediate-early,** which is independent of protein synthesis; **early,** in whose RNAs are encoded proteins required for DNA replication; and **late,** whose processed transcripts are translated into viral structural proteins. Synthesis of viral structural proteins depends on DNA synthesis. Viral replication is relatively slow as compared to herpes simplex viruses: viral DNA synthesis begins slowly after about 12 hours, but it is not maximally replicated until 24 to 36 hours after infection; virion-structural proteins are synthesized from 36 to 48 hours, and infectious virus is first detected 48 to 72 hours postinfection. Cytochemical and immunofluorescent techniques best reveal the synthesis of viral proteins and the development of the cytologic lesions that accompany viral multiplication, namely, focal lesions, followed by generalized cytopathic changes including rounding of cells and the appearance of large intranuclear eosinophilic inclusion bodies (see Fig. 53–12).

De novo biosynthesis of DNA and accumulation of early and late viral proteins are detected initially in the nucleus. Electron microscopic studies show that viral particles, like herpes simplex virions, are assembled in the nucleus (see Fig. 53–11), attain their envelope at the nuclear membrane, and migrate through reduplications of the nuclear membrane into the cytoplasmic endoplasmic reticulum. The maturation of viral particles appears to be inefficient: only rare, completely assembled virions can be detected among many incomplete particles (many of these are noninfectious dense bodies formed by enveloped viral proteins without DNA or assembled capsids). Hence, the yield of infectious virus in cell cultures is low, and as many as 10^6 particles are needed to initiate infection of a new culture. Most infectious virus remains cell associated, and the addition of intact infected cells to a culture therefore initiates viral propagation most efficiently.

Unlike herpes simplex viruses, CMV does not interrupt host macromolecular synthesis but rather stimulates host RNA and DNA synthesis in parallel with viral DNA replication. Accordingly, infected cells are not usually killed, and this explains why latent infections are frequently established.

PATHOGENESIS

CMVs are the most common cause of intrauterine infections. An increasing number of pregnant women excrete CMV as gestation proceeds, apparently owing to increasing levels of certain hormones (e.g., cortisol). Primary infection occurs in 19% to 20% of seronegative pregnant women, with subsequent infection of about 50% of the fetuses. In contrast, only 10% to 20% of pregnant women with latent infection transmit CMV to their fetuses. A mother with a primary or latent infection may transmit CMV to the fetus, either by transplacental transfer during pregnancy or by excreting virus into the genital tract at the time of birth. Protracted viral shedding may follow (although neutralizing Abs develop), and the infant carrier may serve as a source for dissemination (in different studies 15% to 60% of healthy children were shown to excrete virus during the first year of life). Although prolonged viral persistence is common following congenital infection, indefinite persistence is rare, and viral shedding is unusual in adults with Abs. Virus may also become latent in the newborn or during early childhood. Manifest illness infrequently occurs in newborns and infants up to 4 months of age, but if it appears, it usually has a relentless progression, with hepatic and renal insufficiency, pneumonia, neurologic symptoms, and eventual death.

Patients with neoplastic diseases and recipients of organ transplants, subjected to corticosteroids or other immunosuppressive drugs, or those with acquired immune deficiency diseases (e.g., AIDS) are particularly susceptible to viral activation or exogenous infection that results in localized or disseminated disease or inapparent infection. A syndrome resembling infectious mononucleosis (see below) may be observed in recipients of multiple transfusions of blood from latently infected donors. The syndrome has most frequently been reported in patients who have undergone open heart surgery, probably because of the large volumes of blood they receive. Reactivation of a latent infection or occurrence of primary disease, such as following multiple transfusions, is usually associated with an increase in suppressor-cytotoxic T cells (OKT8) and a decrease in helper cells (OKT4).

Like adenoviruses (see Chap. 52), CMVs can be isolated from explants of apparently normal adenoids and salivary glands cultured *in vitro* (see Pathogenesis, Chap. 52). Similar findings have been made with CMVs of mice, rats, hamsters, and guinea pigs.

The pathologic lesion is characterized by necrosis and pathognomonic cellular alterations. The affected cells are greatly increased in size; the nucleus is enlarged and contains a brightly stained eosinophilic inclusion body up to 15 μm in diameter (see Fig. 53–12), larger than that produced by any other virus infecting humans. In addition, the cytoplasm may be swollen and vacuolated and may show up to 20 minute basophilic and osmiophilic structures 2 μm to 4 μm in diameter; these contain DNA and polysaccharide and are therefore positive with Feulgen and periodic acid–Schiff stains.

LABORATORY DIAGNOSIS

Infection can be identified by viral isolation, immunologic assays, and exfoliative cytologic techniques:

Isolation of virus in cultures of human embryonic fibroblasts is the most sensitive method to detect infection in the newborn. Viral replication can be detected within 24 to 36 hours by means of **immunofluorescence, DNA–DNA hybridization** (or *in situ* DNA–RNA hybridization), or *ELISA* techniques. Thus, diagnosis can be made more rapidly, with high sensitivity and accuracy, and less expensively than with actual isolation and immunologic identification of the isolates.

Identification of characteristic cytomegalic cells with intranuclear and cytoplasmic inclusions (particularly in urinary sediment and bronchial and gastric washings) is an inexpensive diagnostic procedure. Detection of viral DNA or Ags by the techniques noted above can rapidly confirm the cytologic diagnosis.

Immunologic assays are valuable for diagnosis by demonstrating an increase in Ab titers. A variety of techniques are available: immunofluorescence, which can also distinguish a baby's IgM Abs from maternal IgG Abs; ELISA; indirect hemagglutination and latex particle agglutination, in which either tanned red blood cells or latex particles are coated with viral Ags; and CF. These techniques are rapid and inexpensive.

EPIDEMIOLOGY AND CONTROL

Infection with human CMV appears to be worldwide and common despite the relative rarity of clinical disease. About 5% to 10% of congenitally infected newborns display signs and symptoms of CMV disease, and from 10% to 18% of all stillborns show characteristic lesions at autopsy. In adults above 16 years of age, typical inclusions in salivary glands are rare, but in a sample taken in the United States, CF Abs were found in 53% of the population between 18 and 25 years old and in 81% of those over 35 years of age. Transplacental passage and infection at birth, during nursing, during blood transfusion, and during sexual intercourse are the most apparent mechanisms for transmitting virus, but person-to-person spread in urine and respiratory secretions seems likely. Latent virus has been detected in lymphocytes, a fact that highlights at least one means by which virus is transmitted while apparently noninfectious. Among pregnant women, 10% to 15% excrete virus during their third trimester, and at least 1% of newborns enter the world with viruria. Children and adults with immunologic deficiencies, naturally or iatrogenically acquired, are particularly susceptible to active disease. Vaccines consisting of attenuated strains and antiviral drugs (e.g., acylovir and interferon) are being studied, but effective measures for prevention and control will probably not be available until more is known of viral characteristics and transmission.

EB (Epstein-Barr) Virus and Infectious Mononucleosis

In a search for the cause of Burkitt lymphoma, Epstein and Barr, in 1964, observed herpeslike viral particles (termed *EB virus*; Fig. 53–13) in a small proportion (0.5%–10%) of the lymphoma cells repetitively cultured *in vitro*. The relation of EB virus to the malignancy is discussed in Chapter 64.

EB virus, however, has been shown to be the cause of infectious mononucleosis. The first indication of this relationship came 4 years after the discovery of EB virus, when G. and W. Henle discovered that lymphocytes from their technician, who had just recovered from infectious mononucleosis, could be serially cultured *in vitro*, unlike normal lymphocytes; and a small number of these cultured cells were found to contain EB virus. Further observations supported the conclusion that EB virus is the etiologic agent of infectious mononucleosis: (1) Transfusion of blood from such persons produces the disease in recipients devoid of Abs. (2) The virus persists in lymphocytes cultured from patients long after recovery from infectious mononucleosis. (3) Patients with infectious mononucleosis develop neutralizing and CF Abs, as well as Abs that react with EB virus–infected cells in an immunofluorescence assay. (4) The Abs persist for years, and their presence or absence is correlated with resistance or susceptibility to infectious mononucleosis. Finally, (5) EB virus has been isolated from pharyngeal secretions of patients with infectious mononucleosis.

PROPERTIES

Virions purified from cultured Burkitt lymphoma cells (Fig. 53–14) are structurally similar to those of other herpesviruses (see Fig. 53–2 and Characteristics of Members

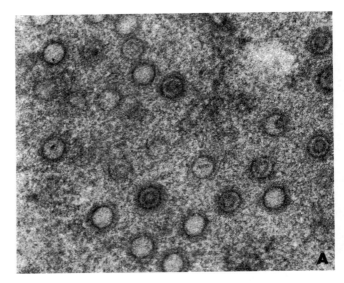

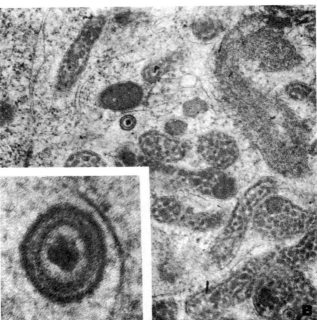

Figure 53–13. Structure of EB virus in cells cultured from Burkitt lymphoma. (*A*) Numerous developing immature particles in a thin section of a lymphoblast nucleus. (Original magnification ×76,500) (*B*) Mature viral particle with envelope, capsid, and nucleoid in the cytoplasm. (Original magnification ×42,000; *inset,* ×213,500; Epstein MA et al: J Exp Med 121:761, 1965)

of the Herpesvirus Family). The virion contains about the same number and size proteins as herpes simplex virus. The structure of the viral DNA isolated from infectious mononucleosis patients or from the transformed cell

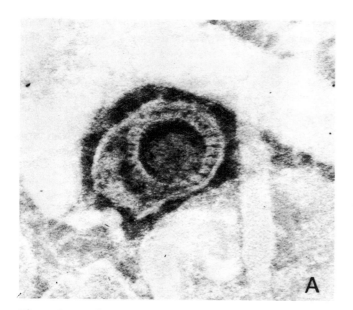 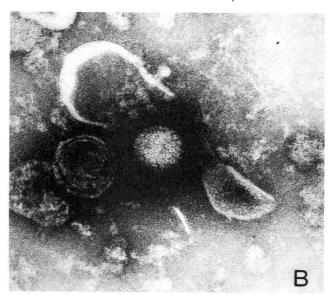

Figure 53—14. Structure of EB virus obtained from cultured Burkitt lymphoma cells. Electron micrographs are of purified, negatively stained virus. (*A*) Empty capsid enclosed in an envelope. (Original magnification ×200,000) (*B*) Viral particle with a disrupted envelope. A capsid (about 75 nm in diameter) similar to that of herpesviruses is clearly visible. (Original magnification ×120,000; Courtesy of K. Hummler)

lines is unique as compared with other herpesviruses (see Fig. 53–1). Indeed, EB virus DNA consists of five unique regions of DNA and seven regions of repeated DNA sequences, including the ends of the genome, which has six to 12 multiple tandem copies. It is also noteworthy that EB virus DNA shows considerable polymorphism among its isolates from patients. However, viral DNA from lymphoid cell lines derived from patients with infectious mononucleosis, though of the same size, differs in about 35% of the sequences when compared with EB virus DNA derived from Burkitt lymphomas. EB virus is immunologically unrelated to other herpesviruses.

The virus selectively infects human B-lymphocytes by means of a receptor related or identical to the receptor for the third component of complement, C3d. The hallmark of infection is **transformation** of the infected cell to a continuously dividing B-lymphocyte cell line—so-called **immortalized** cells. Infection primarily results in **latent infection** in which only limited, noncontiguous regions of the genome are transcribed, yielding at least three **nuclear Ags (EBNAs)** and a **membrane Ag** the precise functions of which are unknown. The most abundant transcripts detected during latent infections are two small, nonmessenger RNAs that appear to be transcribed by polymerase III. Following the immediate early events of infection, including production of a virus-encoded DNA polymerase and a DNase, the linear virion DNA is converted to a covalently closed circular mole-cule via its large terminal repeated nucleotide sequences (see Fig. 53–1). The viral DNA replicates and primarily exists in this episomal form, although small numbers of integrated viral genomes have been detected in some cell lines and in tumor cells.

Latent infection can be converted to the **replicative** phase by several means. *In vitro* cultivation of lymphoid cells infected *in vivo* (e.g., lymphoid cells from patients with infectious mononucleosis) induces production of infectious virus in a small proportion of the cells. The replicative cycle can also be induced by a variety of chemical agents such as uridine analogues or the phorbol ester tumor promoter. The replicative cycle is marked by extensive transcription, increased DNA replication, and production of late proteins, which include capsid antigens and glycoproteins of which at least two (gp 350 and gp 220) are incorporated into the viral envelope and contain epitopes that induce neutralizing Abs. However, the induced cells produce relatively small amounts of infectious virus.

INFECTIOUS MONONUCLEOSIS
Pathogenesis

Primary infection ordinarily occurs via the oral route, and the virus initially replicates in the oropharyngeal epithelium and in the epithelium of salivary gland ducts. The intimate association of lymphoid and epithelial cells in the oropharynx permits ready infection of B-lymphocytes at this site and ensures entrance of EB virus–infected B cells into the circulation. In the largest proportion of the world population these events of primary infection occur during the first few years of life and usually produce only subclinical infection. In the developed countries, apparently owing to the improved hygienic conditions, primary infection is delayed until the age of 15 years or older, and this delayed infection results in the **clinical disease infectious mononucleosis** in up to 50% of the persons infected.

Infectious mononucleosis is an acute infectious disease, primarily affecting lymphoid tissue throughout the body. It is characterized by the appearance of enlarged and often tender lymph nodes, an enlarged spleen, and abnormal lymphocytes in the blood (from which the disease derives its name). In addition, fever and sore throat are common. Occasionally, other diverse manifestations are observed, including mild hepatitis, signs of meningitis or other central nervous system involvement, hematuria, proteinuria, thrombocytopenic purpura, and hemolytic anemia.

Following acute disease, EB virus persists for weeks to months in the oropharyngeal secretions of patients. In a small number of these patients, particularly in young adults (25–35 years), persisting illness and fatigue accompany the continuing viral replication. Whether EB virus causes the chronic illness or is an epiphenomenon is still unclear.

In addition to the EB virus–induced, polyclonal activation of B-lymphocytes, patients with infectious mononucleosis mount a specific Ab response to the viral infection. IgM and IgG Abs are directed against Ags produced during viral replication: early Ags, the viral capsid Ag, and the glycoprotein membrane Ags. The appearance of Abs to the EB virus nuclear Ags is a sign of the development of latent infection. Infection also induces a marked cell-mediated immune response, which is probably mainly responsible for the outcome of the acute EB virus infection. It is the marked proliferation of T-cell (T8) lymphoblasts that produces the characteristic atypical mononuclear cells that are prominent in the blood of patients with infectious mononucleosis. The expanded T-cell population consists to a large extent of EB virus–specific cytotoxic T cells. The increase in cytotoxic T cells results in a reversal of the T4/T8 ratio for 4 to 8 weeks. An increased activity of natural killer cells directed at a variety of Ags also appears early in infection.

Laboratory Diagnosis

Specific diagnosis depends on (1) detection of viral DNA, infectious virus, or viral Ags, (2) Ab response, and (3) observation of abnormal lymphocytes in the peripheral blood.

Intracellular viral DNA can readily be detected by DNA:DNA hybridization with *in vitro* labeled EB virus DNA probes. **Infectious virus** is isolated from peripheral blood, saliva, and lymphoid tissue with immortalization of human B-lymphocytes used as the assay (those from umbilical cord blood are most sensitive). **Viral Ags** are most easily identified by the **indirect immunofluorescence assay;** acute infection is diagnosed with Abs directed against either the capsid Ag in a cell line producing EB viruses or early Ag in nonproducer cell lines superinfected with EB virus. **ELISA** will probably also soon be available for diagnosis. **Serologic tests** can be used but are difficult to interpret owing to the long persistence of Abs following infection. Serologic diagnosis of acute infectious mononucleosis can best be accomplished by demonstrating IgM Abs to capsid Ag or development of Abs to EB virus nuclear Ags, since the nuclear Ags do not appear until late during primary infection.

These techniques for detection of virus or serologic response are specific and highly sensitive, but they are highly specialized and not yet suitable for the general hospital laboratory. A **practical laboratory diagnosis** can be made based on two unique findings:

Abnormal, large lymphocytes with deeply basophilic, foamy cytoplasm and fenestrated nuclei appear, often accounting for 50% to 90% of the circulating lymphocytes. Initially there is a leukopenia, but by the second week of disease the count may rise to 10,000 to 80,000 cells/mm³.

Heterophil Abs (i.e., agglutinins for sheep red blood cells) develop in 50% to 90% of patients during the course of the disease.

The immunogen eliciting the heterophil Abs is unknown, but it appears to be distinct from the EB virus. Thus, a small percentage of patients develop EB virus Abs but no heterophil Abs; heterophil Abs can be adsorbed from serum by bovine red blood cells without reducing the Ab titer to EB virus; and heterophil Abs are transient, but Abs to EB virus persist for years, perhaps for life.

Heterophil Abs also appear during serum sickness following injection of horse serum, and they may be present in serum from healthy persons. However, a differential adsorption test of a patient's serum distinguishes these Abs. Thus, heterophil Abs of infectious mononucleosis are adsorbed by bovine red blood cells but not by guinea pig kidney (which contains Forssman Ag); serum sickness agglutinins are adsorbed by both bovine red blood cells and guinea pig kidney; and normal serum heterophil agglutinins are adsorbed by guinea pig kidney but not usually by bovine red blood cells.

Epidemiology and Control

Infectious mononucleosis appears to be primarily a disease of relatively affluent teenagers and young adults (such as college students), in whom it causes proved disease in about 15% of the susceptible population. (About 75% of entering college students in the United States are free of detectable Abs). The peak incidence is at 15 to 20 years of age.

Successful infection appears to require extensive exposure, or unknown cooperating factors, for multiple cases in families are infrequent despite the presence of virus in pharyngeal secretions of infected persons. In cases of infectious mononucleosis the source of the infecting virus is rarely obvious, perhaps because persons with latent infections produce the virus in pharyngeal and oral secretions for prolonged periods after recovery. The common association of infectious mononucleosis with intensive kissing may reflect a requirement for a large inoculum. No adequate control procedures are available.

Selected Reading

Gershon AA, Steinberg SP, LaRussa P, Ferrara A: The National Institutes of Allergy and Infectious Diseases Varicella Vaccine Collaborative Study Group: Live attenuated varicella vaccine: Efficacy for children with leukemia in remission. JAMA 252:355, 1984

Klein RJ: Initiation and maintenance of latent herpes simplex virus infections: The paradox of perpetual immobility and continuous movement. Rev Infect Dis 7:211, 1985

Pass RF: Epidemiology and transmission of cytomegolavirus. J Infect Dis 152:243, 1985

Rapp F: Persistence and transmission of cytomegalovirus. In Fraenkel-Conrat H, Wagner RR (eds): Comprehensive Virology, Vol 16, p 193. New York, Plenum, 1980

Rickinson AB, Yao QY, Wallace LE: The Epstein-Barr virus as a model of virus–host interactions. Br Med Bull 41:75, 1985

Roizman B (ed): Herpesviruses, Vols I and II. New York, Plenum, 1982

Wagner EK: Individual HSV transcripts: Characterization of specific genes. In Roizman B (ed): Herpesviruses, Vol 3. New York, Plenum, 1984

Wathen MW, Stinski MF: Temporal patterns of human cytomegalovirus transcription: Mapping the viral RNAs synthesized at immediate early, early, and late times after infection. J Virol 41:462, 1982

54

Harold S. Ginsberg

Poxviruses

The smallpox was always present, filling the churchyards with corpses, tormenting with constant fears all whom it had striken, leaving on those whose lives it spared the hideous traces of its power, turning the babe into a changeling at which the mother shuddered, and making the eyes and cheeks of the bethrothed maiden objects of horror to the lover.
—T. B. Macaulay: *The History of England From the Accession of James II,* Vol. IV

It now becomes too manifest to admit of controversy that the annihilation of the smallpox, the most dreadful scourge of the human species, must be the result of this practice (of vaccination).
—Edward Jenner: *The Origin of the Vaccine Inoculation,* 1801

Since the beginning of history, smallpox* has left its indelible mark on the medical, political, and cultural affairs of man. Records show severe epidemics from earliest times. Indeed, by the 18th century the disease had become endemic in the major cities of Europe. Terror of its presence was such that "no man dared to count his children as his own until they had had the disease."†

Because virulent smallpox strains appear to have no animal reservoir, and because vaccination is very effective, complete eradication of this scourge—long a dream of public health authorities—has now become a reality as the result of a vigorous World Health Organization (WHO) program of confinement and immunization initiated in 1966. Whereas in 1945 most of the world's inhabitants lived in endemic areas, October 1977 saw the last reported case of natural infection, detected in Somalia. Immunization has eliminated smallpox from the rest of the world (including India and Pakistan, where severe epidemics still occurred in 1973), and no documented case has been reported in the United States since 1949.

---

* A term initially employed to distinguish the disease from "large pox" (syphilis)

† The Comte de la Condamine, an 18th century French mathematician and scientist

In 1980 the WHO declared that smallpox had been globally eradicated. This remarkable feat has resulted in the saving of over $1 billion annually in global health expenditures, and in the United States it has brought a saving of about $150 million annually in the cost of vaccination and quarantine measures.

With the disappearance of the disease, and with the decline in enforced vaccination, susceptibility will slowly return, bringing a liability to massive epidemics if a source of infection should appear. Hence, it is still too early to be certain that this virus is buried in the graveyard of extinct organisms. There is no absolute certainty that a persistent, unrecognized focus of variola (smallpox) virus does not exist, that a member of another species of poxviruses (e.g., monkeypox or cowpox viruses) may not undergo genetic variation to virulence for humans, or that biological warfare—a horrifying possibility—would not reintroduce smallpox virus. To reduce the risk of escape of this virus from a laboratory (such as occurred in Birmingham, England in 1978, resulting in two cases), the WHO led the campaign to destroy stocks of smallpox virus throughout the world except in four designated laboratories. Thus, in the United States only a single reference stock is preserved, under stringent controls, at the Center for Disease Control in Atlanta.

It would clearly be unwise to ignore the smallpox virus entirely: continued knowledge and constant vigilance are still required. This chapter briefly presents the characteristics of poxviruses and how they replicate. Particular emphasis is placed on the properties of smallpox virus that permitted eradication of the disease it produces, and the methods employed in the eradication program.

General Characteristics

The smallpox (variola) virus is representative of the **poxviruses,** a group of agents that infect both humans and lower animals and produce characteristic vesicular skin lesions, often called **pocks.** Poxviruses (**family Poxviridae**) are the largest of animal viruses (see Characteristics of the Poxviruses): they can be seen with phase optics or in stained preparations with the light microscope. The viral particles (originally called **elementary bodies**) are somewhat rounded, brick-shaped, or ovoid and have a complex structure consisting of an internal central mass—the nucleoid—surrounded by two membrane layers (Figs. 54–1 through 54–3). The surface is covered with ridges that may be tubules or threads (see Fig. 54–2). Poxviruses contain DNA, protein, and lipid. They are relatively resistant to inactivation by common disinfectants and by heat, drying, and cold, characteristics that made their spread so easy in susceptible populations.

The **genus *Orthopoxvirus*** consists of viruses of certain mammals, including variola, vaccinia, monkeypox,

CHARACTERISTICS OF THE POXVIRUSES

Size: 250–390 nm × 200–260 nm
Morphology: Brick-shaped to ovoid (see Figs. 54–1, 54–9, 54–10)
Protein and lipid content: Present (vaccinia: 91.6% and 5%, respectively)
Stability: Relatively resistant to inactivation by chemicals (disinfectants) or by heat, cold, or drying; inactivated by chloroform; variably inactivated by ether
Nucleic acid: Double-stranded DNA*
Molecular weights (daltons) and base ratio: Vaccinia 150 × 10^6 (3.2% of the virion; 231 kilobase pairs); AT/GC = 1.67†
Fowlpox 200 × 10^6 (307 kilobase pairs); AT/GC = 1.84
Antigenicity: Common family and genus Ags
Multiplication‡: In cytoplasm of cells but requires host RNA polymerase II subunit from nucleus
Cytopathogenicity: Predilection for epidermal cells; eosinophilic inclusion bodies produced

* Complementary strands of vaccinia DNA are covalently crosslinked at or near the termini. Other poxviruses have not been examined for this structure.

† DNAs of cowpox, rabbitpox, and mousepox are similar.

‡ A DNA-dependent RNA polymerase in virion.

cowpox, ectromelia of mice, rabbitpox, camelpox and gerbilpox viruses. **Other genera** include viruses specific for birds (*Avipoxvirus*), ungulates (*Capipoxvirus*), and arthropods (*Entomopoxvirus*) and the tumor-producing (fibroma and myxoma) viruses of rabbits (*Leporipoxvirus*). Viruses of a sixth genus, which resemble other poxviruses in structure but not immunologically (hence termed *Parapoxviruses*), include **contagious pustular dermatitis (orf), paravaccinia (milker's nodules),** and

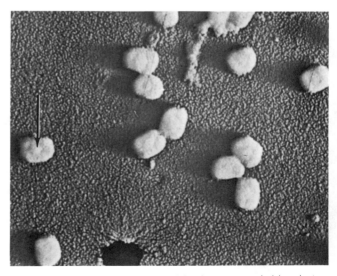

Figure 54–1. Morphology of vaccinia virus as revealed by electron microscopic examination of a shadowed preparation of purified viral particles. Note the central core surrounded by a depression (*arrow*). (Original magnification ×28,000; Sharp DG et al: Proc Soc Exp Biol Med 61:259, 1946)

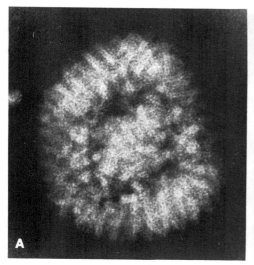

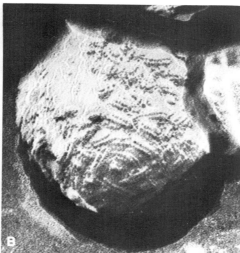

Figure 54–2. Fine structure of a mature vaccinia virion from a purified suspension of viral particles. (*A*) Virion negatively stained with phosphotungstate. The double tract of ridges and a suggestion of the subunit structure can be seen. The protrusion and sawtooth effect of the ridges are noticeable at the periphery. (Original magnification ×224,000) (*B*) Freeze-etched virion, showing subunits, length, and random orientation of the double ridge. (Original magnification ×224,000; Medzon EL, Bauer H: Virology 40:860, 1970)

bovine papular stomatitis viruses. Some poxviruses, such as the **molluscum contagiosum virus** and **Yaba monkey tumor virus,** cannot be classified immunologically in any of these genera.

All poxviruses studied are related immunologically by a common internal Ag extractable from viral particles. They can be divided into genera on the basis of their more specific Ags, nucleic acid homology, morphology, and natural hosts.

Genetic recombination may occur, but only between two viruses of the same immunologic group. The proportion of recombinants seems to be related to the degree of homology between the viral DNAs and parallels the degree of antigenic cross-reactivity.

Poxviruses vary widely in their ability to cause generalized infection, but they share a **predilection for epidermal cells,** in which they multiply in the cytoplasm and produce **eosinophilic inclusion bodies** (termed

Figure 54–3. Thin section of an intact mature vaccinia virus particle showing the inner nucleic acid core and surrounding membranes. The elliptical body (*EB*) on each side of the nucleoid causes a prominent central bulging of the virion. The viral particle is lodged between two cells (*arrows*). (Original magnification ×120,000; Dales S: J Cell Biol 18:51, 1963)

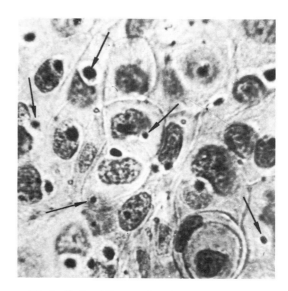

Figure 54–4. Eosinophilic cytoplasmic inclusion bodies (Guarnieri bodies) in corneal epithelial cells of a rabbit eye infected with vaccinia virus. (Original magnification ×900); Coriell LL et al: J Invest Dermatol 11:313, 1948)

Guarnieri bodies; Fig. 54–4). Fibroma and myxoma viruses have, in addition, a great affinity for subcutaneous connective tissues. Most poxviruses also multiply readily in epidermal cells of the chorioallantois of chick embryos, where they produce characteristic **nodular focal lesions,** termed *pocks* (Fig. 54–5). These lesions reflect a second characteristic of poxviruses: the **propensity to cause cellular hyperplasia** before cell necrosis. With myxoma, fibroma, and Yaba viruses the hyperplasia predominates, and tumors develop.

Variola (Smallpox) and Vaccinia

PROPERTIES

The virulence and contagiousness of smallpox virus have understandably limited its laboratory investigation. On the other hand, the closely related and much less dangerous vaccinia virus is one of the most thoroughly investigated animal viruses. Since the two viruses are very similar, they are discussed together.

Morphology

The morphology of vaccinia virions is revealed by various techniques of electron microscopy (see Figs. 54–1 through Fig. 54–3). Viral particles from smallpox crusts and vesicle fluids are morphologically indistinguishable from vaccinia particles. Thin sections (see Fig. 54–3) disclose a **central nucleoid** with a dumbbell-shaped dense coil composed of the viral DNA. The nucleoid is surrounded by **lipoprotein membranes,** and between the nucleoid and the outer viral coat is an **ellipsoidal body.**

In addition to the viral DNA, the viral cores contain several enzymes, primarily for transcription and modification of **immediate early mRNAs:** e.g., DNA-dependent RNA polymerase, polyadenylate polymerase, methyltransferase, and guanylyltransferase.

Chemical and Physical Characteristics

The chemical and physical characteristics of vaccinia virus, the first animal virus to be prepared in sufficient purity and quantity for detailed analysis, are summarized under Characteristics of the Poxviruses. The complexity of the virion is made apparent by the presence of more than 80 distinct species of polypeptides. Their precise localization in the core, lipoprotein membranes, and virion surface structures is only partially known. The majority of the polypeptides, however, are core components, either basic proteins needed for tight folding of the large DNA molecule or virion enzymes essential for uncoating and initiating viral multiplication.

Variola and vaccinia viruses are **relatively stable;** they resist drying and retain their infectivity for many months at 4°C and for years at −20°C to −70°C. Thus, exudates or crusts taken from patients with smallpox may yield infectious virus after almost a year at room temperature, and diagnostic specimens do not need refrigeration. The persistence of infectious variola virus on bedclothes caused a hazard not only for medical personnel but even for laundry workers. The relative resistance of the virus to dilute phenol and other common disinfectants complicated the decontamination of clothing, instruments, furniture, and so forth. However, variola and vaccinia viruses are inactivated by apolar lipophilic solvents (e.g., chloroform), by autoclaving, or by heating at 60°C for 10 minutes.

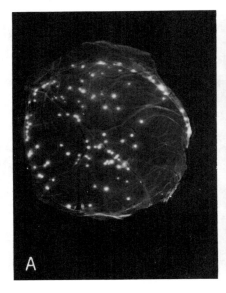 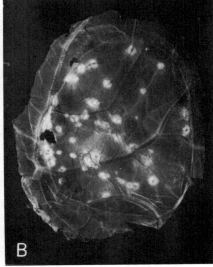

Figure 54–5. Pocks produced by variola virus (*A*) and vaccinia virus (*B*) 3 days after viral inoculation of the chorioallantoic membranes of chick embryos. Note the small gray-white pocks produced by variola virus and the large pocks made by vaccinia virus. (Downie AW, MacDonald A: Br Med Bull 9:191, 1953)

Antigenic Structure and Immunity

The antigenic structure of poxviruses has been examined largely with vaccinia virus, but smallpox virus is very similar. Viral strains from cases of severe smallpox (**variola major**) and those from cases of **variola minor (alastrim)** are immunologically indistinguishable. Moreover, it is striking that the antigens of smallpox viruses have not changed in their immunologic reactivity over the centuries that this virus has been a scourge throughout the world.

The Ags of poxviruses can be measured by all the usual immunologic techniques, including hemagglutination inhibition and neutralization of infectivity.

Stepwise dissection of purified virions with trypsin or chymotrypsin, 2-mercaptoethanol, and a nonionic detergent (e.g., Nonidet P40) permits localization of some of the Ags and polypeptides: (1) Two Ags responsible for neutralizing Abs are present in the tubules of the outer surface membrane; (2) Ags in the viral envelope evoke neutralizing Abs that protect against dissemination of virions throughout the body; (3) one large and three small Ags are identified in the core; and (4) one large Ag has been located between the core and the surface structures. The neutralizing Abs induced by the two surface Ags neutralize only viruses from the homologous subgroup. Neither of these sets of Ags is found on inactivated virus, a fact that explains the ineffectiveness of inactivated virus vaccines. A **family Ag,** which is a component of the core and one of the constituents of the NP fraction, can be extracted from virions with weak alkali; it cross-reacts, in complement fixation (CF) and precipitin assays, with a similar Ag from all poxviruses. Although it is not yet possible to identify the Ags with specific polypeptide chains, most of the polypeptides, have been topographically located in the virion.

Unlike other viruses, the **hemagglutinin** of variola and vaccinia viruses is not a component of the viral particle, and its Abs do not neutralize viral infectivity. It is a lipoprotein embedded in the membranes of infected cells and probably corresponds to a new surface Ag.

Antibody Response

Following infection or immunization, Abs to each of the viral Ags develop. The variations observed in their time of appearance and persistence depend on the nature and quantity of the Ag. Thus, neutralizing and hemagglutination-inhibiting Abs are first detected about the sixth day after onset of illness in the unvaccinated person, whereas CF Abs ordinarily appear 2 to 3 days later. Neutralizing and hemagglutination-inhibition Abs persist at least 20 years after infection, but CF Abs remain less than 2 years. In persons previously immunized, the various Abs generally appear 2 to 3 days sooner after onset of illness, reach higher titers, and persist longer.

Following natural infection, immunity to smallpox is long-lasting, if not persistent for life. In the rare reinfections that have been reported, the disease is usually atypical, very mild, and often without skin rash (variola sine eruptione). Immunity following vaccination, however, may last no longer than a year, and when smallpox was extant, immunization was recommended immediately following exposure regardless of when last vaccinated. If infection occurred in vaccinated persons, the clinical disease was also usually milder than in unimmunized neighbors.

The relative importance of **humoral** and **cellular immunity** remains largely unexplored, although cell-mediated immunity appears to be critical for recovery from infection. Thus, immunization with live vaccine has led to disease in persons with congenital defects of T-lymphocytes (see Chap. 50), as discussed under Complications of Vaccination, below.

Host Range

Variola virus has a much more limited host range than do vaccinia and other poxviruses. Monkeys are the only animals, other than man, known to be naturally infected; also, when variola virus is placed onto monkeys' scarified skin or inoculated intradermally, local lesions and fever follow. A few **animals** (chick embryos, rabbits, mice) and **cultured cells** (e.g., human embryonic kidney, monkey kidney, HeLa cells) are susceptible to experimental infection, and these have been valuable for diagnosis and research.

Multiplication

Although poxviruses contain DNA, **biosynthesis of the viral components and their assembly into viral particles take place entirely within the cytoplasm of the cell.** However, although most biosynthetic reactions can occur in enucleated cells, a large subunit of the nuclear host RNA polymerase II joins with a virus-encoded RNA polymerase subunit to transcribe a special set of mRNAs.

Virus attaches to uncharacterized host cell receptors, and it enters the cytoplasm primarily by the process of engulfment (see Chap. 48). After penetration, viral DNA is released by a **two-stage uncoating process** (see Chap. 48). The first stage is initiated almost immediately after engulfment by preexisting host cell enzymes, which break down the viral membrane part of the protein coat of the viral particle and the membrane of the endocytic vesicle to free the nucleoprotein core into the cytoplasm. The second stage results in breakdown of this core to liberate viral DNA. At the onset of this second stage, a DNA-dependent RNA polymerase present in the intact core transcribes about 25% of the viral genome. The resulting transcripts are processed within the core, and functional **immediate early mRNAs** emerge. They code for proteins required for the final uncoating events and for the enzymes necessary to produce the RNA for a second set of mRNAs, **delayed early.** Finally, after **viral DNA replication** begins, **late mRNAs** appear (derived from about 60% of the genome), while at least some early messengers continue to be made. Transcription continues until about 7 hours after infection. The kinetics of viral macromolecular biosynthesis are presented in Figure 54–6, and the sequential replication events are summarized diagrammatically in Figure 54–7.

Synthesis of **specific enzymes** and of a few **viral structural proteins** begins early in the biosynthetic process, before replication of viral DNA. The products include the second-stage uncoating proteins, three proteins associated with the nucleoprotein core, a protein essential for initiation of viral DNA replication, and enzymes related to DNA biosynthesis.

Viral DNA begins to be synthesized 1.5 to 2 hours after infection and attains its maximal concentration by the time newly made infectious virus is first detected (see Fig. 54–6). The **mode of DNA replication** is imposed by the unique covalent crosslinking of the two strands (see Synthesis of DNA-Containing Viruses, Chap. 48). Synthesis is initiated at either end of the genome. Large circular and forked replicating forms are found, indicating that an endonuclease cleaves the single-stranded crosslinks during replication. **Late viral proteins** are first detected about 4 hours after infection, and **infectious virus** is formed about 1 hour later by packaging viral DNA randomly selected from the preformed pool (see Figs. 54–6, 54–7). **Posttranslational modifications** of several proteins (i.e., cleavage, glycosylation, and phosphorylation) are essential to virion maturation.

Concomitant with the biosynthesis of virus-directed mRNA and viral DNA, biosynthesis of host cell macromolecules is inhibited. Production of host proteins stops because initiation of polypeptide chain synthesis is blocked, and host cell polyribosomes are disrupted; host DNA ceases replicating; and the host mRNAs cannot leave the nucleus, although their synthesis continues unaltered for about 3 hours. How these controls of host cell biosynthesis are induced by the virus and whether they relate to cell injury are still unknown.

The morphologic counterparts of the foregoing biochemical events have been observed in thin sections of infected cells (Fig. 54–8). As viral DNA synthesis increases, regions of dense fibrous material appear in the cell cytoplasm. About 3 hours after infection, some of the early proteins form membranelike structures, which begin to enclose patches of viral components and proceed to form immature particles into which DNA enters. (Fig. 54–8A and B). After the envelope is completed, the nucleoid begins to take shape within the immature particle; an additional membrane encloses the condensing DNA; the lateral bodies differentiate; and, finally, the outer coat structures are laid down on the previously formed membrane, completing the assembly of mature virions (see Fig. 54–7).

Lysis of infected cells is not a prerequisite for liberation of newly formed virions. The viral particles seem to be released through cell villi. Radioautography and immunofluorescence reveal that viral materials may also be transmitted directly from cell to cell through villi.

SMALLPOX
Pathogenesis

Two basic forms of smallpox are recognized: **variola major,** which has a case fatality rate of approximately 25%, and **variola minor,** or alastrim, a less virulent form with a mortality rate below 1%. Although a variety of factors may influence the mortality rate in any epidemic, the epidemiologic evidence is convincing that severe and mild smallpox exist as distinct entities. Nevertheless, it is impossible to distinguish the viruses responsible for these two forms of the disease.

Virus multiplies first in the mucosa of the upper respiratory tract and then in the regional lymph nodes. A **transient viremia** then disseminates virus to internal organs (liver, spleen, lungs), where the virus propagates extensively. A **second viral invasion of the bloodstream** terminates the incubation period (about 12 days) and initiates the **toxemic phase,** characterized by prodromal macular rashes, fever, generalized aching, head-

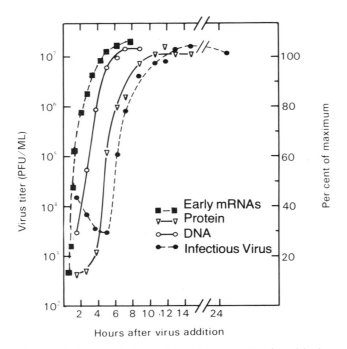

Figure 54–6. Biosynthetic events in the multiplication of vaccinia virus. (Modified from Salzman NP et al: Virology 19:542, 1963)

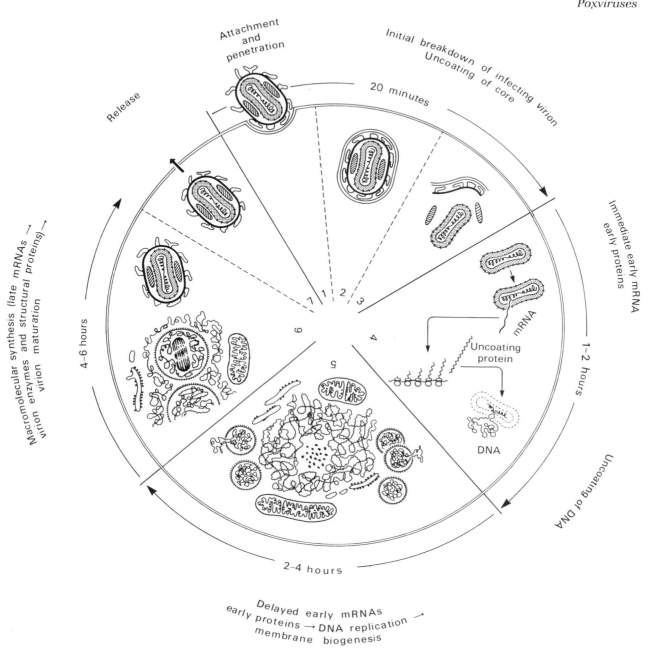

Figure 54—7. Diagram of the sequential events of vaccinia virus infection of a single cell.

ache, malaise, and prostration. Virus spreads to the skin and multiplies in the epidermal cells; the characteristic skin eruption follows in 3 to 4 days. Macular at onset, the rash progresses from papular to vesicular, finally becoming pustular in the second week of illness. In severe cases the rash may become hemorrhagic or confluent. The course of variola minor is similar but shorter, and the rash and other symptoms are less severe.

The **inclusion bodies,** or Guarnieri bodies, surrounded by a clear halo, characteristically develop in cells of the skin and mucous membranes infected with variola or vaccinia virus (see Fig. 54–4). Each inclusion body consists of an accumulation of viral particles and viral Ags. (Similar masses are observed in cells infected with other poxviruses).

A hypersensitivity response to the viral Ags may con-

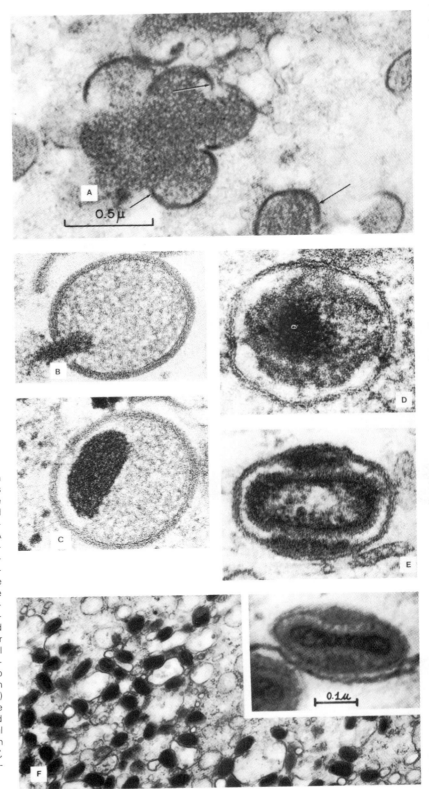

Figure 54–8. Developmental stages in the maturation and assembly of a vaccinia virus: electron micrographs of cells taken at intervals following infections. (*A*) Dense trilaminar viral membranes forming within and around clumps of dense fibrillar material in the cytoplasm. (Original magnification ×48,000) (*B*) Insertion of viral DNA and associated proteins into incomplete particles. (Original magnification ×150,000) (*C*) Condensation of nucleoprotein within the immature particle. (Original magnification ×150,000) (*D*) Maturation of a virion. The complete trilaminar lipoprotein envelope encloses the particle. Within is a dense nucleoid of fibrous nucleoprotein surrounded by a less dense homogeneous substance, thought to be material of the lateral bodies and core. (Original magnification (×170,000) (*E*) A further stage in a maturing virion. The core and two lateral bodies are clearly differentiated within the viral envelope. (Original magnification ×170,000) (*F*) A large group of mature vaccinia virions, smaller and denser than in their formative stages. (Original magnification ×24,000) The inset shows the internal structure of a single particle at higher magnification. The dense dumbbell-shaped core is surrounded by a zone of lower density. (Original magnification ×150,000) (*A* and *F,* Dales S, Siminovich L: J Biophys Biochem Cytol 10:475, 1961; *B* and *C,* Morgan C: Virology 73:43, 1976; *D* and *E,* Dales S, Mosbach EG: Virology 35:564, 1968)

tribute to the eruptive lesions of smallpox, for when the Ab response and hypersensitivity are inhibited in infected rabbits by x-irradiation or cytotoxic drugs, the characteristic pustules do not form, although viral multiplication is unrestricted. The toxinlike properties of the viral particles may also play a role in the cell necrosis. It is clear, however, that pustule formation does not result from secondary bacterial infection, since bacterial cultures of the pustule fluid are ordinarily sterile.

Laboratory Diagnosis

In the event of a recurrence of smallpox, the recognition of the disease will increasingly depend on laboratory procedures as physicians become less familiar with the clinical manifestations. These procedures either identify or isolate the virus, detect viral Ags, or measure a rise in titer of specific Abs. The recommended procedures and the specimens to be collected depend on the stage of the suspected disease (Table 54–1). The stability of the infectious virus and its Ags facilitates laboratory diagnosis because materials can be transported without danger of inactivation.

An early presumptive diagnosis can be made most rapidly by visualizing viral particles (elementary bodies) by electron microscopy, or even by light microscopy in a Giemsa-stained smear of scrapings from the bases of skin lesions. Virus is abundant in lesions during the active disease, but it is rarely detected during the incubation period and prodromal stage before the rash appears. A positive smear is decisive, but a negative result calls for further tests. Identification can be confirmed by viral isolation or by immunologic technique (see Table 54–1).

Epidemiology

Smallpox is confined to man and is **spread chiefly by person-to-person contact.** Although smallpox is considered to be highly contagious, spread is slow, and the probability of infection from a single exposure appears to be low.

Initially the virus is transmitted from the lesions of the upper respiratory tract in droplet secretions or by contamination of drinking or eating utensils; later, when the vesicles or pustules rupture and crusts are formed, the skin lesions also become a source of contagion. Dissemination by fomites is important because the virus is resistant to ordinary temperatures and drying. Airborne transmission of variola virus is unusual but can occur, as has been demonstrated epidemiologically and experimentally.

Although any person infected with variola virus is potentially contagious, the most dangerous disseminators are persons with unrecognized disease, e.g., the partially

TABLE 54–1. Diagnosis of Variola by Laboratory Tests

	Pre-eruptive	Maculopapular and Papular			Vesicular		Pustular		Crusting		
Method of Detection of Virus	Blood	Blood	Skin Lesions	Saliva	Blood	Skin Lesions	Blood	Pustular Fluid	Blood	Crusts	Time Required for Test
Microscopic examination (electron microscopy [most sensitive] or Giemsa-stained smears)			+			+		±		−	1 hour
Viral isolation (culture on chorioallantoic membrane of 12- to 14-day-old chick embryos or in tissue culture)	±	±	+	+	±	+		+		+	1–3 days
Antigen detection* (complement fixation, enzyme-linked immunosorbent assay [ELISA], agar-gel precipitation, immunofluorescence, or radioimmunoprecipitation)	±	±	+			+		+		+	3–24 hours
Detection of antibodies (hemagglutination inhibition, CF, ELISA, neutralization, or radioimmunoprecipitation)	−	±			+		+		+		3 h to 3 days

* Probably the most useful, economical, and efficient of diagnostic procedures. Positive results indicate a rise in Ab titer.

+, test usually positive; ±, test may or may not be positive; −, results usually negative; open spaces also indicate negative results.

(Modified from Downie AW, MacDonald A: Br Med Bull 9:191, 1963)

immune patient who has relatively few lesions. Such cases, easily overlooked or misdiagnosed, have been primarily responsible for introducing smallpox into countries free of the disease.

Prevention and Control

From the systematic beginnings by Jenner in England at the close of the 18th century, artificial immunization has become increasingly effective.* The worldwide eradication of smallpox by the WHO demonstrates the ultimate effectiveness of vaccination. This remarkable achievement was possible, however, primarily because of basic characteristics of the virus: the genetic stability of its Ags responsible for neutralizing Abs; the viremic stages of pathogenesis in which the virus is maximally exposed to Abs; and the lack of an animal reservoir. The eradication campaign was based on the principal of surveillance and containment, i.e., isolation of cases and early immunization of all their contacts. The alternative, eradication by immunizing the entire population of a country, was discarded because it proved impossible to reach all susceptible persons.

PREPARATION OF VACCINE. Since its original use for immunization against smallpox, cowpox virus has been propagated in many different laboratories under diverse conditions, and it is now believed, on the basis of its antigenic structure, to have been inadvertently replaced with an attenuated smallpox virus. The vaccinia virus used today is distinctly different from the cowpox virus encountered in nature.

Successful immunization requires the use of infectious (attenuated) virus, because of the marked lability of the protective Ag. The virus infects the skin at the site of inoculation and ordinarily does not produce viremia. The virus most commonly employed in the vaccine is a dermal strain of uncertain origin. It is prepared from scrapings of vaccinial lesions on the skin of calves or sheep, with 1% phenol added to kill contaminating bacteria and 40% glycerol added to increase the stability of the virus. WHO successfully used lyophilized vaccines to overcome the problem of inactivation of infectivity in hot climates.

* Variolation to protect against smallpox was practiced long before infectious agents and concepts of immunization were understood. The Chinese powdered old crusts and applied them to the nostrils; Brahmins in India preserved crusts and inoculated them into the skin of the unscarred; Persians ingested crusts from patients; and in Turkey, fluid from pocks was inoculated. It was this last practice that Lady Mary Wortly Montague, wife of the British ambassador to Turkey, introduced into England in 1718. Crusts and vesicle fluids were selected from patients during epidemics of mild disease (alastrim). The practice spread to the colonies, where it was more widely used than in the British Isles, but it never became popular because of the risks involved. Jenner introduced the use of attenuated (cowpox) virus in 1776, prompted by the clinical observation of milkmaids who acquired cowpox usually escaped smallpox, even when the disease was rampant in the community.

ADMINISTRATION OF VACCINE AND RESULTS. Classically, vaccine was administered intradermally by gently breaking the epidermis under a drop of vaccine; air jet has been particularly effective for immunization of large numbers, and this was the technique that the WHO employed in the eradication program. Puncture or scarification permits infectious virus to enter the skin, where it multiplies in the deeper layers of the epidermis. The extent of multiplication and spread of virus, and thus the type of reaction that ensues, depend on the state of immunity (and hypersensitivity) of the host. One of three **responses** is seen: (1) **primary** response; (2) **accelerated (vaccinoid)** response; and (3) **early immediate** response (Table 54–2).

Failure to elicit any dermal response is sometimes seen, but it is never the result of complete immunity; it simply indicates that the vaccination technique was faulty or the vaccine inadequate.

Eradication was possible because immunity developing 7 to 10 days after vaccination can protect those contacts of persons with smallpox who are vaccinated shortly after exposure (the incubation period is about 12 days). Protection lasts for 3 to 7 years, but mild smallpox may occur only 1 year after known successful vaccination.

COMPLICATIONS OF VACCINATION. Though vaccination is relatively safe, it gives rise to rare but occasionally fatal complications affecting the skin or central nervous system, especially with initial vaccinations (Table 54–3). Probably the most alarming complication is progressive spread of a primary vaccination response with extensive necrosis of skin and muscle (vaccinia gangrenosa) in those rare persons with thymic dysplasias, who cannot develop cellular immunity (about 1.5 cases per million primary vaccinees). It is essential that physicians and public health officers be aware of these complications and not attempt to vaccinate persons to treat unrelated diseases; e.g., physicians still unsuccessfully treat recurrent herpes simplex virus lesions by vaccination, occasionally resulting in severe complications including spread to unsuspecting contacts. Despite the eradication

TABLE 54–2. Immunologic Status Affecting Response to Vaccination

Response	Day of Appearance (Mean)	Interpretation
Primary	4	No immunity
Accelerated	3	Partial immunity; delayed hypersensitivity
Early or immediate	1	Delayed hypersensitivity; may or may not have immunity

TABLE 54–3. Incidence of Complications Associated With Smallpox Vaccination in the United States

Complication	Complications per 10⁶ Primary Vaccinations (by Age, in Years, at Vaccination)					Complications per 10⁶ Revaccinations (All Ages)
	<1	*1–4*	*5–19*	*20+*	*All Ages*	
Death (from all complications)	5	0.5	0.5	Unknown	1.0	0.1
Postvaccinial encephalitis	6	2	2.5	4	2.9	0.0
Vaccinia gangrenosa	1	0.5	1	7	0.9	0.7
Eczema vaccinatum	14	44	35	30	38	3
Generalized vaccinia	394	233	140	212	242	9
Accidental vaccinia infection	507	577	371	606	529	42

(Modified from Center for Disease Control Morbidity and Mortality Weekly Report 20:340, 1971)

of smallpox, it is still necessary to be aware of the reactions to and complications of vaccination because of the continued use of vaccination in the Armed Forces of the United States owing to the fear that smallpox virus might be used for biological warfare, and the proposed use of vaccinia virus as an expression vector for other immunization programs and gene therapy.

Other Poxviruses That Infect Humans

Several diseases other than variola and vaccinia are caused by poxviruses: monkeypox, cowpox, molluscum contagiosum, contagious dermatitis, milkers' nodules (paravaccinia), and tanapox.

MONKEYPOX

Human infection with monkeypox is clinically indistinguishable from smallpox. The disease is a rare zoonosis and was unrecognized in humans until smallpox was eradicated in the equatorial rain forest areas of west and central Africa. Most cases of human monkeypox have characteristic clinical and epidemiologic features. There is a 2-day prodrome followed by a typical smallpox rash, which evolves over 2 to 4 weeks; the lymphadenopathy is more prominent than in smallpox. The fatality rate is about 15%, and about 13% of cases are mild or very atypical, suggesting the possible occurrence of unrecognized cases. The interhuman transmission rate is much less than with smallpox, and cases resulting from tertiary transmission have not been observed. Thus, although there is concern that monkeypox virus might replace variola virus as a dangerous human pathogen, extensive genetic changes would be required in the monkeypox virus genome before it posed such a danger.

Seroepidemiologic surveys suggest that forest-dwelling monkeys, squirrels, and porcupines are involved in the natural cycle of viral transmission. The means by which humans are infected, however, have not been determined.

COWPOX

Cowpox is a self-limiting occupational disease of humans acquired from the udders and teats of infected cows. The vesicular inflammatory lesions are usually localized on the fingers, but the virus may accidentally be implanted on the face or other parts of the body.

Cowpox virus has properties similar to those of variola and vaccinia virus, but its antigenic structure differentiates it from the other agents in the subgroup. The host ranges of cowpox and vaccinia viruses are similar, but cowpox virus differs in several respects: (1) pocks appear more slowly on chorioallantoic membranes; (2) the virus has a tendency to invade mesodermal tissue, involving capillary endothelium and thus producing hemorrhagic ulcers in the pocks; (3) the inclusion bodies are larger and more eosinophilic than classic Guarnieri bodies; and (4) keratitis is produced slowly in rabbits, in comparison with the rapid development effected by vaccinia virus.

MOLLUSCUM CONTAGIOSUM

The molluscum contagiosum virus produces an uncommon skin disease affecting mainly children and young adults. The lesion is a chronic, proliferative process, restricted to the epithelium of the skin of the face, arms, legs, back, buttocks, and genitals. The virus has been shown to be sexually transmitted, producing inflamed or ulcerated lesions confused with those produced by herpes simplex virus. Electron microscopic observations reveal that the molluscum body (a large cytoplasmic inclusion body) is composed of virions indistinguishable from those of other poxviruses. Mature viral particles develop by a process resembling the formation of vaccinia virions. In addition to the relatively uncommon clinical infections, molluscum contagiosum virus has been transmitted experimentally to humans, and infections have been achieved in cultures of HeLa cells and primary human amnion and foreskin cells. Although virus could not be serially propagated in cultures, viral particles developed, as revealed by electron microscopic examination of thin sections, and cytopathic changes appeared.

The tendency to induce proliferative lesions is even more striking with molluscum contagiosum virus than with other poxviruses. This virus thus appears to provide a link between the common pathogenic viruses of man and tumor-inducing viruses. However, it should be noted that infected cells do not continue to synthesize DNA; it is the neighboring uninfected cells whose rate of cell division is stimulated. The mechanism of this stimulation is unknown.

MILKERS' NODULES (PARAVACCINIA)

Jenner recognized the existence of two diseases affecting the udder and teats of cows: classic cowpox and a second condition consisting predominantly of vesicular lesions. The latter disease is also transmitted to humans, producing painless smooth or warty "milkers' nodules" on the hands and arms. The lesions rarely become pustular. The infected cells contain eosinophilic cytoplasmic inclusion bodies and elementary bodies characteristic of poxvirus infection. Disease is associated with only mild constitutional symptoms and enlargement of regional lymph nodes.

Infection does not confer immunity to either cowpox or vaccinia viruses. Paravaccinia virus cannot be propagated on the chorioallantoic membrane of chick embryos or in laboratory animals usually susceptible to cowpox. The virus, which was isolated from a milkers' nodule of humans, has been serially cultured in fetal bovine kidney tissue cultures, as well as in diploid bovine conjunctival cells and human embryonic fibroblasts. In contrast to many poxviruses, it cannot be propagated serially in continuous human cell lines, such as HeLa cells. Its cytopathic effects resemble those produced by vaccinia virus,

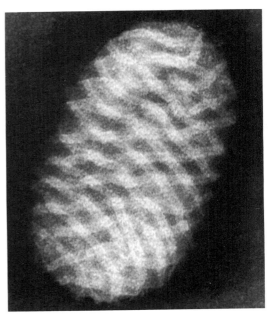

Figure 54–10. Mature contagious pustular dermatitis virus particle, negatively stained with phosphotungstic acid. The woven pattern of threads or tubules can clearly be seen. The apparent crisscrossing of the tubules results from the visualization of both the front and the back faces of the particle. (Original magnification ×600,000, reduced; Nagington J, Horne RW: Virology 16:248, 1962)

and infected cells prepared with Giemsa's stain show metachromatic cytoplasmic inclusion bodies.

Paravaccinia virus has stability characteristics similar to those of poxviruses. In thin sections the average viral particle measures 120 nm × 280 nm and has the typical morphology of a poxvirus. Electron microscopic observations of preparations stained with sodium phosphotungstate (Fig. 54–9) reveal ovoid virions whose size and surface structures are identical to those of contagious pustular dermatitis (Fig. 54–10) and bovine pustular stomatitis viruses.

CONTAGIOUS PUSTULAR DERMATITIS (ORF)

Although a natural affliction of sheep, mainly affecting lambs, contagious pustular dermatitis occurs rarely as an occupational disease of man. Sheep characteristically develop vesicles in the oral mucosa; these become encrusted and heal slowly after several weeks. In humans the infection usually causes a single lesion on a finger, beginning as a small, painless vesicle, which becomes pustular, encrusts, and finally heals. Transmission of infection from man to man has not been recorded.

The causative agent, contagious pustular dermatitis virus, can be isolated in various animal cell cultures or on the chick chorioallantoic membrane. It produces

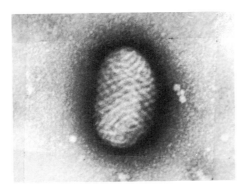

Figure 54–9. Electron micrograph of a viral particle isolated from a lesion of a milker's nodule. The negatively stained viral particle shows the characteristic morphology of a poxvirus. (Original magnification ×114,000; Friedman-Kien AE et al: Science 140:1335, 1963)

characteristic cytopathic changes but rarely gives rise to eosinophilic inclusion bodies. Electron microscopic study of the viral particles (see Fig. 54–10) reveals prominent tubule-like structures characteristic of poxviruses; the virion, in contrast to vaccinia and variola virions, is ovoid (160 nm × 260 nm) and is encircled in a regular pattern by the surface tubules.

TANAPOX VIRUS

A newly identified virus with characteristic poxvirus morphology has been isolated from two epidemics affecting several hundred tribesmen along the Tana River in Kenya. The disease is characterized by one or two pocklike lesions on the exposed upper part of the body and a febrile illness accompanied by severe aching and prostration. The lesion begins with a papule that develops into a raised vesicle and then umbilicates. Initially the pock resembles that of smallpox, but pustulation never follows. The virus has a host range limited to human and monkey cell cultures. It is serologically distinguishable from other orthopoxviruses, including the Yaba poxvirus of monkeys. The tanapox virus resembles a monkeypox virus that has affected monkeys in captivity in the United States, but in nature simian outbreaks have not been detected. Tanapox virus is probably a monkeypox virus that has been transmitted to man in Africa, where the natives use monkey meat and skins.

Selected Reading

BOOKS AND REVIEW ARTICLES

Behbalani AM: The smallpox story: Life and death of an old disease. Microbiol Rev 47:455, 1983

Dales S, Pogo BGT: Biology of poxviruses. Virol Monogr 18:1, 1981

Foege WH, Eddins DL: Mass vaccination programs in developing countries. Prog Med Virol 15:205, 1973

Fenner F: Portraits of viruses: The poxviruses. Intervirology 11:137, 1979

Mack TM: Smallpox in Europe, 1959–1971. J Infect Dis 125:161, 1972

Marsden JP: Variola minor: A personal analysis of 13,686 cases. Bull Hyg 23:735, 1948

Moss B, Winters E, Jones EV: Replication of vaccinia virus. In Cozzarelli N (ed): Proceedings of the 1983 UCLA Symposium on Mechanics of DNA Replication and Recombination, p 449. New York, Alan R. Liss, 1983

SPECIFIC ARTICLES

Arita I: Virological evidence for the success of the smallpox eradication programme. Nature 279:293, 1979

Arita I, Gramyko A: Surveillance of orthopoxvirus infections, and associated research, in the period after smallpox eradication. Bull WHO 6:367, 1982

Bladen RV: Mechanisms of recovery from a generalized viral infection: Mousepox. II. Passive transfer of recovery mechanisms with immune lymphoid cells. J Exp Med 133:1074, 1971

Councilman WT, MacGrath GB, Brinckerhoff WR: The pathological anatomy and histology of variola. J Med Res 11:12, 1904

Downie AW, Taylor-Robinson CH, Count AE et al: Tanapox: A new disease caused by a pox virus. Br Med J 1:363, 1971

Gerhelin P, Berns KI: Characterization and localization of the naturally occurring cross-links in vaccinia virus DNA. J Mol Biol 88:785, 1974

Immunization Practices Advisory Committee: Smallpox vaccine. Morbidity and Mortality Weekly report. 34:341, 1985

Kates J, Beeson J: Ribonucleic acid synthesis in vaccinia virus. I. The mechanism of synthesis and release of RNA in vaccinia cores. J Mol Biol 50:1, 1970

Morrison DK, Moyer RW: Detection of a subunit of cellular pol II within highly purified preparations of RNA polymerase isolated from rabbit poxvirus virions. Cell 44:587, 1986

Harold S. Ginsberg

Picornaviruses

History

Until the 1900s poliomyelitis (Gr. *poli*, gray, and *myelos*, spinal cord) was a disease primarily of infants (hence the name "infantile paralysis"), and this is still the pattern where sanitation is primitive. But with improved sanitation in many countries, in the 75 years prior to widespread immunization in 1960s, epidemics increased, the age distribution advanced, and the disease showed increasing severity as it appeared in young adults. This paradoxical response to improved sanitation was eventually explained by the findings that (1) practically everyone became infected, though the paralytic disease was rare, and (2) the consequences were usually negligible if infection was acquired early in life but might be serious when infection was postponed.

Clinically severe poliomyelitis was never very prevalent. In 1953 there were 1450 deaths and about 7000 cases with residual paralysis in the United States (versus about 500 deaths from measles, which was considered hardly more than a nuisance). However, the visibility of the crippled survivors caused even small epidemics to be terrifying. The problem was dramatized by the severe handicap of Franklin D. Roosevelt, who acquired poliomyelitis as an adult. The public's generous financial support of research (through the March of Dimes) led within 20 years to essentially complete control by immunization. Though the poliovirus was one of the most difficult to work with at the start of this program, it became a model for investigation of many other animal viruses.

In 1909 Landsteiner and Popper transmitted poliomyelitis to monkeys by intracerebral inoculation of a spinal cord filtrate from a patient, and the responsible agent was shown to be a virus.* However, progress depended

---

* Karl Landsteiner was also distinguished for his profound contributions to the understanding of immunologic specificity and for the discovery of human blood groups.

on the development of improved techniques for cultivating the virus. For example, as long as monkeys had to be used for experimentation, epidemiologic studies could demonstrate only that three antigenically distinct polioviruses exist. Adaptation of polioviruses to the cotton rat by Armstrong, in 1939, was a substantial step forward. The turning point came in 1945, however, when Enders, Weller, and Robbins showed that polioviruses can be isolated and readily propagated in cultures in nonneural human or monkey tissue. The incisive investigations that followed soon led to control of the disease.

Many related viruses were discovered as accidental by-products of the intensive pursuit of polioviruses. Thus, **coxsackieviruses*** were isolated in 1948 from the intestinal tract of children by intracerebral inoculation of newborn mice. The subsequent introduction of tissue culture techniques revealed the **echoviruses** (enteric *c*ytopathic *h*uman *o*rphan viruses), a third group of viruses in the gastrointestinal tract of man. They were called orphans because initially they were not clearly associated with disease. A fourth group of related viruses, designated **rhinoviruses** (Gr. *rhino*, nose), was discovered in 1956 during studies of mild upper respiratory infections fitting the description of the common cold.

Classification and General Characteristics

Polioviruses, coxsackieviruses, and echoviruses are similar in epidemiologic pattern, in physical, chemical, and biologic characteristics, and in infecting the human gastrointestinal tract. They were originally given the name **enteroviruses,** but this term seemed inadequate when some coxsackieviruses and echoviruses were also found to produce acute respiratory infections. With the discovery of rhinoviruses, which have similar chemical and physical characteristics but produce primarily acute respiratory infection, **Picornaviridae** (vernacular, **picornaviruses**) was coined as the family designation (*pico*, implying small, and RNA, the nucleic acid component). However, because of differences in certain physical, chemical, and biological characteristics, human picornaviruses were classified into two genera: **Enteroviruses** (which occasionally cause respiratory rather than intestinal or neurologic disease) and **Rhinoviruses** (see Table 55–1 and "Characteristics of Picornaviruses"). To simplify classification and to avoid confusion caused by overlap of host range characteristics, newly isolated enteroviruses are no longer divided into coxsackieviruses and echoviruses. From type 68 upward only the species designation *enterovirus* is employed.

* Named after Coxsackie, NY, the town from which the initial isolates were obtained

TABLE 55–1. Classification of Picornaviruses Affecting Humans

Genus	Species	No. of Types
Enterovirus	*Poliovirus*	3
	Coxsackievirus	
	Coxsackievirus A	23*
	Coxsackievirus B	6
	Echovirus	32†
	Enterovirus	5‡
Rhinovirus	*Rhinovirus*	113

* Type 23 was shown to be identical to echovirus type 9; A23 has been dropped, and the number is unused.

† Type 10 has been reclassified as reovirus 1, and type 28 as rhinovirus 1; the numbers are now unused.

‡ Type 68–72

CHARACTERISTICS OF PICORNAVIRUSES*

Size: 22–30 nm
Morphology: Icosahedral
Capsomers: Probably 32
Nucleic acid: Single-stranded RNA
Reaction to lipid solvents: Resistant
Stability at room temperature: Relatively stable
pH stability: Enteroviruses: Stable at pH 3–9
        Rhinoviruses: Unstable below pH 6
Stability at 50°C: Enteroviruses: Relatively unstable
        Rhinoviruses: Relatively stable
Density in CsCl: Enteroviruses: 1.33–1.34 g/cm^3
        Rhinoviruses: 1.38–1.41 g/cm^3

* Viruses similar to human picornaviruses have been found in several species of lower animals: the agent of foot-and-mouth disease in cattle (a member of the genus *Aphthovirus* that is physically similar to the rhinoviruses), Teschen disease viruses of pigs, and Mengo and encephalomyocarditis viruses of mice (similar to enteroviruses) are members of the genus *Cardiovirus*.

The physical and chemical properties of picornaviruses are summarized under "Characteristics of Picornaviruses." They are small, contain RNA, and do not contain lipid. Polioviruses, described in detail in the following section, will serve as the prototype of the family.

Polioviruses

PROPERTIES

MORPHOLOGY. Electron micrographs of purified virus in thin sections of virus-infected cells reveal small particles with a dense core. Negative staining shows a capsid with a subunit arrangement consistent with icosahedral symmetry (Fig. 55–1). There appear to be 32 capsomers per virion, although clear capsomers are not discernible. Po-

form a smaller peak ringed by promontories. Broad valleys surround the peaks at the fivefold axes, and shallow valleys separate the peaks at the threefold axes. VP4 is associated internally with the inner surface of the capsid and the viral RNA.

The virion RNA has the polarity of the viral mRNA (positive strand), it is infectious, and it can be translated *in vitro*. At its 3' terminus is a poly(A) track of about 90 nucleotides, which is necessary for its infectivity. In addition, the virion RNA has two unusual features: (1) its 5' end is not capped but terminates in pUp, and (2) a protein (**VPg**) of about 7000 daltons is covalently attached to its 5' end. VPg is always covalently linked to the virion RNA; its presence is essential for initiation of RNA replication. However, the RNA remains infectious if VPg is

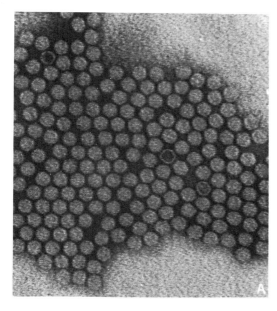

Figure 55–1. Electron micrograph of a purified preparation of poliovirus negatively stained. (*A*) Icosahedral symmetry of viral particles is evident. (Original magnification ×150,000) (*B*) Higher magnification of a viral particle printed in reverse contrast. Capsomers measure approximately 6 nm in diameter; their fine structure is not apparent. (Original magnification ×600,000) (*C*) Same particle as in *B*, marked to display two clear axes of fivefold symmetry (*white lines*). (Mayor HD: Virology 22:156, 1964)

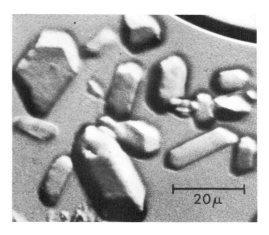

Figure 55–2. Crystals of purified type 1 poliovirus particles. (Schaffer FL, Schwerdt CE: Proc Natl Acad Sci USA 41:1020, 1955)

lioviruses and viruses of the other three picornavirus groups present only minor differences in size and structure.

PHYSICAL AND CHEMICAL CHARACTERISTICS. The characteristics of the polioviruses are given under "Characteristics of Polioviruses." Poliovirus was the first animal virus to be obtained in crystalline form (Fig. 55–2). A single molecule of single-stranded RNA constitutes about 30% of the virion; the remainder consists of four major (**VP1–4**) and one minor (**VPg**) species of proteins. Each surface subunit of the capsid, termed a **protomer,** is composed of three unique intimately associated polypeptide chains (VP1, VP2, and VP3; Fig. 55–3). At each fivefold axis of the capsid a pronounced tilt of the VP1 core forms a large ribbed peak, and at each threefold axis the cores of VP2 and VP3 alternate around the axis to

CHARACTERISTICS OF POLIOVIRUSES

Diameter of virion: 27–30 nm
Diameter of internal core: 16 nm
Diameter of capsomer: 6 nm
Molecular weight of RNA: 2.5×10^6 daltons (7.7 kilobases)
Base composition (G + C): 46 moles %*
Molecular weight of virion proteins†
  VP1: 35×10^3 daltons
  VP2: 28×10^3 daltons
  VP3: 24×10^3 daltons
  VP4: 6×10^3 daltons
  VPg: $\sim 7 \times 10^3$ daltons
Sedimentation coefficient of virion: 157–160 S_{20}
Particle mass of virion: 1.1×10^{-17} g
Molecular weight of virion: $8–9 \times 10^6$ daltons

* Composition is very similar for the three types.
† Virion proteins 1 through 4 are present in equal molar amounts.

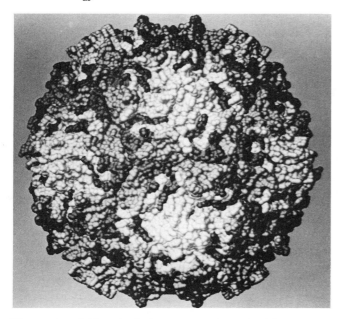

Figure 55–3. Folding of poliovirus major capsid proteins (VP1, VP2, and VP3) to form the outer surface of the virion (derived from computer analysis of x-ray crystallographic data). VP1 is light gray, VP2 is medium gray, and VP3 is dark gray. The three proteins form a protomer. VP4 is present in the inner surface of the capsid and is not visible. (Hogle JM, Filman DJ, Chow M: In Brown F, Chanock RM, Lerner RH [eds]: Vaccines 86: New Approach to Immunization, p 3. Cold Spring Harbor, NY, Cold Spring Harbor Laboratory, 1986)

removed by pronase because newly synthesized VPg is synthesized from the infecting genome.

There are three serotypes; their physical properties are identical, and their base compositions are very similar; the RNAs share 36% to 52% of their nucleotide sequences. Moreover, considerable nucleotide sequence homology exists between the RNAs of polioviruses, other enteroviruses, and rhinoviruses, as detected by cross-hybridization between RNA from virions and replicative forms of RNA from infected cells. This homology appears to be located in the viral RNA polymerase genes (at least between type 1 poliovirus and type 2 rhinovirus).

Polioviruses are more stable than many viruses (e.g., those with lipid envelopes). Hence, their transmission is facilitated because they can remain infectious for relatively long periods in water, milk, and other foods. However, polioviruses are readily inactivated by pasteurization and by many other chemical and physical agents. Magnesium chloride (1M) appears to stabilize their intercapsomeric bonds and hence markedly increases thermal stability.

IMMUNOLOGIC CHARACTERISTICS. The three distinct immunologic types of polioviruses can be recognized by neutralization, complement fixation (CF), gel-diffusion precipitation, and other immunoprecipitation reactions with type-specific sera.

There is no common poliovirus group Ag, but antigenic relations between types do exist. The cross-reactions are particularly prominent when heated virus is employed in CF titrations. The cross-reactivity, however, can be demonstrated only when sera are obtained from humans who have been infected with more than one type of poliovirus; i.e., after an initial infection the Ab response is strictly type-specific, but upon infection with a second type, Abs develop to two or all three of the viruses. Immunologic cross-reactivity between types 1 and 2 is also demonstrated in neutralization titrations by cross-absorption experiments and by the development of heterotypic Abs following natural infections or immunization. Slight cross-reactivity between types 2 and 3 can be detected by neutralization, but not between types 1 and 3. The immunologic kinship between types 1 and 2 Abs is epidemiologically substantiated: possession of type 2 Abs confers significant protection against the paralytic effects of subsequent type 1 infection.

Just as VP1, 2, and 3 polypeptides are closely associated on the surface of the virion (see Fig. 55–3), so each of these proteins can induce neutralizing Abs. VP1, however, is the **immuno-dominant Ag** for neutralizing Abs, and this surface protein contains at least four epitopes capable of inducing neutralizing Abs. VP2 and VP3 each contain a single epitope for neutralizing Abs.

Antigenic variants of types 1 and 2 viruses have been detected by precise plaque-reduction and CF studies with cross-absorbed sera. Through the use of monoclonal Abs, mutants resistant to neutralization, due to point mutations, can be isolated; when polyclonal sera are used, however, these antigenic differences do not affect the capacity of Abs induced by one strain to protect against infection by all other strains of the same type. Despite these minor intratypic differences, **polioviruses** actually **show marked antigenic stability,** both in nature and in laboratory manipulations.

Neutralizing Abs appear early in the course of poliovirus infection, and they have usually reached a high titer by the time the patient is first seen by a physician (Fig. 55–4). They attain a maximum titer 2 to 6 weeks after the onset of disease, decrease to about one fourth that level in 18 to 24 months, and then seem to persist indefinitely. Their presence confers clear protection against subsequent infection. CF Abs appear during the first 2 to 3 weeks after infection, reach maximum titers in about 2 months, and persist for an average of 2 years. Inapparent and nonparalytic infections result in Ab levels as high as those present after severe paralytic disease. Second attacks of paralytic poliomyelitis are rare and are invariably due to a different viral type from that producing the first illness.

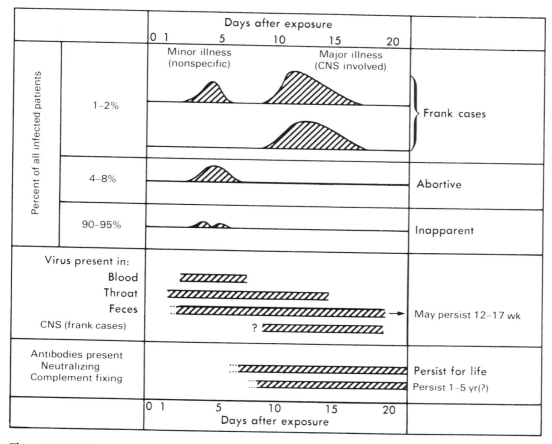

Figure 55—4. Diagram of the times at which the clinical forms of poliomyelitis appear, correlated with the times at which virus is present in various sites and with development and persistence of Abs. The high incidence of subclinical poliovirus infection is also noted. (Horstmann DM: Yale J Biol Med 36:5, 1963)

HOST RANGE. Humans are the only natural hosts for polioviruses. Abs are present in some monkeys and chimpanzees studied in captivity, but there is evidence that the infection is acquired only after capture.

Old World monkeys and chimpanzees are susceptible to infection (by the intracerebral, intraspinal, and oral routes) with fresh isolates as well as with laboratory strains. In contrast, nonprimates are relatively insusceptible. However, by serial passage, strains of poliovirus were adapted to cotton rats and mice. Strains of type 2 virus have also been adapted to suckling hamsters and the chick embryo.

An important development was the discovery that these viruses, hitherto considered purely neurotropic, can multiply and produce cytopathology in human extraneural tissues cultured *in vitro.* It then rapidly became evident that many tissues from primates can furnish cells susceptible to cytopathic changes (Fig. 55–5). Cultures of primate tissues are now widely used for isolation

and identification of polioviruses and other picornaviruses, for production of vaccines, and for experimental studies.

MULTIPLICATION. To initiate infection, polioviruses attach rapidly to specific host cell receptors (composed of lipid and glycoproteins), which are much more prevalent in susceptible than in nonsusceptible tissues. Such **adsorption** to susceptible cells is independent of temperature but depends on the concentration of electrolytes. Infectious particles, but not empty particles, can adsorb, an indication that the conformation of the capsid is critical for attachment. Very soon after attachment the viral capsids are altered by loss of VP4, and about 50% of these particles also lose their RNA; these become noninfectious and elute from the cells. The remaining particles **penetrate** into the host cells, probably in endosomes, and rapidly **uncoat their RNA,** which becomes susceptible to RNase within 30 to 60 minutes after infection.

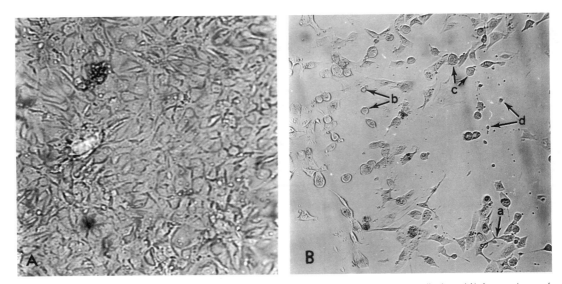

Figure 55–5. Cytopathic changes in monkey kidney cell cultures infected with type 1 poliovirus. (*A*) A monolayer of uninfected cells, unstained. (Original magnification ×200) (*B*) Advanced cytopathic changes in infected cultures, unstained. Polioviruses, like most enteroviruses that infect monkey kidney cells, produce marked cell retraction (*a*), rounding (*b*), and occasionally ballooning of cells (*c*), followed by rapid lysis, leaving a granular debris (*d*). (Original magnification ×200) (Ashkenazi A, Melnick JL: Am J Clin Pathol 38:209, 1962)

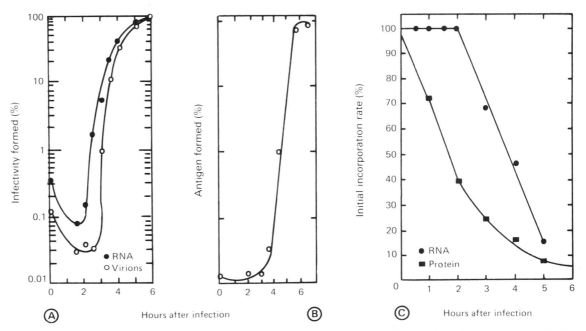

Figure 55–6. Biosynthetic events in poliovirus-infected cells. (*A*) Time course of viral RNA synthesis (measured by its infectivity) and of maturation of virions. (*B*) Biosynthesis of viral capsid proteins measured by incorporation of ^{14}C-labeled amino acids into Ab-precipitable material. (*C*) Rate of total RNA and protein synthesis in poliovirus-infected cells measured by the incorporation of ^{14}C-uridine or ^{14}C-L-valine into acid-precipitable material at the indicated times after infection. (*A*, Darnell JE Jr et al: Virology 13:271, 1961; *B*, Scharff MD, Levintow L: Virology 19:491, 1963; *C*, Zimmerman EF et al: Virology 19:400, 1963)

The **replication** of infectious virus follows the general pattern of viruses with positive-strand RNA, which initially serves as mRNA for the synthesis of viral proteins, including an RNA replicase. The RNA replicative intermediates then develop (see Synthesis of RNA Viruses, Chap. 48), serving first for synthesis of complementary negative strands and then for positive strands. The small **VPg** is covalently linked to the 5'-terminal oligonucleotide of all nascent RNAs (positive and negative strands); it is required for viral RNA synthesis, apparently serving as a primer for the formation of replication complexes. The VPg, however, is cleaved from about half of the plus strands, and these are destined to become uncapped messengers. Hence, VPg distinguishes virion RNAs from viral mRNAs of identical nucleotide sequences.

Replication of viral RNA is independent of biosynthesis of host cell DNA. The production of viral RNA commences within 15 minutes after viral uncoating is completed but all the early molecules, which are copied from the nascent complementary RNA templates, become messengers on very large cytoplasmic polyribosomes. Since internal initiation of protein synthesis on this message does not occur, this polygenic RNA of about 7000 nucleotides serves as a monocistronic message; it is **translated into a single long polypeptide,** termed a **polyprotein,** which is subsequently cleaved into the four individual viral capsid proteins plus VPg and the nonvirion proteins (see Chap. 48). Progeny RNA first appears in viral particles about 3 hours after infection (Fig. 55–6A); once **virion assembly** has started, production of capsid proteins (Fig. 55–6B) and RNA replication are closely coupled, and newly made viral RNA is incorporated into virions within 5 minutes after synthesis. The final step in morphogenesis (Fig. 55–7) appears to be the combination of viral RNA with a shell of viral proteins (VP0, VP1, VP3) termed the **procapsid** (see Chap. 48), during which one of the procapsid proteins (VP0) is cleaved to yield two of the final capsid structures (VP2 and VP4). Since complete viral replication occurs in cells enucleated with cytochalasin B, host cell nuclear functions are not required.

Final assembly of infectious particles is accomplished rapidly. Approximately 500 virions per cell are produced. Initially, virions are released through vacuoles, but after several hours they escape in a burst, accompanied by death and lysis of the host cell.

Synthesis of host cell proteins is inhibited very shortly after viral infection (Fig. 55–6C) owing to inactivation of initiation factors responsible for forming the host cap-binding protein complex. The cessation of host protein synthesis is accompanied by disruption of the host cell polyribosomes. Synthesis of normal host cell RNA ceases about 2 hours after infection, shortly after biosynthesis of viral RNA begins (see Fig. 55–6).

The **cytopathologic changes** accompanying these

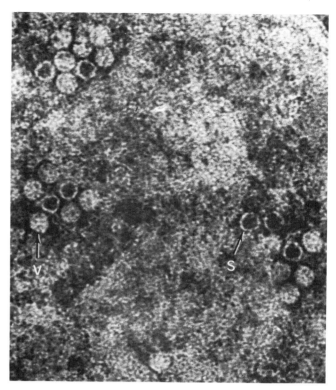

Figure 55–7. Development of poliovirus particles in pieces of the cytoplasmic matrix of artificially disrupted cells. Particles in various stages of assembly from empty shells (*s*) to complete virions (*v*) can be seen. (Original magnification ×200,000; Horne RW, Nagington J: J Mol Biol 1:33, 1959. Copyright by Academic Press, Inc. [London] Ltd.)

biosynthetic events are diagrammed in Figure 55–8. Intranuclear alterations, consisting of rearrangement of chromatin material with condensation at the nuclear membrane, are the first changes detected. One or more small intranuclear eosinophilic inclusion bodies of unknown nature form, and the nucleus becomes distorted and wrinkled and gradually shrinks. These events are probably related to the inhibition of synthesis of the host cell's protein and nuclear RNA (see Fig. 55–6C). The cytoplasm then develops a large eosinophilic mass, which is the site of replication and assembly of viral subunits (Figs. 55–8 and 55–9), and knobs appear on the cell membrane as the result of cytoplasmic bubbling associated with the release of virus. Finally the nucleus becomes pycnotic, the nuclear chromatin becomes fragmented, and the cell becomes rounded and dies.

GENETIC CHARACTERISTICS. The demonstration in 1953 that poliovirus contains only RNA stimulated great interest because it identified **RNA as genetic material.** The finding that polioviruses undergo mutations and re-

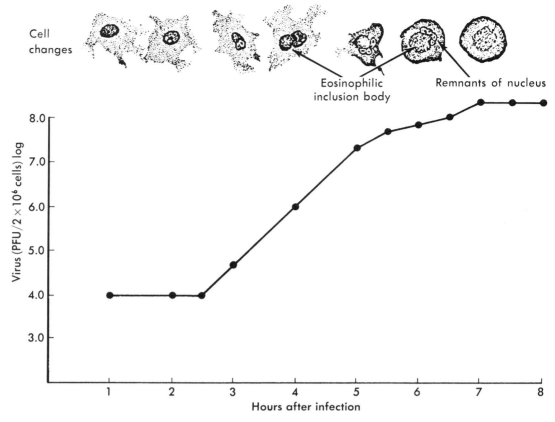

Cell changes

Eosinophilic inclusion body

Remnants of nucleus

Figure 55–8. Diagram of the multiplication of poliovirus and the accompanying pathologic changes in infected cells. A perinuclear cytoplasmic eosinophilic mass (inclusion body) develops as viral multiplication reaches its maximum; the inclusion body impinges on the nucleus, which degenerates as the cell dies. *PFU,* plaque-forming unit. (Adapted from Reissig M et al: J Exp Med 104:289, 1956)

Figure 55–9. Cytoplasmic changes and assembly of virions in a poliovirus-infected cell. Large numbers of membrane-enclosed pieces of cytoplasm (*B*) accumulate in the central region of the cell, pushing the nucleus to one side (the nucleus is not shown in this photograph). Large cytoplasmic vacuoles (*Va*) also develop. A large number of virions (*arrows*) are present in the cytoplasmic matrix, both between and within the membrane-enclosed bodies (*B*). Two large crystals of virions are present (*C*). (Original magnification ×25,000; Dale S et al: Virology 26:379, 1965)

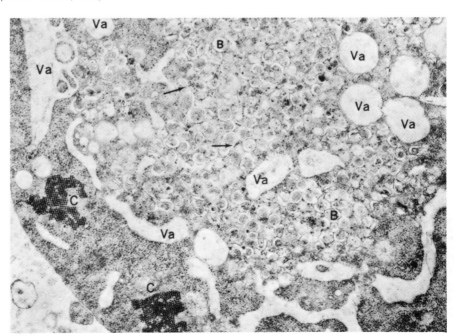

combination just as DNA viruses do provided a theoretical basis for developing attenuated mutants for use as a live virus vaccine. The number of mutant phenotypes observed (Table 55–2) is much larger than the number of viral genes, so that many phenotypes must arise from different mutations of the same gene.

Early studies revealed that recombination could occur between some of the mutants. With conditionally lethal, temperature-sensitive mutants several genes have been mapped: two genes for replication of viral RNA, one for synthesis of capsid proteins, and two for regulation of cell functions. Genetic maps have been obtained by the use of pactamycin to inhibit the initiation of protein synthesis; these maps locate the genes for virion proteins close to the 5' end of the RNA genome and in the order (5'→3') VP4→VP2→VP3→VP1. The nonstructural proteins are encoded toward the 3' end of the genome (see Chap. 48). Nucleotide sequence analysis of the entire genome and amino acid sequence determination of isolated gene products confirm the genetic map.

Genetic studies have also revealed that many mutations are **pleiotropic;** i.e., two phenotypic traits are changed by the same mutation, although the two phenotypes can also be changed separately. The *d* and *e* phenotypes (see Table 55–2), for example, can arise by a single mutation. Hence, mutations affecting neurotropism can be found among mutant phenotypes that are easily detectable *in vitro*, whereas their direct detection in primates would be much more limited and costly.

Several of the mutant phenotypes that are frequently associated with attenuation (see Table 55–2) affect the viral capsid, and others multiply preferentially at or below usual body temperatures. Thus, neurovirulence appears to depend on the ability of the virus to interact with certain cells and to replicate in febrile patients (rct/40 mutants). Furthermore, attenuated viruses induce the synthesis of more interferon than virulent viruses do, and are more readily inhibited by it (see Interferon, Chap. 49). However, the relatively common neurovirulent revertant of the type 3 Sabin vaccine strain does not conform to these patterns because it is due to a mutation changing the uridine at position 472 to cytidine in the small 5' noncoding region.

The ability to make infectious cDNA copies of the viral RNA genome will permit much more detailed genetic studies. In addition, intertypic and intratypic recombinants are being constructed to identify gene functions.

TABLE 55–2. Characteristics of Some Poliovirus Mutants

Class of Mutants	Characteristic(s)	Marker Name
Factors that affect cell–virus interaction and therefore viral multiplication	Ability to multiply at 40°C	rct/40
	*Inability to multiply at 40°C	rct/40⁻
	*Ability to multiply at 23°C	rct/23⁺
	*Heat defectiveness (inability to multiply at 40°C, but usual multiplication at 36°C)	hd
	Plaque size	s
	*Resistance to heating in AlCl₃	a
	Resistance to heat inactivation of virion	t
	Inability to grow in MS cells	ms
Variants distinguished by presence or absence of inhibitory substance in media	*Sensitivity to agar inhibitor at acid pH	d
	Sensitivity to agar inhibitor at neutral pH	m
	Cystine inhibition of multiplication	cy⁺
	Cystine dependence	cyᵈ
	Tryptophan dependence	
	Adenine resistance	
	Guanidine resistance	gʳ
	Guanidine dependence	gᵈ
	Hydroxybenzylbenzimidazole (HBB) resistance	HBBʳ
	Resistance to normal bovine serum inhibitor	bo
	Resistance to normal horse serum inhibitor	ho
Mutants whose markers are physical characteristics of the virus	*Poor elution from Al(OH)₃ gel	Al(OH)₃
	*Greater adsorbability to DEAE-cellulose	e
Immunologic variants	Intratypic antigenic variants	

* Phenotypes associated with attenuated viruses

PATHOGENESIS

The major sequence of events in the multiplication and spread of polioviruses was revealed by studies in chimpanzees and man, as well as in cell cultures. In humans the progression of infection culminates in invasion of the target organs, the brain and spinal cord (Fig. 55–10).

Infection is initiated by the ingestion of virus and its **primary multiplication** in the oropharyngeal and intestinal mucosa. It is not known, however, whether virus multiplies in epithelial or lymphoid cells of the alimentary tract. The tonsils and Peyer's patches of the ileum are invaded early in the course of infection, and extensive viral multiplication ensues in these loci, so that as much as 10^7 to 10^8 infectious doses (i.e., 10^7 to 10^8 times the mean tissue culture infectious dose [see Chap. 44]) of virus per gram of tissue may accumulate. From the primary infectious sites of propagation the virus drains into deep cervical and mesenteric lymph nodes, but since its titer there is relatively low, these nodes may not be im-

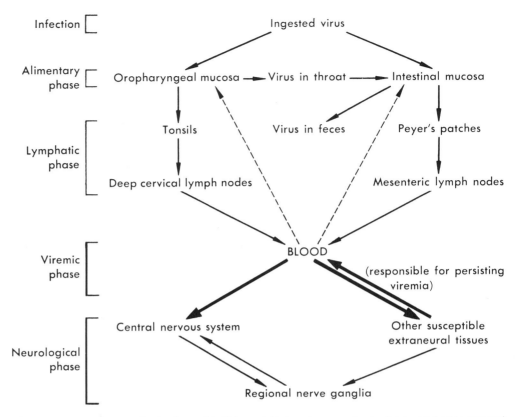

Infection

Alimentary phase

Lymphatic phase

Viremic phase

Neurological phase

Ingested virus

Oropharyngeal mucosa → Virus in throat → Intestinal mucosa

Tonsils   Virus in feces   Peyer's patches

Deep cervical lymph nodes   Mesenteric lymph nodes

BLOOD   (responsible for persisting viremia)

Central nervous system   Other susceptible extraneural tissues

Regional nerve ganglia

Figure 55–10. Pathogenesis of poliomyelitis. This model is based on a synthesis of data obtained in man and chimpanzees. (Adapted from Sabin AB: Science 123:1151, 1956; Bodian D: Science 122:105, 1955)

portant sites of viral replication for progressive infections. From the nodes the virus drains into the blood, resulting in a **transient viremia** that disseminates virus to other susceptible tissues, such as the brown fat (axillary, paravertebral, and suprasternal) and the viscera (probably in reticuloendothelial cells). In these extraneural sites the virus replicates, and it is continually fed back into the bloodstream to establish and maintain a **persistent viremic stage.** In most natural infections, even in nonimmune persons, only transient viremia occurs; the infection does not progress beyond the lymphatic stage, and clinical disease does not ensue.

Viral spread to the central nervous system (CNS) requires persistent viremia, which implies that direct invasion through capillary walls is the major pathway of penetration into the central nervous system. Therefore, the presence of specific Abs in the blood, even at the relatively low levels obtained by passive immunization, effectively halts viral spread and prevents invasion of the brain and spinal cord. However, transmission of virus along nerve fibers from peripheral ganglia may provide an additional route for entry into the CNS, because po-

lioviruses are found in these ganglia during the progression of infection, and virus can spread along nerve fibers in both peripheral nerves and the CNS.

Poliomyelitis generally conjures up the picture of a severe, crippling, and occasionally fatal paralytic disease. However, probably no more than 1% of infections culminate in that syndrome (see Fig. 55–4). A moderate number of infections induce transient viremia, resulting in a mild febrile disease, or so-called summer grippe.

The **course of classic paralytic disease** is initiated by a **minor disease,** which is associated with the viremia and is characterized by constitutional and respiratory or gastrointestinal signs and symptoms (see Fig. 55–4). There follows, after 1 to 3 days or often without any interval, the **major disease,** characterized by headache, fever, muscle stiffness, and paralysis associated with cell destruction in the CNS. Lesions causing **paralysis** occur most frequently in the anterior horn cells of the spinal cord (**spinal poliomyelitis**); similar lesions may occur in the medulla and brain stem (**bulbar poliomyelitis**) and in the motor cortex (**encephalitic poliomyelitis;** Fig. 55–11). Bulbar poliomyelitis is often fatal because of respira-

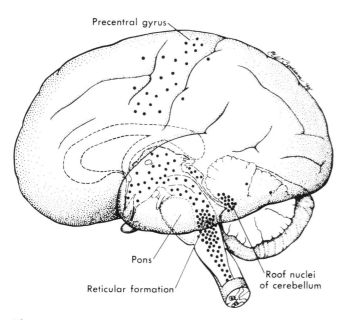

Figure 55–11. Lateral view of the human brain and the midsagittal surface of the brain stem. *Black dots*, usual distribution of lesions. Spinal lesions usually occur in the anterior horn cells; lesions in the cerebral cortex are largely restricted to the precentral gyrus; those of the cerebellum are largely found in the roof nuclei; lesions of the brain stem centers are widespread. (Bodian D: Papers and Discussions Presented at the First International Poliomyelitis Conference. Philadelphia, JB Lippincott, 1949)

tory or cardiac failure; the other forms may result in survival with highly variable patterns of residual paralysis. Paralysis becomes maximal within a few days of onset and is often followed by extensive recovery, which may be aided by physiotherapy. The recovery represents, in part, compensatory hypertrophy of muscles that have not lost their innervation. In the aging this compensatory hypertrophy may be lost; and marked weakness, even paralysis, may return. In addition, although the virus-infected cells in the lesions are irreversibly damaged, neighboring uninfected neurons may also contribute to the paralysis through edema and by-products of necrosis, but paralysis from this cause is reversible. The temporal relation among clinical manifestations, distribution of virus, and the appearance of Abs is summarized in Figure 55–4.

Several host factors may alter the course of infection: fatigue, trauma, injections of drugs and vaccines, tonsillectomy, pregnancy, and age. These factors do not increase the incidence of infection but do affect the frequency and severity of paralysis. Trauma, including hypodermic injections, tends to localize the paralysis to the traumatized muscles. Tonsillectomy, recent or of long standing, markedly increases the incidence of bul-

bar poliomyelitis. The mechanisms of action of these localizing host factors are not certain. They may increase infection of peripheral nerve ganglia or transmission of virus along peripheral nerves associated with the affected area, or they may increase the permeability of blood vessels in corresponding areas of the CNS. The greater severity of paralysis in adults, and even more in pregnant women, may be related to endocrine factors: steroids, for example, greatly heighten the severity of infection in experimental animals.

LABORATORY DIAGNOSIS

Laboratory methods for the diagnosis of poliomyelitis are simple and efficient, because of the ready availability of excellent tissue culture methods and immunologic techniques. Virus is isolated most readily from feces or rectal swabs for about 5 weeks after onset, and from pharyngeal secretions for the first 3 to 5 days of disease. Multiplication of virus is detected by the development of characteristic cytopathic changes (see Fig. 55–5B) and by failure of the infected tissue cultures to become acid in 1 to 4 days after inoculation, revealed by incorporating phenol red in the medium (since this indicator shifts from red to yellow on acidification). The latter criterion arises from the inability of the dying infected cells to produce organic acids from glucose. Final identification of the virus is accomplished by neutralization tests by means of a standard serum for each of the three types. If an enterovirus other than poliovirus is also present, which is not uncommon, identification is more complex and requires multiple pools of antisera.

For serologic diagnosis, Ab levels are compared in sera obtained during the acute phase of the disease and 2 to 3 weeks after onset, by means of neutralization or, less often, CF titrations. Titration endpoints in neutralization are based on the inhibition of cytopathic effects or on continued acid production; the latter assay is especially convenient, since it can be carried out in disposable plastic trays and the endpoints can be determined rapidly without microscopic observations. CF tests are usually less dependable, because the different Ags from intact and incomplete virions are present in viral preparations, and Abs directed against each Ag appear at different times.

EPIDEMIOLOGY

Serologic surveys show that polioviruses are globally disseminated. In densely populated countries with poor hygienic conditions, practically 100% of the population over 5 years of age has Abs to all three types of poliovirus, epidemics do not occur, and paralytic disease is rare. In countries with improved sanitation, in contrast, the young are shielded from exposure, and prior to the effective use of vaccines, many who reached adulthood had escaped infection and were therefore without protective Abs. Because the incidence and severity of paralytic disease increase with age, if infection is delayed until susceptible persons are above age 10 to 15 years, more severe crippling disease occurs.

The widespread use of vaccines has strikingly altered the epidemiologic picture. In some countries (Sweden, Finland, Denmark, the Netherlands), paralytic poliomyelitis appears to have been eradicated. In the United States, epidemics have been eliminated except in pockets of lower socioeconomic groups, among whom immunization has not been widespread (although available free of charge); a recent study indicated that there were at least 20 million unimmunized children. Among those unprotected, small epidemics are again occurring in the very young (reviving the picture of "infantile paralysis") because, in contrast to the situation in countries with poor sanitation, polioviruses are not widely disseminated while babies are still protected by maternal Abs. However, the few cases of poliomyelitis that occur are also seen in older children and young adults.

TRANSMISSION. Poliomyelitis occurs primarily in the summer, like the common summer diarrheal diseases. This finding first suggested transmission by the fecal route. Indeed, a large amount of virus is excreted in the feces for an average period of 5 weeks after infection, even in the presence of a high titer of circulating Abs (see Fig. 55–4). A patient is maximally contagious, however, during the first week of illness, when pharyngeal excretion of virus also occurs. Multiple modes of infection probably account for the fact that infections can occur in any season of the year.

Person-to-person contact is the primary mode of spread, and transmission within families and schools appears to be the major mechanism of dispersion throughout a community. Presently, however, occurrence of the rare case in immunized populations does not produce secondary cases owing to the insufficient numbers of susceptible persons. Prior to widespread immunization, flies occasionally served as accidental vectors, but they were not an important mode of distribution. Water- and milk-borne epidemics caused by fecal contamination have also been reported. Dissemination of virus is rapid and extensive in nonimmune members of a family or in other contact groups, but the ratio of paralytic disease to inapparent infections is low, i.e., about 1:200 in temperate zones.

PREVENTION AND CONTROL

Until vaccine became available in 1954, the only approaches to the control of infection were passive immunization and nonspecific public health measures (isola-

tion of patients; closing of such gathering-places as schools and swimming pools; widespread spraying of insecticides); none of this proved successful in preventing or stopping an epidemic.

The present era of successful control can be attributed to three major discoveries: (1) Protection is required against the three distinct antigenic types of poliovirus. (2) Multiplication of poliovirus to a high titer in cultures of nonnervous tissues affords a practicable procedure for preparing large quantities of virus free of the nervous tissue that may induce demyelinating encephalomyelitis. (3) As the role of viremia in pathogenesis suggested, the infection can be interrupted before the CNS is infected. In addition, the protection of monkeys, mice, and man by passive immunization with immune serum or pooled γ-globulin proved that even low titers of Abs can be effective in preventing paralytic poliomyelitis.

The development of poliomyelitis vaccine proceeded by two different approaches: the preparation of an **inactivated virus vaccine,** based on evidence that poliovirus inactivated with formalin could immunize monkeys; and the development of a **live attenuated virus vaccine,** modeled on the successful control of smallpox and yellow fever by such vaccines.

INACTIVATED VIRUS VACCINE. Salk demonstrated that all three types of polioviruses could be inactivated in about 1 week by 1:4000 formalin, pH 7, at 37°C, with retention of adequate antigenicity. When purified virus is used, the inactivation follows pseudo–first-order kinetics. However, in crude viral preparations, aggregation of the viral particles results in a complex inactivation curve: the exponential rate of viral inactivation is not constant, and the inactivation curve tails off markedly. Failure to recognize this complication led to some serious initial difficulties in vaccine production, exemplified by an incident in which residual infectious virus in several lots of commercial vaccine induced 260 cases of poliomyelitis with 10 deaths. Fortunately, the errors were soon rectified, and a safe, highly effective vaccine was developed.* Extensive controlled studies showed an effective protection against paralytic poliomyelitis in 70% to 90% of those immunized, and subsequent use of vaccine in the general population confirmed its protective ability.

The inactivated vaccine has been supplanted by the live attenuated vaccine (see below) in the United States, but inactivated vaccine is still popular elsewhere. When

---

* In the presence of 1M $MgCl_2$, inactivation by formalin shows much less tailing off. Moreover, this procedure, which can be carried out at 50°C, inactivates adventitious viruses present in monkey cell cultures. These viruses, which include SV40 virus (see Chap. 64), are more resistant to formalin inactivation than poliovirus but are not stabilized by $MgCl_2$ against heat inactivation. Because 1M $MgCl_2$ also selectively reduced heat inactivation of infectious poliovirus, heating in its presence may similarly be used with infectious virus preparations to eliminate extraneous viruses such as SV40.

the inactivated vaccine is used in the United States (for example, for immunodeficient children), it is recommended that the vaccine containing all three poliovirus types be administered in three intramuscular or subcutaneous injections over a 3- to 6-month period and that a fourth injection be given after 6 to 12 months. In the early, extensive studies of this vaccine, Ab levels for all three types appeared to fall to approximately 20% of their maximum titer within 2 years, and thereafter to decline at a slower rate. The actual persistence of Abs was difficult to evaluate, however, because of the uncontrolled occurrence of reinfections. Booster injections of vaccine every 5 years are therefore recommended. More concentrated vaccines are being tested abroad, and it is believed that these preparations will not require more than two primary inoculations.

Immunization does not prevent reinfection of the alimentary tract unless serum Ab levels are very high (which is unusual except shortly after booster doses). However, infection of the oropharyngeal mucosa and tonsils is generally prevented, eliminating transmission by pharyngeal secretions. This effect may explain the decreased incidence of infection observed after the widespread use of inactivated virus vaccine.

LIVE ATTENUATED VIRUS VACCINE. Infection of the alimentary canal with attenuated live viruses offers several hypothetical advantages: (1) long-lasting immunity, similar to that following natural infections, (2) prevention of reinfection of the gastrointestinal tract and therefore elimination of this route for transmission of the virus, and (3) inexpensive mass immunization without the need for sterile equipment.

Three different sets of attenuated viruses were independently selected by Cox, Koprowski, and Sabin by multiple passage in a foreign host, most frequently tissue culture. The strains developed by Sabin were chosen by the U.S. Public Health Service for commercial production of vaccines. These strains lack neurovirulence for susceptible monkeys inoculated both intramuscularly and intracerebrally, but they occasionally cause paralysis following intraspinal inoculation. The type 1 and 2 strains are genetically stable, probably because they contain several mutations that decrease virulence (see Table 55–2), although a minor increase in neurovirulence may occur during passage in humans. Fortunately, however, the alimentary tract of humans does not offer a marked selective advantage to mutants with increased neurovirulence. In contrast, the type 3 vaccine strain reverts more frequently; it is estimated to produce approximately one case of paralytic poliomyelitis for every 3×10^6 vaccinated people, and the frequency is much greater in children with immunodeficiency diseases and in adult males. (Oligonucleotide mapping and the genetic markers in mutant strains used for immunization are of particular value in determining whether or not the vaccine was responsible for the rare postimmunization case of disease.)

In most persons, oral administration of a single type in a dose of 10^5 to 3.2×10^5 TCID$_{50}$ produces infection of the gastrointestinal canal, excretion of virus in high titer for 4 to 5 weeks, and development of Abs to a titer of approximately 1:128 in 3 to 4 weeks. Serologic conversion occurs in over 95% of those without Abs to any of the three types at the time the vaccine is administered. During the period of relatively high Ab titers, natural reinfection of the alimentary tract is prevented, but the duration of this protection has not been clearly defined.

Since Abs and immunity persist following natural infections, a similar persistence was expected to follow immunization with infectious attenuated virus. In fact, however, Abs decrease at approximately the same rate as after immunization with inactivated viruses, i.e., a diminution in 2 years to about 20% of the maximum titer. Antibody levels are generally higher, however, following vaccination with live rather than with inactivated vaccines.

The three viruses are fed together, and interference, with multiplication of one or more types, which was initially feared, is minimized by adjusting the viral concentrations so that type 1 is present in the greatest quantity and type 2 in the lowest. For maximum Ab response, immunization with three doses of the trivalent vaccine during the first year of life, preferably from 3 to 18 months of age, is recommended; a booster is also advised for all children at the time of entrance to elementary school. Further vaccine administration is believed unnecessary unless one is exposed to a known case of poliomyelitis or anticipates travel to a region where poliomyelitis is endemic.

Preexisting infection of the alimentary canal with other enteroviruses may interfere with successful implantation of the poliovirus vaccine strains. Hence, community immunization programs are usually carried out in the winter or early spring, when enteroviruses are less prevalent.

CRITIQUE OF POLIOVIRUS VACCINES. Each class of the vaccines has advantages and disadvantages.

Inactivated virus vaccine, which is now of high potency and moderate purity, has the distinct **advantage** that it is safe and remarkably effective when properly employed. For example, the exclusive use of the inactivated vaccine in Finland and Sweden has apparently eliminated paralytic poliomyelitis in these countries. Indeed, there has not been a single case for more than 12 years; and in Finland, despite constant surveillance, no poliovirus has been isolated during this period. However, the inactivated vaccine has the following **disadvantages:** (1) logistic problems of administration by sterile injection to large numbers of people, especially children; (2) greater cost, both for administration and for several doses of vaccine; (3) requirement for booster immunizations every 5 years; and (4) failure to eliminate intestinal reinfection and fecal excretion. This last feature, however, could also be an advantage if immunity is not long-lasting, since natural infection could then occur at the usual rate, inducing immunity without producing paralytic disease.

The **live attenuated virus vaccine** has clear **advantages:** (1) it is easily administered; (2) it is relatively inexpensive; (3) it results in synthesis and excretion of IgA Abs into the gastrointestinal tract, thus producing alimentary tract resistance, decreasing spread within the population, and therefore conferring **"herd immunity"** as well as **individual immunity** (however, as noted above, it is unclear how long this persists); and (4) its effectiveness approaches 100%. The **disadvantages** of this vaccine as presently constituted are (1) reversion to increased virulence of the viruses employed, particularly type 3, and (2) dissemination of virus to unvaccinated contacts. The latter process might be advantageous by increasing the resistance of members of a group; it is, however, a potential hazard because the transmission is uncontrolled and may infect immune-deficient persons, and the viruses may be mutants of increased virulence.

Despite the safe immunization of millions with live virus vaccine in many countries, its acceptance for general use was slow in the United States, owing to the ear-

lier accident with the inactivated virus vaccine and the fear of reversion of the attenuated strains to neurovirulence. The initial hesitancy has been overcome, however, and at present only the live virus vaccine is routinely used in the United States.

Whatever the advantages of either kind of vaccine may be, both have been used in the United States with remarkable effects (Fig. 55–12): In 1955, when the inactivated virus vaccine was approved for general use, 28,985 cases of poliomyelitis were reported in the United States; in the following year there were 15,140 cases; in 1964 there were only 122 cases; and in 1969 (after the shift to the live vaccine) a mere 20 cases of paralytic disease and no deaths were reported. From 1969 to 1981, 203 cases of paralytic poliomyelitis occurred, in four general categories: (1) 43 cases in three well-defined epidemics, affecting 22 unimmunized preschool children of Mexican–American parentage in South Texas in 1970, 11 unimmunized students in a Christian Science boarding school in Connecticut in 1972, and 10 cases in 1979 produced by a wild-type virus that spread from unimmunized persons in the Netherlands through Canada into four states; (2) 22 cases imported into the United States, most commonly from Mexico; (3) 41 endemic cases, unassociated with travel or vaccine; and (4) 100 vaccine-associated cases, occurring in persons who had either received the live virus vaccine or were in contact with recipients; of these, 14 cases were in persons with immunodeficiency diseases. **Indeed, the majority of**

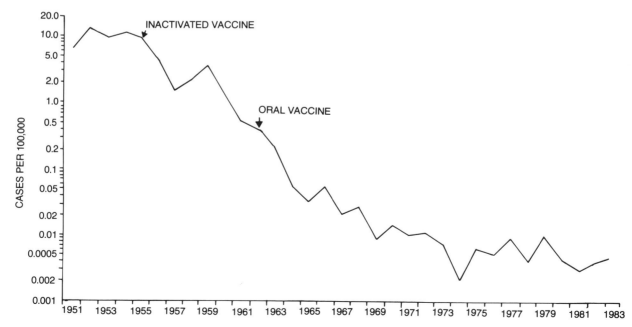

Figure 55–12. Number of cases of paralytic poliomyelitis after administration of killed virus vaccine (1955–1965) and live virus vaccine (1961–1983). Figures in parentheses give the total case rate per 100,000 population. (Data from United States Centers for Disease Control)

cases of paralytic disease in the United States now occur from vaccine administration, in either recipients or contacts. This disadvantage of the live virus vaccine could be revealed only as a consequence of its impressive effectiveness.

Because there is no reservoir for polioviruses other than humans, it is theoretically possible to eradicate the disease by global immunization. Owing to the practical problems of such a program, however, its fulfillment seems unlikely. Therefore, as sanitation improves and immunization decreases the prevalence of infection, as well as of disease, more and more people could reach adulthood, when disease is most dangerous, without ever encountering the virus, and therefore without protective Abs. Hence, the constant threat that virus will be introduced into a poliovirus-free, nonimmune population will probably make vaccination, at least of children, necessary indefinitely. It is depressing, however, that social and economic factors hamper the attainment of even this limited goal: as noted above, many of the economically disadvantaged are not vaccinated, and if the number of susceptible persons increases (for lack of early natural infection), the threat of devastating epidemics may once again arise.

Coxsackieviruses

Following the demonstration that yellow fever virus and other togaviruses (see Chap. 60) are more infectious and pathogenic for newborn than for adult mice, Dalldorf and Sickles attempted to utilize this unique host for studies of poliomyelitis. Instead they isolated a new virus from the feces of two children from Coxsackie, N.Y., in 1948. This new development offered the first major clue that many viruses other than polioviruses infect the intestinal tract of man.

Coxsackieviruses are distinguished from other enteroviruses by their much greater pathogenicity for the suckling than for the adult mouse. They are divided into two groups on the basis of the lesions observed in suckling mice: **group A viruses** produce a diffuse myositis with acute inflammation and necrosis of fibers of voluntary muscles; **group B viruses** evoke focal areas of degeneration in the brain, focal necrosis in skeletal muscle, and inflammatory changes in the dorsal fat pads, the pancreas, and occasionally the myocardium.

PROPERTIES

PHYSICAL AND CHEMICAL CHARACTERISTICS. Those few coxsackieviruses that have been appropriately examined are similar to polioviruses in physical properties and chemical composition (see "Characteristics of Polioviruses" and "Characteristics of Picornaviruses") but differ significantly in RNA base composition. Hybridization analyses also show only about 5% nucleotide sequence homology. However, the replicase genes of the Coxsackie B3 virus and type 1 poliovirus show marked similarity.

IMMUNOLOGIC CHARACTERISTICS. Each of the 23 group A and six group B coxsackieviruses is identified by a type-specific Ag, measured by neutralization and *in vitro* assays (e.g., CF, enzyme-linked immunosorbent assay [ELISA]). In addition, all from group B and one from group A (A9) share a group Ag, which is detected by agar-gel diffusion. Cross-reactivities have also been observed between several group A viruses, but no common group A Ag has been found. A type-specific virion Ag of a few types (see Laboratory Diagnosis) causes agglutination of group O human red blood cells (RBCs) at 37°C (maximum titers are obtained with RBCs from newborns).

Type-specific Abs usually appear in the blood within a week after onset of infection in humans, and they attain maximum titer by the third week. Neutralizing Abs persist for at least several years, but CF Abs decrease rapidly after 2 to 3 months. Resistance to reinfection, according to epidemiologic data, appears to be long-lasting. Monoclonal Abs have detected at least five epitopes responsible for neutralizing Abs; their precise locations are not yet known, but purified VP2 can induce neutralizing Abs. In patients infected with group B or A9 viruses, the Ab directed against the so-called group-specific Ag appears earlier and persists longer than the type-specific Ab (suggesting a secondary response resulting from prior infection by a related virus).

HOST RANGE. Suckling mice inoculated by the intracerebral, intraperitoneal, or subcutaneous route are employed for propagation and isolation of coxsackieviruses. Mice 4 to 5 days old are still susceptible to infection by group A viruses, but group B viruses multiply best in mice 1 day old or less. Even adult mice can be rendered susceptible to group B viruses by cortisone administration, x-irradiation, continuous exposure to cold (4°C) during the period of infection, or severe malnutrition prior to and during infection. Denervation of the limb of an adult mouse increases susceptibility to group A viruses, with resulting myositis and muscle necrosis limited to the affected extremity.

The striking susceptibility of newborn mice may be partially explained by their failure to produce interferon when infected by coxsackieviruses. The increased susceptibility produced by cortisone may also be accounted for by inhibition of interferon synthesis (see Chap. 49).

Newborn mice infected with **group A viruses** develop a total flaccid paralysis, resulting from severe and extensive degeneration of skeletal muscles; there are no significant lesions elsewhere. Muscle necrosis may be so extensive that a marked liberation of myoglobulin results, causing renal lesions similar to those developing in the crush syndrome.

Group B viruses produce quite different manifestations in suckling mice, including tremors, spasticity, and spastic paralysis. Degeneration of skeletal muscle is focal and limited. The most prominent pathologic lesions are necrosis of brown fat pads, encephalomalacia, pancreatitis, myocarditis, and hepatitis. Adult as well as suckling mice develop pancreatitis, but in adult mice most of the other lesions do not appear or are so minimal that the mice survive. Necrosis of the myocardium, however, is often noted and is markedly increased by cortisone. Cortisone or pregnancy in adult mice transforms an inapparent infection into a fatal one.

Intracerebral inoculation of rhesus monkeys with A7 and A14 viruses produces widespread degeneration of ganglionic cells of the CNS, followed by flaccid paralysis similar to that caused by poliovirus. (Because of this behavior, A7 virus was initially mistaken for a new type of poliovirus.)

Tissue culture techniques have become increasingly valuable for study and isolation of coxsackieviruses and for obtaining attenuated strains by repeated passage of wild-type virus. The group B and A9 viruses multiply readily in various cell cultures, but most group A viruses do not (Table 55–3). The similar tissue culture host range of group B and A9 viruses parallels their antigenic relation and further suggests that these viruses are very closely related despite their dissimilar pathologic effects in suckling mice.

TABLE 55–3. Multiplication of Coxsackieviruses in Cell Cultures

Group and Type	Cell Culture		
	Monkey Kidney	HeLa	Human Amnion or Embryonic Kidney
GROUP A			
Types 1, 2, 4–6, 19, 22	−	−	−
Type 7	±*	±*	−
Type 9	+	±*	+
All others	±*	±†	±‡
GROUP B			
Types 1–6	+	+	+

* Not readily isolated in cell culture, but strains have been adapted to multiply in indicated cells.

† Types A13, 15, 18, and 21 multiply readily on first passage.

‡ Types A11, 13, 15, 18, 20, and 21 grow in human embryonic kidney cells.

(Adapted from Wenner HA, Lenahan MF: Yale J. Biol Med 34:421, 1961)

MULTIPLICATION. The multiplication cycle of coxsackieviruses is very similar to that of polioviruses. However, the assembled virions tend to remain within the cell rather than to be released rapidly into the culture medium. The cytopathic changes are also similar to those caused by other enteroviruses, but those produced by group A viruses develop much more slowly.

PATHOGENESIS

Most coxsackievirus infections in humans are mild; infections mimicking those of man have not been produced in laboratory animals. Hence, we have very little knowledge of the pathogenesis of human infections or the pathology of the lesions. The marked diversity of clinical syndromes, however, indicates that virus enters through either the mouth or the nose and follows a pathogenic course, from local multiplication through viremic spread, that is akin to that demonstrated in poliovirus infections (see Fig. 55–10). In biopsies obtained from a few patients with coxsackievirus A infections, focal necrosis and myositis were noted, but the lesions were not distinctive. In children who died of **myocarditis of the newborn**, a highly fatal disease caused by group B coxsackieviruses, the myocardium showed edema, diffuse focal necrosis, and acute inflammation; focal necrosis with inflammatory reaction also occurred in liver, adrenals, pancreas, and skeletal muscle, and occasionally there was diffuse meningoencephalitis. Group B coxsackieviruses also appear to cause mild interstitial focal myocarditis and occasionally valvulitis in infants and children.

The coxsackieviruses can produce a remarkable variety of illnesses (Table 55–4), and even the same virus may be responsible for quite different types of disease. Still, a number of group A viruses have not been definitely implicated as causative agents of any human disease. Some viruses in each group are associated with at least one distinctive syndrome, which can usually be diagnosed on clinical grounds alone. Thus, **herpangina*** is caused by certain group A viruses and **epidemic pleurodynia†** and myocarditis of the newborn by certain group B viruses. Other syndromes present no clinical features distinctive for coxsackieviruses: rarely, illness simulating paralytic poliomyelitis can be induced, particularly by A7

* Herpangina is an acute disease with sudden onset of fever, headache, sore throat, dysphagia, anorexia, and sometimes stiff neck. The diagnosis depends on recognition of the pathognomonic lesions in the throat: at the onset, small papules are present, but these soon become circular vesicles that ulcerate.

† Epidemic pleurodynia (epidemic myalgia; Bornholm disease) is an acute febrile disease with sudden onset of pain in the thorax (a "stitch in the side"), which is aggravated by deep breathing (simulating pleurisy) and by movement. The pain may be chiefly abdominal or associated with other muscle groups and may be accompanied by muscle tenderness.

TABLE 55—4. Clinical Syndromes Commonly Associated With Coxsackieviruses

Clinical Syndrome	Coxsackieviruses	
	Group	Predominant Types
Aseptic meningitis	A	2, 4–7, 9, 10, 12, 16
	B	All
Paralytic disease	A	4, 7, 9,
	B	3–5
Herpangina	A	1–6, 8–10, 16, 21, 22
Fever, exanthema	A	2, 4, 9, 16
	B	4
Acute upper respiratory infection (cold)	A	2, 10, 21, 24
	B	2–5
Hand-foot-and-mouth disease	A	16 (4, 5, 9, and 10 [rarely])
Epidemic pleurodynia or mylagia	B	1–5
	A*	4, 6, 8, 9, 10
Myocarditis of the newborn	B	2–5
Interstitial myocarditis and valvulitis in infants and children	B	2–5
Pericarditis	B	1–5
Undifferentiated febrile illness	All	All

* Much less common than group B viruses

virus; a few group A and B viruses cause an acute upper respiratory illness; and pancreatitis, nephritis, and hepatitis have occasionally been associated with group A and B coxsackievirus infections.

Group B viruses may produce myocarditis of human newborns by intrauterine infection, as can certain group A viruses in mice. These findings suggest that coxsackieviruses may, like rubella virus, be responsible for some cases of congenital heart disease. Indeed, women with coxsackievirus infections during the first trimester of pregnancy have been shown to give birth to newborns with twice the normal incidence of congenital heart lesions.

LABORATORY DIAGNOSIS

The etiologic diagnosis of group A coxsackievirus infections depends on isolation of the causative agent from feces, throat secretions, or cerebrospinal fluid by inoculating suckling mice. However, for the initial isolation of group B and A9 viruses, inoculation of cell cultures is more suitable (see Table 55–3). In autopsies of patients with myocarditis and valvulitis, the Ag of group B viruses has been demonstrated by immunofluorescence.

A newly isolated virus is grouped as A or B on the basis of the lesions produced in suckling mice. Type identification is considerably more cumbersome, owing to the large number of types. One aid in identification is based on the fact that relatively few coxsackieviruses in-

duce hemagglutination of human group O RBCs at 37°C; these types (B1, B3, B5, A20, A21, and A24) are rapidly distinguished from other coxsackieviruses and can readily be identified by hemagglutination-inhibition titrations. With group A viruses, because of the large number of types, identification is initiated by neutralizing titrations by means of pools containing several type-specific sera, and final identification is accomplished with the individual type-specific sera.

While serologic diagnosis without viral isolation is not practicable because of the large number of possible viruses, **identification of an isolated virus as the cause of a particular illness requires serologic confirmation of infection** (by neutralization, immunofluorescence, CF, ELISA, or hemagglutination-inhibition titrations) because many enteroviruses appear to be present as harmless inhabitants of the intestinal tract rather than as etiologic agents of a current disease.

EPIDEMIOLOGY AND CONTROL

Coxsackieviruses are widely distributed throughout the world, as demonstrated by the occurrence of proved epidemics and by the results of serologic surveys. The type prevalent in any locality varies every few years, probably owing to the development of immunity in the population. For example, in 1947–1948, coxsackievirus B1 was predominant in epidemics observed in New York and New England, but by 1951 the B3 virus produced epidemics throughout the world, replacing the B1 virus.

Coxsackieviruses are highly infectious within a family or the closed population of an institution (about 75% of susceptible persons are infected). However, the mechanism of spread may vary with the strain of virus and the clinical syndrome. Most clinical infections and epidemics occur in summer and fall, and the viruses are frequently present in the feces, suggesting a fecal–oral spread. However, viruses may also be isolated from nasal and pharyngeal secretions and produce acute respiratory disease, suggesting spread by the respiratory route as well.

No effective control measures are yet available. Immunization is not practical because of the large number of viruses that induce human disease and the relative infrequency of epidemics caused by any single virus.

Echoviruses

The first echoviruses were accidentally discovered in human feces, unassociated with human disease, during epidemiologic studies of poliomyelitis. Viruses were termed **echoviruses** (an acronym for *e*nteric, *c*ytopathic, *hu*man, *o*rphan viruses) if they were found in the gastrointestinal tract, produced cytopathic changes in cell cul-

TABLE 55–5. *Diseases Associated With Infection by Echoviruses*

Clinical Syndrome	Common Epidemic	Common Endemic	Uncommon Epidemic	Uncommon Sporadic
Aseptic meningitis	4, 6, 9, 30		3, 7, 11, 16, 18, 19	1, 2, 5, 13–15, 17, 20–22, 25, 31–34
Neuronal injury				
Paralysis			4, 6, 30	1, 2, 9, 11, 16, 18
Encephalitis			3	2, 4, 6, 7, 9, 18, 19
Rash, fever	4, 9		16, 18	1–7, 14, 19
Acute upper respiratory infection		20?	19	4, 8, 9, 11, 22, 25
Enteritis	6		11, 14, 18	8, 12, 19, 20, 22–24, 32
Pleurodynia				1, 6, 9
Myocarditis				1, 6, 9, 19
Neonatal infections			11	4, 9, 17–20, 22, 31

tures, did not induce detectable pathologic lesions in suckling mice, and had the properties listed under "Characteristics of Picornaviruses." Most echoviruses, however, are no longer "orphans" in the world of human diseases but have been associated with one or more clinical syndromes ranging from minor acute respiratory diseases to afflictions of the CNS (Table 55–5).

Initially, 34 viruses were assigned echovirus serotype designations. However, once they were characterized, echoviruses 10 and 28 were reclassified (see Table 55–1).

PROPERTIES

Data on the characteristics of echoviruses are exceedingly fragmentary, and the viruses cannot be adequately compared.

PHYSICAL AND CHEMICAL CHARACTERISTICS. The morphology and general chemical characteristics are similar to those of polioviruses and coxsackieviruses. Infectious RNA has been extracted from a number of echoviruses, but it has not been studied in detail.

Echoviruses are generally stable, but there are marked variations, and some of these viruses are considerably less stable than polioviruses. Like polioviruses, heating at 50°C inactivates infectivity and alters the virion antigenic specificity; and as with polioviruses, 1M $MgCl_2$ stabilizes echoviruses to heat inactivation.

HEMAGGLUTINATION. Of the 32 echoviruses, 12 show **hemagglutinating activity** with human group O erythrocytes.* Maximum titers are obtained with RBCs from newborn humans (as with some coxsackieviruses), but the opti-

mum temperature for the reaction varies with the virus.† The hemagglutinin is an integral part of the viral particle. Some types (3, 11, 12, 20, and 25) elute spontaneously from agglutinated RBCs at 37°C, but, unlike orthomyxoviruses and paramyxoviruses (see Chaps. 56 and 57), they do not remove the receptors from the cells, which are still agglutinable by the same or by other echoviruses.

IMMUNOLOGIC CHARACTERISTICS. The type designation of each echovirus depends on a specific Ag in the viral capsid, and neutralization titration is the most discriminating method for its identification. There is no group echovirus Ag, but heterotypic cross-reactions occur between a few pairs,‡ causing major difficulties in the identification of freshly isolated viruses and in the serologic diagnosis of infections.

Immunologic studies can also be carried out by CF, ELISA, and hemagglutination-inhibition titrations. The CF titrations have the advantage of simplicity but the disadvantage of increased cross-reactivity among echoviruses; it is also difficult to obtain satisfactory Ag for this assay from some isolates.

HOST RANGE. The original notion that echoviruses were not pathogenic for experimental animals has proved to be incorrect for at least 14 of the known viruses.§ Intra-

---

* Types 3, 6, 7, 11–13, 19–21, 24, 29, and 30.

† Maximum titers are obtained at 4°C for types 3, 11, 13, and 19, and at 37°C for types 6, 24, 29, and 30. Titers for types 7, 12, 20, and 21 are independent of temperature.

‡ Types 1 and 8 show a major antigenic overlap by neutralization titrations, and type 12 cross-reacts to a lesser extent with type 29. Antibodies directed against type 23 neutralize type 22 virus, but the reciprocal reaction does not occur. Minor reciprocal cross-neutralization occurs between types 11 and 19 and between types 6 and 30.

§ Types 1–4, 6–9, 13, 14, 16–18, and 20.

spinal or intracerebral inoculation of virus into rhesus and cynomolgus monkeys initiates viremia, neuronal lesions, and meningitis, occasionally associated with detectable muscle weakness. Some strains of types 6 and 9 produce lesions in newborn mice similar to those induced by group B and A coxsackieviruses, respectively.

Cultures of kidney cells from rhesus or cynomolgus monkeys are most suitable for isolation and propagation of all echoviruses (Table 55–6). The final cytopathic changes produced by most echoviruses are similar to those induced by polioviruses and coxsackieviruses (see Fig. 55–5B).

MULTIPLICATION. Judging from the fragmentary data available, the multiplication of echoviruses resembles that of polioviruses.

Echoviruses replicate in the cytoplasm of infected cells. The virions appear to assemble and become oriented in columns supported by a fine filamentous lattice distinct from the endoplasmic reticulum (Fig. 55–13), a procedure similar to the assembly process of coxsackieviruses. Crystalline viral arrays may form. Viral particles are subsequently dispersed in the cytoplasm and released from the host cell through small rents in the plasma membrane or in cytoplasmic protrusions that are shed from the cell. Eventually, infected cells disrupt. Types 22 and 23 echoviruses, in contrast, appear to have a different mode of replication: they cause characteristic nuclear changes, and unlike all other echoviruses, they are not inhibited by 2(α-hydroxybenzyl)-benzimidazole or guanidine (see Chap. 49).

PATHOGENESIS

Echoviruses usually enter humans by the oral route; a few probably infect through the respiratory tract. The majority of infections probably remain limited to the primary cells infected in the alimentary or respiratory tract. It is obvious, however, from the clinical manifestations elicited (see Table 55–5), that virus occasionally disseminates beyond the initial organs infected, causing fever, rash, and symptoms of CNS infections. In fact, virus can be isolated from the blood in several of the syndromes

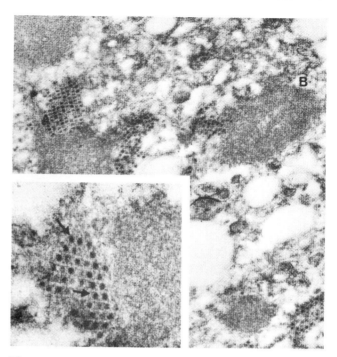

Figure 55–13. Assembly of type 9 echovirus particles in the cytoplasm of infected cells. Viral particles appear to differentiate into columns on a fine filamentous lattice at cytoplasmic template sites (A). Another crystal-like array of virions associated with finely granular masses is shown (B). (Original magnification ×50,800) (*Inset*) A mass at higher magnification. (Original magnification ×112,800) Transected fibrils lie between the particles (*arrows*). (Rifkind RA et al: J Exp Med 114:1, 1961)

listed, and from the cerebrospinal fluid in aseptic meningitis.

The pathologic effects of echovirus infection are still unknown, owing to the general mildness of the diseases. Cerebral edema and some focal destructive and infiltrative lesions have been noted in the CNS of the rare fatal cases examined, but the pathologic findings are not distinctive. Similar neurologic injury has been produced in monkeys and chimpanzees following intracerebral or intraspinal inoculation of the viruses.

LABORATORY DIAGNOSIS

Viral isolation in rhesus monkey kidney cells offers the most sensitive and reliable procedure for diagnosis of an echovirus infection. Feces and throat secretions are the most abundant sources of virus; infectious virus persists in feces longer than in any other body excretion or fluid. Use of kidney cell cultures from patas and rhesus monkeys is valuable for identification of the specific virus type isolated (see Table 55–6). The differential host sus-

---

TABLE 55–6. Susceptibility of Cell Cultures to Echoviruses

Culture	Virus
Rhesus and cynomolgus monkey kidneys	All
Patas monkey kidneys	Types 7, 12, 19, 22–25
Human amnion and kidneys	All
Continuous human cell lines	Poor until adapted

ceptibility noted and the limited number of viruses (12) possessing hemagglutinating activity afford convenient tools for preliminary grouping of an unknown echovirus and reduce the expense of the immunologic identification of a freshly isolated agent. Neutralization titrations provide the final criterion for identification because of their greater specificity.

Diagnosis solely by serologic analysis of the patient's paired sera is cumbersome* and expensive and is usually employed only during an epidemic caused by a single virus type.

EPIDEMIOLOGY AND CONTROL

The epidemiologic features of echovirus infections resemble those of other enteroviruses, especially coxsackieviruses. But for those echoviruses that cause respiratory infections (particularly for those, such as type 9, that produce extensive waves of infection), respiratory secretions may be a more significant route of viral transmission than feces. This route is also suggested by the rapid and pervasive spread of virus within the family unit.

Immunization does not appear practicable or warranted because of the large number of viruses and the relative infrequency of epidemics produced by a single agent.

New Enteroviruses

Newly identified picornaviruses that are not polioviruses but conform to the characteristics of enteroviruses are no longer separated into the species coxsackievirus and echovirus because of ambiguities presented by overlapping host range variations. Of the five such enteroviruses isolated (enteroviruses 68–72), two of these viruses, types 70 and 72, merit special attention. **Enterovirus 70,** which has the typical physical and chemical characteristics of other enteroviruses (see "Characteristics of Picornaviruses"), was isolated from many patients during epidemics of **acute hemorrhagic conjunctivitis** that swept through Africa, Asia, India, and Europe from 1969 to 1974. The disease is characterized by sudden swelling, congestion, watering, and pain in the eyes accompanied by subconjunctival hemorrhages. Symptoms subside rapidly, and recovery is usually complete within 1 to 2 weeks. In a rare patient the virus is neurovirulent and produces a poliomyelitis-like disease. The virus is readily isolated in diploid human embryonic lung fibroblasts and KB cells, but it can easily be adapted to propagation in primary monkey kidney cells. Enterovirus 70 multi-

plies best at 33°C but not at all at 38°C, usually a property of a rhinovirus rather than an enterovirus. This property correlates with its preferential infection of the conjunctivae.

Enterovirus 72 is the designation assigned to **hepatitis A virus,** which, after it could be propagated *in vitro,* was shown to have the physical and chemical characteristics of enteroviruses (discussed in detail in Chap. 63 to compare it with the other major viruses causing hepatitis).

Enterovirus 68 has been isolated from patients with acute respiratory infections. **Enterovirus 71,** which appears to be highly pathogenic, has been associated with epidemics of a variety of acute diseases, including aseptic meningitis, encephalitis, paralytic poliomyelitis-like disease, and hand-foot-and-mouth disease.

Rhinoviruses

Acute afebrile upper respiratory diseases, grouped clinically as the **common cold,** are the most frequent afflictions of man. Although the diseases are not serious, they cause much discomfort as well as the loss of more than 200 million man-days of work and school each year in the United States alone.

There have been many attempts to discover the etiology of this syndrome. Kruse, in 1914, showed that the common cold could be transmitted to man by a filterable agent, but subsequent studies in every conceivable animal failed to isolate the virus; only man seemed susceptible! From extensive human transmission experiments by Andrewes and his colleagues in England and by Dowling and Jackson in the United States, the notion emerged that the common cold was caused by a large number of viruses, rather than by a single agent, as had commonly been thought. (They also presented evidence that seems to explode the myth that cold, dampness, and thin clothes provoke the onset of a cold.) It is now known that viruses belonging to several different families can cause a common cold syndrome (see Epidemiology, Prevention, and Control, below).

Since the initial isolations of viruses from patients with common cold in 1956, at least 115 immunologically distinct but biologically related viruses have been isolated. Their chemical and physical characteristics led to their classification into the genus designated *Rhinovirus,* of the picornavirus family.

It is not clear why rhinoviruses were not isolated sooner, despite many efforts, since suitable cells were available, and the initial isolations were finally accomplished with methods employed unsuccessfully in earlier studies. The subsequent isolation of numerous types, however, was clearly facilitated by the important discovery that optimal propagation of rhinoviruses, in human

---

* Because of the virtual absence of shared Ags, it would require that each serum be tested with 32 different viruses.

embryo or monkey kidney cells, requires special conditions approximating those in the nasal cavities: an incubation temperature of 33°C and pH of 6.8 to 7.3. The viral multiplication and cytopathology are minimal—with some rhinoviruses, not even detectable—if the infected cell cultures are maintained under the more usual conditions of 37°C and pH 7.6.

PROPERTIES

STRUCTURAL PROPERTIES. Because of the relatively poor viral yield in cell cultures, only a few rhinoviruses have been investigated in any detail. The virion consists of a single molecule of RNA having a molecular weight like polioviruses (see "Characteristics of Polioviruses"). The capsid closely resembles that of other picornaviruses (see "Characteristics of Picornaviruses" and "Characteristics of Polioviruses"), as shown by x-ray crystallographic analysis (see Figs. 55–3 and 55–14), but the

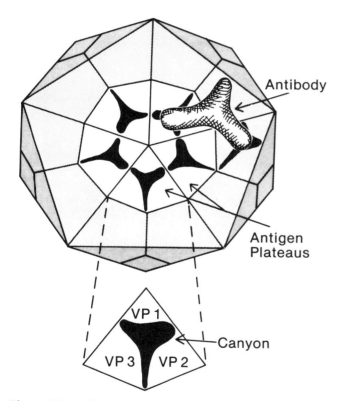

Figure 55–14. (*Top*) Organization of a rhinovirus capsid. The capsid consists of 12 pentamers, each at a fivefold axis (one is highlighted). Each pentamer comprises five wedge-shaped protomers. An Ab molecule is depicted bridging two protomers. (*Bottom*) Parts of the three unique polypeptide chains (VP1, VP2, VP3) are exposed at the protomer's surface (see Fig. 55–3), while the smallest polypeptide (VP4) is buried. The *canyon*, thought to bind host cell receptor near its base, is too narrow (about 20 Å) to accommodate the blunt-nosed binding site of the Ab. (Courtesy of R. R. Rueckert)

subunit structures seem to be more loosely bonded. There is also striking homology between the genomes of type 14 rhinovirus and type 1 poliovirus: from 65% homology of the regions encoding the replicase to 44% homology in the VP1 genes.

Rhinoviruses are sharply distinguished from other picornaviruses because they are **inactivated at low pH, maintain their infectivity at 50°C, and have higher buoyant density in CsCl** (see "Characteristics of Picornaviruses"). When rhinoviruses are held at pH 3 to 5 for 1 hour at 37°C, the virions are disrupted, yielding RNA, empty capsids, and free VP4; more than 99% of their infectivity is lost. The other picornaviruses are not affected by these conditions. Conversely, rhinoviruses are more stable than the other picornaviruses when heated at 50°C at neutral pH.

Not all rhinoviruses have yet been tested for heat stability, but the strains that multiply only in human cells (H strains) appear to be more stable than those that multiply also in monkey cells (M strains). Molar $MgCl_2$ partially stabilizes the M strains. The high buoyant density of rhinoviruses in CsCl is due to the permeability of their capsids to cesium ions (as many as 5000 cesium molecules may reversibly bind to an RNA molecule).

No hemagglutinating activity has been detected for rhinoviruses, but they have the novel ability to inhibit the spontaneous hemagglutination of trypsin-treated human RBCs.

IMMUNOLOGIC CHARACTERISTICS. Each of the distinct rhinoviruses identified possesses a type-specific Ag. There is not a common group Ag, but several consistent reciprocal cross-reactions (e.g., types 2 and 49, types 13 and 14) and more frequent unilateral heterologous neutralization reactions are noted with sera from immunized rabbits. Sequential immunization of rabbits with two or three different types further demonstrates immunologic relationships among rhinoviruses because acquired immunologic memory effects a secondary response to heterologous but related types. Thus several clusters of antigenically related rhinoviruses have been detected (e.g., types 13, 14, and 41; types 5, 17, and 42), a fact that may be utilized for the development of an effective vaccine.

Like polioviruses, the **capsid possesses four clusters of epitopes that induce neutralizing Abs,** with VP1 presenting the dominant antigenic sites (see Fig. 55–14). Natural human infections stimulate the production of type-specific neutralizing Abs (IgM, IgA, and IgG) that confer resistance to reinfection by virus of the same type (see Epidemiology, Prevention, and Control, below). Specific Abs appear in nasal secretions and serum 2 to 3 weeks after infection and continue to increase for 4 to 5 weeks after primary infection. The Ab response appears to be greater to the M than to the H strains.

HOST RANGE. Humans are the natural host for rhino-viruses. The chimpanzee is the only uniformly suscepti-ble laboratory animal; after intranasal inoculation, virus multiplies in the nasal and pharyngeal mucosal cells, and type-specific Abs appear, but disease does not de-velop. Some types infect monkeys, but in these animals also, disease is not produced.

Cell and organ cultures are the only practical experi-mental hosts available. The **H viruses** multiply and pro-duce cytopathic changes in human embryo kidney cells, in certain human diploid cell lines, and in a specially selected HeLa cell line, termed HeLa "R." The **M viruses** multiply and produce cytopathic changes in primary cultures of rhesus and monkey and human embryo kid-ney cells, human diploid cells, and continuous human cell lines (e.g., KB and HeLa cells). Organ cultures of hu-man embryonic nasal and tracheal epithelium are partic-ularly sensitive hosts for multiplication of rhinoviruses. A few recently isolated rhinoviruses multiplied only in these organ cultures, whereas others could be propa-gated in human cell cultures after preliminary isolation and several passages in organ cultures. Whether these rhinoviruses with limited host ranges to highly differenti-ated cells are new types or only uniquely fastidious strains of previously isolated types is unknown.

The **cytopathic changes** observed in cell cultures under optimal conditions qualitatively resemble those produced by other picornaviruses, but they are slower to develop and are usually incomplete. In infected organ cultures, ciliary activity diminishes, and superficial epi-thelial cells begin to shed 18 to 22 hours after infection. As in infections in humans, only the fully differentiated outer epithelial cells are injured; deeper cells appear un-affected.

MULTIPLICATION. The biochemical events in the replica-tion of rhinoviruses do not differ significantly from those of other picornaviruses. For maximum replication, as for viral isolation, a temperature of 33°C is optimal. Higher temperatures restrict a temperature-sensitive step late in viral multiplication. A pH lower than usual (pH 7–7.2) also optimizes viral multiplication in many cell cultures.

PATHOGENESIS

Human infections appear to be confined to the respira-tory tract and generally fit the **syndrome termed the common cold.** Virus infects and replicates in the cili-ated epithelial cells lining the nose, and during the first 2 to 5 days of illness, viruses can be isolated from nasopha-ryngeal secretions but not from other secretions or body fluids. A small number of the infected epithelial cells are shed into the nasal secretions, but the nasal mucosa is not denuded. Symptoms appear to result from an inflam-matory response triggered by the infection; the mecha-nisms of the response that effects increased mucus pro-duction is unknown. Rhinoviruses have also been associated with some exacerbations of chronic bronchi-tis and a few cases of bronchopneumonia in children and young adults (i.e., so-called primary atypical pneu-monia; see Chap. 40). The absence of a satisfactory exper-imental animal has hindered detailed studies of patho-genesis.

LABORATORY DIAGNOSIS

Isolation of the etiologic agent from nasopharyngeal se-cretions is the only practical method of establishing the diagnosis. Organ cultures of human embryonic nasal or tracheal epithelium are the most sensitive hosts and are required for isolation of some rhinoviruses. Such cul-tures, however, are inconvenient for routine diagnostic purposes; therefore, monolayer cultures of primary hu-man embryo kidney cells, human diploid cells, or HeLa "R" cells are generally used for primary isolations. The isolated virus can be typed by neutralization titrations with standard sera. Because of the existence of at least 115 distinct immunologic types, the number of titrations required is first narrowed by preliminary neutralization by means of pools containing several type-specific sera.

Routine serologic diagnosis is not practical, owing to the absence of a family cross-reacting Ag, the existence of so many distinctive types, and the small amounts of virus obtained in cell cultures.

EPIDEMIOLOGY, PREVENTION, AND CONTROL

The epidemiology of specific rhinovirus infections is that of the common cold. Seroepidemiologic surveys demon-strate that Abs to several prototype viruses are prevalent in many parts of the world and that rhinovirus infections are geographically widespread. Antibodies are found in relatively few infants and children, whereas the majority of adolescents and adults have high titers of Abs to one or more of the viruses studied. School children fre-quently introduce the virus into a family, where it spreads readily, particularly to those whose nasal secre-tions lack IgA Abs. The secondary attack rate may be as high as 70% in a family if the primary patient manifests symptoms of a common cold. Successful viral transmis-sion appears to depend primarily on direct contact be-tween infected and susceptible hosts. The high potential infectivity of rhinoviruses is further evident from a study of military recruits during their initial 4 weeks of training: 90% became infected with one or more different viruses, and 40% had two or more rhinovirus infections.

The incidence of isolation of rhinoviruses corre-sponds to the occurrence of minor respiratory infec-tions, which is greatest in fall, winter, and early spring. Rhinoviruses play a significant role as causative agents of common colds, but clearly they are not solely responsible for production of these illnesses. For example, in one 2-year study of college students, rhinoviruses were isolated from 24% of patients with common colds, from 2% of those convalescing from a cold, and from 1.6% of healthy students; in a study of families with young children, 20% of upper respiratory infections were associated with rhinoviruses; and in a study of military recruits, rhinoviruses were isolated from 31.5% of those with common colds. These studies have also demonstrated that inapparent infections occur.

It is common knowledge that the same person may have repeated episodes of the common cold, even five or six times in a single year. One reason is the prevalence of a large number of immunologically unrelated viruses that cause this syndrome, including viruses other than rhinoviruses, e.g., A21 coxsackievirus and coronaviruses (see Chap. 58).

When volunteers were successively infected with four or five different rhinoviruses, the infection conferred specific immunity for at least 2 years to homologous but not to heterologous rhinoviruses. The degree of protection appears to depend on the Ab levels present at the time of reexposure, particularly the level of IgA Abs in nasal secretions. How long specific immunity persists is unknown.

It should theoretically be possible to prepare a vaccine that could induce immunity for any single virus or for all rhinoviruses. In fact, an inactivated type 13 vaccine has proved effective when administered intranasally to volunteers. However, many rhinovirus types are widespread (in contrast to the few predominant special types seen with other organisms), and several may be prevalent concurrently. A vaccine containing 113 different rhinoviruses appears to be impractical, but if the antigenically related clusters of rhinoviruses are sufficiently broad and encompass enough types, the development of an effective vaccine may yet be possible.

Selected Reading

POLIOVIRUSES
Books and Review Articles

Cooper PD: Genetics of picornaviruses. Compr Virol 9:133, 1977
Melnick JL: Portraits of viruses: The picornaviruses. Intervirology 20:61, 1983
Rueckert RR: On the structure and morphogenesis of picornaviruses. Compr Virol 6:131, 1976
Sangar DV: The replication of picornaviruses. J Gen Virol 45:1, 1979
Symposium: Biology of poliomyelitis. Ann NY Acad Sci 61:737, 1955

Specific Articles

Bodian D: Histopathologic basis of clinical findings in poliomyelitis. Am J Med 6:563, 1949
Enders JF, Weller TH, Robbins FC: Cultivation of the Lansing strain of poliomyelitis virus in cultures of various human embryonic tissues. Science 109:85, 1945
Hogle JM, Chow M, Filman DJ: Three-dimensional structure of poliovirus at 2.9 A resolution. Science 220:1358, 1985
Horstmann DM, McCallum RW, Mascola AD: Viremia in human poliomyelitis. J Exp Med 99:355, 1954
Horstmann DM: Control of poliomyelitis: A continuing paradox. J Infect Dis 146:540, 1982
Kim-Farley RJ, Schonberger LB, Nkowane BM et al: Poliomyelitis in the USA: Virtual elimination of disease caused by wild virus. Lancet 2:1315, 1984
Pollarsch MA, Kew OM, Semler BL et al: Protein processing map of poliovirus. J Virol 49:873, 1984
Racaniello VR, Baltimore D: Molecular cloning of poliovirus cDNA and determination of the complete nucleotide sequence of the viral genome. Proc Natl Acad Sci USA 78:4887, 1981
Rekosh D: Gene order of the poliovirus capsid proteins. J Virol 9:268, 1977
Sabin AB: Oral poliovirus vaccine: Recent results and recommendations for optimum use. R Soc Health J 2:51, 1962
Salk JE: A concept of the mechanism of immunity for preventing poliomyelitis. Ann NY Acad Sci 61:1023, 1955
Special Advisory Committee on Oral Poliovirus Vaccine: Report to the Surgeon General, USPHS. JAMA 190:49, 1964
Toyada H, Kohara M, Kataoka V et al: Complete nucleotide sequences of all three poliovirus genomes: Implication of genetic relationship, gene function and antigenic determinants. J Mol Biol 174:561, 1984

COXSACKIEVIRUSES
Books and Review Articles

Melnick JL, Wenner HAA, Phillips CA: The enteroviruses. In Lennette EH, Schmidt NJ (eds): Diagnostic Procedures for Viral and Rickettsial Disease, 5th ed. New York, American Public Health Association, 1980

Specific Articles

Burch GE, Sun S, Chu K et al: Interstitial and coxsackievirus B myocarditis in infants and children: A comparative histologic and immunofluorescent study of 50 autopsied hearts. JAMA 203:1, 1968
Dalldorf G, Sickles GM: An unidentified, filterable agent isolated from the feces of children with paralysis. Science 108:61, 1948

ECHOVIRUSES
Books and Review Articles

Moore M: Enteroviral disease in the United Sates, 1970–79. J Infect Dis 146:103, 1982
Wenner HA, Behbehani AM: Echoviruses. Virol Monogr 1:1, 1968

Specific Articles

Mirkovic RR, Kono R, Yin-Murphy M et al: Enterovirus type 70: The etiologic agent of pandemic acute hemorrhagic conjunctivitis. Bull WHO 49:341, 1973

RHINOVIRUSES
Books and Review Articles

Dingle JH: The curious case of the common cold. J Immunol 81:91, 1958
Gwaltney JM: Rhinovirus. In Evans AS (ed): Viral Infection of Humans: Epidemiology and Control, p 491. New York, Plenum, 1982

Specific Articles

Cooney MK, Fox JP, Kenney GE: Antigenic groupings of 90 rhinovirus serotypes. Infect Immun 37:642, 1982

Fox JP, Cooney MK, Hall CE: The Seattle virus watch. V. Epidemiologic observations of rhinovirus infections, 1965–1969, in families with young children. Am J Epidemiol 101:122, 1975

Pelon W, Mogabgab WJ, Phillips IA, Pierce WE: A cytopathogenic agent isolated from naval recruits with mild respiratory illness. Proc Soc Exp Biol Med 94:262, 1957

Price WH: The isolation of a new virus associated with respiratory clinical disease in humans. Proc Natl Acad Sci USA 42:892, 1956

Rossman MG, Arnold E, Erickson JW et al: Structure of a human common cold virus and functional relationship to other picornaviruses. Nature 317:145, 1985

56

Harold S. Ginsberg

Orthomyxoviruses

History and Classification

In 1918 to 1919 one of the most devastating plagues in history swept the world, killing approximately 20 million persons and afflicting a huge part of the human population. The underlying disease, influenza,* had been known to occur in large epidemics for several centuries. Indeed, the pandemics of 1743 and 1889–1890 were only slightly less disastrous than that of World War I.

The influenza bacillus (*Hemophilus influenzae*; see Chap. 31) was originally named as the primary cause of the disease by Pfeiffer in the great pandemic of 1889–1890. However, in 1933, Smith, Andrewes, and Laidlaw in England found that filtered, bacteria-free nasal washings from patients with influenza produced a characteristic febrile illness when inoculated intranasally into ferrets. The viral etiology was soon confirmed in other laboratories, and it eventually became clear that *H. influenzae* is only one of a number of bacterial pathogens (others include *Staphylococcus aureus* and *Streptococcus pneumoniae*) that may cause severe, often fatal secondary pneumonia in patients with influenza.

Further progress in the investigation of the virus and the disease was accelerated by the fortunate findings that influenza viruses can multiply to high titer in the chick embryo, a convenient and inexpensive laboratory animal, and that they cause **hemagglutination** of chicken red blood cells (RBCs; see Hemagglutination, Chap. 44). This reaction, discovered by chance in 1941 by Hirst and also by McClelland and Hare, proved to be of great practical and theoretical importance: It provided a simple method for detecting and quantitating influenza viruses; its specific inhibition by Abs to the virus pro-

---

* Derived from an Italian form of Latin *influentia* (influence), reflecting the widespread supposition that epidemics resulted from an astrologic or other occult influence such as an unhappy conjunction of the stars

vided a highly sensitive **hemagglutination-inhibition** test for measuring Abs; and its study revealed the mechanism of infection of host cells, since the receptor sites for the virus on the RBCs proved to be the same as those on the susceptible host cells. These **receptors** were shown to be mucoproteins possessing a **terminal N-acetyl-neuraminic acid (NANA) group.** As described earlier (see Hemagglutination, Chap. 44), absorption of virus leads to release of NANA by a viral enzyme, **neuraminidase;** the RBCs thereby become inagglutinable, and soluble mucoproteins present in respiratory secretions become nonreactive with fresh virus.

Of major interest in this family of viruses is the frequent emergence of novel antigenic variants as the source of pandemics, and the analysis of the mechanism responsible for this unusual genetic instability.

The successful investigations of influenza viruses, and the general availability of tissue cultures, led to the discovery of additional viruses (e.g., **parainfluenza viruses**) that agglutinate RBCs and react with similar mucoproteins. These were originally classified together with influenza viruses and termed *myxoviruses* (Gr., *myxo*, mucus), but later discovery of major physical and chemical differences among the viruses led to their separation into two families (Table 56–1): **Orthomyxoviridae** * and **Paramyxoviridae** (vernacular, orthomyxoviruses and paramyxoviruses), whose distinguishing characteristics are listed in Table 56–2. Orthomyxoviruses (influenza viruses) are described in this chapter, and paramyxoviruses are described in Chapter 57.

Influenza Viruses

After the isolation of the causative agent of influenza it soon became evident that a complex group of viruses was involved. The agents isolated from humans in England and the United States were found to be similar but not identical to the swine influenza virus isolated by Shope in 1931, and many viral strains isolated proved to be antigenic variants when compared with the initial isolates. In 1940 Francis and Magill, studying patients with influenza in the United States, independently isolated viruses that were immunologically distinct from the original strains. The agent isolated in 1933 was termed **influenza A virus,** and the second discovered was called **influenza B virus.** A third distinct antigenic type, **influenza C virus,** was subsequently isolated in 1949. Influenza C virus rarely produces clinical disease and has not been responsible for epidemics.

Since the discovery of influenza viruses **major new**

*Orthomyxoviridae, to contrast with **Paramyxoviridae,** is the designation assigned to this family by the International Committee on Viral Nomenclature.

TABLE 56–1. Classification of Orthomyxovirus and Paramyxovirus Families

Family	Genus (Type)	Species (Subtype)*
Orthomyxovirus	*Influenzavirus A*	
		H_1N_1 (A_1 human, $H_{sw}N_1$)†
		H_2N_2 (A_2)
		H_3N_2 (A_{HK}·A_3)
		$H_{eq}N_{eq}$ ($H_{eq2}N_{eq2}$)
		$H_{av}N_{av}$ ($H_{av8}N_{av8}$)
	Influenzavirus B	B‡ (human)
	Influenzavirus C	(human)
Paramyxovirus	*Paramyxovirus*	Parainfluenza 1–4
		Simian (SV5) parainfluenza
		Mumps
		Newcastle disease (NDV)
	Morbillivirus	Measles
		Rinderpest
		Canine distemper
	Pneumovirus	Respiratory syncytial

* Based on the immunologically distinct surface Ags, the hemagglutinin (H), and the neuraminidase (N), which undergo antigenic variation

† $H_{sw}N_1$, a virus isolated from swine in 1931 and humans in 1933 (previously designated H_0N_1), and the first major antigenic variant, H_1N_1, isolated in 1947, have very similar nucleotide sequences and are therefore classified as the same species. *sw,* swine; *eq,* equine; *av,* avian.

‡ Antigenic variations among strains are known, but the information is inadequate to enable division into subtypes.

antigenic variants (i.e., **species,** or **subtypes;** see Table 56–1) of influenza A and B viruses have continually emerged; the new variants are only remotely related to the earlier viruses or to each other. The frequent recurrence of the epidemic disease reflects the genetic variability of influenza viruses.

TABLE 56–2. Characteristics of Orthomyxoviruses and Paramyxoviruses

Characteristic	Orthomyxoviruses	Paramyxoviruses
Particle size	Small (80–120 nm)	Large (125–250 nm)
Diameter of internal helical core (nucleocapsid)	9 nm	18 nm
Localization of nucleocapsid	Nucleus	Cytoplasm
Segmented genome	+	−
Frequent genetic variation	+	−
Virion RNA polymerase	+	+
Separate hemagglutinin and neuraminidase	+	0
Filamentous forms	Common	Observed
Hemolysin	0*	+†
Prominent cytoplasmic inclusions	0	+
Syncytial formation	0	+

* At low pH they produce hemolysis.

† All species except pneumoviruses (respiratory syncytial virus)

Although bacterial types are narrow subgroups within the species, it should be noted that the original so-called **types** of influenza virus (see Table 56–1) are broad groups, now called **genera.**

PROPERTIES
Morphology

Influenza viruses are somewhat heterogeneous in size and shape, but are generally roughly spherical or ovoid (Fig. 56–1). Influenza A viruses have a mean diameter of 90 nm to 100 nm, whereas influenza B viruses are somewhat larger, approximately 100 nm in diameter. Filamentous forms of similar diameter occur in fresh isolates (Fig. 56–2).

Influenza virus particles (like paramyxoviruses) are distinguished by spikes or rods that cover the entire surface (see Fig. 56–1). These are evenly spaced and appear to be arranged in interlocking hexagons so that each rod has six neighbors. (Type C influenza virus particles display areas sparsely covered with spikes, revealing an underlying lattice of hexagons and pentagonal units.) Beneath the outer zone is a continuous membrane (Fig. 56–3). Disruption of the particle by lipid solvents uncovers an inner helical component, the nucleocapsid (see

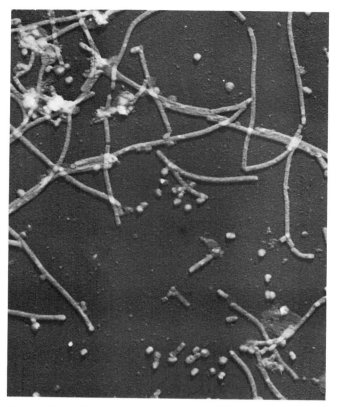

Figure 56–2. Electron micrograph of influenza H_2N_2 virus (third passage) showing filamentous viral particles as well as a few spherical ones. Chromium-shadowed preparation. (Original magnification ×10,400; Choppin PW et al: J Exp Med 12:945, 1960)

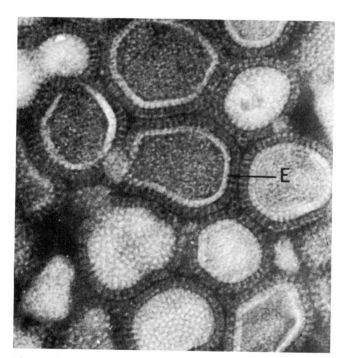

Figure 56–1. Electron micrograph of intact influenza A virions embedded in phosphotungstate. The virions are of variable shape and size and show evenly spaced, short surface projections covering the entire surface of the particles. *E*, envelope. (Original magnification ×300,000; Hoyle L et al: Virology 13:448, 1961)

Fig. 56–3). The purified nucleocapsid (Fig. 56–4) is composed of filaments of variable length (averaging 60 nm), each containing one of the segments of the viral RNA.

Upon complete disruption of the virion with sodium dodecyl sulfate (SDS) the **hemagglutinin** and the **neuraminidase** subunits can be separated by electrophoresis or rate zonal centrifugation. Each of these viral surface projections has a distinct structure (Fig. 56–5). The dispersed hemagglutinins, which are rod shaped, are univalent and therefore attach to RBCs but do not agglutinate them. When the SDS is removed, the hemagglutinins aggregate into clusters of radiating rods (Fig. 56–5C) that are multivalent and therefore produce hemagglutination. The neuraminidase subunits are mushroom-shaped and, after removal of SDS, aggregate into pinwheel-like structures (Fig. 56–5D); both the single and aggregated units have enzymatic activity. These arrangements originate from the mutual adherence of the terminal hydrophobic portions of the proteins, which are normally embedded in the lipid bilayer of the virion envelope.

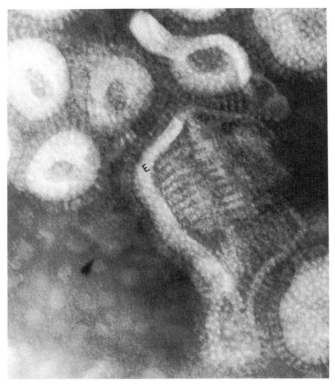

Figure 56–3. Partially disrupted influenza virus particle. The disrupted outer membrane or envelope (*E*), which is 6 nm to 10 nm thick, appears to have collapsed and become distorted, revealing the nucleocapsid folded in parallel repeating bands. (Original magnification ×300,000; Hoyle L et al: Virology 13:448, 1961)

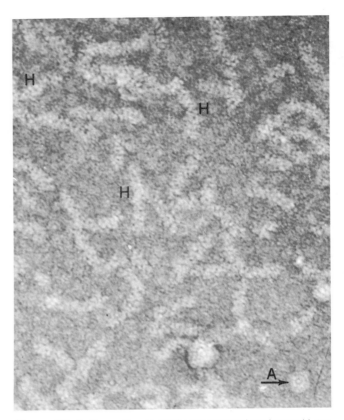

Figure 56–4. Preparation of highly concentrated nucleocapsid prepared by ether disintegration of purified influenza A virus particles and embedded in phosphotungstate. The helical structure of the nucleocapsid is apparent, particularly in regions marked *H*. The particle at *A* is interpreted as part of an elongated structure viewed along the particle axis. (Original magnification ×270,000; Hoyle L et al: Virology 13:448, 1961)

The **filamentous forms** of virus (see Fig. 56–2), often seen in freshly isolated strains, are very pleomorphic and frequently appear to be composed of spherical subunits; they may be as long as 1 μm to 2 μm and can be observed by dark-field microscopy. Evenly spaced spikes, similar to those of the more classic spherical particles, project from the surfaces. Filaments, which are apparently infectious, appear to assemble as a result of some defect in the development of particles and their emergence from the cell membrane. The capacity of an influenza virus strain to produce a predominance of filamentous forms is a stable genetic attribute of the strain.

Physical and Chemical Characteristics

The intact spherical particles contain approximately 0.8% to 1.1% RNA, 70% protein, 6% carbohydrate, and 20% to 24% lipids. A lower proportion of RNA is found in preparations containing numerous defective viral particles or filamentous forms.

Disruption of intact particles with lipid solvents, followed by removal of the hemagglutinin and neuramini-

dase by adsorption onto RBCs, reveals that the RNA is associated entirely with the **inner helical core** (the **S Ag**), which is 5% RNA by weight. In accord with electron micrographs of the nucleocapsid, gentle chemical extraction and physical separation of the nucleocapsid from other viral components yield nucleoproteins of three size classes. The **genome** extracted directly from the virion consists of **eight separate pieces of single-stranded RNA,** corresponding to segments of the nucleocapsid. (The influenza C genome, however, consists of only seven segments.) All the RNA pieces have almost identical 5′ ends consisting of a sequence of about 13 nucleotides ending with an Appp. In addition, the 3′ termini of all the genome RNAs have a high degree of conservation for the first 12 nucleotides (Fig. 56–6). Owing to the partial sequence complementarity between the 3′ and 5′ termini of each RNA, a panhandle holds each RNA in a circular form in virions and in infected

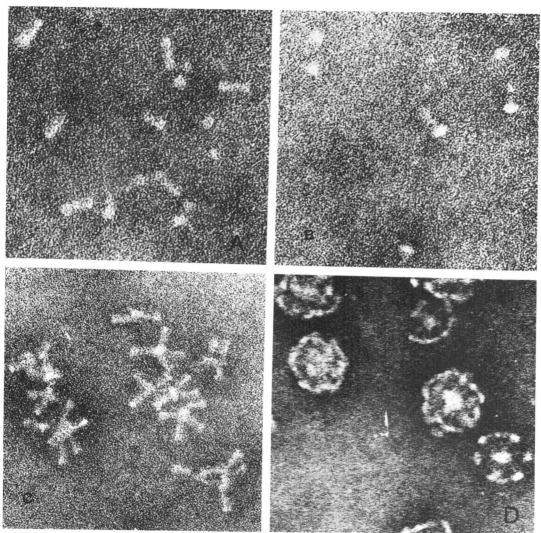

Figure 56–5. Morphology of the hemagglutinin and neuraminidase subunits of influenza A virus. (*A*) Single hemagglutinins appear as thick rods 14 nm × 4 nm in the presence of SDS. (*B*) Individual neuraminidase subunits dispersed in SDS are seen as oblong structures with a centrally located fiber possessing a terminal knob 40 nm in diameter. (*C*) Clusters of hemagglutinins formed by removal of SDS. (*D*) Neuraminidase subunits aggregated by the tips of their tails to form pinwheel-like clusters when SDS was removed. (Original magnification ×500,000; Laver WG, Valentine RC: Virology 38:105, 1969)

cells. It is striking that these conserved sequences at the 3' and 5' ends of the RNAs have extensive homology in all strains of types A, B, and C thus far examined. Since each RNA segment codes for a single viral protein, except segments 7 and 8, which encode two proteins each (Table 56–3), this physical structure is consistent with the marked genetic lability and very high recombination rate of influenza viruses (see Immunologic Characteristics and Genetic Characteristics, below).

Influenza virus RNAs are single-stranded molecules (total mol. wt. 5.9 to 6.3 × 10^6 daltons, assuming one molecule of each RNA segment per virion). Influenza A, B, and C viruses are probably phylogenetically quite distant, for their genomes differ significantly in size and base compositions (the A + U/G + C ratio is about 1.25 for type A, 1.42 for type B, and 1.46 for type C).

Seven distinct **virion proteins** can be separated by polyacrylamide gel electrophoresis (see Table 56–3). Two

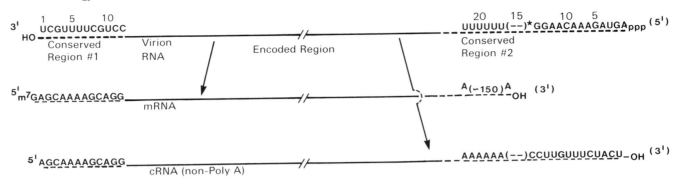

Figure 56–6. Diagram illustrating the nucleotide sequences of the two conserved regions of all orthomyxovirus genome RNAs and the two classes of transcribed cRNAs: polyadenylated incomplete transcripts (i.e., the mRNAs) and nonpolyadenylated complete transcripts, which are the templates for virion RNAs. The sequences 14–16 [(–)*] at the 5' ends are variable. The mRNAs are copied by the virion transcriptase, whereas the complete cRNAs are transcribed by a polymerase modified by newly synthesized viral proteins. (Modified from Hay AJ, Skehel JJ: Br Med Bull 35:47, 1979)

other proteins, **nonstructural components NS$_1$** and **NS$_2$,** are translated from two overlapping reading frames of RNA segment 8. The membrane **matrix (M$_1$) protein** (composed of identical small monomers) is the most abundant virion protein (see Table 56–3; Fig. 56–7); it is associated with the inner surface of the lipid bilayer and confers stability on the **viral envelope.** A second small polypeptide (**M$_2$**) is translated from a spliced RNA transcribed from segment 7, and it is inserted into the surface of infected cells but has not been detected in virions.

The **hemagglutinin** and **neuraminidase** are glycoproteins (containing glucosamine, fucose, galactose, and mannose). The rod-shaped hemagglutinin is synthesized as a single glycoprotein, but for virions to attain infectivity, its precursor must be proteolytically cleaved to two unique polypeptides (HA$_1$ and HA$_2$), which are held together by a single disulfide bond. The functional hemagglutinin consists of three noncovalently linked HA$_1$ and HA$_2$ polypeptides (see Fig. 56–7). The neuraminidase is a mushroom-shaped spike with a boxlike head composed

TABLE 56–3. Influenza A and B Virus Genome Segments and Encoded Proteins*

RNA Segment	Length (Nucleotides)	Encoded Polypeptide	Function
1	2341	PB$_2$	Host cell RNA cap "binding"; component RNA transcript
2	2341	PB$_1$	Initiation of transcription; component of transcriptase; possible endonuclease
3	2233	PA	Component transcriptase; elongation of mRNA
4	1778	HA	Surface glycoprotien; trimer; major antigenic component; CTL target Ag
5	1565	NP	Component of nucleoprotein complex, associated with each RNA segment; component of RNA transcriptase; expressed on cell surface; CTL target Ag
6†	1413	NA	Neuraminidase activity; glycoprotein on surface of viral envelope; tetramer
7	1027	M$_1$	Major component of virion; underlies lipid bilayer of viral envelope
		M$_2$	Nonstructural protein; CTL target Ag
8	890	NS$_1$	Nonstructural protein; function unknown
		NS$_2$	Nonstructural protein; function unknown
Total:	13,588		

* A/PR/8/34 strain (H$_1$N$_1$) used as an example

† Influenza B virus encodes a second protein (NB), which is nonstructural and of unknown funciton.

(Modified from Lamb RA: In Palese P, Kingsbury DW [eds]: Genetics of Influenza Viruses, p 21. New York, Springer-Verlag, 1983)

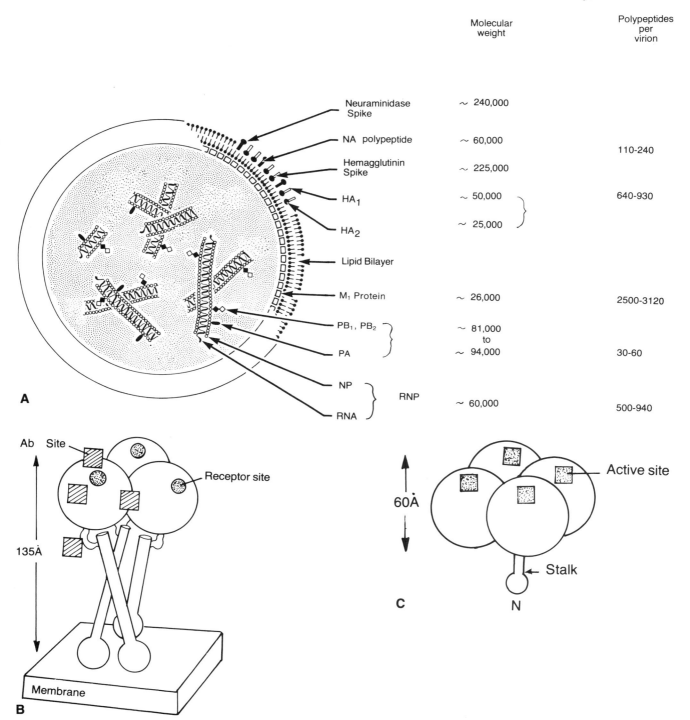

	Molecular weight	Polypeptides per virion
Neuraminidase Spike	∼ 240,000	
NA polypeptide	∼ 60,000	110-240
Hemagglutinin Spike	∼ 225,000	
HA₁	∼ 50,000	640-930
HA₂	∼ 25,000	
Lipid Bilayer		
M₁ Protein	∼ 26,000	2500-3120
PB₁, PB₂	∼ 81,000 to	
PA	∼ 94,000	30-60
NP		
RNA	∼ 60,000	500-940

Figure 56–7. Schematic model of influenza virus particles. (*A*) The nucleocapsid is segmented and gives the appearance of a double helix owing to association of internal proteins (NP, PB₁, PB₂, and PA) with the single-stranded RNAs. Also shown is the estimated number of each polypeptide per virion. (*B*) The hemagglutinin spike is composed of three sets of HA₁ and HA₂ polypeptides. (*C*) The neuraminidase spike consists of four NA polypeptides. (*B*, modified from Wilson IA, Skehel JJ, Wiley DC: Nature 289:366, 1981; *C*, derived from Varghese JN, Laver WG, Colman PM: Nature 303:35, 1983)

of four identical glycosylated polypeptides joined by disulfide bonds (see Fig. 56–7). The **nucleocapsid protein (NP)** is a single phosphorylated polypeptide species. Three large internal polypeptides (PB_1, PB_2, PA), associated with RNA transcription and replication, are present in the nucleocapsid in relatively small numbers (see Fig. 56–7). Influenza B virus RNA segment 6 is transcribed into a bicistronic mRNA, which, in addition to neuraminidase, encodes a small, glycosylated, **nonstructural polypeptide (NB)** whose function is still unknown. The NB protein has not been detected in influenza A virus–infected cells.

Tryptic peptide maps show little difference between the nucleocapsid proteins from different strains of influenza A, but sharp differences between the hemagglutinins are noted. These striking differences are confirmed by nucleotide and amino acid sequences of many different strains. This result is consistent with evidence that within a given viral type (genus) the nucleocapsid proteins of different strains are antigenically similar, whereas the glycoprotein surface Ags are strain specific.

The lipid of the viral particle is two-thirds phospholipid and one-third unesterified cholesterol. The kinds and concentrations of the individual lipids resemble those of the plasma membrane of the host cells. Thus, when cells are labeled with ^{32}P before viral infection, the lipids incorporated into viral particles, except for phosphotidic acid, are seen to be derived from the host cell. At least a portion of the viral polysaccharide is also of host origin. In contrast, the RNA and proteins are specified by the viral genome.

STABILITY. The high lipid content makes these viruses susceptible to rapid inactivation by lipid solvents and surface-active reagents. As with most other viruses with lipid envelopes, the infectivity of influenza viruses is relatively labile on storage at $-15°C$ or $4°C$ but is retained for long periods at $-70°C$.

Immunologic Characteristics

VIRAL ANTIGENS. The **hemagglutinin,** which is the major surface glycoprotein Ag (see Figs. 56–5*A* and 56–7), is measured by direct hemagglutination, and its Abs are assayed by hemagglutination-inhibition, neutralization, complement fixation (CF), or enzyme-linked immunosorbent (ELISA) assays. The **neuraminidase,** which is the other surface glycoprotein of influenza A and B virions (Fig. 56–5*B* and 56–7), is assayed by enzyme activity. Antibodies specific for neuraminidase inhibit enzyme activity but do not neutralize infectivity. Influenza C viruses do not have neuraminidase activity.

The major structural **matrix protein of the viral envelop (M_1 protein)** is measured by CF, immunodiffusion, immunoprecipitation, and ELISA; its specific Abs cannot neutralize infectivity or inhibit hemagglutinin and neuraminidase activities. The **internal** or **nucleocapsid (NP) Ag** corresponds morphologically to the internal helical component (the RNA-protein core; see Fig. 56–4) and is immunologically identical to the soluble Ag that is present in infected cells. Surprisingly, the NP protein is expressed on the surface membrane of infected cells and is a major target for cytotoxic T cells (CTLs). This Ag is assayed by CF titration or ELISA.

IMMUNOLOGIC GROUPING. On the basis of their nucleocapsid and M protein Ags the many influenza viruses are divided into three distinct **immunologic types (genera):** A, B, and C. The Ags of each type are unique and do not cross-react with those of the other two.

Within types A and B immunologic variants are distinguished by antigenic differences of the hemagglutinin (H) and neuraminidase (N). The antigenic variations of these two proteins, however, are genetically independent. Over the past few decades a major variant (i.e., **subtype**) of type A has emerged after varying intervals. The variant was either a new subtype ($H_1N_1 \rightarrow H_2N_2 \rightarrow H_3N_2$)* or a reemergence of an old one (H_1N_1 in 1977).

ANTIGENIC VARIATION. Influenza A virus undergoes two distinct forms of antigenic variation: **Antigenic drift** reflects minor antigenic changes in either the hemagglutinin or the neuraminidase, or both. **Major antigenic shift** occurs infrequently and reflects the appearance of viral strains with surface Ags that are immunologically only distantly related to those on earlier strains. The antigenic shift may involve either the hemagglutinin alone or the neuraminidase as well. Influenza A viruses have undergone three major antigenic shifts since 1933, detected by immunologic studies on sera from persons of different ages and on the viruses isolated. Such "seroarcheological" studies suggest that an H_2 virus was probably responsible for the large epidemic in 1890 and that a large epidemic in 1900 was caused by an H_3 virus. A virus similar to swine influenza virus ($H_{sw}H_1$)† presumably accounted for the human infections between 1918 and 1929, since sera from persons born during that period contain Abs to the swine agent; nucleotide sequence data show it to be an H_1N_1 subtype influenza A virus.

The first human influenza A virus (subtype H_1N_1)‡ was isolated in 1933; it was responsible for all influenza A infections until 1957; moreover, in sera from persons born during this period, regardless of their age when

---

* With the recognition that the hemagglutinin (H) and neuraminidase (N) glycoproteins vary independently and determine the antigenic characteristics of a viral strain, the subtypes are now named accordingly (see Table 56–1).

† Isolated by Shope in 1931 from pigs with a severe respiratory infection.

‡ The genus is termed type A; the subtype representing the first human influenza A virus to be isolated was originally called A_0.

tested, the highest influenza Ab titers are the H_1N_1 subtype prevalent at that time. In 1947 there emerged an H_1N_1 virus whose hemagglutinin's antigenicity had drifted to the extent that it was not efficiently neutralized by Abs to the 1933 H_1N_1 virus (see Table 56–1). The 1947 virus supplanted all prior strains, as indicated by isolations and by the appearance of Abs. In 1957 the H_2N_2 (Asian) influenza virus became prevalent. In 1968 another relatively large antigenic shift occurred, and the Hong Kong H_3N_2 virus emerged; the neuraminidase molecules are antigenically similar to those of the original H_2N_2 virus, but the hemagglutinin is chemically and immunologically unique. In 1976 swine influenza virus (H_1N_1) unexpectedly appeared at a U.S. Army post at Fort Dix, N.J., but it again disappeared after a brief encounter with about 200 soldiers. In the late fall of 1977 an H_1N_1 virus, another old acquaintance, emerged in the Soviet Union and Hong Kong, but this H_1N_1 was antigenically closer to the variants isolated in 1950 than to the original strains.

Influenza B viruses also undergo antigenic variations, but these are neither so extreme nor so frequent as those of A viruses, and some immunogenic cross-reactivity occurs among all the B variants. Hence, influenza B variants have not been classified into distinct subtypes. The originally isolated B virus was prevalent from 1936 to 1948, and the second variant appeared in 1954. Antigenic drift frequently occurred and eventually (in 1962) yielded an only distantly related antigenic variant.

The continual antigenic variation of influenza viruses is of considerable practical importance and theoretical interest. Each major shift has found a large proportion of the world population immunologically defenseless against the newly emerged virus. Furthermore, the neutralizing Abs induced by the vaccine current at the time did not react with the variant strain. The minor antigenic drifts that more frequently occur between pandemics, in contrast, may reduce the effectiveness of the vaccine but do not make it useless.

BASIS FOR ANTIGENIC VARIATIONS. The **minor antigenic variations** of antigenic drift result from mutations in the hemagglutinin and neuraminidase genes. The hemagglutinin mutations effecting antigenic drift are primarily confined to the four Ab combining sites in the H_1 polypeptide (see Fig. 56–7). The new variants always have mutations in two or more of the reactive epitopes. Therefore, the mutants emerge by selection of viruses that are less susceptible to neutralizing Abs prevailing in the population. In fact, experimental passage of viruses in the presence of small amounts of Ab in mice or chick embryos leads to similar selection of new variants.

A **major antigenic shift,** in contrast, cannot be explained by a simple mutation because the peptide maps of the hemagglutinins from different viral subtypes differ greatly, indicating extensive diversity in amino acid sequences. The change probably results from **recombination** (i.e., **gene reassortment**) between a human and an animal strain, both influenza A viruses. Such recombinants between different viral species have been produced experimentally in animals and have been selected by passage in immunized animals.

Each new major variant results from the adding of new major antigenic determinants while some of the previous ones are retained. Hence, **primary immunization** with H_2N_2 virus induces formation of neutralizing and hemagglutination-inhibiting Abs that react with H_2N_2 virus itself and with H_1N_1 subtype viruses, although the H_1N_1 virus does not elicit neutralizing or hemagglutination-inhibiting Abs able to react with the H_2N_2 virus. The complexity of the immunologic reactivities reflects the independent changes of the hemagglutinin and the neuraminidase, which are encoded in different genes (see Table 56–3). With a major antigenic change (a new subtype) the chemical changes of the hemagglutinin and neuraminidase, or of the hemagglutinin alone, are of such magnitude as to add antigenic reactivities that do not cross-react with the prior surface Ags. However, the antigenic specificities of the major internal NP and M proteins may not change even with major antigenic shifts.

On successive exposures to influenza viruses, whether by infection or by artificial immunization, the **Ab response is predominantly directed against the Ags of the viral strain with which one was initially infected.** Thus, if a child were infected first with an H_1N_1 virus in 1933 and then with an H_1N_1 virus in 1947, his Ab response in 1947 would be greatest to the 1933 H_1N_1 viral Ags, although he would also develop 1947 H_1N_1 Abs. With advancing age and an increased number of infections the Ab response to infection becomes broader, but the titer of Abs against the Ag of the original infecting virus remains the highest. This phenomenon, termed the **doctrine of original antigenic sin,** is reflected in the Ab levels of persons in different age groups (Fig. 56–8); it suggests that the initial encounter with influenza virus elicits a primary response and that with subsequent meetings a secondary response induces higher Ab titers owing to generation of an enlarged population of cross-reactive memory cells persisting since the primary antigenic response.

The prominent antigenic shift, resulting in the appearance of the major antigenic variants described, has led to speculation about whether an almost limitless number of major antigenic changes can occur. However, continued studies of "serologic archeology," which led to the concept of original antigenic sin, imply that the number of subtypes is limited, and the appearance and reappearance of viruses from H_1N_1 to H_3N_2 between 1889 and the present confirms the hypothesis that the repertoire of influenza A subtypes is limited (see Fig. 56–8).

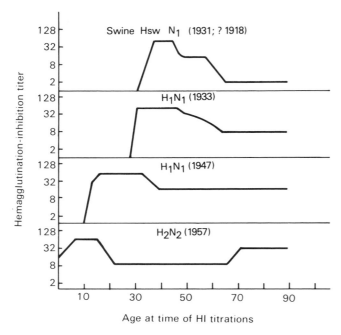

Figure 56–8. Age distribution, in the general population, of mean hemagglutination-inhibition Abs to subtypes of influenza A viruses. Sera were obtained in 1952 and 1958, and each was assayed with the viruses noted. The highest Ab titers for each virus tested were in those persons who were probably infected when that subtype first became prevalent (dates in parentheses). (Modified from Davenport FM et al: J Exp Med 98:641, 1953, with addition of data for H_2N_2 virus)

The observations, as well as the appearance first of the swine influenza H_1N_1 virus in 1976 and then of a similar H_1N_1 virus in 1977, strengthen the thesis that only a limited number of antigenic variations are possible. However, since the swine H_1N_1 did not flourish and spread and the 1933 H_1N_1 virus did not emerge after the H_3N_2 virus, it appears that the subtypes are not compelled to reappear in a set order, although the immune status of the population must serve as a major selective mechanism. It is striking that from 1977 to the present, for the first time since the influenza virus was isolated, two major subtypes, H_1N_1 and H_3N_2, have both persisted in the population. Moreover, this dual existence has resulted in the development of reassortment of RNA segments between the two subtypes as they spread in the population.

IMMUNITY. Either clinically evident or inapparent infection leads to immunity, and the immunity to viruses of the same antigenic structure appears to be long lasting. However, reinfections can be caused by variants with minor antigenic differences. **Immunity is induced by the hemagglutinin,** since it can be evoked by infection of purified hemagglutinin. Neuraminidase probably also plays a role, however, for antineuraminidase Abs effectively reduce viral spread from infected cells and therefore diminish the impact of infection. Abs directed against the nucleocapsid Ag or other internal proteins, in contrast, do not confer immunity.

Since humoral immunity is highly type specific, and generally subtype specific, infection or artificial immunization with one influenza H_1N_1 virus affords immunity against infection with other H_1N_1 viruses but not against H_2N_2 viruses (nor, of course, against influenza B viruses). Circulating antihemagglutinin and antineuraminidase Abs are present for many years after infection, and disease is usually mild when due to reinfection by a subtype similar to one previously experienced.

CELL-MEDIATED IMMUNITY. In human and mouse infections virus-specific T-lymphocytes appear that are cytotoxic for infected cells, reacting with virus-specific glycoproteins (both the HA_1 and HA_2 hemagglutinin subunits) as well as with the broader-reacting M_2 and NP proteins. Thus, the immune cytotoxicity exhibits cross-reactivity between influenza viruses of the same genus (e.g., influenza A) as well as subtype specificity. The reaction requires that the virus-induced cytotoxic T cells and the infected cells share the host major class I histocompatibility complex. Cell-mediated immunity, and its persistence in man, plays an important role in recovery from influenza virus infections, as it does in many other viral infections (see Chaps. 50 and 51).

ARTIFICIAL IMMUNIZATION. Artificial immunization is limited not only by the marked antigenic variation of the viruses but also by the restriction of the infections to the respiratory mucous membranes, where secretory IgA Abs are required, and where Ab concentrations are only approximately 10% of those in the blood. Hence, minor antigenic modifications of the infecting virus permit it to escape neutralization more readily than it could if viremia were an essential part of the infectious process. The situation is analogous to the outgrowth of drug-resistant bacterial mutants when the drug concentration is borderline.

Genetic Characteristics

The remarkable genetic variability of influenza viruses involves not only antigenic subtypes but also other genetic markers. These include affinity for Ab, reactions with RBCs from animals of different species, virulence, reactions with soluble mucoprotein inhibitors, heat resistance, host range, and morphology. Only a few of these mutations have an obvious bearing on the behavior of influenza viruses in nature, but they illustrate the ease with which these agents vary and the potential types of selective pressures at work in nature.

Variation in the affinity of viruses for specific Ab (see

Chap. 50) is frequently noted. Thus, viruses isolated in the course of a single epidemic may vary in their susceptibility to neutralization by antiserum prepared with homologous virus (isolated during the same epidemic) or with heterologous strains. Strains isolated during the height of epidemics commonly react to high titer only with homologous Abs. Oligonucleotide mapping and nucleotide sequence analysis of such selected viruses reveal the potential frequency of mutations affecting the hemagglutinin epitopes responsible for neutralizing Abs and the number of antigenic variants that may circulate during an epidemic. The emergence of similar variants in nature may permit the persistence of virus in the population during interepidemic periods, but new subtypes appear to emerge from reassortment of RNA segments between two distinct viruses rather than sequential mutations.

Numerous mutants with increased virulence for a given host or organ system have been isolated. Conversely, **temperature-sensitive (ts) mutants** unable to multiply effectively at temperatures above 37°C and **cold-adapted mutants** (multiply best at 32°C) are less virulent and may prove valuable for live virus vaccines. The finding that such mutants revert to a wild-type phenotype at a relatively high frequency may, however, prove to be an insurmountable handicap to their use.

Genetic recombination, more properly termed **genetic reassortment,** has been extensively studied in influenza viruses because of its special epidemiologic and clinical implications for such a variable virus, the opportunity (which was initially unique) to investigate RNA as genetic material, and the numerous markers available.

The first evidence of recombination between animal viruses was obtained by Burnet in 1949, using influenza viruses: infections with mixtures of neurotropic and nonneurotropic strains of different antigenic identity yielded recombinants in which neuropathogenicity from one strain was combined with an antigenic character from the other. Subsequently, genetic recombination has been observed with many other naturally occurring strains carrying various markers and with conditionally lethal ts mutants. Extraordinarily high recombination frequencies between influenza viruses have been reported (up to 50%). The new genotypes, however, are not the consequences of true recombination (see Chap. 48) within an RNA molecule; rather, they emerge as the result of an independent assortment and segregation (i.e., reassortment) of separate segments of the viral genomes. The high frequency of reassortment recombination between influenza A viruses of different subtypes has made it possible to identify the viral protein encoded in a particular RNA segment by associating a given segment with a given polypeptide in recombinant viruses. This approach is possible because each RNA segment encodes only one or two proteins (Table 56–3) and because RNA and proteins

from different subtypes have unique electrophoretic mobilities. To complete the RNA segment assignments biochemical and biophysical analyses of the viral RNAs and proteins were required.

Reassortment is detected only within a genus, and not between influenza A and B viruses. When a mixed infection is initiated with high multiplicities of influenza A and B viruses, however, viral particles appear that have surface Ags of both parent viruses and are therefore neutralizable by Abs to either. This property results from **phenotypic mixing** (see Chap. 48) and is not passed on to the progeny.

Host Range

Strains of human influenza viruses are best propagated experimentally in the amniotic cavity of chick embryos (the most sensitive and the most convenient host) or in the respiratory tract of ferrets or mice. Many strains also multiply readily in cultures of monkey kidney, calf kidney, and chick embryo cells. Viruses can be readily adapted to propagation in the allantoic cavity of the chick embryo, as well as in the respiratory tracts of monkeys and many rodents.

Multiplication

ENTRY. Infection is initiated with the attachment of virions to susceptible host cells by reactions between the hemagglutinin spikes and specific *N*-acetylneuraminic acid–containing mucoprotein receptors.* The host receptors are similar to or identical with those on RBCs and with the soluble mucoprotein inhibitors in human and animal secretions. After attachment the viral particles, through the process of **endocytosis,** are engulfed into coated pits and vesicles, finally entering endosomes where they are exposed to about pH 5.0 (Fig. 56–9). This acidic pH activates the fusion function of the hemagglutinin, which permits the viral nucleocapsid to enter the cytoplasm. With this process infectivity is rapidly lost **(viral eclipse).**

REPLICATION. The initial steps in viral replication following entrance of the viral genome into the cell are still unclear. Unlike other RNA-containing viruses (e.g., picornaviruses), influenza viruses cannot replicate in enucleated cells. Moreover, ultraviolet irradiation, dactinomycin, or mitomycin C blocks viral multiplication if administered during the first 2 hours of infection but not thereafter (i.e., before synthesis of viral RNA is established). However, chemical inhibitors of DNA biosynthesis (e.g., arabinosylcytosine) do not reduce propagation

---

* Although the neuraminidase-containing spikes can also react with cell surface receptors, virions remain infectious after these spikes are removed by trypsin. Antibodies to neuraminidase cannot neutralize infectivity, although they block enzyme activity.

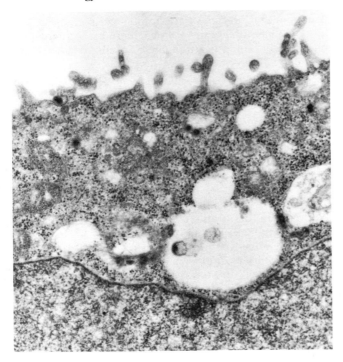

Figure 56—9. Attachment and phagocytic engulfment of influenza virus particles into clathrin-coated vesicles and endosomes. Some nucleocapsids have entered the cytoplasm as result of fusion of the viral envelope with the endosomal membrane. (Courtesy of A. Yoshimura and S. Ohnishi)

of infectious virus. Hence, **functioning but not replicating host DNA is essential for early events in multiplication of influenza viruses.** Amantadine, which inhibits RNA polymerase II, also blocks viral production. These phenomena ensue because the host continually supplies transcripts whose 5′ ends are cannibalized to provide caps for the 5′ termini of the viral mRNAs and to serve as primers for viral transcription.

Influenza viruses are **negative-strand (antimessenger) RNA viruses,** and therefore contain an **RNA-dependent RNA polymerase (RNA transcriptase)** within the virion to transcribe the virion RNA segments into mRNAs. The PB_1, PB_2, and PA virion proteins (see Table 56–3 and Fig. 56–7) in concert fulfill the RNA polymerase functions. Studies using temperature-sensitive mutants indicate that the PB_2 protein attaches to the cap structure of a nascent host mRNA, which is then cleaved by a viral endonuclease (apparently one of the viral P proteins). The cap structure then acts as a primer for transcription to produce the viral mRNA. PB_1, which is initially found at the first nucleotide added onto the primer, appears to catalyze the addition of each nucleotide. PB_2 dissociates from the capped primer, after the addition of the first 11 to 15 nucleotides, after which it associates with PB_1 and PA to move down the growing mRNA chain. Two classes of complementary RNAs (**c**RNAs) are made in infected cells: polyadenylated incomplete transcripts of the virion RNAs (terminated about 17 nucleotides from the conserved 5′ ends of the

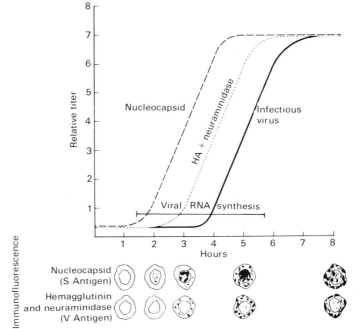

Figure 56—10. Multiplication of influenza A virus. Diagram of the biosynthesis of virions and viral subunits measured by titrations and immunofluorescence.

templates) that are associated with polysomes and serve as the mRNAs; and nonpolyadenylated complete transcripts, which are the templates for virion RNAs (see Fig. 56–6).

Primary virion transcription, detected with radiolabeled probes within the first hour of infection, can occur in the absence of protein synthesis and yields only mRNAs. Replication of the nonpolyadenylated cRNAs requires synthesis of viral proteins, apparently to modify the RNA polymerase, permitting production of complete transcripts. The complementary RNAs are predominant during the first 2 to 3 hours of infection, and virion RNAs predominate thereafter. However, the mechanism for regulating the synthesis of full-length transcripts (cRNAs) and virion RNAs, rather than mRNAs, is unknown. Both classes of complementary RNAs and the virion RNAs are made separately for each viral RNA segment through the usual intermediary replicative forms (see Chap. 48). Both the mRNAs and virion RNAs are synthesized in the nucleus.

The nucleocapsid protein (NP), which is about 90% of the protein associated with RNA-protein fragments, is associated with polymerase activity but probably only in a structural role. The NP (detected with fluorescein-labeled or ferritin-labeled Abs) is synthesized in the cytoplasm and is rapidly transported into the nucleus, where a significant proportion of the viral RNA is also found (Figs. 56–10 and 56–11). The newly assembled nucleocapsids subsequently move into the cytoplasm and migrate to the cell membrane. The hemagglutinin and neuraminidase proteins remain in the cytoplasm throughout replication (Fig. 56–10).

ASSEMBLY. About 4 hours after infection with influenza A virus the virion M_1 protein becomes associated with the inner surface of the cell plasma membrane, and discrete patches of the membrane thicken and incorporate hemagglutinin and neuraminidase molecules, which gradually replace the host proteins in these segments (Figs. 56–12 and 56–13). As segments of the helical nucleocapsid impinge on the altered membrane, it buds and forms viral particles, which are released as they are completed; virions cannot be detected within the cell.

Although the assembly of virions is an imperfect process, yielding virions of considerable morphologic heterogeneity and many noninfectious particles, it must have an effective control for packaging the appropriate set of nucleocapsid segments. Thus, **reassortants** derived from two subtypes containing distinguishable RNA segments never contain two copies of the same gene from different parents (e.g., a single virus does not contain the hemagglutinin genes from two parents).

Viral particles are released over many hours, without lysis of the infected cells, but eventually the cells die. The mechanism for releasing the budding virions is unclear.

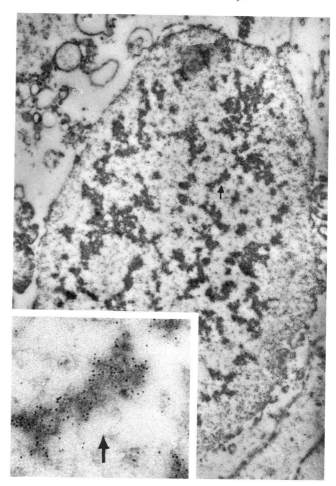

Figure 56–11. Nucleus of influenza virus–infected cell showing aggregates of dense material labeled with ferritin-conjugated specific Ab (*arrow*). The chromatin is sparse, and the nuclear membranes are disrupted. (Original magnification ×26,000) (*Inset*) Higher magnification of the portion of the nucleus marked by the arrow. Ferritin-conjugated Ab is present within the aggregates of dense material. The intervening nuclear matrix is nearly devoid of ferritin, i.e., nucleocapsid Ag. (Original magnification ×97,000; Morgan C et al: J Exp Med 114:833, 1961)

Neuraminidase may serve this function, since specific Abs to neuraminidase decrease viral release, though it cannot neutralize infectivity. However, univalent Fab fragments of this Ab do not reduce viral release, though they neutralize neuraminidase activity *in vitro*. The bivalent Ab may thus block viral release by binding virions to the membrane rather than by inhibiting specific enzyme activity.

The sequence of events with influenza B virus is similar, but the latent period (identical with the eclipse period for viruses that are assembled at the cell membrane) is 1 to 2 hours longer.

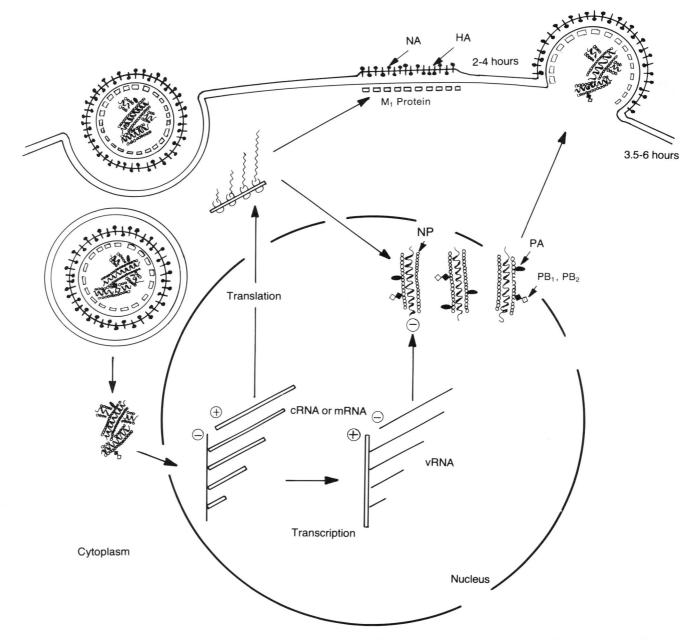

Figure 56–12. Schematic representation of the steps in the biosynthesis of influenza viruses, from adsorption, endocytosis, fusion of the viral envelope, and penetration of the nucleocapsid into the cytoplasm to insertion of newly made viral proteins into the cell's plasma membrane, assembly, and budding of the virion. The site of viral RNA transcription and replication is unknown but is probably the nucleus.

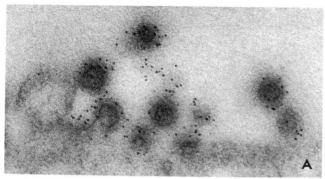

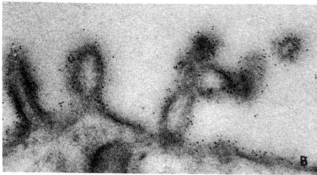

Figure 56–13. Development of influenza virus particles at the surface of an infected cell. Viral particles and components are labeled with ferritin-conjugated specific Ab. (*A*) In addition to fully formed viral particles, ferritin–Ab complexes have tagged one particle (*left of center*) presumed to be in the process of budding and two others (*below, right*) probably at early stages of differentiation. (Original magnification ×140,000) (*B*) Surface of an infected cell with several cytoplasmic protrusions with which ferritin–Ab complexes have combined as a result of the virus-specific antigenic change in the surface. No mature viral particles are evident. (Original magnification ×97,000; Morgan C et al: J Exp Med 114:825, 1961)

PATHOGENESIS

Influenza, an acute respiratory disease associated with constitutional symptoms, results from infection and destruction of cells lining the upper respiratory tract, trachea, and bronchi. Virus enters the nasopharynx and spreads to susceptible cells, whose membranes contain the specific mucoprotein receptors. The virus must first pass through respiratory secretions; and though these contain mucoproteins that can also combine with viral particles, infection is not blocked because the viral neuraminidase hydrolyzes the mucoproteins, rendering them ineffective as inhibitors.*

* The evolutionary selection of viruses containing neuraminidase is understandable. Were influenza virus devoid of its surface hydrolytic enzyme, the secretory mucoprotein would be as effective as Abs, and infection would be difficult to establish.

During acute illness the ciliated epithelial cells of the upper respiratory tract are primarily involved. Viral multiplication is followed by necrosis of infected cells and extensive desquamation of the respiratory epithelium, which is directly responsible for the respiratory signs and symptoms of the acute infection. Indeed, there is a direct correlation between the severity of disease and the quantity of virus produced and shed in nasopharyngeal secretions (peak titers range from 10^3 to 10^7 $TCID_{50}$/ml secretions). Those shedding less than 10^3 $TCID_{50}$/ml have only minor illness or are asymptomatic.

In nude mice or in mice treated with anti-θ serum, experimental infections are mild, implying that most of the cell damage is caused by virion-activated cytotoxic T-lymphocytes that recognize viral Ags on infected cells. Such activated T cells are also cytotoxic to cells infected with other influenza viruses of the same genus, probably because the related viral M_2 and NP proteins are the Ag signals.

Early in the course of the uncomplicated disease, constitutional symptoms—fever, chills, generalized aching (particularly muscular), headache, prostration, and anorexia—are more prominent than would be expected from a local infection of the respiratory tract. Viremia, however, is not an essential event in the pathogenesis of influenza infection, and although it has been detected on rare occasions, the constitutional symptoms are probably due to breakdown products of dying cells absorbed into the bloodstream. The liberation of endogenous pyrogen from polymorphonuclear leukocytes (PMNs), with which influenza viruses can react and enter, is probably not important in producing fever, since few such cells are found in the upper respiratory tract. However, the reaction of viral particles with PMNs *in vitro* inhibits the leukocytes from phagocytizing bacteria favoring their spread. Moreover, in experimentally infected mice, alveolar macrophages are greatly hampered in their capacity to clear bacteria from the lungs.

Normally, influenza is a self-limited disease lasting 3 to 7 days. About 10% of patients with clinical influenza have small areas of lobular pulmonary consolidation. **Secondary bacterial pneumonias** are the major cause of death. Although **fatal primary influenza virus pneumonia** without bacterial invasion is rare, it accounts for the unusual deaths from primary influenza. It occurs most frequently in persons with diminished respiratory function (e.g., those with mitral stenosis or chronic pulmonary disease, and women during the later stages of pregnancy). Viruses isolated from such fatal cases seem no more virulent in experimental animals than those causing minor illness.

Fatal nonbacterial pneumonia was more common in the 1918–1919 pandemic, and postmortem descriptions of the rare deaths in more recent epidemics are identical. The lungs appear unforgettably huge, distended with edema fluid and blood that pour out if the lung is cut. Microscopically, the tracheal and bronchial epithelium shows marked destruction and often denudation. The bronchioles and alveoli are distended with cell debris, blood, and edema fluid, but purulent exudate is absent. Fibrin-free, hyaline-like membranes often line the alveoli. Apparently, the pathogenesis of influenza pneumonia evolves in two stages: from bronchial and bronchiolar epithelial necrosis to hemorrhage and edema.

The highly fatal pneumonia during the pandemic of 1918–1919 was predominantly characterized by secondary bacterial complications. The most prominent invaders were *Staphylococcus aureus, Hemophilus influenzae*, and β-hemolytic streptococci. In recent epidemics, coagulase-positive *S. aureus* and pneumococci have been most frequent. In addition to the usual severe epithelial injury of influenza, the secondary bacterial pneumonias show **deeper invasion** of the walls of the bronchi and bronchioles by bacteria, destruction of alveolar walls, purulent exudate, abscess formation, and vascular thrombi.

The influence of influenza virus infection on bacterial invasion has been studied experimentally. Introduction of pneumococci, streptococci, or staphylococci into the lungs of healthy mice results in little if any inflammatory response, but extensive bacterial pneumonia develops in animals with pulmonary edema induced by influenza virus or irritant chemicals. This effect of influenza viruses is probably promoted by their capacity to inhibit phagocytosis of bacteria by alveolar macrophages as well as PMNs.

Extrapulmonary lesions have also been observed in fatal cases of influenza pneumonia, including central nervous system involvement such as the Guillain-Barré syndrome (ascending myelitis) or encephalitis; Reye syndrome (encephalopathy and fatty liver), which inexplicably has been particularly associated with influenza B epidemics; hemorrhage into the adrenals, pancreas, and ovaries; and renal tubular degeneration. Viruses have only rarely been isolated from such lesions, however, and the relation of viral infection to the lesions is obscure.

LABORATORY DIAGNOSIS

A presumptive diagnosis of influenza can often be made from clinical and epidemiologic considerations. Laboratory confirmation of a clinical diagnosis is generally too costly for individual or sporadic cases, but it is used to establish the presence of the agent in the community, to determine its specific type, and to carry out epidemiologic studies. Diagnosis may be established by viral isolation, demonstration of specific Ab increase, and immunofluorescent demonstration of specific Ags in epithelial cells present in nasal secretions or sputum.

Virus is usually **isolated** by inoculating nasal and throat washings or secretions into the amniotic sac of 11- to 13-day-old chick embryos or onto monolayers of monkey kidney cell cultures. The latter have the advantage of also supporting multiplication of many respiratory viruses other than influenza. Fresh isolates of influenza virus may fail to produce cytopathic changes in monkey kidney cells, and the presence of virus is best detected by the **hemadsorption** technique, employing guinea pig RBCs. With embryonated eggs, virus can be detected in the amniotic fluid, after 2 to 3 days' incubation, by **hemagglutination,** by means of guinea pig or human RBCs for influenza A and B viruses and chick RBCs for influenza C virus. The newly isolated virus from either chick embryos or cell cultures is usually typed by **hemagglutination inhibition** with standard antisera.

Immunologic methods are used most frequently for diagnosis of influenza infection. Because the majority of persons already have influenza virus Abs at the time of infection, it is essential to demonstrate an increase by comparing titers in serum specimens obtained during both the acute and the convalescent phases of the disease (paired sera). **Hemagglutination-inhibition (HI)** techniques are most often employed for this purpose, although they are handicapped by the troublesome presence in serum of **nonspecific** mucoprotein or protein **inhibitors.** These may be eliminated by treating the serum, after heat inactivation (56°C for 30 min), with either the receptor-destroying enzyme (RDE) from *Vibrio cholerae* (a neuraminidase), a mixture of trypsin and potassium periodate, or the adsorbent kaolin. **ELISA** and **solid-state radioimmunoassays** are at least as sensitive as the HI, but they are not used generally in hospital diagnostic laboratories.

The **complement-fixation (CF)** assay is equally sensitive, and it circumvents the difficulties presented by nonspecific serum inhibitors. With crude preparations, which contain the nucleocapsid antigen, CF has broad specificity and can identify only viral type, but if one utilizes the hemagglutinin separated from virions, the assay is just as strain-specific as hemagglutination inhibition. An increase in specific Abs can also be measured by **neutralization** titrations, but because of its greater expense and the time required, this procedure is employed only for special purposes.

Immunofluorescent techniques furnish a method for establishing the diagnosis of influenza while the patient is still acutely ill: fluorescein-conjugated Abs reveal the presence of virus-specific Ags in desquamated cells from the nasopharynx. This technique permits diagnosis of about 74% of the infections detected by viral isolation or by immunologic assays.

EPIDEMIOLOGY

Influenza occurs in recurrent epidemics that start abruptly, spread rapidly, and are frequently distributed worldwide. An influenza A epidemic generally appears every 2 to 4 years and an influenza B epidemic every 3 to 6 years, but the patterns have not been completely predictable. Although epidemics occur periodically in any given geographic locality, outbreaks occur somewhere every year. Epidemics of influenza A viruses are usually more widespread and more severe than those of influenza B. Influenza C virus has not caused epidemics, and it usually produces inapparent infections.

The incidence is highest in the age group of 5 to 9 years; above 35 it gradually declines with increasing age. The very young and the very old suffer the highest mortality, with about three-quarters of all influenza deaths occurring in those over 55 years of age. Indeed, even without virologic or serologic evidence an influenza epidemic is recognizable by the increased mortality due to pneumonia in the elderly. Other special groups showing elevated mortality include pregnant women and persons with chronic pulmonary disease or cardiac insufficiency. A striking exception to the usual age-related mortality rate, however, was noted in the severe 1918–1919 pandemic: The majority of the 20 million deaths occurred in young adults, probably because older persons had previous exposure to and hence some immunity against this unusually virulent influenza virus.

Epidemics are common from early fall to late spring. Outbreaks often develop in many places in a country at almost the same time and spread rapidly to neighboring communities and countries; with the common use of air travel, intercontinental spread has also become rapid. The rapid dissemination is not entirely due to the speed of modern transportation, however, for this characteristic was also noted when man could travel no faster than the speed of his horse. The pattern of epidemic spread may be related to the occurrence of sporadic cases and the probable seeding of the virus in the population several weeks prior to an explosive outbreak. Even during an epidemic, a **high ratio of infection to disease,** from 9:1 to 3:1, can be demonstrated serologically.

The pathway of widespread dissemination of the virus has now been exemplified by several well-studied episodes, such as the 1957 pandemic caused by a new variant, H_2N_2 (Asian) subtype, which apparently emerged from central China in February of 1957 (Fig. 56–14). The arrival of the virus in the United States was detected in naval personnel in Newport, RI, on June 2 and shortly thereafter in San Diego, Calif, without a traceable connection between the two episodes. The first civilian outbreak was observed in a conference in Davis, Calif, on June 20, followed by several similar small episodes else-

where in California. From the conference the virus was carried directly by some of the more peripatetic members to a meeting of young people in Iowa, and from this location it was seeded throughout the country. This initial dissemination resulted in small, sporadic outbreaks until September, when epidemics occurred in almost all parts of the country. Similar spread along paths of travel occurred throughout the world.

In the summer of 1968, after an appropriate period of antigenic drift (11 years), another influenza A variant (H_3N_2) appeared in Hong Kong and produced a mild but widespread pandemic whose spread was strikingly similar to that of 1957. Although the 1968 Hong Kong virus had antigenic characteristics clearly different from the previously isolated H_2N_2 viruses, there was considerable immunologic relatedness, owing to the cross-reacting neuraminidase. The H_3N_2 viruses isolated from epidemics in 1969, 1970, and thereafter showed further antigenic drift.

When swine influenza $(H_{sw}N_1)$ virus (considered an H_1N_1 subtype) was isolated from a fatal illness and from approximately 200 nonfatal infections at Fort Dix, NJ, in the spring of 1976, it was postulated that this virus would be the next pandemic subtype. This prediction was based on the concept that the genetic variation of influenza A virus is limited (see Basis for Antigenic Variations, above) and that the emergence of subtypes is cyclic. However, the feared $H_{sw}N_1$ (H_1N_1) virus did not spread, and the previously prevalent H_3N_2 subtype (A Victoria) remained epidemiologically viable. The emergence of a different H_1N_1 virus in China and Russia in 1977 and its rapid spread to other continents heralds this subtype as a prevalent species and again suggests the limited variations of influenza viruses.

Many questions concerning the epidemiology of influenza remain unanswered. Not the least puzzling among the unknowns are the following: Why does the virus not spread rapidly at the time of the initial infections in a community? Where is the virus during interepidemic intervals? How does the virus become "masked" or "go underground" in the interval between its seeding and the occurrence of an epidemic? What provocative factors induce the epidemic?

After seeding, or during the interval between epidemics, virus may simply be transmitted slowly, producing inapparent infections or sporadic cases; remain latent in the persons previously infected; or reside, active or latent, in an animal reservoir. Influenza virus has rarely been isolated in nonepidemic periods, which speaks against the first possibility. The intriguing ecology of swine influenza virus, which is activated by cold weather in the presence of *H. influenzae suis* (see Latent Viral Infections, Chap. 51), offers one example of the second mechanism. Finally, human strains of influenza A virus

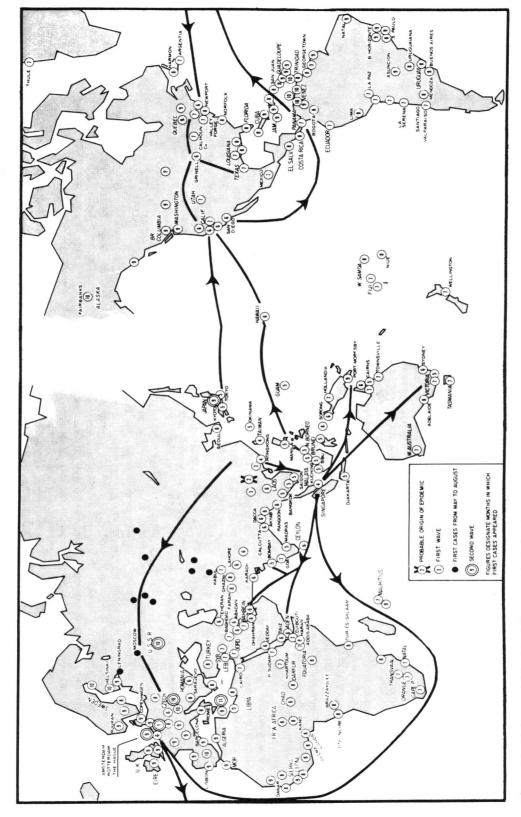

Figure 56–14. Progress of the Asian influenza pandemic from its probable origin in central China, Feb. 1957 to Jan. 1958. (Langmuir AD: Am Rev Resp Dis 83:1, 1961)

show immunologic and genetic relations to influenza viruses of horses, ducks, chickens, and pigs,* which supports the third mechanism and suggests that animals may serve as a source of new variants by reassortment with human strains.

The appearance of influenza viruses after a silent interval no doubt frequently depends on the development of a new antigenic variant that can escape an immunologic barrier existing in nature. However, "old" strains, having only minimal antigenic changes, also initiate epidemics, presumably because a sufficient number of previously uninfected persons enter the population, and the general Ab level falls below that necessary to prevent infections. The nature of the provoking factors that initiate an epidemic remains a mystery. It is also important to note that new variants, or even a different virus (e.g., type B), commonly emerge in a **heralding wave** of infections the spring before they produce fall and winter epidemics.

PREVENTION AND CONTROL

The high incidence of inapparent infections, the short incubation period, and the high infectivity preclude the successful use of isolation or quarantine procedures to control influenza. Quarantine of travelers entering a country can delay but not prevent the entrance of virus. In South Africa in 1957, for example, where ships were quarantined and the passengers and crew forbidden to land, infection did not enter through the ports, but the virus finally entered from the north, probably being carried by immigrant laborers traveling overland.

Artificial immunization can prevent influenza to a significant extent (reducing the incidence 60% to 80%), but not completely. Viruses propagated in chick embryos, partially purified, and inactivated by formalin or chemically disrupted, can provide a highly effective vaccine if the viruses utilized include a strain whose hemagglutinin and neuraminidase glycoproteins are closely related immunologically to the currently prevalent strain. This requirement is not always easy to satisfy. For example, although influenza vaccines containing an early H_1N_1 virus were highly effective in 1943 and 1945, the vaccine

employed in 1947 failed because it did not include the newly emerged H_1N_1 antigenic variant. Hence, influenza must be under constant global surveillance, including accurate antigenic characterization of isolated strains.† Vaccines currently employed contain a mixture of several strains of influenza A and B viruses in order to cover the known antigenic spectrum. New major antigenic variants are added as they appear.

Despite the proved value of the available inactivated viral vaccines, several factors have limited their use and possibly their effectiveness. (1) Pyrogenic reactions, accompanied by constitutional symptoms (not unlike the manifestations of mild influenza) and by local reactions, have been common, particularly in infants and young children; an incidence of 10% to 20% is not unusual in children less than 6 years of age, even when partially purified whole virus is used. Hence, the disrupted, purified component virus vaccine is recommended for children less than 12 years old. (2) Secretory IgA Abs in respiratory secretions are probably critical for successful protection, but subcutaneous injection of inactivated virus induces only low levels of such Abs in the respiratory tract. (3) Abs begin to decrease about 3 months after immunization, and immunity is often lost within 6 months.

Generally, a single subcutaneous injection of 0.25 ml to 0.5 ml containing 0.75 μg to 15 μg of each hemagglutinin subtype will confer immunity in 2 to 4 weeks. Persons immunized with a new subtype, especially children, show a primary immunologic response, whereas those who have had previous exposure to the Ags in the vaccine exhibit a secondary response. Therefore, if the vaccine contains a new major antigenic variant to which most individuals have no detectable Abs, such as H_2N_2 in 1957, a second injection is recommended a month after the first.

Influenza virus immunization is generally effective for all age groups. However, owing to the reactions to immunization and the limited nature of the disease in the general population, the vaccine is generally recommended for broadly defined high-risk groups and persons who provide essential general community services. Special groups recommended to be targeted for annual immunization are adults and children with chronic heart diseases, bronchopulmonary diseases, renal diseases, and metabolic disorders such as diabetes; residents of nursing homes and other chronic care facilities; persons more than 65 years of age; and medical personnel, policemen, and firemen.

New methods of preparation and administration of vaccines are being tested with encouraging results. (1) **Subunit vaccines,** consisting of hemagglutinin and neuraminidase components of disrupted virions, are be-

---

* The N_1 in human viruses extant during the late 1930s is immunologically related to the neuraminidase present in at least four of the avian influenza viruses as well as to that in the swine virus; the N_2 present in the H_2N_2 and H_3N_2 viruses is similar to the neuraminidase of a turkey influenza virus isolated in 1966; and the hemagglutinin of the H_3N_2 virus is immunologically similar to the hemagglutinin of an equine influenza virus ($H_{eq2}N_{eq2}$) as well as an avian hemagglutinin (H_{av7}). Indeed, every subtype of hemagglutinin (13 subtypes) and neuraminidase (nine subtypes) has been identified in avian influenza viruses. Moreover, viruses isolated from seals and whales, which have infected humans through laboratory contact, have been traced to be of avian origin. It therefore seems possible that new epidemic influenza viruses arise by reassortment with avian viruses.

† The World Health Organization has established centers throughout the world for this purpose.

ing used rather widely. This formulation permits administration of a greater antigenic mass, therefore effecting greater Ab responses, with fewer major toxic reactions. (2) **Attenuated infectious viruses** (selected by serial passage at low temperatures) have been widely used in the Soviet Union as well as in other countries, and those selected as cold-adapted mutants are being employed experimentally in the United States. The attenuated viruses have been shown to produce IgA Abs in respiratory secretions. With live vaccines, however, it is impossible to select and test rapidly a suitable derivative of the viral subtype that has recently emerged. (3) **Intranasal aerosol administration** of inactivated viral vaccine induces an adequate response of IgA Abs in nasal secretions as well as of specific IgA, IgM, and IgG Abs in serum, but this method has not yet been adequately evaluated. (4) **Recombinant DNA vaccine** in which the hemagglutinin gene is inserted into a heterologous viral vector (e.g., vaccinia virus) and expressed as the viral vector replicates.

A new approach to immunization has developed from the ability to "tailor-make" variants of influenza viruses by **genetic reassortment** of a new antigenic variant with an avirulent cold-adapted mutant, or with an established strain, in order to ensure propagation of the newly emerged subtypes to high titers. (The swine A influenza H_1N_1 vaccine virus used for nationwide immunization in 1976 was prepared by this method from the Fort Dix isolate.)

Widespread immunization has revealed an additional source of concern, at least with inactivated intact virus vaccine. Following immunization of more than 35 million persons with inactivated swine influenza (H_1N_1) virus in the fall of 1976, the Guillain-Barré syndrome occurred in 354 recipients, with 28 deaths. Most illnesses occurred 2 to 4 weeks after immunization. The incidence was about 1 in 100,000 persons who received the vaccine, almost six times greater than that in the nonimmunized population in the same period. Influenza virus vaccine produced after 1977 has not been associated with neurologic complication. Moreover, the etiology of the Guillain-Barré syndrome is not specifically related to influenza immunization. It has also been reported following other immunization procedures (e.g., rabies, smallpox), but the scale and the surveillance of the 1976 immunization program was never previously equaled so that an etiologic association could not be established with other immunogens. Such complications, in a mass program that retrospectively proved unnecessary, not only brought criticism on the use of influenza vaccine but unfortunately also discouraged public acceptance of other immunization programs.

CHEMOTHERAPY. The discovery that **amantadine (1-adamantanamine)** can inhibit an early step in the multiplication (uncoating) of some influenza viruses (see Chap. 49) reawakened hopes for the successful chemical control of viral diseases. Amantadine and rimantadine (an amantadine derivative) are about 70% effective in protecting against proven infections. Amantadine is also effective therapeutically in that virus is cleared more rapidly, and symptomatic improvement occurs about 1 day earlier if the drug is given within 24 to 48 hours after onset. Its clinical usefulness is limited, however, because its therapeutic value has been less striking than its prophylactic effect, its effectiveness is restricted to influenza A viruses (it does not affect influenza B), and it has neurologic toxic effects (particularly in the aged).

Ribavirin (1-β-ᴅ-ribofuranosyl-1,2,4-triazole-3-carboxamide, virazole), which inhibits synthesis of viral RNA by blocking guanine biosynthesis, has been more effective than amantadine in preventing experimental influenza A infection in cell cultures and animal models. Oral administration in clinical trials, however, has shown ribavirin to have only a small, inconsistent prophylactic or therapeutic effect.

Selected Reading

BOOKS AND REVIEW ARTICLES

Brachiale TJ, Brachiale VL: CTL recognition of transfected H-2 gene and viral gene products. In Notkins AL, Oldstone MBA (eds): Concepts in Viral Pathogenesis, Vol II, p 174. New York, Springer-Verlag, 1987

Burnet FM: Portraits of viruses: Influenza virus A. Intervirology 11:201, 1979

Lamb RA, Choppin PW: The gene structure and replication of influenza virus. Ann Rev Biochem 52:467, 1983

Mitchell DM, McMichael AJ, Lamb JR: The immunology of influenza. Br Med Bull 41:80, 1985

Murphy BR, Webster RG: Influenza viruses. In Fields BM (ed): Virology. New York, Raven Press, 1985

Palese P, Kingsbury DW (eds): Genetics of Influenza Viruses. New York, Springer-Verlag, 1983

Sweet C, Smith H: Pathogenicity of influenza virus. Microbiol Rev 44:303, 1980

Webster RG, Laver WG, Air GM, Schield GC: Molecular mechanisms of variation in influenza viruses. Nature 296:115, 1982

SPECIFIC ARTICLES

Broom J, Ulmanen I, Krug RM: Molecular model of a eucharyotic transcription complex: Functions and movements of influenza P proteins during capped RNA-primed transcription. Cell 34:609, 1983

Buonogurio DA, Nakada S, Parvin JD et al: Evolution of human influenza A viruses over 50 years: Rapid, uniform rate of change in NS gene. Science 232:980, 1986

Colman PM, Varghese JN, Laver WG: Structure of the catalytic and antigenic sites in influenza virus neuraminidase. Nature 303:41, 1983

Davenport FM, Minuse E, Hennessy AV, Francis T Jr: Interpretations of influenza antibody patterns of man. Bull WHO 41:453, 1969

Lamb RA, Zebedee SL, Richardson CR: Influenza virus M_2 protein is an integral membrane protein expressed on the infected-cell surface. Cell 40:627, 1985

Plotch SJ, Bouloy M, Krug RM: Transfer of 5′ terminal cap of globin mRNA to influenza viral complementary RNA during transcription in vitro. Proc Natl Acad Sci USA 76:1618, 1979

Practices Advisory Committee: Prevention and Control of Influenza. Morbidity and Mortality Weekly Report 34:261, 1985

Smith GL, Hay AJ: Replication of the influenza virus genome. Virology 118:96, 1982

Townsend ARM, McMichael AJ, Carter NP et al: Cytotoxic T cell recognition of the influenza nucleoprotein and hemagglutinin expressed in transfected mouse L cells. Cell 39:13, 1984

Wilson IA, Skehel JJ, Wiley DC: Structure of the hemagglutinin membrane glycoproteins of influenza virus at 3A resolution. Nature 289:366, 1981

Wiley DC, Wilson IA, Skehel JJ: Structural identification of the antibody-binding sites of Hong Kong influenza haemagglutinin and their involvement in antigenic variation. Nature 289:373, 1981

Yoshimura A, Ohnishi S: Uncoating of influenza virus in endosomes. J Virol 51:497, 1984

57

Harold S. Ginsberg

Paramyxoviruses

The paramyxoviruses (**Paramyxoviridae**) differ widely pathogenically. **Parainfluenza** and **respiratory syncytial viruses** produce **acute respiratory diseases, measles virus** causes a **generalized exanthematous disease,** and **mumps virus** initiates a **systemic disease** of which **parotitis** is a predominant feature. However, on the basis of chemical and several biological properties the paramyxoviruses are relatively homogeneous (Table 57–1). There are also sufficient differences in their characteristics to permit their classification into three genera: **Paramyxovirus, Morbillivirus,** and **Pneumovirus.** Respiratory syncytial virus, a pneumovirus that cannot hemagglutinate or cause hemolysis of red blood cells (RBCs), differs most sharply from the other paramyxoviruses.

General Properties

The characteristics that are similar for all paramyxoviruses are discussed in this section; the distinctive properties are described in the following sections on the individual viruses.

MORPHOLOGY

The virions are **roughly spherical enveloped particles** of heterogeneous sizes (see Table 57–1), larger than influenza viruses (see Table 56–2). Electron microscopic examination of negatively stained virions discloses that they appear similar to orthomyxoviruses. The intact viral particle has a well-defined **outer envelope,** about 10 nm thick, covered with short (8-nm to 12-nm) **spikes** that are more or less regularly arranged (Fig. 57–1). Disruption of the envelope reveals an inner **helical nucleocapsid** and serrations with a regular periodicity of about 5 nm (Figs. 57–2 through 57–4). The nucleocapsid is distinctly differ-

TABLE 57–1. *Characteristics of Human Paramyxoviruses*

Common Properties		Distinguishing Properties				
			Parainfluenza	Mumps	Measles	RSV
Average size	125–250 nm (range 100–800 nm)	Hemagglutinin*	+	+	+	−
Nucleocapsid diameter	18 nm (except RSV = 14 nm)	Hemadsorption†	+	+	+	−
Viral genome	5 × 10⁶ daltons (17–20 Kb); single negative-strand molecules	Hemolysin	+	+	+	−
Virion RNA polymerase	+	Neuraminidase	+	+	+	−
Reaction with lipid solvents	Disrupts	Antigenic types	4	1	1	1
Syncytial formation	+	Antigenic relationships	Mumps	Parainfluenza	−	−
Cytoplasmic inclusion bodies	+ (Measles virus nuclear)	Genus	Paramyxovirus	Paramyxovirus	Morbillivirus	Pneumovirus
Site of multiplication	Cytoplasm					

* Chicken and guinea pig RBCs

† Infected cells adsorb guinea pig RBCs.

+, present; −, absent

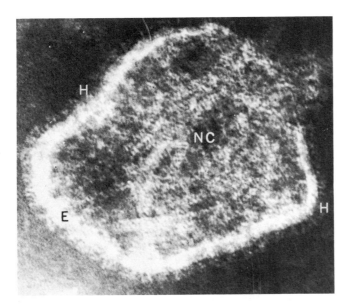

Figure 57–1. Electron micrograph of type 3 parainfluenza virus embedded in phosphotungstate. The envelope (*E*), peripheral short projecting spikes (*H*), and internal nucleocapsid (*NC*) are apparent. The helical nucleocapsid can also be seen escaping from a small break in the envelope. (Original magnification ×210,000; Courtesy of A. P. Waterson, St. Thomas's Hospital Medical School, London)

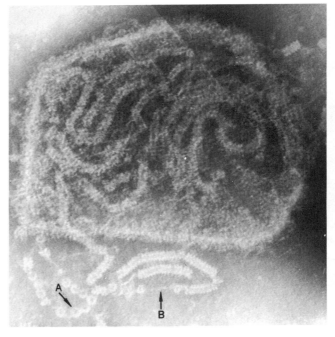

Figure 57–2. Partially disrupted mumps virus particle showing the envelope and the hollow helical strands forming the nucleocapsid. Several broken strands have been released, revealing periodic structures (*A*) that make up the helical nucleocapsid. Fine threads connecting separated pieces are visible (*B*). (Original magnification ×250,000; Horne RW, Waterson AP: J Mol Biol 2:75, 1960. Copyright by Academic Press, Inc. [London] Ltd.)

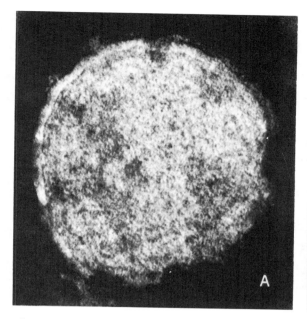
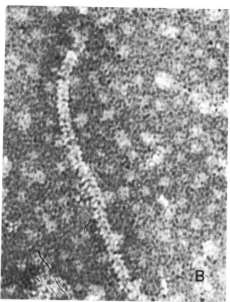

Figure 57–3. Fine structure of measles virus, revealed by negative staining with sodium phosphotungstate. (*A*) Particle showing the characteristic envelope and peripheral projections. The nucleocapsid is tightly packed and shows an appearance of concentric rings toward the periphery. (Original magnification ×280,000) (*B*) The portion of the helical nucleocapsid released from a disrupted virion. (Original magnification ×240,000; Horne RW, Waterson AP: J Mol Biol 2:75, 1960. Copyright by Academic Press, Inc. [London] Ltd.)

ent from that of influenza viruses (see Table 56–2): its diameter is approximately twice as great, its characteristic periodic serrations are more discrete, and it can be isolated from the virion as a single long helical structure. Filamentous virions are observed in thin sections of infected cells but not in negatively stained preparations, because they are apparently disrupted during fixation.

PHYSICAL AND CHEMICAL CHARACTERISTICS

The paramyxoviruses that have been purified and analyzed all have a similar protein, lipid, and RNA composition. The nucleocapsid consists of a single species of protein (NP) and **one large molecule of single-stranded RNA** of **negative polarity** (see Table 57–1). The viral envelope contains three proteins. Two glycoproteins form the surface projections (Fig. 57–5). One glycoprotein varies in viruses of the three genera: in parainfluenza and mumps virus it has both **hemagglutinin and neuraminidase activities** (termed **HN**); in measles virus it lacks neuraminidase activity (hence, termed **H protein**); and in respiratory syncytial virus it has neither hemagglutinin nor neuraminidase functions and is called the **G protein.** The other surface glycoprotein, which consists of two disulfide-linked subunits (F_1 and F_2), is responsible for the virion's **cell fusion** activity of all paramyxoviruses as well as the **hemolytic** function of

parainfluenza, mumps, and measles viruses (see Table 57–1). The third envelope protein (M), which is nonglycosylated, forms the **inner layer of the envelope,** maintaining its structure and integrity. The HN protein of parainfluenza and mumps virus is a dimer consisting of disulfide-linked, identical polypeptide chains.

Because the viral envelope has a high lipid content, organic solvents or surface-active agents rapidly inactivate the virions by dissolving their envelopes, thus liberating the envelope proteins and the nucleocapsids from the disrupted particles.

The virions are relatively unstable, losing 90% to 99% of their infectivity in 2 to 4 hours when suspended in a protein-free medium at room temperature or at 4°C.

MULTIPLICATION

The multiplication cycles do not appear to differ for each paramyxovirus, except for the great variations in the length of various phases. For example, the eclipse period is 3 to 5 hours for parainfluenza viruses, 16 to 18 hours for mumps virus, and 9 to 12 hours for measles virus. The basic biosynthetic events (predominantly derived from studies of parainfluenza and Newcastle disease viruses) are similar to those for other enveloped viruses that possess a negative single-stranded RNA and contain an RNA-dependent RNA polymerase within the virion (see Syn-

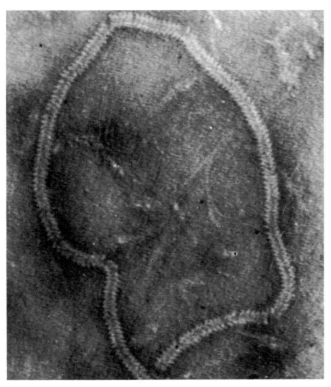

Figure 57–4. Strand of helical nucleocapsid of type 1 (Sendai) parainfluenza virus embedded in phosphotungstate. (Examination by tilting through large angles reveals the sense of the helix to be left-handed.) (Original magnification ×200,000; Horne RW, Waterson AP: J Mol Biol 2:75, 1960. Copyright by Academic Press, Inc. [London] Ltd.)

thesis of RNA Viruses, Chap. 48), but there are several unique features. In contrast to orthomyxoviruses, paramyxoviruses multiply without restraint in the presence of actinomycin D. The unsegmented virion RNAs of all paramyxoviruses consist of a linear series of linked genes. Because each gene produces a single mRNA, each contains conserved nucleotide sequences designating start and termination sites, the latter containing a poly(A) signal for its mRNA (Fig. 57–6). The genes are linked by a highly conserved **intergenic sequence,** GAA. The **P** and **L proteins** function for **RNA synthesis.** The **NP protein,** like its counterpart in orthomyxoviruses, does not have catalytic activity, but appears to provide the genomic RNA with the appropriate configuration for its **transcription and replication.** Since the genes in the viral genomes differ somewhat in each genus of the paramyxovirus family the genome maps also vary, but the basic structures are similar to that shown (see Fig. 57–6).

Virion infectivity, as well as hemolytic and cell fusion properties, requires **maturation of the F glycoprotein** by proteolytic cleavage of a larger precursor (F_0) by a cellular enzyme to produce two polypeptide chains, F_1 and F_2, which are linked by disulfide bonds into the functional dimer. With most paramyxoviruses the NP proteins of the nucleocapsid, as well as the HN (hemagglutinin–neuraminidase) protein, are detected only in the cytoplasm of infected cells, but the measles virus nucleocapsid is also present in the nucleus.

After viral RNA is synthesized in the cytoplasm, it is rapidly associated with the newly made nucleocapsid protein, but only a relatively small proportion of the nucleocapsids thus formed is assembled into virions. Occasionally, positive strands, owing to their excess, may be accidentally assembled into virions as well as nucleocap-

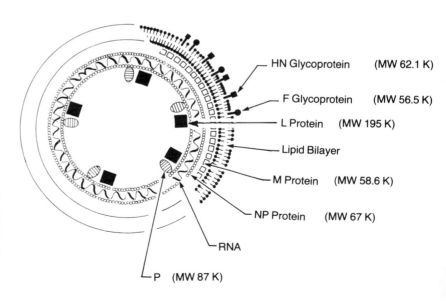

Figure 57–5. Diagram of a paramyxovirus. (Different forms are shown for the *F* and *HN* glycoproteins to indicate their chemical differences, although they are not distinguishable in electron micrographs.) The F and HN glycoproteins are anchored in the lipid bilayer, the F by its F_1 C terminus and the HN by its N terminus. The M protein forms the inner layer of the envelope and maintains its structure and integrity. The actual arrangements of the NP and L proteins in the nucleocapsid are unknown. *MW,* molecular weight; *K,* kilodaltons.

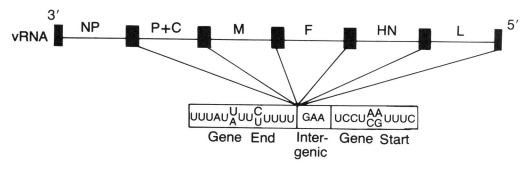

Figure 57–6. Diagram of the parainfluenza 3 virus RNA genome depicting the order of the genes (NP, P+C, M, F, HN, and L), the gene-end nucleotide sequences, the start sequences, and the intergenic sequences. The gene-end sequences contain the transcription termination sequences and the mRNA complementary poly(A) sequences. The 12 gene-end nucleotides are semiconserved, whereas the ten gene start nucleotides and the three intergenic nucleotides (GAA) are relatively well conserved for all five gene junctions. The nucleotide position of the translational start sites from the 5' end of the genes varies from 33 to 35 nucleotides for the M protein gene to 194 to 196 nucleotides for the F gene. (Data from Spriggs MK, Collins PL: Human parainfluenza virus type 3: Messenger RNAs, polypeptide coding assignments, intergenic sequences and genetic map. J Virol 59:646, 1986)

sids. The cytoplasmic inclusion bodies (Fig. 57–7) are predominantly accumulations of the excess nucleocapsids.

Electron microscopic examination of infected cells reveals the remarkable assembly and maturation of paramyxoviruses at the plasma membrane (Fig. 57–8). Strands of viral nucleocapsid can be seen to associate with the viral M protein lining the regions of thickened cell membranes containing virus-specific glycoproteins: these differentiated cell membranes are destined to become the viral envelopes. Intact viral particles are noted only at the cell membrane, where they are assembled and released from the cell by budding. The final assembly of the nucleocapsid and the specifically altered plasma membrane are especially prominent with these viruses (see Fig. 57–8).

Parainfluenza Viruses

Parainfluenza viruses were recognized in 1957 as important causes of acute respiratory infections of man and were initially termed *hemadsorption virus*, *croup-associated viruses*, and *influenza D*. At least four antigenic types infect man, and one infects monkeys.

The major characteristics of parainfluenza viruses infecting man are listed in Table 57–1.

PROPERTIES
Immunologic Characteristics

Each parainfluenza virus possesses three distinct Ags from which it derives its type specificity: The **HN (hemagglutinin–neuraminidase)** and the **F (fusion–hemolysin)** surface Ags, and an internal nucleocapsid Ag, the **NP protein** (see Fig. 57–5). Neutralizing Abs are directed against the HN glycoprotein, and at least four epitopes function in the neutralization and hemagglutinin-inhibition reactions. Distinct epitopes have also been identified with neuraminidase activity, indicating that neuraminidase and hemagglutinin functions are associated with unique regions of the HN glycoprotein. Parainfluenza viruses are immunologically unrelated to influenza viruses. Progressive major antigenic alterations have not been detected; only type 4 parainfluenza virus has subtypes, A and B.

Although there is not a single Ag common to all

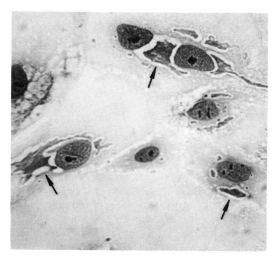

Figure 57–7. Eosinophilic cytoplasmic inclusions (*arrows*) in dog kidney cells infected with type 2 parainfluenza virus. Nuclei appear unaffected. (Original magnification ×450; Brandt CD: Virology 14:1, 1961)

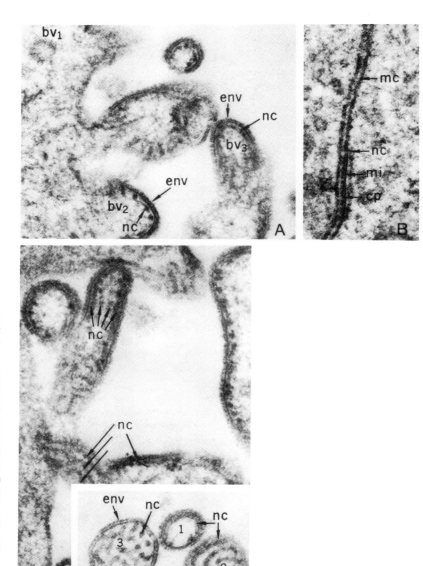

Figure 57–8. Parainfluenza virions forming at the cell membrane of a chick embryo cell and being released by budding. (*A*) Viral buds (*bv*) at various stages of development. The altered cell membrane forming the viral envelope (*env*) and the nucleocapsid (*nc*) beneath the envelope can be clearly distinguished. The nucleocapsid is cut transversely in two of the buds (*bv₁* and *bv₂*) and longitudinally in a third (*bv₃*). (Original magnification ×105,000) (*B*) Differentiation of the viral envelope at the cell membrane. The cell membranes (*mc*) of two adjacent cells are shown. On contact with the nucleocapsid (*nc*) the cell membrane differentiates into the internal membrane of the envelope (*mi*) and the outer layer of short projections (*cp*). (Original magnification ×105,000) (*C*) Arrangement of the nucleocapsid (*nc*) impinging on the altered cell membrane that is participating in the formation of the viral envelope. Notice the regularity of the arrangement of the nucleocapsid, seen in cross-sectional and longitudinal views. (Original magnification ×105,000) (*D*) Thin section of virions. Note the structure of the envelope covered with short projections (*env*) and the various arrangements of the nucleocapsids (*nc*). In particle 1 the nucleocapsid is arranged parallel to the envelope. In particles 2 and 3 the nucleocapsid is arranged more irregularly. (Original magnification ×92,000; Berkaloff A: J Microsc 2:633, 1963)

parainfluenza viruses, there are immunologic relations among the parainfluenza viruses (at least types 1, 2, and 3) and with mumps viruses (see Table 57–1). These are noted by heterotypic Ab responses to infection in those who have had prior infections with one or more of the other viruses. The serologic response to the initial infection (the **primary response**) in humans, as well as in lower animal hosts, is strictly **type-specific.** Heterotypic Abs, particularly between types 1 and 3, subsequently appear in humans, probably as a result of repeated infections with different members of the group; each addi-

tional infection broadens the Ab response. The human heterotypic responses to infections are sufficiently frequent to permit the following conclusions: these agents constitute a group of viruses with cross-reactive Ags, the qualitative and quantitative characteristics of the heterotypic Ab response to infection depend on prior immunologic experience, and immunologic diagnosis of infection by a specific virus may be unreliable except following the initial infection.

Most adults have a relatively high titer of circulating Abs to all antigenic types but usually lack the critical neutralizing IgA Abs in nasal secretions. These appear following acute infections, but they decrease substantially within 1 to 6 months, although specific IgG serum Ab levels remain relatively high. Recurrent infections occur despite the presence of neutralizing Abs in the serum, although the severity of disease is reduced (perhaps owing to a secondary response of IgA Abs). The initial infection with a given type is the most severe and usually occurs in children; infections in adults are commonly afebrile and minor.

Reactions With Erythrocytes

The hemagglutinins and neuraminidases of parainfluenza viruses resemble those of influenza viruses in most biological characteristics (see Hemagglutination, Chap. 44). Maximum hemagglutination titers are obtained with chicken RBCs at 4°C (types 1 and 2) or with guinea pig RBCs at 25°C (types 1 and 3). The **hemolysin (F) glycoprotein** is inhibited by type-specific antiserum and is similar to the hemolysins of mumps and Newcastle disease viruses. For hemolytic activity to occur, intact mucoprotein receptors on RBCs are required, and therefore **anti-HN Abs** also inhibit hemolysis.

Host Range

Primary cultures of cells from monkey and embryonic human kidneys are the hosts of choice for primary isolations, neutralization titrations, and investigations of the biological properties of parainfluenza viruses. Organ cultures of human tracheal and nasal epithelium have similarly proved to be sensitive for primary isolations. Types 2 and 3 also multiply well in human continuous epithelium-like cell lines (e.g., HeLa cells). Type 1 virus has been adapted to HeLa and human diploid cells, but neither cell line is satisfactory for initial viral isolation.

In cell cultures cytopathic changes develop very slowly (particularly with type 4); infection is most quickly detected by hemadsorption of guinea pig RBCs. The cytopathic changes that eventually appear consist of stringiness or rounding of cells (types 1 and 4) or formation of large **syncytia** containing eosinophilic **cytoplasmic inclusion bodies** (types 2 and 3; see Fig. 57–7). Syncytia are produced by the viral surface F glycoprotein, which causes fusion and then dissolution of the fused membranes of infected cells (see Cytopathic Effects, Chap. 51). This remarkable capacity of parainfluenza viruses (even after ultraviolet or heat inactivation) to induce fusion of cells from many animal species has had great general utility for studying the genetics of eukaryotic cells as well as for virology.

Intranasal inoculation of very small amounts of virus into hamsters, guinea pigs, and cotton rats yields relatively high titers of virus in the lungs, but pulmonary lesions only develop in a particular species of cotton rats (*Sigmadon fulviventer*) **after type 3 infection.** The animals develop type-specific Abs and resist intranasal challenge with homologous virus for 1 to 3 months, but then susceptibility returns. This pattern mimics the relation of Abs and recurrent susceptibility in humans.

Type 3 virus is highly infectious for cattle. It appears to be harmless, however, except under conditions of stress, such as the herding of cattle together for transportation, when it may induce an acute febrile upper respiratory disease (hence the term *shipping fever*).

Parainfluenza viruses have been adapted to propagation in chick embryos.

PATHOGENESIS

Parainfluenza viruses cause a spectrum of illnesses, primarily in infants and young children, ranging from mid-upper-respiratory infections to croup or pneumonia (Table 57–2). In infants type 1 and 2 viruses are particularly prone to produce laryngotracheobronchitis (croup) and type 3 virus to cause bronchiolitis and pneumonia. Type 4, which consists of two subgroups (4A and 4B), is difficult to isolate, and apparently is a less common etiologic agent. The occasional infections of adults usually evoke a subclinical illness or a mild "cold." Even in children the majority of infections with parainfluenza viruses appear to be clinically inapparent. Conclusions concerning pathogenesis and development of lesions are derived from studies of infections in volunteers, observations of natural infections, and investigation of experimental infections in cotton rats, the best animal model. The virus enters by the respiratory route, and in most adults it multiplies and causes inflammation only in the upper segments of the tract. In infants and young children, however, the bronchi, bronchioles, and lungs are occasionally involved. **Viremia is neither an essential nor a common phase of infection.**

Parainfluenza viruses, like all other paramyxoviruses, can readily establish a persistent infection *in vitro* (see Chap. 51). Scattered findings suggest that similar chronic infections may follow acute diseases *in vivo*. Thus, a parainfluenza-like virus has been isolated by cocultivation of susceptible cells and brain tissue from a patient with multiple sclerosis, and structures resembling para-

TABLE 57–2. Clinical Syndromes Associated With Parainfluenza Viruses

Disease	Virus Type
Minor upper respiratory disease	1, 3, 4*
Bronchitis	1, 3
Bronchopneumonia	1, 3
Croup	1, 2

* Clinical disease uncommon

myxovirus nucleocapsids have been observed in electron micrographs of tissues from patients with various collagen diseases (e.g., systemic lupus erythematosus). However, no conclusive data exist indicating that a parainfluenza virus is the cause of any chronic disease, as measles virus has been shown to produce **subacute sclerosing panencephalitis** (see below and Chap. 51).

LABORATORY DIAGNOSIS

An etiologic diagnosis of parainfluenza virus infection requires laboratory procedures; the lack of distinctive clinical features precludes etiologic diagnoses on this basis. Measurement of a **rise in serum Abs** by enzyme-linked immunosorbent assay (ELISA), hemagglutination-inhibition, complement-fixation (CF), or neutralization titration permits diagnosis conveniently and economically. However, serologic techniques alone cannot reliably establish the specific type of virus that is responsible, because of the frequency and the degree of heterotypic Ab responses, as discussed earlier under Immunologic Characteristics.

The specific parainfluenza virus responsible for an infection can be identified by **viral isolation**, although the marked instability of the virions makes isolation difficult. For this purpose nasopharyngeal secretions containing antibiotics are added to primary tissue cultures of monkey or embryonic human kidney. Although cytopathic changes may not be detectable except with type 2 virus, or may develop very slowly, viral infection can be recognized rapidly and conveniently by **immunofluorescence** or by **adsorption of guinea pig RBCs** to infected cells. Hemadsorption to cells infected with types 1 and 3 viruses can usually be detected within 5 days after inoculation of the patient's secretions, but types 2 and 4 often require 10 days or more. The specific type can be identified by hemadsorption-inhibition techniques utilizing standard sera. Immunofluorescence can yield comparable diagnostic results in only 24 to 48 hours. It should be remembered that the simian parainfluenza virus SV5 is a common latent agent in monkey kidney cultures, and its emergence from this tissue must not be confused with its primary isolation from man.

EPIDEMIOLOGY

Parainfluenza viruses produce disease throughout the year, but the peak incidence is noted during the "respiratory disease season" (late fall and winter). Most infections are endemic; but sharp, small epidemics occasionally occur with types 1 and 2, and at present these epidemics usually occur simultaneously every other year. Parainfluenza virus infections are **primarily childhood diseases:** type 3 infections occur earliest and most frequently, so that 50% of children in the United States are infected during the first year of life, and almost all are infected by 6 years of age; 80% of children are infected with types 1 and 2 by 10 years of age. Type 4 viruses induce few clinical illnesses but infections are common: By 10 years of age 70% to 80% of children have Abs.

Parainfluenza viruses are **disseminated in respiratory secretions.** Type 3 shows the most effective spread, and during outbreaks in closed populations (e.g., in institutions or hospitals) all children who are free of neutralizing Abs become infected. Under similar circumstances only about 50% of children are infected with type 1 or type 2 virus.

The epidemiologic patterns and the clinical manifestations of parainfluenza virus infections, in children and adults, emphasize the protective effect of neutralizing Abs as well as the lack of complete or long-lasting immunity (probably owing to an inadequate level of IgA Abs in the respiratory secretions). In contrast, young infants appear to be partially protected by passively acquired, maternal serum Abs. Febrile and severe illness is observed only with the initial infection. Reinfection may be produced by the same virus within as little as 9 to 12 months, but it results in a much milder disease.

PREVENTION AND CONTROL

Reducing the attack rate of respiratory diseases is an important social and economic goal. However, prevention of parainfluenza virus infections would probably reduce the incidence of acute respiratory illnesses by only about 15% in children less than 10 years old, and by much less in adults. Nevertheless, an effective vaccine, evoking an Ab response in the respiratory tract, would be of value for young children, especially in hospitals and institutions. Unfortunately, however, an effective vaccine is not available, although newer recombinant DNA techniques have raised hopes that one will be developed in the near future.

Mumps Virus

The unique clinical picture of mumps can be recognized in writings of Hippocrates from the fifth century B.C. The etiologic agent was not isolated and identified as a virus, however, until 1934, when Johnson and Goodpasture produced parotitis in monkeys by inoculating bacteria-free infectious material directly into Stensen's duct. No further major progress was made until the virus was propagated in the chick embryo and was found (in 1945) to agglutinate chicken RBCs, as do influenza viruses. The subsequently isolated parainfluenza viruses were found to have an even closer relation to mumps virus.

PROPERTIES

Immunologic Characteristics

Mumps virus, like parainfluenza viruses, contains the surface **hemagglutinin–neuraminidase (HN) glycoprotein** (also termed the **V [virion] antigen),** which induces protective Abs, **the hemolysis–cell fusion (F) glycoprotein Ag,** and the internal **RNA-protein nucleocapsid (NP),** which is immunologically identical with the soluble Ag from cells (called the **S antigen).**

Antigenically, mumps virus exists as a single type; no immunologic variants have been detected. It cross-reacts significantly, however, with parainfluenza and Newcastle disease viruses. Therefore, a rise in heterotypic Abs to parainfluenza viruses is seen in the serum from mumps patients.

Antibodies against the nucleocapsid Ag appear within 7 days, and high titers are attained within 2 weeks after the onset of clinical illness. The HN Abs appear later (2–3 weeks after onset), attain maximum titers in 3 to 4 weeks, and persist longer than the nucleocapsid Abs. The HN Abs are measured by neutralization, hemagglutination-inhibition, ELISA, or CF assays with purified HN glycoprotein; they appear to reflect the degree of immunity. Nucleocapsid Abs, which are conveniently assayed by CF titrations, do not afford protection against subsequent infection. Humoral and cellular immunity develop after subclinical as well as clinical infections and usually persist for many years, although second infections have been reported.

Mumps infection induces delayed hypersensitivity that can be observed by a skin test with infectious or inactivated virus. A positive skin reaction correlates roughly with immunity, providing a useful epidemiologic tool, but false reactions (both positive and negative) occur.

Reactions With Erythrocytes

The viral particle **agglutinates RBCs** from several animal species (see Hemagglutination, Chap. 44). Mumps virus, however, has only weak neuraminidase activity against soluble mucoproteins, and therefore hemagglutination is highly susceptible to these inhibitors in serum or culture fluids. The hemagglutinin present on the surface of infected cells can also be detected in tissue culture by **hemadsorption.**

Mumps virus, like parainfluenza viruses, hemolyzes susceptible RBCs at 37°C by interacting with the same specific receptors as are involved in hemagglutination. Several viral particles per cell are required to produce hemolysis, and even under optimal conditions only about 50% of the hemoglobin is released. Calcium inhibits hemolysis. Viral hemagglutination, hemadsorption, and hemolysis are inhibited by Abs to the virion surface Ags (the HN and F glycoproteins).

Host Range

Humans are the only natural hosts for mumps virus, but the virus can infect monkeys and 6- and 8-day-old chick embryos (amniotic cavity or yolk sac); no pathologic lesions of the embryos are noted. After adaptation by serial passage in the allantoic sac of the chick embryo the virus can infect guinea pigs, suckling mice, hamsters, and white rats but has lost its virulence for man and monkey.

In culture, chick embryo and many types of mammalian cells (including monkey and human kidney cells) support multiplication of mumps virus. Infected cultures can be recognized by the appearance of viral particles (hemadsorption or agglutination of RBCs) and soluble Ag (CF), as well as by the slow development of cytopathic effects (cytolysis and development of giant cells with cytoplasmic inclusions; Fig. 57–9). Mumps virus can also replicate in human T-lymphocytes.

PATHOGENESIS

Mumps typically has an acute onset of parotitis (with painful swelling of one or both glands) 16 to 18 days after exposure. The virus is **transmitted in saliva and respiratory secretions** and its **portal of entry** in man is the **respiratory tract. Primary viral multiplication** also takes place in the respiratory tract epithelium and cervical lymph nodes; the salivary glands, the primary target organs, are infected via the bloodstream (Fig. 57–10). Thus, **viremia** begins several days before the development of mumps and before virus is present in the saliva.

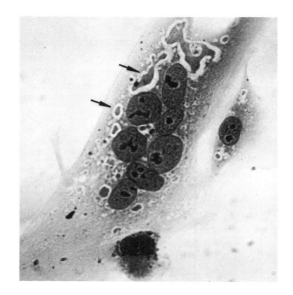

Figure 57–9. Eosinophilic cytoplasmic inclusions and giant cell formations in monkey kidney cells infected with mumps virus. Inclusions (*arrows*) develop in close proximity to the nuclei, which are unaffected. (Original magnification ×450; Brandt CD: Virology 14:1, 1961)

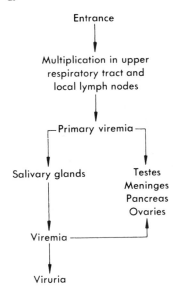

Entrance

↓

Multiplication in upper
respiratory tract and
local lymph nodes

↓

Primary viremia

Salivary glands   Testes
                         Meninges
                         Pancreas
                         Ovaries

Viremia

↓

Viruria

Figure 57–10. Schematic representation of the pathogenesis of mumps.

Virus is present in the blood and in saliva for 3 to 5 days after the onset of the disease, and since the kidney is frequently infected, **virus in the urine** is common for 10 days or more after the onset. The long incubation period reflects the time required for virus to establish a local infection, spread in the blood to the target organs, and multiply sufficiently to damage cells and induce inflammation. The occasional complications also suggest widespread dissemination of the virus.

The acute onset of fever and of salivary gland inflammation may be followed in 4 to 7 days by orchitis (in 20% to 30% of males past puberty),* by meningitis or meningoencephalitis (0.2% to 0.5% of cases, although cerebrospinal fluid pleocytosis occurs in over 50%), and occasionally by pancreatitis, oophoritis, myocarditis, or presternal edema. Even more rarely, nephritis or paralytic manifestations may appear. When infection occurs during the first trimester of the pregnancy, fetal wastage may also occur. The complications may develop at the same time as parotitis or even in the absence of salivary gland involvement.

The most complete pathologic descriptions of mumps virus lesions have been those of affected parotid glands and testes. The parotid lesions consist of interstitial inflammation of the gland and degeneration of the epithelium of the ducts. No characteristic inclusion bodies have been observed in patients' tissues, in contrast to tissue cultures (see Fig. 57–9). Testes show dif-

* Despite the commonly told tales, sterility rarely if ever results from mumps orchitis.

fuse degeneration, particularly of the epithelium of the seminiferous tubules, as well as edema, serofibrinous exudate, marked congestion, and punctate hemorrhages in the interstitial tissue. Meningoencephalitis or meningitis has been studied pathologically only in rare cases, and proof of etiology has often been lacking. The findings described are typical of postinfectious encephalitis, with perivascular demyelination.

LABORATORY DIAGNOSIS

Diagnosis can usually be made solely from clinical observations. However, to establish the diagnosis of mumps in atypical or subclinical infections (about one-third of infections) **viral isolation** and assays of **Ab response** are required. Virus can be isolated by inoculation of saliva, secretions from the parotid duct, or spinal fluid into the amniotic cavity of 8-day-old chick embryos or appropriate cell cultures; the latter appear to be more sensitive for primary isolation. Infection of test cells can be detected earliest by **immunofluorescence** or **hemadsorption** techniques. Since the virion is unstable at 4°C or at room temperature, specimens must be used immediately or stored at −70°C.

Immunologic diagnosis is made by demonstrating an increase in Abs after infection. **Complement fixation** with the **nucleocapsid (soluble) Ag** can detect a rise within the first 7 to 10 days after onset of illness. After 14 days serologic diagnosis is accomplished most readily with the **viral particles (HN Ag)** in **hemagglutination-inhibition, ELISA,** or **CF** titrations. The antigenic cross-reactions between the surface (HN) antigens of mumps and parainfluenza viruses complicate immunologic diagnosis, making it necessary to test all the Ags of related viruses; the highest Ab rise occurs in response to Ags of the infecting virus. When diagnosis is of critical importance, the **plaque neutralization** assay can detect Ab rises that may not be evident in other assays owing to the cross-reactive CF Abs and the nonspecific inhibitors of hemagglutination. If acute and convalescent sera are not available, serologic diagnosis may be made by testing for IgM Abs that appear early and do not persist.

Though the **skin test** may be useful in surveys for estimating immunity, it frequently leads to erroneous conclusions in individual cases and is not reliable for diagnosis. Indeed, because skin testing itself may induce a rise in Abs and confuse the interpretation of immunologic reactions, skin tests should not be used during the course of illness.

EPIDEMIOLOGY

Mumps is predominantly a **childhood disease** spread by droplets of saliva. Salivary secretions may contain infectious virus as early as 6 days before and as long as 9 days after the appearance of glandular swelling. Virus is also present in the saliva of patients with meningitis or orchitis, even in the absence of clinical involvement of the salivary glands. Although viruria develops, there is no evidence for transmission from this source.

Mumps is not nearly so contagious as the childhood exanthematous diseases (e.g., measles and chickenpox), and many children escape infection; hence, **disease in adults** is not uncommon (prior to artificial immunization about 50% of U.S. Army recruits were without Abs). **Subclinical infections,** detectable immunologically, are also much more frequent than in other common childhood diseases.

Mumps appears most often during the winter and early spring months, but it is endemic throughout the year. Large epidemics have been observed about every 3 to 5 years.

PREVENTION AND CONTROL

Because subclinical infections are common, control of infection by isolation is not effective. The disease can be prevented by immunization. **Infectious attenuated virus,** inoculated subcutaneously, induces development of Abs in about 90% of Ab-free subjects (both children and adults). Although vaccine induces infection, viremia and viruria are not detectable, clinical reactions do not occur, and virus does not spread to exposed contacts. The Ab response is not as great as that accompanying natural infections, but Ab levels and protection against clinical mumps persists for at least 12 years after immunization.

The live virus vaccine is of considerable value for susceptible adults in whom the disease is more severe and the complications more frequent. If widespread immunization of children is to be employed, the vaccine should induce persistent immunity, similar to that which follows the natural disease. It is not clear whether the infectious attenuated virus vaccine satisfies this requirement because sufficient data are not yet available.

Formalin-inactivated virus vaccine has been shown to induce a good Ab response and to be clinically effective. However, Abs decline 3 to 6 months after immunization, and neither the effectiveness nor the persistence of immunity appears to be as satisfactory as following the live virus vaccine.

Passive immunization with **γ-globulin from convalescent serum** has been employed to prevent infection after exposure, particularly in men, for whom orchitis is a relatively frequent and severely discomforting complication. Its effectiveness, however, has not been clearly demonstrated. Convalescent whole human serum was once used for similar purposes, but it is not recommended because of the considerable danger of contamination with hepatitis viruses (see Chap. 63) or the human immunodeficiency virus (HIV; see Chap. 65).

Measles Virus

Measles (**morbilli**)* is one of the most infectious diseases known, and it is almost universally acquired in childhood; fortunately, immunity is essentially permanent. It was first described as an independent clinical entity by Sydenham in the seventeenth century. Although measles was demonstrated as early as 1758 to be transmissible in volunteers, and was transferred to monkeys in 1911, it was not until 1954 that the virus, through Enders' persistent and careful search, was isolated reproducibly from patients and was shown to produce cytopathic changes in tissue cultures. This achievement led to a rapid advance in knowledge of the virus and to the development of an effective vaccine.†

PROPERTIES
Immunologic Characteristics

All measles strains studied belong to a **single antigenic type.**‡ Specific Abs are produced to each of the major viral Ags: the **hemagglutinin (H),** the **hemolysin–cell fusion (F),** and the **nucleocapsid (NP),** as well as the **matrix (M),** and the internal **P** and **L** proteins, all of which exist free in cell extracts as well as assembled in virions. Antibodies to the **virion surface glycoproteins** (hemagglutinin and hemolysis–cell fusion factor), in the presence of specific complement components (the alternate complement pathway), lyse infected cells, which contain these viral glycoproteins in their plasma membranes. These **cytotoxic Abs** probably play a role in certain aspects of the pathogenesis of the disease as well as in the elimination of infected cells during recovery. In the absence of complement, however, these same Abs cause the viral glycoproteins to accumulate ("cap") in a limited region of the infected cell surface, and finally to shed from the cell (see Chap. 51). It is likely that this phenomenon permits infected cells to escape immunologic destruction and to persist, possibly causing chronic infections such as **subacute sclerosing panencephalitis** (see Chap. 51).

Circulating Abs are detected 10 to 14 days after infection (i.e., when the rash appears or shortly thereafter) and reach maximal titer by the time the exanthem disappears. Antibody titers (neutralizing, CF, and hemagglutination-inhibiting) remain high following infection, and

---

* **Rubeola** is often employed as a synonym; unfortunately, this term has also been used as a synonym for rubella (German measles).

† This brief account illustrates only in part the unique role Enders played in leading the way to the recent control of two important diseases, poliomyelitis and measles.

‡ Measles virus, however, is related to the viruses of canine distemper and rinderpest (of cattle) in antigenic, physical, and biological properties. Considerable nucleotide sequence homology also exists between the genomes of measles and canine distemper viruses.

immunity persists for life (as shown by epidemiologic investigations; see Persistence of Antibodies, Chap. 51). In monkeys, reinfection can be produced 3 to 6 months after a primary infection, but clinical disease does not ensue. Abs directed against the hemagglutinin (H) and fusion (F) glycoproteins neutralize virus *in vitro;* passive immunity in experimental animals only requires the H Abs.

Like most orthomyxoviruses and paramyxoviruses, measles virus (or its separated hemagglutinin) agglutinates monkey RBCs. In contrast to orthomyxoviruses and other paramyxoviruses, measles virus does not elute spontaneously from agglutinated cells, the RBC receptors are not destroyed by *Vibrio cholerae* neuraminidase, and the virions do not contain neuraminidase molecules.

Cell-mediated immunity is also demonstrable by the time the rash appears. Human T-lymphocytes from normal as well as immune persons possess receptors for measles virus: the lymphocytes are agglutinated by purified virus, and they form rosettes upon reaction with infected cells. Following rosette formation, the virus-infected cells are killed. Although lymphocytes from both immune and susceptible persons are cytotoxic, the immune lymphocytes kill more effectively.

Host Range

Man is the natural host for measles virus, but this virus is highly contagious for both humans and monkeys. Monkeys in captivity commonly develop spontaneous measles, with humans probably serving as the source. Experimental measles infections have been produced in many animals other than primates; hamsters and mice have proved to be particularly useful.

Although unmodified measles virus from patients replicates poorly *in vitro*, multiplication has been achieved in a variety of mammalian as well as chick embryo cells. Both primary and continuous mammalian cell cultures are commonly employed.* Viruses adapted to propagation *in vitro* can also multiply in the amniotic sac or in the chorioallantoic membrane of chick embryos, and one strain has been adapted to propagation in brains of newborn mice.

The development of **large syncytial giant cells** is generally the major cytopathic effect produced by measles infection of cultured cells (Fig. 57–11). Eosinophilic inclusion bodies develop in both the nuclei and the cytoplasm of syncytial cells (see Fig. 57–11). The inclusions are composed of dense, highly ordered arrays of viral nucleocapsids (Figs. 57–12 and 57–13). Immunofluores-

* **Primary cultures:** human embryonic kidney, human amnion, monkey or dog kidney, chick embryo cells, bovine fetal tissue; **continuous cell lines of human origin:** HeLa, KB, Hep-2, amnion, heart, nasal mucosa, bone marrow, kidney. Primary cultures of human embryonic or monkey kidneys are most susceptible to unadapted viruses.

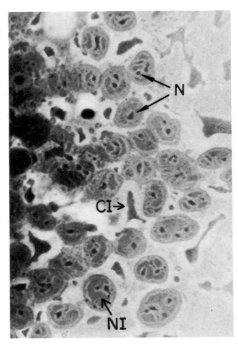

Figure 57–11. Inclusion bodies of cells infected with measles virus. A giant cell consisting of a large syncytium of cells is illustrated; each large round body is a nucleus. Large eosinophilic cytoplasmic inclusions are indicated by *CI* and numerous intranuclear eosinophilic inclusion bodies by *NI.* The nucleoli (*N*) are intact. (Original magnification ×750; Kallman F et al: J Biophys Biochem Cytol 6:379, 1959)

cence reveals specific viral Ags in both the nuclear and cytoplasmic inclusions.

PATHOGENESIS

Measles is a highly contagious, acute, febrile, exanthematous disease. The pathogenesis in man resembles the general pattern described for smallpox (see Chap. 54) and mumps (see Fig. 57–9), with **local multiplication** followed by **hematogenous dissemination.** Virus **transmitted in respiratory secretions** enters the upper respiratory tract, or perhaps the eye, and multiplies in the epithelium and regional lymphatic tissue. Virus may also disseminate to distant lymphoid tissue by a brief primary viremia. Viral multiplication in the upper respiratory tract and conjunctivae causes, after an incubation period of 10 to 12 days, the prodromal (i.e., pre-rash) symptoms of coryza, conjunctivitis, dry cough, sore throat, headache, low-grade fever, and Koplik spots (tiny red patches with central white specks on the buccal mucosa in which are noted characteristic giant cells containing viral nucleocapsids). Viremia occurs toward the end of the incubation period, permitting further wide-

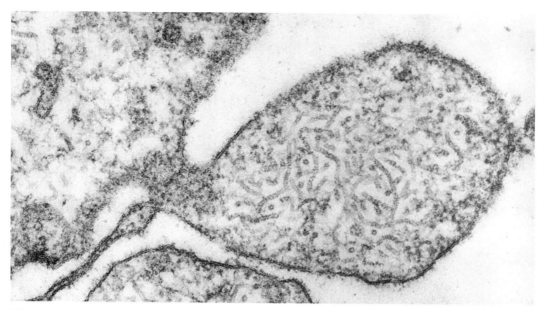

Figure 57–12. Measles virus particles budding from the surface membrane of an infected cell. Viral nucleocapsids are seen within the forming particles; the fuzzy structures on the surfaces of the virions probably correspond to the surface projections seen by negative staining (see Fig. 57–3). (Original magnification ×98,000; Nakai T et al: Virology 38:50, 1969)

spread dissemination of virus to the lymphoid tissue and skin. With the diffuse secondary multiplication of virus the prodromal symptoms are intensified and the typical red, maculopapular rash appears, first on the head and face and then on the body extremities.

Viral Ags and nucleocapsids are present in endothelial cells of the subcutaneous capillaries but usually infectious virus is not detectable in the affected superficial epidermal cells. The characteristic rash appears to result predominantly from interactions of immune T-lymphocytes with infected cells. It is striking that when children who are T-cell deficient owing to thymic dysplasia develop measles, they do not display a rash but manifest extensive giant-cell pneumonia.

Virus is excreted in the secretions of the respiratory tract and eye, and in urine, during the prodromal phase and for about 2 days after the appearance of the rash. **This early shedding of virus, before the disease can be recognized, promotes its rapid epidemic spread.** The blood, lymph nodes, spleen, kidney, skin, and lungs also contain detectable virus during this period. Measles virus can multiply in and has been isolated from human macrophages and lymphocytes, suggesting that these cells may play a role in its dissemination in the body and in the pathogenesis of the disease. The leukocytic involvement may also be responsible for the leukopenia observed during the prodromal stage as well as for depression of delayed-type hypersensitivity reactions (e.g.,

the tuberculin test). Measles virus can induce striking aberrations in the chromosomes of leukocytes during the acute disease. Although the chromosomal pulverization produced is probably lethal to the cell, a possible relation between some of the changes and the initiation of leukemia has been suggested.

The characteristic **viremia** in measles, in contrast to the more localized respiratory infections produced by influenza and parainfluenza viruses, probably contributes to the notably effective immunity conferred by the disease.

Complications

Bronchopneumonia and **otitis media,** with or without a bacterial component, are frequent complications of the disease. **Encephalomyelitis** is the most serious complication, appearing about 5 to 7 days after the rash. Its incidence in most epidemics is about 1 in 2000 cases (higher in children over 10 years); but in some outbreaks, particularly in the widespread infection of malnourished infants in Africa, the incidence has been much higher. However, cerebrospinal fluid pleocytosis may occur in as many as 30% of the cases. The mortality rate of encephalomyelitis is about 10%, and permanent mental and physical sequelae have been reported in 15% to 65% of survivors.

There is no evidence that measles encephalomyelitis, which occurs in certain epidemics, is due to viral strains

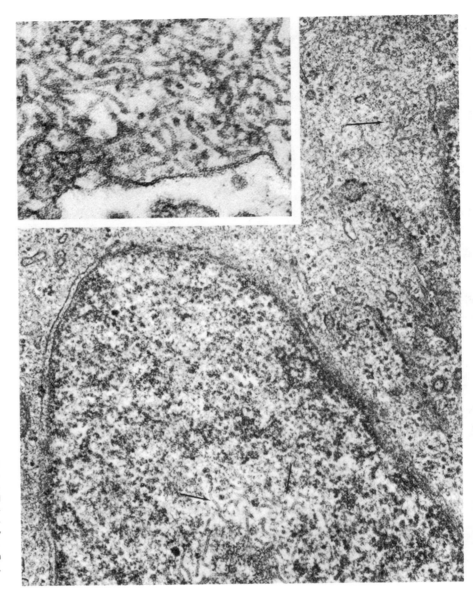

Figure 57–13. Electron micrographs of cell infected with measles virus, showing intranuclear and intracytoplasmic matrices containing viral nucleocapsids (*arrows;* original magnification ×60,000). (*Inset*) Higher magnification of an extensive accumulation of nucleocapsids. Where the tubules are favorably oriented, cross-striations of the nucleocapsids can be seen. (Original magnification ×140,000; Nakai T et al: Virology 38:50, 1969)

with increased virulence. Indeed, measles virus cannot be isolated from the brain, but lymphocyte infiltration and demyelination are prominent pathologic features, reminiscent of allergic encephalitis. These findings suggest that measles encephalomyelitis is a hypersensitivity response either to the measles virus or to virus-altered host tissue (i.e., an autoimmune phenomenon).

Giant cell pneumonia, a rare disease of debilitated children, or of those with immunodeficiency disease, was proved to be due to measles by isolation of virus.

Subacute sclerosing panencephalitis (SSPE), a progressive, fatal complication of measles, is an excellent example of a single virus inducing an acute disease and a chronic illness separated by a long interval during which there is restricted synthesis and expression of viral genes (see Slow Viral Infections, Chap. 51, for a complete description).

Pathology

The development of very large multinuclear **giant cells** (Warthin–Finkeldey syncytial cells) is the predominant and characteristic feature of the pathology of measles. These distinctive cells are found in nasal secretions during the prodromal stage of the disease, as well as in lymphoid tissue of the gastrointestinal tract, particularly the appendix. Giant cells are also often observed in sputum from patients with bronchopneumonia and may contain eosinophilic nuclear and cytoplasmic inclusions, similar to those seen in infected cell cultures; and such cells are characteristic of the rare **giant cell pneumonia.** These giant cells are presumably produced by cell fusions, like syncytial cell formation in infected cell cultures.

In **measles encephalomyelitis** the brain shows perivascular hemorrhage and lymphocytic infiltration early in the disease; areas of demyelination later appear in the brain and spinal cord.

Brains from patients with **subacute sclerosing panencephalitis** display a degeneration of the cortex and especially the underlying white matter. They contain characteristic intranuclear and intracytoplasmic inclusion bodies not unlike those noted in acute measles (see Figs. 57–11, 57–12, and 57–13), perivascular infiltration of plasma cells and lymphocytes, scattered degeneration of nerve cells, hypertrophy of astrocytes, microglial proliferation, and demyelination.

LABORATORY DIAGNOSIS

The epidemiologic and clinical features of measles are usually so characteristic that laboratory confirmation of the diagnosis is unnecessary except for investigative purposes. During the prodromal stage of the disease a rapid and simple presumptive diagnosis can be made by demonstrating specific viral immunofluorescence and characteristic giant cells in smears of the nasopharyngeal mucosa. Definitive diagnosis can be accomplished by isolation of virus from nasal or pharyngeal secretions, blood, or urine; viral isolations are best achieved in primary cultures of human embryonic or monkey kidneys. Serologic diagnosis can be made by comparing acute and convalescent sera using hemagglutination-inhibition, ELISA, or CF titrations, as described for other paramyxoviruses. Finding measles Abs in the cerebrospinal fluid is suggestive of injury to the blood–brain barrier, probably from an immunologic reaction, and is often a sign of severe neurologic disturbance such as acute encephalomyelitis or SSPE.

EPIDEMIOLOGY

Measles is a **highly contagious disease** in which virus is **spread in respiratory secretions.** It is predominantly a childhood affliction that occurs in epidemics during the winter and spring in rural areas. Since measles virus does not have a reservoir in nature other than humans, and since long-lasting immunity follows infection, persistence of the virus in a community depends upon endemic infections in a continuous supply of susceptible persons. It has been estimated that in urban areas 2500 to 5000 cases per year are required for continued transmission. Hence, it is clear why in the more highly developed countries epidemics tend to appear in 2- to 3-year cycles, as a sufficient number of nonimmune children arises in the population, and why the disease disappears from small, isolated communities. However, even in a highly immunized population exogenous introduction of virus can initiate a limited epidemic; for example, epidemics are now appearing in high schools and universities in which over 95% of the students were immunized as children.

In the United States the highest incidence is in children 5 to 7 years of age, and the disease is relatively mild. However, since 1973, after widespread immunization began, the age-specific incidence has steadily increased to older children and teenagers. Moreover, the disease is more severe in adults and in young children (but transplacental immunity usually protects newborns and babies up to the age of 6 months). In communities having primitive and crowded living conditions (e.g., in West Africa) most cases occur in infants less than 2 years old; the illnesses tend to be severe and often fatal (probably owing to malnutrition and a high rate of secondary infections), and epidemics occur yearly because the close contacts probably decrease the proportion of susceptibles required. If the virus is introduced into isolated, unimmunized communities, where the exposure is rare and measles has not struck in many years, the incidence is very high, and the illness is severe and frequently is fatal to very young children and to the elderly (e.g., the Faroe Islands; see Persistence of Antibodies, Chap. 51).

PREVENTION AND CONTROL

Public health measures alone, such as isolation, have not successfully prevented or even limited measles epidemics. Prior to the development of vaccines for active immunization, passive immunization with pooled γ-globulin* was used to furnish temporary protection. Because of the long incubation period, large doses, even when administered shortly after exposure, prevent the disease, and small doses reduce its severity. This procedure is still effectively employed for exposed susceptibles, particularly adults.

---

* Because the majority of adults in most countries have had measles, and because levels of circulating Abs remain high, pooled normal adult γ-globulin is effective in passive immunization.

After the successful cultivation of measles virus in tissue culture, vaccines were developed and effective control became possible. Following the methods that had proved so successful in the control of poliomyelitis both **attenuated live virus vaccine** and **formalin-inactivated virus vaccine were prepared.**

The initial "attenuated" virus induced mild measles with fever in about 80% of recipients, so that γ-globulin containing measles Abs was administered at a different site to reduce the reactions. Viruses of greater attenuation were eventually obtained, however, and they are now used without an accompanying injection of γ-globulin. Babies immunized prior to their first birthday often have a poor Ab response, and it is recommended that they be reimmunized when about 15 months old. In children older than 1 year, the live attenuated virus selected for general immunization induces an Ab response in almost 100% of those who previously lacked Abs, but the vaccine still produces reactions in 15% to 20% of children. The Ab titers are about 10% to 25% of the levels observed after the natural disease. Neutralizing Abs endure without significant decline for at least 2 years after immunization but decrease twofold to threefold by 5 years; CF Abs begin to decline in 6 to 8 months. Although effective immunity has been demonstrated at least 10 years after immunization, measles does occasionally occur in vaccinated children, particularly in those immunized during the first year of life. How long adequate protection persists after vaccination is still unclear, but the increasing number of measles outbreaks reported among adolescents and young adults who were previously immunized as young children suggests that the immunity may be less lasting than that following natural measles.

The vaccine does not cause acute neurologic complications and virus does not spread from vaccinees to susceptibles in the same family. Immunization with live virus vaccine has proved highly effective in large studies, and during epidemics it appears to prevent measles in more than 95% of children. In the United States, where widespread immunization has been employed, large epidemics have been eliminated and the incidence of measles has been reduced from about 450,000 reported cases per year prior to 1960 to 2543 in 1984 (Fig. 57–14).

The attenuated virus vaccine is distributed in a lyophilized form, usually mixed with other attenuated viruses. The most popular combination is measles virus mixed with mumps and rubella viruses. It is comforting that the multiple-virus formulation has neither increased the reaction rate to measles attenuated virus nor decreased the Ab response to any of the viral immunogens present.

A **formalin-inactivated** virus vaccine has been shown to elicit Abs in about 75% of the recipients after a course of three intramuscular injections. This vaccine has not received acceptance, however, because protection is only temporary, neutralizing Abs as well as CF Abs begin to decline rapidly within 3 to 6 months, and unanticipated severe disease has been reported in vaccinees who were subsequently infected naturally or reimmunized with live virus vaccine. (It should be noted, however, that this **"atypical" measles** is also observed in some recipients of live virus vaccine who subsequently develop natural measles.)

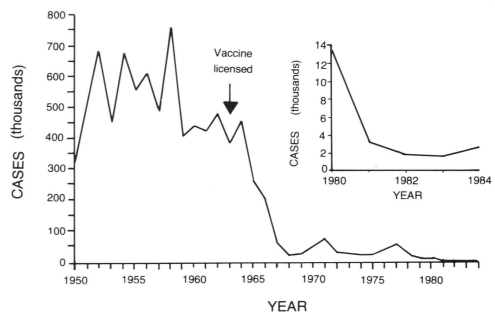

Figure 57–14. Reported cases of measles in the United States from 1950 to 1984. (Recreated from MMWR 34:308, 1985)

Respiratory Syncytial Virus

From a chimpanzee with coryzal illness, and from a laboratory worker who had been in contact with the animal, Morris and co-workers isolated a new virus in 1956. The following year Chanock isolated similar viruses from two infants with pneumonia and croup. Subsequent studies have indicated that **the virus is a major cause of lower respiratory tract disease during infancy and early childhood** throughout the world. Since it characteristically introduces formation of large syncytial masses in infected cell cultures, it was named **respiratory syncytial (RS) virus.**

PROPERTIES

The RS virus is related to parainfluenza, measles, and mumps viruses, but several distinctions led to its classification in a separate genus, *Pneumovirus* (see Table 57–1). In particular, hemagglutination, hemadsorption, hemolytic, and neuraminidase activities are not detectable despite the presence of regularly spaced clublike projections (peplomers) on the virion's surface (Fig. 57–15). Moreover, the virions and the nucleocapsids are extremely fragile, which makes preservation of infectivity difficult. In addition, filamentous virions are present in purified preparations (see Fig. 57–15) and in thin sections (Fig. 57–16); the filaments, unlike those of influenza viruses (see Fig. 56–2, Chap. 56), are often much narrower than the spherical virions, and they have the appearance of an elongated nucleocapsid rather than a folded helix covered with an envelope. Finally, the spherical virions are somewhat less variable in size and are slightly smaller (see Table 57–1) than the typical paramyxoviruses; the nucleocapsid has a slightly smaller diameter, and the helix has a regular periodicity slightly larger than for other paramyxoviruses.

The single-stranded RNA genome of RS virus is similar in size and organization to that of the other paramyxoviruses (see Fig. 57–6). However, the RS virus RNA encodes four additional proteins: two nonstructural proteins (NS_1 and NS_2); a third (SH), whose functions are unknown; and a second envelope membrane protein (22K).

Infectivity of RS virus is completely destroyed during storage at $-15°C$ to $-25°C$ for only several days. Viral suspensions can be preserved without complete inactivation by adding protein (5% to 10% normal serum or albumin), freezing rapidly, and maintaining at $-70°C$.

Immunologic Characteristics

Antigenic variants have been noted among the RS strains studied, but the immunologic cross-reactivity between variants is too great to permit division into distinct types. It is noteworthy that the antigenic differences do not appear to be progressive with successive epidemics.

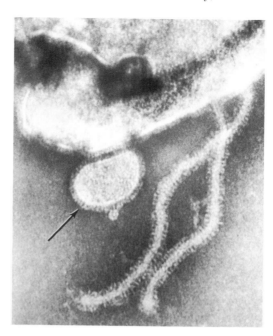

Figure 57–15. Intact respiratory syncytial virus particle (*arrow*) and two filaments negatively stained with phosphotungstate; note the regularly arranged clublike peripheral projections. (Original magnification ×110,000; Bloth B et al: Arch Gesamte Virusforsch 13:582, 1963)

Each variant has a specific surface Ag that can be detected by plaque neutralization assays in cell cultures with standard sera. Purified virions and extracts of infected cells contain the specific viral **surface Ags** and the other viral protein including the **major nucleocapsid (NP) Ag,** detectable by CF titration, that is common to all RS viruses but not to other paramyxoviruses.

Neutralizing Abs are directed primarily against the **fusion (F_1 and F_2) proteins.** Monoclonal Abs that react with the large **G surface glycoprotein** also neutralize virus and provide passive immunity to experimental animals.

Protection from reinfection is brief following RS virus infections, probably because the titer of IgA Abs in nasal secretions declines rapidly. Serum Abs persist, but even a level of neutralizing Abs as high as 1:256 does not provide adults with effective protection against reinfection. Babies during the first 6 months of life are most often infected and most seriously affected, despite the high level of serum IgG Abs derived from their mothers; and infants with neutralizing Abs may even develop bronchopneumonia on reinfection.

Host Range

RS virus has been observed to cause illness only in humans, chimpanzees, owl monkeys, and cotton rats. In addition, intranasal inoculation of virus into mice, fer-

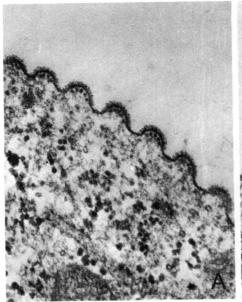
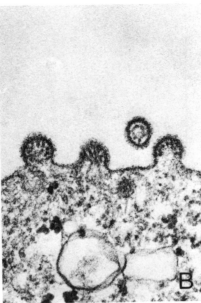
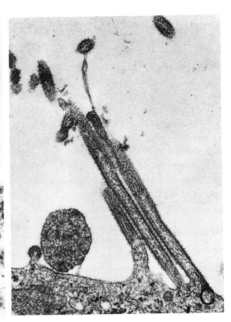

Figure 57—16. Electron micrographs of thin sections of a cell culture infected with respiratory syncytial virus. Different stages are seen in the morphogenesis of the viral particles at the plasma membrane. (*A*) Early stages of budding show thickening of the cell membrane with the appearance of fringelike projections. Submembranous accumulation of the nucleocapsid is also noted. (Original magnification ×75,000) (*B*) Later stages of viral budding and a single free virion are seen. The circular arrangement of the nucleocapsid in the spherical particles suggests an organized packing of this component. (Original magnification ×76,000) (*C*) Filamentous forms of viral particles are developing at the cell membrane. The linear arrangement of the nucleocapsid is visible within the filamentous particles. (Original magnification ×42,000; Norrby E et al: J Virol 6:237, 1970)

rets, monkeys, and many other mammals produces inapparent infections, from which virus can be recovered.

RS virus multiplies and produces cytopathic changes in continuous human epithelium-like cell lines, in human diploid cell strains, and in primary simian and bovine kidney cell cultures. About 10 hours after infection, virus-specific Ags, detected by immunofluorescence, appear in the cytoplasm of infected cells but not in their nuclei. As the quantity of Ag increases, it accumulates at the cell membrane, where virions assemble and bud from the altered cell membrane like other paramyxoviruses (see Fig. 57—16).

The formation of syncytia and giant cells, as in parainfluenza and measles virus infections, is the predominant and characteristic cytopathology (Fig. 57—17). Prominent eosinophilic cytoplasmic inclusions, consisting of densely packed material having a granular or threadlike appearance, are commonly found in infected cells, particularly in the syncytia (see Fig. 57—17). These inclusion bodies may be considered "scars" of the infection, for they do not contain RNA, viral particles, or accumulated virus-specific Ags.

PATHOGENESIS

The pathogenesis of RS virus infections has been largely surmised on the basis of clinical observations. The **initial infection** results from **viral multiplication in epithelial cells of the upper respiratory tract;** it most often ends at this stage in adults and older children. In about 50% of infected children less than 8 months old, **virus spreads into the lower respiratory tract**—the bronchi, the bronchioli, and even the pulmonary parenchyma—causing febrile upper respiratory disease, bronchitis, bronchiolitis, bronchopneumonia, and croup.

The dissemination of virus into the lower respiratory tract, with concomitant production of more serious disease, is probably due at least partially to the absence of IgA Abs in the respiratory secretions and to the slow development of Abs in the previously uninfected infant. With increasing age, and therefore repeated exposure and greater immunologic response to RS virus, infections are likely to be milder and confined to the upper respiratory tract and are most likely to be inapparent. In adults, because of the absence or limited quantity of virus-specific secretory (IgA) Abs in the respiratory tract, infections do occur despite the presence of circulating Abs; most frequently the disease is afebrile and clinically resembles the common cold.

Autopsy examination of infants dead of RS virus infection shows severe necrotizing lesions of the epithelium of the bronchi and bronchioles, and interstitial pneumonia consisting of mononuclear infiltration, patchy atelectasis, and emphysema. The exact mechanism of epithe-

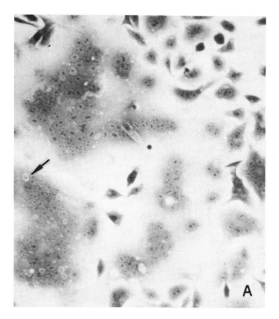
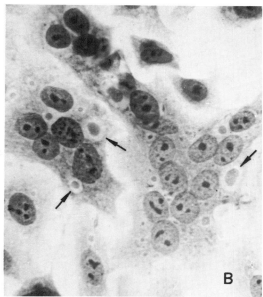

Figure 57–17. Syncytium formation and cytoplasmic inclusions (*arrows*) in Hep-2 cells infected with respiratory syncytial virus. Syncytia are characterized by the very large aggregates of intact nuclei and extensive cytoplasmic masses devoid of cell membranes. (*A*, original magnification ×125; *B*, original magnification ×540; Bennett CR Jr, Hamre D: J Infect Dis 110:8, 1962)

lial cell injury is unclear, but direct damage owing to viral replication rather than an immunologic response is probably the major cause. Cytoplasmic inclusion bodies are noted, but syncytial formation is not observed. These changes represent one end of the spectrum of disease produced by RS virus. In infants who die with bronchiolitis the lungs contain small amounts of virus and intracellular Ags that can be detected by immunofluorescence, whereas in infants with pneumonia large quantities of infectious virus and viral Ags are present.

LABORATORY DIAGNOSIS

Detection of RS viral Ags in exfoliated cells of the respiratory tract by **specific immunofluorescence** offers the most rapid and economic means for diagnosing acute infections. **A precise diagnosis of RS** virus infection, however, requires either **isolation of the virus** or demonstration of a **rise in Abs.** Serologic diagnosis is ordinarily the most reliable and easiest, using ELISA or CF (the most convenient and economical) or neutralization titrations. Very young infants, however, may not produce a detectable Ab response, in which case a laboratory diagnosis depends upon isolation of virus.

Because RS virus is highly labile, isolation is most efficient when nasal or pharyngeal secretions are inoculated directly from the patient into cultures of human continuous cell lines (Hep-2, HeLa, or KB). Characteristic giant cells develop within 2 to 14 days, but infection can be detected more rapidly by immunofluorescence. Virus can be isolated in about 70% of patients during the first week of illness. Newly isolated RS virus can be identified by CF, ELISA, or neutralization titrations with standard antisera.

EPIDEMIOLOGY

RS virus is a **major cause of respiratory disease in young children,** and the **most common cause of nosocomial infections** in pediatric wards. Infections have a worldwide distribution, and they occur in yearly epidemics of varying magnitude. Virus spreads rapidly through the susceptibles in a community, so that epidemics are sharply circumscribed and relatively brief. The outbreaks occur primarily in infants and children between late fall and early spring. Many adults may be infected during the episode, but they have mild disease or inapparent infections.

RS infections uniformly occur early in life. Indeed, RS virus is the only recognized virus that preferentially produces severe respiratory disease and has its maximum impact during the first 6 months of life. Severe pneumonia and bronchitis occur in infants between 6 weeks and 6 months after birth with the peak incidence in those 2 months old. Serious infections occur 30% more frequently in male babies and more commonly in white infants. Although infants have a poor Ab response to infection, approximately one-third of infants in the United States develop Abs in the first year of life, and 95% by 5 years of age. However, reinfections are common, and about 80% of previously infected children are reinfected in the second year of life.

PREVENTION AND CONTROL

Because RS virus ranks so high on the list of causes of respiratory disease in young children, preventive methods are highly desirable. Isolation or general public health measures are not adequate to control the spread of infection: it is difficult to recognize the disease early, and inapparent cases are frequent.

Experience with an alum-precipitated, formalin-inactivated vaccine was discouraging. Though the serum Ab response was good, the clinical response to subsequent RS virus infection was startling and paradoxic: both the **incidence and the severity** of disease **increased** strikingly, particularly in infants. These results resembled the severe reactions in children immunized with inactive measles virus and then naturally infected with measles virus. Animal models indicate that this frightening response was probably due to a cell-mediated hypersensitivity as well as to Ab–virus complexes producing an Arthus-type reaction. This dramatic failure of immunization was primarily the result of formalin damage to the neutralizing epitopes of the viral envelope F and G proteins.

Development of a live RS virus vaccine using selected attenuated viral variants, particularly temperature-sensitive mutants, was unsuccessful owing to the mutants' high reversion rates. Rapid progress is being made, however, in developing a vaccine consisting of a live viral vector, such as vaccinia virus or adenovirus, in which the F and G protein genes of RS virus are inserted to permit their expression. This approach has been successful in animals. Although the disease does not afford effective, lasting immunity, an effective viral vector vaccine administered early in infancy should prevent or reduce the serious effects of primary infection.

Newcastle Disease Virus

Newcastle disease virus (NDV) is primarily a respiratory tract pathogen of birds, particularly chickens, but it occasionally produces accidental infections in man. Human infections are almost exclusively confined to poultry workers and laboratory personnel. The disease is characteristically mild and limited to conjunctivitis without corneal involvement. Although predominantly of veterinary interest, this virus merits brief mention because of the prominent role it has played in the investigation of paramyxoviruses.

NDV possesses the characteristic properties of paramyxoviruses listed in Table 57–1. This virus particularly resembles parainfluenza and mumps viruses in morphology, chemistry, and reactions with RBCs. Moreover, many patients with mumps virus infection develop hemagglutination-inhibiting and CF Abs to NDV.

Selected Reading

PARAINFLUENZA VIRUSES
Books and Review Articles

Chanock RM, McIntosh K: Parainfluenza viruses. In Fields BN (ed): Virology. New York, Raven Press, 1985
Choppin PW, Compans RW: Reproduction of paramyxoviruses. Compr Virol 4:95, 1975

Specific Articles

Chanock RM: Association of a new type of cytopathogenic myxovirus with infantile croup. J Exp Med 104:555, 1956
Deshpande KL, Portner A: Monoclonal antibodies to the P protein of Sendai virus define its structure and role of transcription. Virology 140:125, 1985
Storey DG, Dimock K, Kang CY: Structural characterization of virion proteins and genomic RNA of human parainfluenza virus 3. J Virol 52:761, 1984
Yewdell J, Gerhard W: Delineation of four antigenic sites on a paramyxovirus glycoprotein via which monoclonal antibodies mediate distinct antiviral activities. J Immunol 128:2670, 1982

MUMPS VIRUS
Specific Articles

Julkunen I, Vaananen P, Penttinen K: Antibody responses to mumps virus proteins in natural mumps infection and after vaccination with live and inactivated mumps virus vaccines. J Med Virol 14:209, 1984
Levitt LP, Mahoney DH, Casey HL, Bond JO: Mumps in a general population: A seroepidemiologic study. Am J Dis Child 120:134, 1970

MEASLES VIRUS
Specific Articles

Fine PEM, Clarkson JA: Measles in England and Wales I: An analysis of factors underlying seasonal patterns. Int J Epidemiol 11:5, 1982
Frank JA Jr, Orenstein WA, Bart KJ et al: Major impediments to measles elimination. The modern epidemiology of an ancient disease. Am J Dis Child 139:881, 1985
Girandon P, Weld TF: Correlation between epitopes on hemagglutinin of measles virus and biological activities: Passive protection by monoclonal antibodies is related to their hemagglutination inhibiting activity. Virology 144:46, 1985
Norrby E: Measles. In Fields BN (ed): Virology. New York, Raven Press, 1985

Panum PL: Observation made during the epidemic of measles on the Faroe Islands in the year 1846. New York, American Publishing Association, 1940 (Reprint)

RESPIRATORY SYNCYTIAL VIRUS
Books and Review Articles

McIntosh K, Chanock RM: Respiratory syncytial virus. In Fields BN (ed): Virology. New York, Raven Press, 1985

Scott EJ, Taylor G: Respiratory syncytial virus. Brief review. Arch Virol 84:1, 1985

Specific Articles

Chanock RM, Roizman B, Myers R: Recovery from infants with respiratory illness of a virus related to chimpanzee coryza agent (CCA). I. Isolation properties and characterization. Am J Hyg 66:281, 1956

Collins PL, Dickens LE, Buckler-White A et al: Nucleotide sequences for the gene junctions of human respiratory syncytial virus reveal distinctive features of intergenic structure and gene order. Proc Natl Acad Sci USA 83:4594, 1986

Morris JA, Blount RE, Savage RE: Recovery of cytopathic agent from chimpanzees with coryza. Proc Soc Exp Biol Med 92:544, 1956

58

Harold S. Ginsberg

Coronaviruses

After the discovery that rhinoviruses are major etiologic agents of the common cold (see Chap. 55), more than 50% of illnesses still could not be associated with known causative agents. However, when Tyrrell and Bynoe in 1965 introduced the use of ciliated human embryonic tracheal and nasal organ cultures, they revealed a new group of viruses (and also improved the isolation of known agents). The unique properties of the new group included their distinctive club-shaped surface projections (Fig. 58–1), which give the appearance of a solar corona to the virion. Hence the family name **Coronaviridae** (L. *corona;* vernacular, *coronaviruses*, crown) was proposed to include the agents isolated initially as well as viruses propagated in human embryonic cell cultures by Hamre and Procknow from patients with acute respiratory diseases, and similar viruses from a variety of lower animals, presenting a total of 11 viral species.*

Properties

MORPHOLOGY

Electron microscopic examinations of negatively stained preparations reveal moderately pleomorphic spherical or elliptical virions. The surface is covered with distinctive, widely spaced, pedunculated projections (**peplomers**), 20 nm long, with narrow bases and club-shaped ends (see Fig. 58–1). Thin sections show virions with an outer membrane (envelope), an electron-lucent intermediate zone, and an inner nucleocapsid consisting of an electron-dense shell and a central zone containing

---

* Avian infectious bronchitis virus, calf neonatal diarrhea virus, murine hepatitis virus, porcine transmissible gastroenteritis virus, porcine hemagglutinating encephalitis virus, rat pneumotropic virus, rat sialodacryoadenitis virus, turkey bluecomb disease virus, bovine coronavirus, canine coronavirus, and feline infectious peritonitis virus

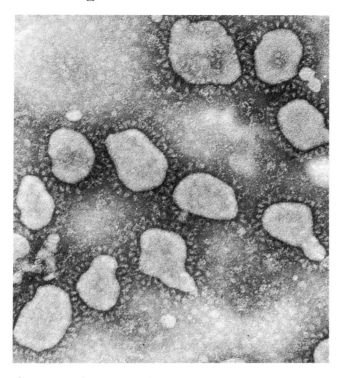

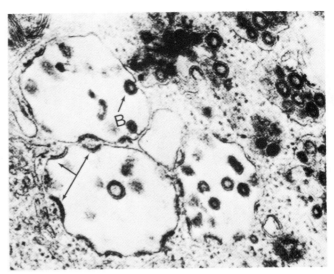

Figure 58—2. Development of a coronavirus in a human diploid cell. Thickened membranes (*arrows*) indicate sites of early morphogenesis. A particle budding into a vacuole (*B*) and several mature particles within vacuoles are also present. Mature particles show inner and outer shells with a translucent zone between them. (Original magnification ×50,000; Becker WB et al: J Virol 1:1019, 1967)

Figure 58—1. Coronaviruses. The negatively stained particles show the distinctive corona effect produced by the pedunculated surface projections, which are approximately 20 nm long and have a club-shaped end about 10 nm wide. The marked pleomorphism of the virions may be noted. (Original magnification ×144,000; Kapikian AZ: Diagnostic Procedures for Viral and Rickettsial Infections, 4th ed. New York, American Public Health Association, 1969)

amorphous material of variable density (Fig. 58—2). The nucleocapsid appears to be a loosely wound helix.

PHYSICAL AND CHEMICAL CHARACTERISTICS

The properties of all species studied are similar, but the most detailed characterization was done with the murine hepatitis virus owing to its ease of propagation and greater viral yield in cell cultures. Whenever comparable studies have been possible, the human coronaviruses proved to be comparable. The virion is enveloped in a lipid bilayer membrane (which projects two glycoproteins, E_1 and E_2) (Fig. 58—3). The most prominent projection, the **peplomer** or E_2 **spike,** consists of two polypeptide chains of equal sizes derived by proteolytic cleavage of a precursor protein. The E_2 protein and the hemagglutinin of influenza virus (see Chap. 56) are comparable. The E_1 **glycoprotein** spans the envelope; a small, glycosylated amino-terminal domain projects on the external surface, but the major portion of the protein (greater than 85%) is in the membrane and on the envelope's

internal surface. Like the **matrix (M) protein** of orthomyxoviruses (see Chap. 56) and paramyxoviruses (see Chap. 57) this internal domain interacts with the nucleocapsid (see Fig. 58—3). The long, flexible, helical nucleocapsid consists of a single molecule of capped and polyadenylated infectious RNA (i.e., plus-stranded) associated with many molecules of a **basic phosphoprotein, the N protein** (see Fig. 59—2). The presence of lipid in the envelope makes the virion sensitive to lipid solvents. Deviations of as little as 0.5 units from the virion's optimal pH stability, which varies with viral strains, inactivate the virus, probably reflecting the lability of its envelope.

IMMUNOLOGIC CHARACTERISTICS

Coronaviruses contain three major antigens that are components of the virions: the peplomers (E_2), the matrix protein (E_1), and the nucleocapsid protein (N). Antibodies directed against these proteins divide human coronaviruses into two antigenic groups, of which the viruses 229E (isolated in cell culture) and OC43 (isolated in organ culture) are the prototypes. However, although 229E and OC43 viruses appear antigenically distinct by *in vitro* neutralization assays and other immunologic reactions carried out *in vitro*, infected children and adults develop Abs that react with both viruses following a single illness. Monoclonal Abs elicited by the three virion Ags indicate that the E_2 peplomer induces neutralizing

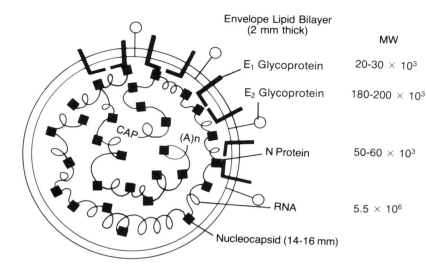

Envelope Lipid Bilayer
(2 mm thick)

	MW
E₁ Glycoprotein	$20\text{-}30 \times 10^3$
E₂ Glycoprotein	$180\text{-}200 \times 10^3$
N Protein	$50\text{-}60 \times 10^3$
RNA	5.5×10^6

Nucleocapsid (14-16 mm)

Figure 58–3. Model of coronaviruses schematically shows the helical nucleocapsid, consisting of a single-stranded RNA of positive polarity associated with N protein (only a few of the associated N [■] proteins are indicated); the association of the nucleocapsid with the cytoplasmic domains of the E₁ glycoproteins; and the surface projecting E₂ glycoprotein peplomer.

Abs, which are directed against three distinct epitopes. The E₁ membrane protein also induces Abs that neutralize, but this activity requires complement (see Chap. 51).

The peplomers of some coronaviruses effect hemagglutination with red blood cells (RBCs) from humans, mice, rats, and chickens. Unlike influenza and parainfluenza viruses, coronaviruses do not elute spontaneously from agglutinated RBCs, and treatment of susceptible RBCs with neuraminidase does not reduce hemagglutination, indicating that coronaviruses attach to different cellular receptors.

The viruses from humans appear antigenically unrelated to coronaviruses from other animals in neutralization tests, but complement fixation (**CF**) assays indicate a partial immunologic relatedness of some human types with mouse hepatitis virus.

HOST RANGE

Coronaviruses affecting man appear able to multiply in only a very limited range of host cells. Viruses isolated in organ cultures of ciliated human embryonic tracheal or nasal tissues (the OC type viruses) do not multiply well in monolayer cultures of human diploid or embryonic kidney cells, and vice versa; hence, both culture systems must be employed for viral isolations. The strains related to the 229E virus have the broadest host range in cell culture and can be propagated in primary or secondary human embryonic kidney cells, in human diploid fibroblast lines, and in heteroploid human embryonic lung cell lines. The highest yields of both types of coronaviruses are obtained in a human rhadomyosarcoma cell line, which is therefore probably the best for isolation of virus from clinical specimens. Two of the OC strains have been adapted to growth in the brains of

suckling mice, but other laboratory animals have not proved to be susceptible to infection.

MULTIPLICATION

Coronaviruses infecting humans cannot be studied biochemically owing to poor replication and hence low yields. Their similarities of virion structure, including their genomes, to mouse hepatitis virus (MHV), however, suggest that MHV's method of multiplication also applies to human coronaviruses. Upon entry of virus into the cell via endocytosis, the viruses replicate entirely in the cytoplasm (Fig. 58–4). Shortly after entry a portion of the viral RNA genome, probably at its 5′ end, is translated into an "early" RNA-dependent RNA polymerase, which then transcribes the genome into a full-length complementary strand. This negative strand is transcribed into a set of seven viral mRNAs, transcribed by two different "late" RNA polymerases. These capped and polyadenylated mRNAs are uniquely arranged as a **nested set** of progressively decreasing size so that each smaller RNA contains all of the 3′ sequences except those of the gene translated from the next larger RNA (see Fig. 58–4). A second feature is that all the mRNAs, including the genomic RNA, have an identical 5′ leader sequence of about 72 nucleotides that is encoded only at the 5′ terminus of the genome RNA. This finding implies that all the viral mRNAs are formed by joining two noncontiguous RNAs. A study of the replicative intermediate RNAs, which are of genome length, indicates that the joining of the common 5′ leader to the body of each mRNA is not accomplished posttranscriptionally by splicing but occurs during transcription, an unusual biosynthetic process. The findings suggest that the leader RNA is synthesized independently and sequestered to prime transcription of

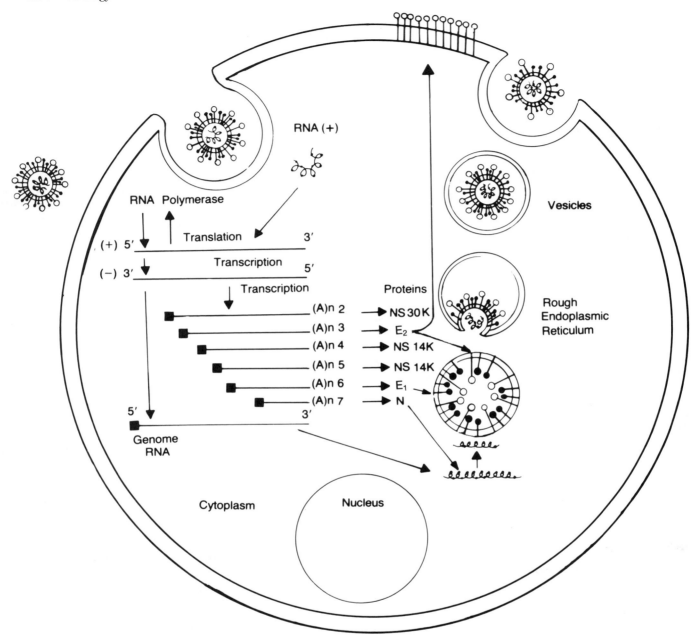

Figure 58–4. Schematic representation of the steps in replication of coronaviruses. 🍄, E_1 glycoprotein; 🍄, E_2 peplomer; ■, the 5' common capsid, leader sequence on vested mRNAs; (A)*n*, 3' poly(A) mRNA sequences; *NS*, nonstructural proteins. (Modified from Sturman LD, Holmes KV: The molecular biology of coronaviruses. Adv Virus Res 28:35, 1983)

each mRNA by binding to complementary intergenic initiation sites on the negative-strand RNA.

The translation of the structural proteins (E_1, E_2, and N) is associated with **assembly of virions** by budding into vesicles of the rough endoplasmic reticulum and the Golgi apparatus. The E_2 protein, which is cotransla-

tionally glycosylated, and the E_1 matrix protein, which is glycosylated in the Golgi apparatus, are inserted in the vesicle membranes and serve as the signals for association with the N-phosphoprotein–genome RNA complex, the **nucleocapsid** (see Fig. 58–4). Although the E_2 peplomers are also inserted into the plasma membrane, virions are not assembled at the cell surface. Rather, the virions are **released** by fusion of virion-filled vesicles with the plasma membrane, the cellular secretory process of exocytosis.

Viral multiplication is relatively slow compared with a number of other enveloped viruses such as orthomyxoviruses or togaviruses: the eclipse period lasts at least 6 hours, and maximum viral yield is not attained until about 24 hours after infection. Viral multiplication is optimal at 32°C to 33°C, as with rhinoviruses (see Chap. 55), and infectivity is rapidly lost at higher temperatures or with prolonged incubation.

Pathogenesis

Coronaviruses constitute another group of viruses responsible for common colds and pharyngitis. Patients from whom the viruses were recovered, and volunteers who were inoculated intranasally to establish the etiologic relation of coronaviruses to disease, exhibited signs and symptoms of acute upper respiratory infections: coryza, nasal congestion, sneezing, and sore throat were common; less frequent symptoms were headache, cough, muscular or general aches, chills, and fever. Pulmonary involvement has not been noted. In volunteers inoculated intranasally, viral multiplication occurred in superficial cells of the respiratory tract; viral excretion was detectable at the time symptoms were first noted, about 3 days after infection; and the mean duration of illness was 6 to 7 days. Although coronaviruses have many distinctive viral characteristics, the diseases they cause cannot be distinguished clinically from common colds produced by rhinoviruses or even from those occasionally initiated by influenza viruses.

Unfortunately, details of pathogenesis cannot be experimentally studied since an animal model for human coronaviruses does not exist. Other coronaviruses infect a variety of animals, however, producing hepatitis, transmissible gastroenteritis, encephalomyelitis, infectious peritonitis, and pneumonitis. Indeed, despite its being an enveloped virus whose infectivity is quite labile, especially at alkaline pH, it is transmitted by the fecal–oral route in many animals. Moreover, coronaviruses can be detected by electron microscopy in stools of infants with necrotizing enterocolitis, and two of these viruses have been isolated and passed in human fetal intestine organ cultures.

A number of coronaviruses infecting animals, particularly murine hepatitis virus, produce persistent infections. Moreover, some rodent coronaviruses infect the central nervous system, producing diverse pathology such as acute fatal encephalitis or slowly progressive, chronic demyelination, depending upon viral strain, as well as rodent age, strain, and route of inoculation. No

such disease has been attributed to coronaviruses in humans, but some human viral strains can infect human and mouse brain cells, which suggests that neurotropism may have clinical implications in man.

Laboratory Diagnosis

Coronaviruses are so fastidious in their host requirements that isolation is not practicable for routine diagnosis. In principle, the diagnostic serologic procedures of CF, enzyme-linked immunosorbent assay (ELISA), neutralization, and hemagglutination-inhibition titrations can be used, but CF and ELISA titrations are the only practical procedures. Neutralization titrations are difficult and expensive, since they require human organ cultures and cell cultures; moreover, neutralizing Abs are present in 50% to 80% of the population, and although reinfection is common, an increase in titer is difficult to demonstrate. Hemagglutination has been obtained with only two viruses, both belonging to the same immunologic type of OC viruses.* For seroepidemiologic surveys, ELISA titrations are more informative than CF assays because CF Abs decline rapidly after infection. However, neutralization and hemagglutination-inhibition assays are more Ag-specific, although less sensitive than ELISA.

Epidemiology

Acute upper respiratory infections generally have their greatest incidence in children, particularly those between 4 and 10 years of age. It is therefore especially striking that coronaviruses are frequently associated with common colds in adults, and it is a member aged 15 years or older who most frequently introduces infection into a family. Virus is not commonly isolated from children, but neutralizing Abs have been detected in up to 50% of children 5 to 9 years old. Moreover, infection (diagnosed by an increase in Ab titer) is more common in older children and adults, although up to 70% of adults have Abs.

Available data indicate that coronavirus infections occur throughout all regions of the world studied. Infections occur predominantly during the winter and early spring and vary markedly from year to year. During a particular respiratory disease season only a single immunologic type appears to be the causative agent. In a seroepidemiologic study of adults covering a 4-year period in Washington, DC, coronavirus infections occurred in 10% to 24% of those with upper respiratory illnesses. A study

---

* Cultured human and monkey cells infected with OC43 viruses adsorb rat and mouse RBCs, so the hemadsorption-inhibition technique can therefore be used.

of Michigan families showed all age groups to be infected, from 0 to 4 years (29.2%) to over 40 years (22%), with the highest incidence at 15 to 19 years. These findings contrast sharply to the situation with other respiratory viruses (e.g., respiratory syncytial virus), where there is a marked decrease in disease incidence with increasing age. It is noteworthy that rhinovirus infections are uncommon during the periods of prevalent coronavirus infections. During epidemics all segments of the population are affected, but virus tends to spread preferentially within families.

Coronavirus-like particles have been found in stools of children and adults in many parts of the world. However, the frequency of viral detection has been similar in the sick and the healthy. The appearance of these viral particles in stools of newborns in hospital nurseries is poorly understood; we understand neither the cause of illness nor the way it is spread.

Prevention and Control

Additional data on duration of immunity following natural infection, on antigenic structure and immunogenic potential of the virions, and on incidence of infections are essential before the need for or the feasibility of a vaccine is established. Indeed, the high frequency of reinfection accompanied by disease suggests that successful immunization may not be possible.

Selected Reading

BOOKS AND REVIEW ARTICLES

Siddell SG, Anderson R, Cavanagh D et al: Coronaviridae. Intervirology 20:181, 1983

Siddell SG, Wege H, Ter Meulen V: The biology of coronaviruses. J Gen Virol 64:761, 1983

Sturman LS, Holmes KV: The molecular biology of coronaviruses. Adv Virus Res 28:35, 1983

SPECIFIC ARTICLES

Baric RS, Stohlman SA, Razavi MK, Lai MMC: Characterization of leader-related small RNAs in coronavirus-infected cells: Further evidence for leader-primed mechanism of transcription. Virus Res 3:19, 1985

Hamre D, Procknow JJ: A new virus isolated from the human respiratory tract. Proc Soc Exp Biol Med 121:190, 1966

Hamre D, Kindig DA, Mann J: Growth and intracellular development of a new respiratory virus. J Virol 1:810, 1967

Tyrrell DAJ, Bynoe ML: Cultivation of a novel type of common-cold virus in organ cultures. Br Med J 1:1467, 1965

Harold S. Ginsberg

Rhabdoviruses

Were the basis for evolutionary development not so indelibly imprinted on scientific thought, a modern virologist might consider the emergence of the striking bulletlike morphology of rhabdoviruses (Gr. *rhabdos*, rod) a reflection of the violence of our times. This unique form, which was first described for vesicular stomatitis virus (VSV), a virus of cattle and horses, is also associated with rabies virus (Fig. 59–1) and at least 25 other viruses that infect a variety of mammals, fish, insects, and plants. Properties of the agents in this group are summarized under "Characteristics of Rhabdoviruses." Several rhabdoviruses replicate in arthropods as well as in mammals (e.g., VSV, Hart Park virus, Flanders virus, Kern Canyon virus) and hence were previously considered to be arboviruses. Rabies virus is the only member of the group known naturally to infect and produce disease in humans; it will be considered in detail. The Marburg virus,* a simian virus that only accidentally infects humans, has some similar features and will be discussed briefly.

Rabies Virus Group†

The terrifying change of a docile, friendly dog into a vicious, rabid (L. *rabidus*, mad) beast, often with convulsions, struck terror in those in its vicinity and was long considered the work of supernatural causes. The infectious nature of rabies was recognized in 1804, but it was Pasteur, in the 1880s, who suggested that the responsible etiologic agent was not a bacterium. He used his knowledge of the properties of infectious agents and his great intuition to demonstrate for the first time that the patho-

---

* The Marburg virus and the related Ebola viruses are officially classified in the family Filoviridae.

† Lyssavirus (Gr. *lyssa*, rage) is the official designation for this genus of the family Rhabdoviridae.

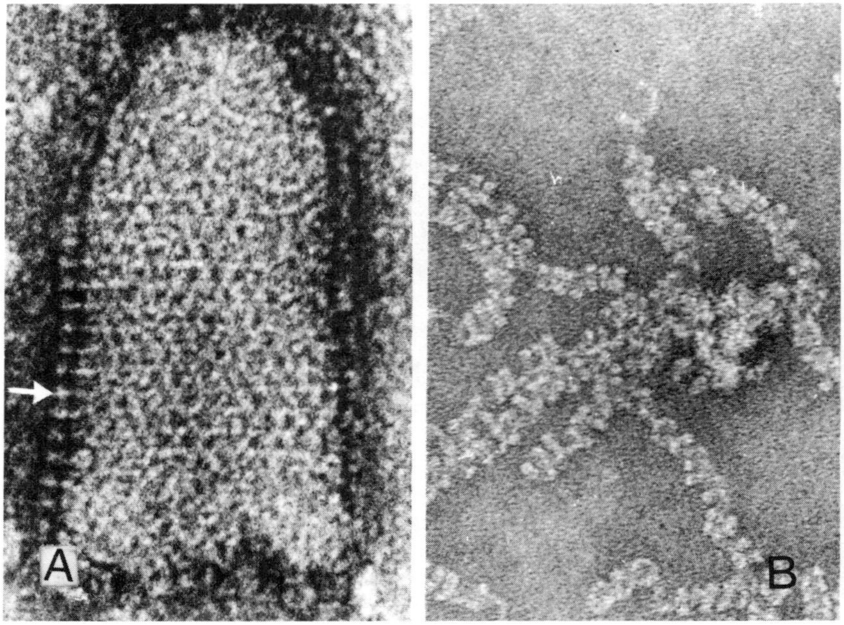

Figure 59–1. Morphologic characteristics of rabies virus. (*A*) Intact rabies virus particle embedded in phosphotungstate and viewed in negative contrast. On the left are well-resolved surface projections 6 nm to 7 nm long (*arrow*). (Original magnification ×400,000) (*B*) Helical nucleocapsid isolated from disrupted rabies virions. Note the tightly coiled and partially uncoiled regions of the single-stranded helix. (Original magnification ×212,000; *A*, Hummeler K et al: J Virol 1:152, 1967; *B*, Sokol F et al: Virology 38:651, 1969)

CHARACTERISTICS OF RHABDOVIRUSES

Morphology: Bullet-shaped; 130–240 nm × 70–80 nm
Nucleic acid: Single-stranded RNA of negative polarity; $3.5–4.6 × 10^6$
    daltons (11–14 Kb); noninfectious
Virion enzyme: RNA transcriptase
Nucleocapsid: Helical; 18 nm wide
Effect of lipid solvents: Disrupt virions; inactivate infectivity
Maturation: Budding at cytoplasmic membranes
Hosts: Wide variety of mammals, fish, invertebrates, and plants
Common antigens: None

---

genicity of a virus (before viruses had actually been identified) could be modified by serial passage in an animal other than its natural host. Fifty serial intracerebral passages in rabbits yielded a modified virus, **fixed virus** (as contrasted with the **wild-type** or **street virus**), which was used for immunization.*

Upon discovery of the filterable causative agent in 1903, Negri described the presence of prominent cytoplasmic inclusion bodies (**Negri bodies**) in the nerve cells of infected human beings and animals. Their characteristic appearance and easy recognition made possible the rapid pathologic diagnosis of infection.

Although rabies was one of the first diseases of man to be recognized as caused by a virus, the agent was stud-

---

*Since the fixed virus strain has residual pathogenicity for humans and other animals, it should be considered to be only partially attenuated.

ied very little until the late 1960s, when methods for propagating attenuated viruses in cell cultures overcame the dangers encountered in handling the virus and the difficulties involved in growing it to high titer.

PROPERTIES
Morphology

The virions (which average 180 × 75 nm) are cylindrical, resembling a bullet (see Fig. 59–1), with one rounded and one planar end, the latter probably arising by collapse of the region where the budding particle is sealed. Regularly spaced projections, each with a knoblike structure at the distal end, cover the surface of the virion. Shorter bullet-shaped and cylindrical particles, probably defective particles (see Chap. 48), are also occasionally observed in electron micrographs. The helical nucleocapsid is symmetrically wound within the envelope along the axis of the virion, often giving the appearance of a series of transverse striations (see Fig. 59–1A). The purified nucleoprotein is a ribbonlike helical strand, consisting of regular rodlike protein subunits attached to a thread of nucleic acid (see Fig. 59–1B).

Physical and Chemical Properties

The viral envelope consists of a lipid bilayer covered by external surface projections composed of a glycoprotein (**G protein**) and a **matrix nonglycosylated M protein** (**M_2 protein**) that reinforces the membrane internally (Fig. 59–2). The glycoprotein surface projections, which

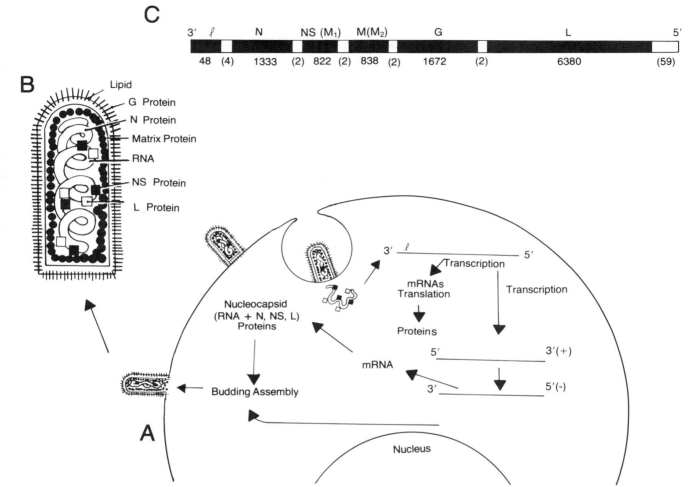

Figure 59–2. Model of rhabdovirus (vesicular stomatitis virus; VSV) virion, its genome, and the steps in viral multiplication. (*A*) The virion and its attachment, uncoating, transcription, translation, RNA replication, assembly, and budding from the infected cell. (*B*) Model of the virion. (*C*) Structure of the RNA genome of VSV, a typical rhabdovirus. The specific coding regions are indicated by the solid boxes; the open boxes indicate the intergenic sequences. The proteins encoded in the genome are noted above each gene, and the number of nucleotides in each gene and in the intergenic regions are noted below. ℓ designates the leader sequences transcribed and individually spliced on each mRNA transcribed. M_1 is the rabies protein comparable to NS, and M_2 is the rabies matrix protein comparable to M of VSV.

act as a hemagglutinin of goose **red blood cells (RBCs),** serve for virion attachment to host cells. The nucleocapsid consists of one molecule of **single-stranded RNA of negative polarity,** many identical copies of a **phosphorylated N protein,** and a few copies of the **RNA-dependent RNA transcriptase,** composed of a large **(L) protein** and a smaller **phosphorylated NS(M_1) protein.**

The rabies virus RNA genome has not been studied in detail. However, VSV, which is readily replicated in cell culture and easily purified, has been exquisitely investigated. Since the morphologic, physical, and chemical characteristics of VSV are so similar to those of rabies

virus, where comparable studies have been done, VSV will be considered the model for discussion of rabies virus structure and replication. Indeed, it is striking that the genome structure and the order of comparable genes of VSV (Fig. 59–2) are so similar to those of paramyxoviruses (see Chap. 57).

As with most enveloped viruses, infectivity deteriorates rapidly at room or refrigerator temperatures in the absence of protein from tissue or added normal serum or albumin. Inactivation is much slower in crude tissue extracts or in infected tissues stored in neutral glycerol. Infectivity is quite stable in frozen or lyophilized tissue extracts.

Immunologic Characteristics

All the rabies viruses isolated from man and other animals, throughout the world, appear to be of a **single immunologic type.** Selected modified (fixed) and wild-type (street) viruses, prepared by many different methods and propagated in different tissues, are also immunologically similar, although minor antigenic variants have been isolated. Several other viruses, immunologically related to rabies virus but distinguishable from it, appear to be limited geographically to regions of Africa and, in host range, primarily to lower animals and insects. However, some of these viruses have been isolated from human illnesses.*

The virion's surface structures, which consist of the G protein, are responsible for the production of neutralizing as well as hemagglutination-inhibiting Abs. Abs to the nucleocapsid proteins, in contrast, are recognized by complement fixation (CF), immunoprecipitation (e.g., enzyme-linked immunosorbent assay [ELISA]), and immunofluorescence. A third class of virus-specific Abs appears after either infection or immunization; in the presence of complement they lyse infected cells whose plasma membranes have incorporated viral Ags. **These cytolytic Abs may play a deleterious rather than a protective role in pathogenesis.** Virus-specific, cytotoxic T-lymphocytes have been demonstrated after infection and immunization, but their role in recovery or protection has not been demonstrated.

Owing to the long incubation period, circulating Abs may be present at the onset of illness. Neutralizing Abs also appear in high titer in the brain and cerebrospinal fluid of patients and can serve as a valuable indicator of infection.

Host Range

Rabies virus can infect all mammals so far tested. Among domestic animals, dogs, cats, and cattle are particularly susceptible. Skunks, bats, foxes, squirrels, raccoons, coyotes, mongooses, and badgers are the principal wildlife hosts. Birds are also susceptible, but less so than mammals.

To establish laboratory infections hamsters, mice, guinea pigs, and rabbits (in order of decreasing susceptibility) are employed. Intracerebral inoculations are more reliable than subcutaneous or intramuscular inoculations.

Rabies virus can be propagated in chick or duck embryos. Attenuated strains developed by multiple passage in embryonated eggs now serve as important sources for vaccines. Cultures of cells from many different animal species can also support viral multiplication. Hamster kidney, human diploid, and chick embryo cell cultures

---

* Duvenhaga virus, isolated from the brain of a man who had been bitten by a bat in South Africa; Lagos bat virus (Nigeria); and Mokola shrew virus (Nigeria)

maintained at 31°C to 33°C are most commonly used. Wild virus is propagated with greater difficulty than modified strains. Cytopathic changes are not usually observed in infected cultures, but intracytoplasmic Ag can be detected by immunofluorescence.

Multiplication

Despite the long incubation period in natural infections, and in experimental animals, the characteristics of viral multiplication in cell cultures are not unusual: the eclipse period is 6 to 8 hours, and the initial cycle of multiplication is completed in 19 to 24 hours. Since cytopathic changes are absent or minimal, carrier cultures or persistent infections (see Chap. 51) are established readily.

The family relationship of rabies virus to VSV suggests that the reactions in their multiplication are essentially identical. The virion attaches to specific receptors of susceptible cells through the viral surface glycoprotein projections and enters the cell in endocytic vesicles. Like influenza viruses, the acidic pH of the vesicle induces viral envelope–plasma membrane fusion and the viral nucleocapsid enters the cytoplasm (see Fig. 59–2). The **virion's RNA-dependent RNA polymerase (transcriptase),** which is composed of the L and NS (M_1) proteins associated with the nucleocapsid, transcribes the viral genome to produce five monocistronic, capped, and polyadenylated mRNAs and the short leader RNA, which is identical on all the mRNAs. The genome has a single promoter and the transcriptase therefore enters the genome at a single site and moves sequentially down it, so that production of each mRNA requires that the preceding 3′ RNA be made first. It is not entirely clear whether the mRNAs are then produced by either terminating and restarting transcription at each intergenic sequence junction or processing a transcript of the entire genome. The rapid production of mRNAs *in vivo* and *in vitro*, the inability to demonstrate the processing of mRNAs from large transcripts, and the presence of conserved polyadenylation signals in the terminal 11 nucleotides of each gene suggests that the first mechanism is used to produce the mRNAs.

During viral replication cytoplasmic "factories" form prominent matrices of helical nucleocapsids. These masses of nucleocapsids appear as cytoplasmic inclusion bodies (i.e., Negri bodies) that can be identified by specific immunofluorescence. The virions then assemble by budding from cytoplasmic membranes (and from the basolateral surfaces of plasma membranes of cultured epithelial cells) (see Fig. 59–2).

PATHOGENESIS

A wound or abrasion of the skin, usually inflicted by a rabid animal, is the major portal of entry into man; the virus enters with the animal's saliva. (A dense population of infected bats may also create an aerosol of infected secretions, by which virus appears to obtain entrance into the respiratory tract.) Virus multiplies in muscle and connective tissue but remains localized for periods that vary from days to months; it then **progresses along the axoplasm of peripheral nerves** to ganglia and eventually to the **central nervous system (CNS),** where it multiplies and produces severe and usually fatal encephalitis. Hematogenous spread of virus to the CNS has been claimed but not established. For transmission by a mammal, the virus must reach the salivary glands, but it apparently does so via efferent nerves rather than through blood and lymph vessels.

The **incubation period** is usually 3 to 8 weeks but can be as short as 6 days or as long as 1 year. It depends on the size of the viral inoculum, the severity of the wound, and the length of the neural path from the wound to the brain, that is, it is shorter following bites on the face and head. (These bites are also usually more extensive.) Illness is ushered in by a **prodromal period,** with irritability, abnormal sensations about the wound site, and hyperesthesia of the skin. **Clinical disease** becomes apparent with the development of the increased muscle tone and difficulty in swallowing, owing to painful and spasmodic contractions of the muscles of deglutition when fluid comes in contact with the fauces. Often the mere sight of liquids will induce such contractions; hence the common name *hydrophobia* (Gr., fear of water). The final stages of the disease result from the extensive damage in the CNS. A fatal outcome has been considered inevitable, but several patients with proved rabies have recovered after being given extensive care for sustaining vital functions. Epidemiologic data further suggest that recognizable rabies follows only 30% to 50% of proven exposures. For example, in an unintentional study about one-half of untreated persons developed clinical rabies and died following severe mutilation by a rabid wolf, which must certainly have effected a viral infection in all those attacked.

Pathologically, rabies is an encephalitis with neuronal degeneration of the spinal cord and brain. **Negri bodies** within affected neurons are characteristic and the only pathognomonic microscopic finding. These cytoplasmic inclusions are sharply defined, spherical or oval, eosinophilic, Feulgen-negative bodies, 2 μm to 10 μm in diameter, containing a central mass of basophilic granules (Fig. 59–3). Several may be found in a single cell. Immunofluorescence studies have demonstrated specific viral Ags in the Negri body, and electron micrographs show (Fig. 59–4) that it consists of a large matrix of viral nucleocapsids (eosinophilic material) and budding virions (the basophilic granules). Virions are also seen in intercellular spaces and in synaptic junctions.

Inclusion bodies are most abundant in Ammon's horn of the hippocampus but may also be found in large numbers in many other sites in the brain and in the posterior

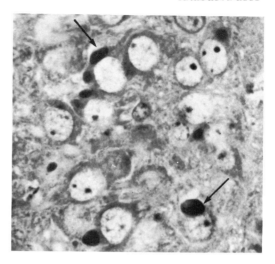

Figure 59–3. Negri bodies in the brain of a mouse infected with rabies virus. Numerous large dark cytoplasmic inclusion bodies (*arrows*) are present. The matrix of the inclusion body is stained intensely; the internal granules appear as light vacuoles. Stained by the dinitrofluorobenzene method for protein-bound groups. (Original magnification ×2000, reduced; courtesy of Dr. H. Koprowski, Wistar Institute, and Dr. R. Love, Jefferson Medical College of Thomas Jefferson University)

horn of the spinal cord. In the absence of identifiable Negri bodies a pathologic diagnosis of rabies cannot be made.

LABORATORY DIAGNOSIS

Definitive diagnosis of infections in humans, and in suspected animal vectors, depends on any one of the following findings: detection of viral Ags in specimens of brain, spinal cord, or skin by immunofluorescence (probably the method of choice, considering speed, accuracy, and cost); identification of Negri bodies in brain tissue; and isolation of virus from brain or saliva. Since the incubation period may be relatively short, the need for an immediate definitive diagnosis makes serologic techniques of little value early in the disease because circulating Abs appear slowly. Serum and cerebrospinal fluid (CSF) Abs eventually reach high levels, however, and if clinical doubts exist during later stages, Ab titrations (neutralization, CF, ELISA, immunofluorescence) can establish the diagnosis in an unimmunized patient.

Negri bodies are detected in impression smears prepared from the region of Ammon's horn. Seller's method, with a stain composed of a mixture of basic fuchsin and methylene blue, is commonly employed. Their distinctive stained appearance (**cherry red with deep blue granules**) differentiates them from other inclusion bodies, particularly those produced by distemper virus in dogs. Fluorescent Ab techniques provide reliable con-

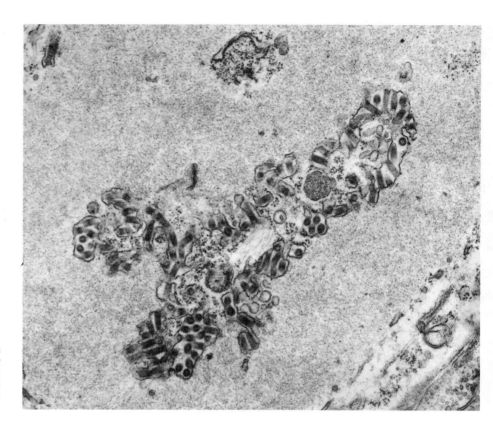

Figure 59–4. Electron micrograph of a Negri body containing several inner bodies composed of developing and mature virions. (Original magnification ×19,500; Matsumoto S: Adv Virus Res 16:257, 1970)

firmation of the specific nature of the inclusion body. The presence of Negri bodies is diagnostic, but **failure to detect them does not exclude rabies** and should be followed by attempts to isolate the virus.

Virus is preferably isolated by inoculating saliva, salivary gland tissue, or hippocampal brain tissue intracerebrally into infant mice. The mice develop paralysis after an incubation period of 6 to 12 days, depending upon the quantity of virus present. The illness is not pathognomonic of rabies, and the virus must be identified by immunologic techniques or by demonstrating Negri bodies in brain tissue from the inoculated animals.

EPIDEMIOLOGY

Although medical interest in rabies centers on infections of humans, epidemiologically this is a dead end to the infectious cycle, since humans contract rabies but do not normally transmit the disease. Dogs are generally the most dangerous source of infection for man, with cats next. In the United States, however, cattle are the most commonly infected domestic animals. The incidence of rabies in humans and dogs has decreased continuously in the United States since the institution of an effective immunization program and leash laws for dogs, and epizootic canine rabies (dog-to-dog transmission) has become rare. But rabies still remains enzootic in wild animals, and therefore persists enzootic in dogs. Thus control measures remain essential. For example, human rabies dropped from 33 cases in 1946 to only 1 to 3 cases per year since 1960, and rabies in animals (detected annually in almost every state) has continued a slight but uneven decline (Fig. 59–5).

Wild mammals serve as a large and uncontrollable reservoir of **sylvan rabies,** which is an increasing threat to people and domestic animals throughout the world. The most frequent wildlife sources in North America are skunks, bats, raccoons, and foxes, in order of their potential danger. Moreover, an epizootic in foxes has reintroduced rabies into Western Europe, which once was relatively free of the disease. Worldwide, rabies in animals, including dogs, steadily increases. A worldwide total of about 1000 fatal human cases is reported annually to the World Health Organization (WHO), and the actual number must be several times greater.

Vampire and insectivorous **bats** are important reservoirs and could be one of the most important links in the ecology of rabies: experimentally, the virus can remain latent in these animals for long periods; and virus has been detected in the nasal mucosa, brown fat, and sali-

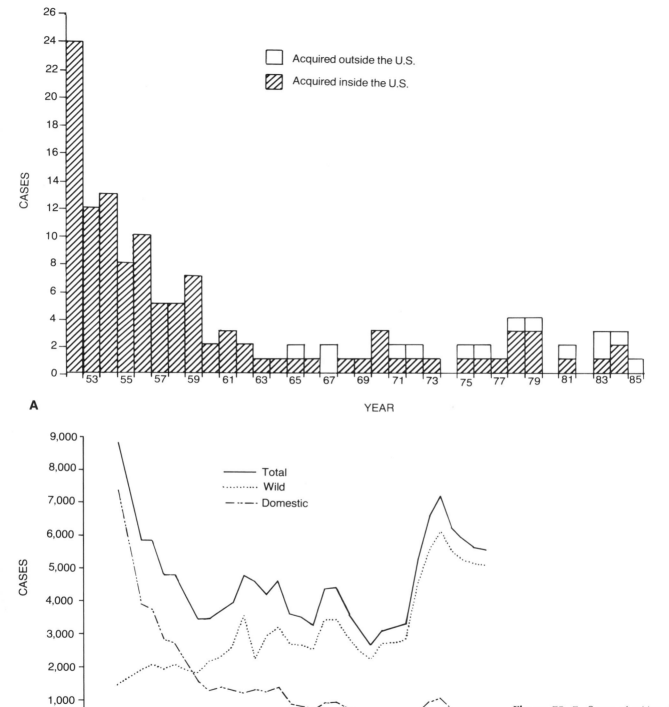

Figure 59–5. Cases of rabies detected in the United States (*A*) in humans, 1951–1985, and (*B*) in wild and domestic animals, 1953–1985. (Data from Centers for Disease Control: Rabies Surveillance, Annual Summary, 1986)

vary glands of apparently healthy bats, permitting transmission to man or other animals without a direct bite.

PREVENTION AND CONTROL

An effective program for rabies control must be directed toward preventing the disease in domestic animals, as well as in man. However, control measures directed against wildlife, which is now the main source of human cases, are impracticable. In fact, the primary rabies problem in the United States and many other countries is no longer human cases but the decision on whether to undertake immunization of humans, with its attendant dangers, after suspected exposure.

Vaccines

Pasteur's original vaccine was prepared by homogenization of partially dried* spinal cords from rabbits infected with modified (fixed) virus. Daily injections were given for 15 to 20 days with cords desiccated for progressively shorter periods. Frequent difficulties resulted, however, from the inexact method of viral inactivation, the lack of quantitative controls, and the complication of demyelinating allergic encephalitis.

Research led to a number of advancements in vaccines with development of **phenol-inactivated vaccines** (e.g., the Semple vaccine, which consisted of a 5% suspension of inactivated infected animal nervous tissue, usually rabbit), which were used in the United States until 1957. More recently, a vaccine consisting of rabies virus **propagated in embryonated duck eggs** and **inactivated by β-propiolactone** was prepared; this vaccine has a lower risk of allergic encephalomyelitis, but it is considerably less immunogenic than the nervous-tissue vaccines.

The low immunogenicity of the duck embryo vaccine and the continuing occurrence of allergic encephalomyelitis spurred efforts to replace it with a modified rabies virus propagated in cell cultures. Such a virus could be partially purified and concentrated in order to prepare highly immunogenic inactivated vaccines. An inactivated virus vaccine prepared from **virus grown in human diploid fibroblasts** is now being used in humans after being successfully tested in monkeys, rodents, dogs, and foxes. Five 1-ml injections (days 0, 3, 7, 14, and 28) elicit Ab levels superior to those following a full course of the phenolized or duck embryo vaccine. The new vaccine reliably produces high titers of Abs, and it has been shown to protect humans bitten by rabid animals; it also appears to be free of encephalitogenic factors. This vaccine is being widely used in Europe and was approved by the Food and Drug Administration in 1980 for use in the United States.

Alternatively, the **purified virion surface glycoprotein** produces neutralizing Abs and may eventually be used for immunization as a purified Ag or produced in a viral expression vector (e.g., vaccinia virus) in which a cDNA copy of the rabies virus G protein gene has been inserted.

Several types of **live attenuated virus vaccines** have been prepared from strains of virus adapted to chick embryos and to canine, feline, monkey, and murine cultured cells. Although considered unsuitable for humans, these vaccines have proved valuable for use in dogs, cats, and cattle. Live attenuated viruses after many passages in canine and feline cell lines have proved effective in dogs and cats, respectively. Inactivated viruses obtained from cultured cells of animal origin are also used in many parts of the world.

Public Health Measures

Primary control of rabies requires restriction of dogs and other domestic animals, as well as limitation of spread from wildlife to the greatest extent possible. The excellent WHO report (1973) recommends compulsory prophylactic immunization of dogs; registration of all dogs, destruction of stray dogs, and isolation and observation of suspect dogs; and attempted control of rabies in wildlife by trapping and other means. The danger of rabies is responsible for current severe restrictions on the transport of dogs across some national boundaries (e.g., Great Britain).

Prophylactic Treatment

Despite the low incidence of human rabies in many parts of the world, the question frequently arises as to the course to follow after an animal bite or scratch. The incidence of demyelinating encephalomyelitis, 1:500 to 1:8000, with a fatality rate of about 1:35,000 has resulted in the elimination of the use of a nervous tissue vaccine. The duck vaccine lacks nervous tissue and is presently recommended in parts of the world in which the cell culture vaccine is not affordable ($200 to $250 per course). Unfortunately, it has not completely eliminated the occurrence of encephalomyelitis (neuroparalytic reactions in about 1 in 25,000 immunized). Systemic reactions (fever, myalgia, etc.) also occur in about one third of the recipients after five to eight doses.

In deciding on the management of a person bitten by an animal suspected of being rabid a number of factors must be considered, but it is not possible to present guidelines for all situations. However, the following is recommended: if exposure to rabies appears definite or if a person was attacked by a nondomestic animal (e.g., skunk, raccoon), treatment with the human diploid cell

---

* Desiccation was employed to reduce infectivity, since Pasteur had noted earlier that dried cultures of chicken cholera bacteria lost their pathogenicity, but not their immunogenicity, for chickens.

vaccine and with human immunoglobulin* should be instituted promptly; vaccine should not be given to a person who has had only minimal contact with a questionable source of infection; and if the responsible animal can be captured alive, it should be observed for at least 10 days for the development of symptoms, since Negri bodies may not be detectable in the brain early in the disease, and the virus can appear in the saliva several days before the onset of symptoms. Suspected bats, however, should be killed and examined at once.

Prophylactic treatment is directed toward confining the virus to the site of entry. Local treatment of the wound consists of thorough cleansing (cauterization is no longer suggested) and infiltration with antiserum. Because of the long incubation period active immunization usually produces an adequate Ab level before the virus has reached its target organs in the CNS. **Combined inoculation with hyperimmune serum and vaccine** is the most effective regimen: serum Abs provide an immediate barrier to passage of virus, which lasts about 14 days, and meanwhile Abs are elicited by the vaccine. The use of hyperimmune serum, however, makes it mandatory to give the full course of vaccine injections in addition to two booster injections, since serum depresses Ab production. Antiserum alone should probably never be used.

Long-Term Prophylaxis

Veterinarians, laboratory workers, dog handlers, wildlife workers, and certain hobbyists (e.g., spelunkers) may be sufficiently exposed to rabies to justify prophylactic immunization. Members of this high-risk group should receive a course of vaccine followed in several months by a booster injection. Subsequent booster doses should be given every 2 to 3 years or following a suspected exposure.

Filoviridae—Marburg and Ebola Viruses

In 1967 in Marburg and Frankfurt, Germany, and in Yugoslavia, an acute febrile illness appeared in laboratory workers handling tissues and cell cultures from recently imported African green monkeys. The 31 cases that occurred, including cases of nosocomial infections in hospital personnel, resulted in seven fatalities. Similar episodes of hemorrhagic fever occurred in simultaneous epidemics in Zaire and the Sudan in 1976, resulting in over 500 cases and 430 deaths. From blood and organ suspensions from fatal cases, viruses, termed the Marburg and Ebola viruses, were isolated.

---

* A single injection of γ-globulin from hyperimmunized humans is recommended.

The viruses contain one molecule of linear, noninfectious RNA (presumably negative-stranded) and at least five major proteins. Virion infectivity is inactivated by lipid solvents. Electron microscopic examinations of negative-stained preparations reveal pleomorphic virions that are filamentous or somewhat rod-shaped, but with a variety of bizarre cylindrical and fishhook-like forms. The virions have a uniform diameter of 80 nm, but virions vary in length from 130 nm to as long as 14,000 nm, with the peak infectivity of Marburg virus at 790 nm and for Ebola virus at 970 nm. Like rhabdoviruses most particles have one rounded end; the other extremity is flat or occupied by a large bleb. Prominent cross-striations and an inner helical structure add to its similarities with rhabdoviruses. Marburg and Ebola viruses are not antigenically related. They are not considered to be rhabdoviruses, to which they are not immunologically related. They are now officially classified as members of a new family termed **Filoviridae.**

The illness produced by Marburg or Ebola virus is that of a severe hemorrhagic fever: a sudden onset with high fever, gastrointestinal upset, constitutional symptoms, and marked prostration, followed by uremia, rash, hemorrhages, and CNS involvement. Fatal cases show necrotic foci in many organs, including the brain; the liver and lymphatic tissues are most severely affected. Medical personnel should take great precautions when handling saliva, urine, or blood from patients.

Serologic surveys in Uganda and Kenya show that Marburg virus is present in monkeys and humans, producing **inapparent infections.** Hemorrhagic fever due to Marburg virus reappeared in South Africa in 1975 (three cases), in Kenya in 1980 (one case and the physicians who cared for the patient), and in Zimbabwe in 1982 (a single case).

Cases of Ebola virus hemorrhagic fever have also been observed since the initial episodes. In 1977 there was a single fatal case in Zaire, but in 1979 there were 34 cases resulting in 22 deaths at the same site in the Sudan as the original cases. The source of the virus has not been identified in any of the episodes, but serologic surveys show Abs in about 5% of the humans in the areas of Zaire where the cases occurred; 26% of the domesticated guinea pigs used as a food source also have Abs. However, there has not been any direct epidemiologic evidence that the guinea pigs were the source of virus infecting humans; rather it is suggested that both the humans and guinea pigs were infected from an unknown source.

The extensive use of primary monkey cell cultures, which may harbor Marburg or an Ebola virus, makes it imperative that physicians and virologists be aware of these simian viruses that on occasion cause disease in man.

Selected Reading

BOOKS AND REVIEW ARTICLES

Baer GM (ed): The Natural History of Rabies. New York, Academic Press, 1975.

Banerjee AK: Transcription and replication of rhabdoviruses. Microbiol Rev 51:66, 1987.

Dean DJ, Evans WM, McClure RC: Pathogenesis of rabies. Bull WHO 29:803, 1963.

Emerson SV: Rhabdoviruses. In Fields BM (ed): Virology, p. 1119. New York, Raven Press, 1985.

Expert Committee Report on Rabies: Sixth Report, WHO Technical Report Series No. 523, World Health Organization, Geneva, 1973.

Shope RE: Rabies. In Evans AS (ed): Viral Infections of Humans: Epidemiology and Control (2nd ed). New York, Plenum Medical Books, 1984.

Specific Articles

Appelbaum E, Greenberg M, Nelson J: Neurological complications following antirabies vaccination. JAMA 151:188, 1953.

Kabat EA, Wolf A, Bezer AE: The rapid production of acute disseminated encephalomyelitis in rhesus monkeys by injections of heterologous and homologous brain tissue with adjuvants. J Exp Med 85:117, 1947.

Kissling RE: Marburg virus. Ann NY Acad Sci 174:932, 1970.

Pasteur L: Methode pour prevenir la rage appres morsure. CR Acad Sci [D] (Paris) 101:765, 1885.

Tierkel ES, Sikes RK: Preexposure prophylaxis against rabies. JAMA 201:911, 1967.

U.S. Government: Recommendation of the Immunization Practices Advisory Committee (ACIP). Public Health Service, U.S. Department of Health and Human Services, Center for Disease Control, Atlanta, 1980

Harold S. Ginsberg

Togaviruses, Flaviviruses, Bunyaviruses, and Arenaviruses

The **arthropod-borne viruses (arboviruses)** multiply in both vertebrates and arthropods. In the cycle of transmission the former serve as reservoirs and the latter mostly as vectors, acquiring infection with a blood meal, but in some instances arthropods can also serve as reservoirs, maintaining the viruses by transovarian transmission. The virus is propagated in the arthropod's gut; and if it attains a high titer in its salivary glands, it can be transmitted when a fresh host is bitten. The viruses often cause disease in humans and other vertebrate hosts, but no ill effects are evident in the arthropods.

Most of the viruses described in this chapter were until recently classified, on epidemiologic grounds, as arboviruses. But increasing knowledge of their chemical and physical characteristics has revealed great heterogeneity among the arthropod-borne viruses. A number of these viruses with somewhat similar chemical and physical characteristics (Table 60–1), and of great medical significance to man, were previously grouped into a single family, **Togaviridae** (vernacular, **togaviruses;** *toga*, coat) comprising four genera. However, detailed characterization of a number of viruses clearly indicated that **flaviviruses,** previously classified as a genus in the family, were sufficiently distinct from other members that they should be designated as members of a separate family, **Flaviviridae** (see Table 60–1). Togaviridae comprise three genera: *Alphavirus* and *Rubivirus* (rubella virus; German measles virus), which are viruses infecting humans, and *Pestivirus*, a genus consisting of viruses that

TABLE 60–1. Comparison of Viral Characteristics of Members of the Families Togaviridae and Flaviviridae

Property	Togaviridae	Flaviviridae
Virion size	60–65 nm	45 nm
Envelope	+	+
Nucleocapsid	35–39 nm; cubic	30 nm; noncubic
RNA	Positive sense; 4.4×10^3 Kd (Sindbis virus, 11,274 nucleotides)	Positive sense; 3.8–4.2×10^3 Kd (yellow fever virus, 10,862 nucleotides)
Envelope proteins	gp E_1: 45–50 Kd gp E_2: 52–59 Kd gp E_3: 10 Kd*	gp E: 53 Kd M: 8.7 Kd
Nucleocapsid (C) protein	Nonglycosylated: 30 Kd	Nonglycosylated: 13.5 Kd
Nonstructural (NS) proteins	89, 82, and 16 Kd	About 7 species

* Present only in Semliki Forest virus

infect only animals (hog cholera virus and bovine diarrhea virus). The unique pathogenesis, epidemiology, and clinical problems of rubella, however, warrant its separate description (see Chap. 61). This chapter will also consider two other medically significant groups of viruses whose characteristics distinguish them sharply from togaviruses and flaviviruses. The family **Bunyaviridae** (previously called the Bunyamwera virus supergroup) includes the largest group of arthropod-borne viruses; it will undoubtedly be subdivided with further characterization. The family **Arenaviridae** (formerly included among the arboviruses) is also discussed in this chapter, although arthropod transmission is not observed.

The frequent association of an enveloped structure with transmission by arthropods may be more than coincidental: enveloped viruses lose infectivity readily (e.g., on drying, on exposure to bile) and must therefore be spread by either intimate contact (e.g., orthomyxoviruses, paramyxoviruses) or insect bite, whereas naked viruses, such as picornaviruses, tend to be more stable and can survive a more circuitous fecal–oral spread.

Arthropod-borne diseases, because of their vectors, depend strongly on climatic conditions: They are endemic in areas of tropical rain forests, and epidemics in temperate areas usually appear after heavy rainfall has caused an increase in the vector population. Apart from agents known to cause human diseases, a large number of additional arboviruses have been isolated from mosquitoes and ticks trapped in forests and from animals, especially monkeys, caged in the jungle as "sentinels" to permit insects to feed on them. These viruses are not known to cause prominent diseases of humans, but they are attracting a good deal of attention because of their threat as the world's expanding and increasingly mobile population impinges progressively on jungles. Over 350 arthropod-borne viruses have now been isolated (including some **rhabdoviruses** [discussed in Chap. 59] and **orbiviruses** [see Chap. 62]).

History

Yellow fever virus was the first arthropod-borne virus to be discovered, through the work of Major Walter Reed. He headed the United States Army Yellow Fever Commission, established in 1901 to try to overcome the disastrous effect of yellow fever on American troops in Cuba during the Spanish–American War. Reed and the members of the commission demonstrated transmission of this disease in bold experiments with human volunteers* built on the astute observations of Carlos Finlay, a Cuban physician, showing the association between yellow fever and mosquitoes. They also demonstrated the filterability of the agent. These studies established, for the first time, a virus as an agent of human disease, and an insect as the vector for a virus.

Their discoveries led to the eventual control of yellow fever, which for more than 200 years had intermittently been one of the world's major plagues and which, in fact, was a deciding factor in France's failure to complete the Panama Canal. This disease was by no means purely tropical; for example, an epidemic in the Mississippi Valley in 1878 caused 13,000 deaths, and substantial epidemics occurred in the nineteenth century as far north as Boston.

Immunologic Classification

The viruses isolated from arthropods may be divided into at least 34 distinct groups; many are still ungrouped. Table 60–2 lists the principal viruses that infect humans, the families and genera to which they have been assigned, and some of their clinical and epidemiologic characteristics.

* It should be noted that such experiments are not permitted today, and accordingly it is exceedingly difficult to establish whether a newly isolated virus is the cause of a disease or merely a fellow traveler with the undiscovered true etiologic agent.

TABLE 60–2. Classification and Description of Arthropod-Borne Viruses and Clinically Related Viruses of Humans

Family	Genus (Group)	Subgroup Complex	Viral Species	Vector	Clinical Disease(s) in Man	Geographic Distribution
Togaviridae*	*Alphavirus*	I	Eastern equine encephalitis (EEE)	Mosquito	Encephalitis	Eastern U.S., Canada, Brazil, Cuba, Panama, Philippines, Dominican Republic, Trinidad
		II	Venezuelan equine encephalitis (VEE)	Mosquito	Encephalitis	Brazil, Colombia, Ecuador, Trinidad, Venezuela, Mexico, Florida, Texas
		III	Western equine encephalitis (WEE)	Mosquito	Encephalitis	Western U.S., Canada, Mexico, Argentina, Brazil, British Guiana
			Sindbis	Mosquito	Subclinical or arthritis, rash	Egypt, India, South Africa, Australia, Sweden, Finland, Soviet Union
			(4 others) Semliki Forest	Mosquito	Fever or none	East Africa, West Africa
		IV	Chikungunya	Mosquito	Headache, fever, rash, joint and muscle pains	East Africa, South Africa, Southeast Asia
			Mayaro	Mosquito	Headache, fever, joint and muscle pains	Bolivia, Brazil, Colombia, Trinidad
		V–VII	Getah (each subgroup contains a single virus)	Mosquito	Subclinical or none known	
Flaviviridae	*Flavivirus*	I	St. Louis encephalitis	Mosquito	Encephalitis	U.S., Trinidad, Panama
			Japanese B encephalitis	Mosquito	Encephalitis	Japan, Guam, Eastern Asian mainland, Malaya, India
			Murray Valley encephalitis	Mosquito	Encephalitis	Australia, New Guinea
			Ilheus	Mosquito	Encephalitis	Brazil, Guatemala, Trinidad, Honduras
			West Nile	Mosquito	Headache, fever, myalgia, rash, lymphadenopathy	Egypt, Israel, India, Uganda, South Africa
			(8 other viruses)	Mosquito		
		II	Dengue (4 types)	Mosquito	Headache, fever, myalgia, prostration, rash (sometimes hemorrhagic)	Pacific islands, South and Southeast Asia, Northern Australia, New Guinea, Greece, Caribbean islands, Nigeria, Central and South America, Republic of China
		III	Yellow fever	Mosquito	Fever, prostration, hepatitis, nephritis	Central and South America, Africa, Trinidad
		IV	Tick-borne group (Russian spring–summer encephalitis group) 15 viruses	Tick	Encephalitis; meningo-encephalitis, hemorrhagic fever	Russian spring–summer encephalitis: U.S.S.R., Canada, U.S.; others: Japan, Siberia, Central Europe, Finland, India, Malaya, Great Britain (louping ill)

(Continued)

TABLE 60–2. (Continued)

Family	Genus (Group)	Subgroup Complex	Viral Species	Vector	Clinical Disease(s) in Man	Geographic Distribution
		V–VII	(11 viruses) Rio Bravo (bat salivary gland) (16 others)	Mosquito Unrecognized Unrecognized	Encephalitis	California, Texas
Bunyaviridae	*Bunyavirus* (Bunyamwera supergroup)	C group	Marituba and 10 others	Mosquito	Headache, fever	Brazil (Belem), Panama, Trinidad, Florida
		Bunyamwera group	Bunyamwera and 17 others	Mosquito	Headache, fever, myalgia, fever only, or none	Uganda, South Africa, India, Malaya, Colombia, Brazil, Trinidad, West Africa, Finland, U.S.
		California group	California encephalitis and 10 others	Mosquito	Encephalitis or none	U.S., Trinidad, Brazil, Canada, Czechoslovakia, Mozambique
		10 other subgroups (3 ungrouped members of genus)	46 viruses			
	Phlebovirus fever group	Phlebotomus	37 viruses	Phlebotomus	Sandfly fever; headache, fever, myalgia	Italy, Egypt
	Nairovirus	Crimean-Congo hemorrhagic fever group (4 others)	19 viruses	Tick	Fever, headache, gastrointestinal and renal symptoms, hemorrhages	Africa, Europe, Asia
	Unkuvirus	Uukuniemi group	Uukuniemi 7 others 8 unassigned viruses	Mosquito Tick		Finland
Arenaviridae	*Arenavirus*	Tacaribe	Tacaribe, Junin, Tamiami, Machupo, Pichinde, and 3 others Lymphocytic choriomeningitis Lassa		Headache, fever, myalgia, hemorrhagic signs	South and Central America
Ungrouped			Silverwater	Tick	None known	Canada
			Rift Valley fever	Mosquito	Headache, fever, myalgia, joint pains, hemorrhagic signs, rash	Africa
			Crimean hemorrhagic fever	Tick	Headache, fever, myalgia, hemorrhagic signs	Southern U.S.S.R.
			36 others	Mosquito	None known	
Others			48 viruses (14 groups of 2–8 viruses)	Mosquito Tick	None known in most cases	

* Rubivirus (rubella, or German measles, virus), also a togavirus, is discussed in Chapter 61.

Alphaviruses and flaviviruses are classified on the basis of hemagglutination-inhibition, neutralization, and complement fixation (CF) assays: members of a subgroup cross-react best with each other, but not with members of other families (Table 60–3). Bunyaviruses are classified into 10 subgroups by CF, which shows greater cross-reactivity among members of the family than does hemagglutination-inhibition, but many bunyaviruses are still unclassified. Species within a subgroup are identified by neutralization with standardized antisera; there is less cross-reaction with this test.

The immunologic cross-reactions seen within the ma-

TABLE 60–3. Results of Hemagglutination-Inhibition Titrations With Alphaviruses and Flavivirus

Immune Serum	Viruses (Antigen)								
	Genus	EEE	VEE	WEE	Sindbis	Chikungunya	Mayaro	Semliki Forest	St. Louis Encephalitis
EEE	Alphavirus	**10,240**	80	160	20	40	20	20	<10
VEE	Alphavirus	160	**640**	80	20	80	80	40	<10
WEE	Alphavirus	80	160	**10,240**	160	40	80	40	<10
Sindbis	Alphavirus	80	10	2560	**1280**	10	40	40	<10
Chikungunya	Alphavirus	20	20	40	<10	**1280***	80	80	<10
Mayaro	Alphavirus	40	40	320	40	640	**1280***	**1280***	<10
Semliki Forest	Alphavirus	40	80	160	10	40	320	**2560**	<10
St. Louis encephalitis	Flavivirus	<10	<10	<10	<10	<10	<10	<10	**2560**

* Note cross-reactions among alphaviruses, but not between the alphaviruses and one flavivirus tested. Also note the cross-reactivity among some viruses forming two subgroups; EEE, VEE, WEE, Sindbis; and Chikungunya, Mayaro, Semliki Forest. When neutralization assays are employed instead of hemagglutination inhibition, EEE appears to be antigenically unique, and VEE shows less cross-reactivity with WEE and Sindbis viruses. At least seven subgroups of flaviviruses have also been identified.

(Modified from Casals J: Ann NY Acad Sci 19:219, 1957)

jor groups of arthropod-borne viruses suggest close phylogenetic relations. However, chemical analysis of the glycoprotein Ags and cross-hybridization of viral RNAs for characterizing genetic relatedness are just beginning to bring sharper taxonomic criteria into this field.

The immunologic cross-reactions among arthropod-borne viruses are of practical as well as theoretic interest. Thus, with the viruses that show cross-reactivity in hemagglutination-inhibition or CF tests, cross-reacting neutralizing Abs may be evident after repeated immunization, though not after primary immunization. Moreover, infection by one virus may confer a demonstrable increase in resistance to subsequent challenge with another. Epidemiologic evidence suggests that such cross-protection may be important in nature; vaccines are being developed to take advantage of these findings.

Togaviruses—Alphaviruses

The major alphaviruses pathogenic for man (see Table 62–2) produce severe encephalitis, particularly western and eastern equine encephalitis viruses (**WEE and EEE,** respectively).

PROPERTIES

Because of their wide distribution and the serious nature of illnesses caused by alphaviruses, particularly encephalitis, these viruses have been studied extensively. Most of them, however, are difficult and dangerous to work with, and the physical and chemical characterization is incomplete except for a few less pathogenic viruses that

can be readily propagated in cell cultures. Sindbis and Semliki Forest viruses, which are not usually pathogenic for humans, have been most extensively used for biochemical studies.

Morphology

The virions are roughly spherical; in thin sections they have an outer membrane (lipoprotein envelope) and a core of electron-dense material (ribonucleoprotein), and they tend to pack in crystalline arrays (Figs. 60–1 and 60–2). Negative staining shows an outer membrane covered with fine projections and a capsid, which appears to consist of 32 capsomers arranged in icosahedral symmetry (Fig. 60–3).

Physical and Chemical Characteristics

The purified virions that have been analyzed each contain one **positive-strand, infectious RNA molecule** (see Table 60–1), which amounts to 4% to 8% of the particle weight. The RNA, like most viral and eukaryotic cell mRNAs, has a polyadenylic acid-rich sequence at its 3′ terminus and a cap at its 5′ end. As with other enveloped viruses, the precise lipid composition reflects that of the host cells' plasma membranes.

Alphaviruses contain a basic nucleocapsid protein and either two or three envelope glycoproteins (Sindbis and Venezuelan equine encephalitis [VEE] viruses, two; Semliki Forest virus, three) that form the surface spikes that span the lipid bilayers to establish association with the capsid (Fig. 60–4).

The E_1 surface glycoprotein spike is the **hemagglutinin.** Virions agglutinate RBCs from newly hatched chicks or adult geese. Maximum hemagglutination (or **hemad-**

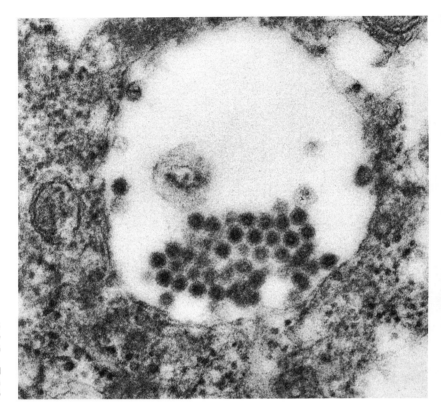

Figure 60–1. Thin section of a mature WEE virus within the cytoplasmic vacuole of an infected cell. The viral particle has a dark central nucleocapsid 30 nm in diameter and a peripheral membrane about 2 nm thick. Note the cubical shape of the virions aggregated in a crystalline-like array. (Original magnification ×100,000; Morgan C et al: J Exp Med 113:219, 1961)

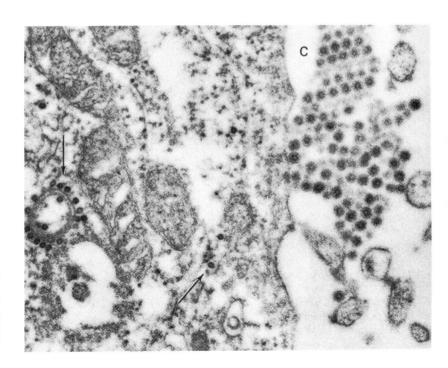

Figure 60–2. Extracellular crystal (*C*) composed of WEE virus particles. Dense precursor particles are scattered in the cytoplasm and are present on opposite sides of two concentric lamellae near the left border (*arrows*). Mature virions are seen only within vacuoles (see Fig. 60–1) or outside of cells. (Original magnification ×86,000; Morgan C et al: J Exp Med 113:219, 1961)

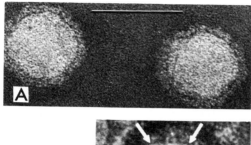

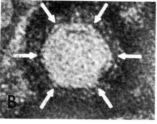

Figure 60–3. Morphology of a negatively stained alphavirus. (A) The spherical particles have a dense nucleocapsid and an envelope covered with fine projections. (Original magnification ×240,000) (B) In an occasional virion a nucleocapsid with a clear polygonal outline and the semblance of an ordered capsomeric structure is seen. Arrows point to the axes of fivefold symmetry of the icosahedron. (Original magnification ×360,000; Simpson RW, Houser RE: Virology 34:358, 1968)

sorption on infected cell cultures) is effected within narrow ranges of pH and temperature.* The E_1 protein also has **hemolytic activity.** The virions become firmly bound to the red blood cell (RBC) surface and do not elute spontaneously. The surface glycoproteins also effect attachment to host cells.

Cell lipids inhibit hemagglutination; hence, detection of hemagglutinating activity in lysates of cells (particularly brain or spinal cord) requires preliminary extraction with lipid solvents. Moreover, treatment of virions with such solvents or with nonionic detergents inactivates infectivity and liberates the glycoproteins that induce production of species-specific neutralizing Abs.

The infectivity of most togaviruses decreases rapidly at 35°C to 37°C *in vitro*; infectivity and hemagglutinating activity have maximum stability at about pH 8.5.

Immunologic Characteristics

Alphaviruses also fall into immunologic subgroups whose members show cross-reactivity of the E_1 glycoprotein. The alphavirus cross-reactivity among subgroups, particularly subgroups I to IV (see Table 60–2), resides in the antigenic properties of the nucleocapsid protein. Monoclonal Abs to specific epitopes on the E_1 and E_2 glycoproteins neutralize virus and inhibit hemagglutination (see Table 62–3). Neutralizing Abs appear about 7 days after onset of disease and exist for many years, probably for life, which correlates with the persistence of solid immunity. Hemagglutination-inhibiting Abs appear at the same time and are easier to assay but less specific; CF Abs also rise early but are not detectable after 12 to 14 months.

Most togaviruses, in contrast to yellow fever and dengue flaviviruses (see below), maintain their immunoge-

* Maximum agglutination is attained at pH 6.4 and 37°C.

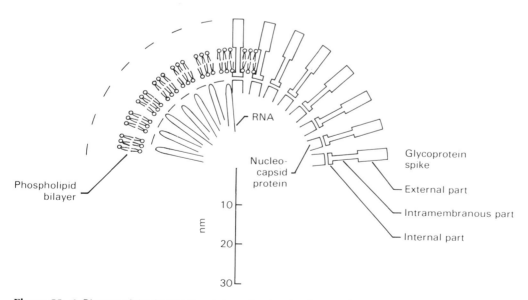

Figure 60–4. Diagram of the Semliki Forest virus, showing the glycoprotein spikes spanning the envelope lipid bilayer and associating with the protein of the nucleocapsid. (Modified from Simons K et al: In Capaldi RA [ed]: Membrane Techniques. New York, Marcel Dekker, 1977)

nicity and immunologic reactivity following destruction of infectivity by formalin, heat, or β-propriolactone.

Cross-reactive hemagglutination-inhibiting Abs develop after infection or artificial immunization; the degree of cross-reactivity increases up to about 1 month after the immunogenic stimulus. For example, an animal infected with EEE virus develops homologous Abs about 7 days after infection, and 3 to 6 weeks later develops relatively high titers of hemagglutination-inhibiting Abs against all alphaviruses. If the same animal is subsequently inoculated with EEE virus or another alphavirus, a rapid increase of hemagglutination-inhibiting Abs against all alphaviruses follows (a secondary group response); the Ab response is considerably greater than that following a single inoculation of either virus alone. Upon subsequent infection, or immunization, with a different but antigenically related virus the highest immunologic response is often to the first virus to which the subject was exposed—another example of the principle of original antigenic sin (see Influenza Viruses, Chap. 56). Proposed immunization procedures take advantage of this broad-ended secondary response.

Host Range

Alphaviruses multiply in a wide range of vertebrates and arthropods. Most of the viruses can also be propagated on a variety of primary and continuous cell cultures, including cultures of mosquito cells. They produce cytopathic effects (except in mosquito cells), and infected cells can also be detected by hemadsorption. A sensitive and reproducible plaque assay can be carried out with susceptible vertebrate cells. Because of their sensitivity and convenience, cell cultures are now the host of choice for experimental and diagnostic work.

The major vertebrate hosts for alphaviruses in nature are birds, rodents, and monkeys. Horses are readily and often fatally infected by the equine encephalitis viruses (whose initial isolation from horses, during epizootics, gave rise to their names). Monkeys are also useful hosts for studying the pathogenesis of infection.

Before its replacement by cell cultures the newborn mouse was the laboratory host of choice; it is highly susceptible to infection by all members of the family. Viruses multiply to high titer in brain, producing extensive pathogenic changes. Some of the viruses also multiply in muscle, lymphoid, or vascular endothelium cells. Resistance increases with age: most mice by age 3 to 6 months are quite resistant to infection by peripheral routes.

Wild birds and domestic fowl, particularly when newly hatched, can be infected artificially or by the bite of an infected mosquito. Embryonated chicken eggs are sensitive, convenient hosts for many studies. Mosquitoes (*Culex, Anopheles,* and *Mansonia*) are the arthropod hosts in nature for alphaviruses.

Multiplication

Togaviruses are positive-strand RNA viruses, which multiply in the cytoplasm. They show considerable variation in the lengths of their multiplication cycles, although the temporal differences are relatively minor within each subgroup. Thus, the duration of the cycle is relatively short for alphaviruses (e.g., WEE virus; Fig. 60–5); in contrast, flaviviruses multiply more slowly (e.g., the maximum yield of type 2 dengue virus is attained 20 to 30 hours after infection).

Virus is adsorbed rapidly by susceptible cells in culture and enters the cell in endocytic vesicles. Eclipse is evident within 1 hour. The uncoated parental genome serves as mRNA for the nonstructural proteins required for replicating RNA. The replication forms of viral RNA superficially resemble those in picornavirus-infected cells (see Chap. 48). However, in **alphavirus** infections two **RNA species** of positive polarity are synthesized: **49S virion RNA and 26S mRNA,** each of which is transcribed from a full-length copy (negative polarity) of the virion RNA (Fig. 60–6). The 26S mRNA is identical with the 3′ OH terminal one-third of the 49S virion RNA and codes for the structural proteins, whereas the 49S RNA serves as messenger for translation of the nonstructural proteins (see Fig. 60–6). Extensive replication of 26S RNAs amplifies the messenger for structural proteins when they are most needed. As with picornaviruses, the 26S RNA acts as a **monocistronic mRNA** to produce a poly-

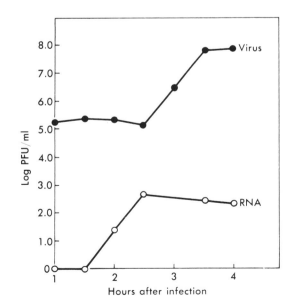

Figure 60–5. Temporal relation of the biosynthesis of infectious (viral) RNA and infectious WEE virus particles. *PFU,* plaque-forming units. (Wecker E, Richter A: Cold Spring Harbor Symp Quant Biol 27:137, 1962)

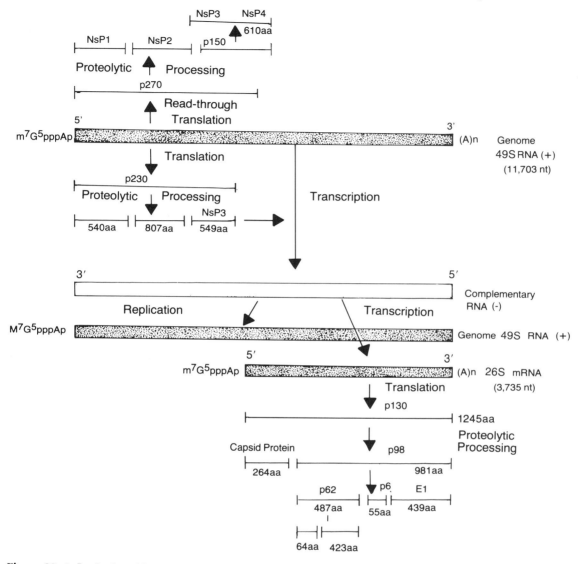

Figure 60–6. Replication of Sindbis virus, a model of alphavirus genome replication, transcription, and translation. The 49S genome (+) RNA is translated into polyprotein that is processed into the nonstructural proteins. The 26S mRNA, which is transcribed from the full-length negative-polarity RNA, is translated into polyprotein that is processed into viral structural proteins. (Recall that Sindbis virus envelope has two glycoproteins, E_1 and E_2, whereas Semliki Forest virus has an additional protein, E_3). *p*, indicates precursor protein; e.g., p 130 signifies a precursor protein of 130,000 daltons. (Modified from Strauss JH, Strauss EG, Hahn CS, Rice CM: In Rowlands DJ, Mayo M, Mahy BWJ [eds]: The Molecular Biology of the Positive Strand RNA Viruses, p 75. New York, Academic Press, 1987)

protein, which appears to have autoprotease activity, to process first the capsid protein and subsequently the other structural, virion proteins (see Fig. 60–6).

The envelope proteins are glycosylated and transported through the endoplasmic reticulum and Golgi apparatus to the plasma membrane, where they become transmembrane proteins. Nucleocapsids recognize the sites of insertion, and final **assembly of the virion** is accomplished by **budding at the cell surface.** The viral particle becomes infectious when it acquires its final coat (i.e., the envelope). Thus maturation of alphavirus particles (e.g., WEE and VEE) and release from the host cell are almost simultaneous (within 1 minute) (Figs. 60–7 and 60–8). Cells infected with Sindbis virus produce virions at the extraordinary rate of 10^4 per cell per hour for approximately 12 hours.

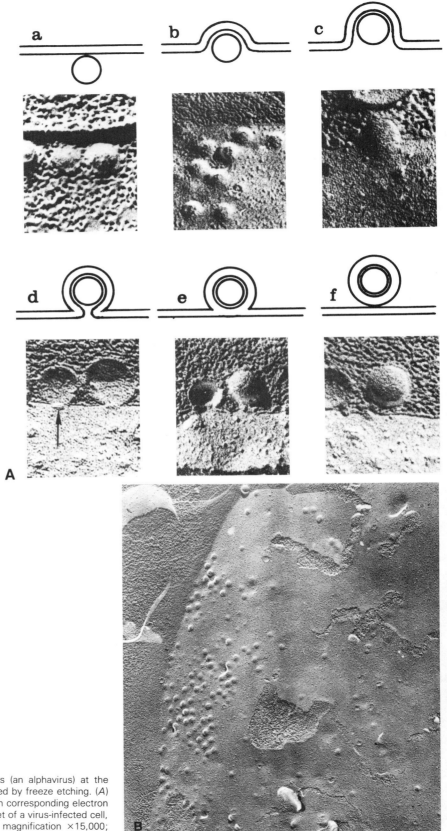

Figure 60–7. Morphogenesis of Sindbis virus (an alphavirus) at the plasma membrane of an infected cell as observed by freeze etching. (*A*) Schematic representation of viral maturation with corresponding electron micrographs. (*B*) Outer surface of the inner leaflet of a virus-infected cell, showing budding and mature virions. (Original magnification ×15,000; Brown DT et al: J Virol 10:524, 1972)

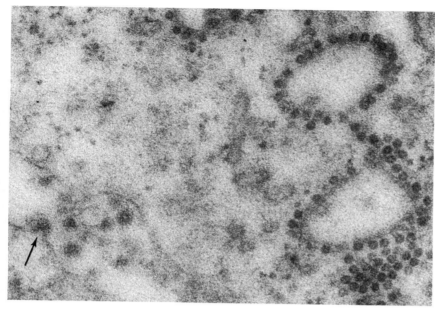

Figure 60–8. Characteristic formation of particles of WEE virus at the membrane of vacuoles in the cytoplasm of an infected cell. Extracellular virus is visible at lower left. One particle seems to be in the process of emerging from an adjacent cell membrane (*arrow*). (Original magnification ×96,000; Morgan C et al: J Exp Med 113:219, 1961)

PATHOGENESIS

When an infected mosquito bites a prospective host it injects virus from its salivary glands into the bloodstream or lymph of its victim. Successful infection depends on the presence of sufficient virus in the saliva of the mosquito and a paucity of neutralizing Abs in the host. Details of the pathogenesis of infections in humans are largely inferred from experimental studies in animals.

All alphavirus infections are similar in the initial stages. Virus is removed from the blood by, and multiplies in, skin and reticuloendothelial cells (mainly in the spleen and lymph nodes). Viremia follows, initiating the systemic phase of the clinical disease. Finally, virus invades various tissues—the central nervous system in the encephalitides; the skin, bone marrow, and blood vessels in the hemorrhagic fevers. The mechanisms by which encephalitis viruses invade the central nervous system are not yet understood: entrance may be effected directly through the blood–brain barrier, at nerve endings, or, less likely, by transmission along nerves.

Despite similarities in certain pathogenic characteristics, considerable diversity exists in the diseases produced by alphaviruses. Two types of **clinical syndromes** are seen (see Table 60–2): (1) EEE, WEE, and VEE, and Semliki Forest viruses produce a **systemic phase** of disease (chills, fever, aches) resulting from the viremia, and an **encephalitic phase** may follow after a variable time; (2) infections by Sindbis, Chikungunya, Mayaro, and Getah viruses are confined to the systemic phase, and the symptoms are primarily fever, arthritis, and rash. Chikungunya virus also rarely produces hemorrhagic fever.

EEE virus may produce severe illnesses, with high mortality, and often with severe residual neurologic damage among survivors. WEE virus, in contrast, usually causes a less severe disease (most often appearing in infants and children), and most patients recover completely. WEE virus also frequently initiates abortive disease (fever and headache) or clinically inapparent infections. On rare occasions, WEE virus produces massive cerebral necrosis in newborns owing to transplacental transmission of the virus from infected mothers (like rubella virus) (see Chap. 61). VEE virus primarily infects horses; transmission to man usually results in a mild disease with variable systemic symptoms and only rarely causes severe encephalitis (e.g., in the summer of 1971 a devastating epizootic killed thousands of horses in Texas and Mexico, with mild disease occurring in over 100 humans as well).

Pathologically, severe alphavirus infections show gross involvement of viscera as well as of brain and spinal cord. The brain lesions caused by EEE virus are scattered in the white and gray matter, particularly in the brain stem and basal ganglia; the spinal cord shows milder changes. WEE infections, on the other hand, chiefly affect the brain, producing lymphocytic infiltration of the meninges and lesions of the parenchyma, predominantly in the gray matter. Lesions generally consist of necrosis of neurons, glial infiltration, and perivascular cuffing. Inflammatory reactions in walls of small blood vessels, and thrombi, may occur.

LABORATORY DIAGNOSIS

Togavirus infection is diagnosed by viral isolation or by serologic procedures. To isolate a virus from a patient, specimens are inoculated intraperitoneally and intracerebrally into newborn and suckling mice, or into susceptible cell cultures. The viremic phase of an alphavirus encephalitis is usually completed by the time a patient seeks medical assistance. Hence, isolation of virus from the blood and spinal fluid of patients during acute disease is unusual. With those viruses that produce fever, arthritis, and rash (e.g., Chikungunya virus), however, the responsible virus may be isolated from blood during the initial stages of illness. To obtain a virus from a fatal case of encephalitis, emulsions of brain and spinal cord are used.

A newly isolated virus is identified by hemagglutination-inhibition titrations with standard antisera, to determine its immunologic group, and by neutralization titrations, to establish its species.

Early serologic diagnosis can be made by detecting IgM Abs with the enzyme-linked immunosorbent assay (ELISA). Final serologic diagnosis is made with serum drawn during the acute illness and during convalescence. Antigens are obtained from infected cell cultures. Definitive diagnosis can be accomplished by showing a four-fold or greater increase in Ab titer by means of ELISA, hemagglutination-inhibition, CF, or neutralization assays. Neutralization titrations are the most specific, but *in vitro* assays are preferred because of their simplicity, speed, and economy. However, within an immunologic subgroup, rises in heterologous Ab may occur and obscure a precise etiologic diagnosis when *in vitro* assays are used. CF titrations are not adequate for epidemiologic surveys, moreover, because CF Abs do not persist as long as those measured by hemagglutination-inhibition or neutralization titrations.

EPIDEMIOLOGY

For most alphaviruses man is merely an incidental host. The mosquito is the common arthropod vector. The cycle of infection has been elucidated best for the equine encephalitis viruses, particularly **WEE** and **EEE,** and can be simply diagrammed as follows:

Despite the names of the diseases, the horse, like man, appears to be a dead end in the chain of infections (Fig. 60–9). In fact, horses are not significant reservoirs in nature for WEE and EEE viruses, probably because viremia does not usually reach sufficiently high levels to infect mosquitoes with regularity. It is noteworthy, however, that equine infections usually appear 2 to 3 weeks prior to the occurrence of disease in humans. Birds are the principal natural hosts of WEE and EEE viruses (see Fig. 60–9). Birds likewise appear to be the most likely hosts in

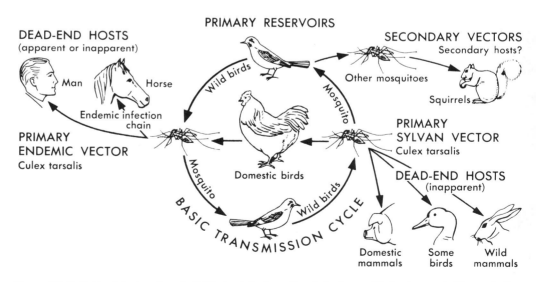

Figure 60–9. Epidemiologic pattern for WEE virus infections. The chains for rural St. Louis encephalitis are similar, except that horses are inapparent, rather than apparent, hosts. EEE infections also have a similar summer infection chain, but a few significant differences exist. (1) The identity of the vector infecting man is unknown. (2) Domestic birds do not appear to be a significant link in the chain. (3) It has a bird-to-bird secondary cycle in pheasants, whose role is unclear. (Hess AD, Holden P: Ann NY Acad Sci 70:294, 1958)

which viruses can persist in nonepidemic periods and during the seasons in which transmission by mosquitoes is not prominent (**overwintering**). Wild snakes and frogs, as well as some rodents, are probably secondary reservoirs for WEE. Hibernating mosquitoes are also a possible reservoir, for infectious virus can persist in them for at least 4 months.

WEE is generally found in the United States west of the Mississippi River, but increasingly the virus is also being isolated along the eastern seaboard. In contrast, EEE is confined to the eastern part of the United States and Canada. Both viruses are also found in the Caribbean, Central America, and South America.

The primary vector of WEE virus is the culicine mosquito, *Culex tarsalis* in central and western United States and *Culiseta melanura* in the northeast. The primary mosquito vector of EEE is not certain, but *Culiseta melanura* is susceptible to experimental infection and has been found infected in nature. Other mosquitoes also appear to be implicated in the epidemiologic cycles involving mammals.

VEE virus is distributed in the Everglades region of Florida (where it is endemic in rodents), in the southwestern United States, and in Central America, northern South America, the Amazon Valley, and southern Mexico. The natural cycle of infections involves mammals and mosquitoes; the natural reservoir is in small mammals rather than birds. **Aedes mansonia** and **psorophora** are the major vectors in epidemic spread of the virus. Horses are invariably infected when human disease occurs and appear to be the major source of virus for mosquitoes. An example, the severe outbreak of 1971, has been mentioned under Pathogenesis.

Chikungunya (African for "that which bends up") virus infections may be a notable exception to this general epidemiologic pattern: man is the only known vetebrate host, with *Anopheles* and *Aedes* mosquitoes the vectors.

PREVENTION AND CONTROL

Control measures are aimed at preventing transmission of the virus by eradicating, or at least reducing, the population of arthropod vectors and at increasing the host resistance by artificial immunization. The former procedure is the only effective means to prevent spread of the viruses, since humans and horses are dead ends, and this method has been relatively successful. Effective vaccines have been prepared against some of the viruses. Despite great effort, however, the viruses continue to exist and cases occur in the United States. In 1984 there were four human cases of EEE (two in Florida and two in Massachusetts) and two cases of WEE in South Dakota. Moreover, 103 EEE equine infections and three WEE equine infections were detected in the same year.

Formalinized chick embryo **EEE and WEE vaccines** produce effective Ab responses in horses. But these vaccines have not been utilized in humans except for protection of laboratory workers; their effectiveness in man has not been established. A live, attenuated **VEE virus vaccine** was successfully used to immunize horses in Texas in 1971; this procedure, along with strict quarantine of the equine population, halted the epidemic. An attenuated VEE virus vaccine has also been used experimentally in humans, but its effectiveness has not yet been demonstrated. The attenuated virus was also formalin-inactivated and proved to be safe and highly immunogenic; its clinical effectiveness, however, is unproved.

Flaviviruses

The unique structural, morphologic, and replicative characteristics of **flaviviruses** led to their being separated into a family distinct from **Togaviridae** (see Table 60–1), although the viruses in both families have similar epidemiologic features (i.e., arthropod borne) and some produce similar clinical syndromes (see Table 60–2). Flaviviruses, however, do cause a greater variety of illnesses (encephalitis, hemorrhagic diseases, and severe systemic illnesses).

PROPERTIES

Despite the serious nature of many of the illnesses, the broad distribution of the viruses, and the recurrence of epidemics (particularly dengue fever), relatively few of the 60 members of the Flaviviridae family have been studied in detail. However, the marked similarity of those viruses investigated, although they are in different subgroups (see Table 60–2), permit generalities to be made.

Morphology

Flaviviruses are spherical and have a thin, unit membrane envelope, surface projections, and a dense core (i.e., nucleocapsid) whose symmetry is unclear. Thus, although their morphology is similar to that of togaviruses, flaviviruses are significantly smaller (see Table 60–1) and do not have distinct icosahedral symmetry of their nucleocapsids. In thin sections, however, flaviviruses frequently form crystalline-type arrays like togaviruses (see Figs. 60–1 and 60–2).

Physical and Chemical Characteristics

The flavivirus genome, like that of the togaviruses, is a single, linear **positive-strand RNA,** but there the similarity ends. The flavivirus genomes are significantly smaller (see Table 60–1); its 5' terminus is capped, but it does not have a 3' poly(A) tail. Moreover, the genome organization of flaviviruses is different in that the virion structural proteins are encoded in the 5'-terminal quarter, and the remainder encodes the nonstructural protein genes (Fig. 60–10). The virion contains only three structural proteins: a single envelope glycoprotein projection, the **E protein;** the envelope **membrane (M) protein;** and the **nucleocapsid C protein** (see Table 60–1). The number of nonstructural viral proteins is uncertain in flavivirus-infected cells; as many as 12 have been described, and at least three have been identified in yellow fever virus–infected cells (see Fig. 60–10). The nonstructural proteins detected appear to function in viral RNA replication, and at least one of the viral proteins must have protease activity.

The envelope E glycoprotein surface spikes are the viral **hemagglutinin** and effect **hemadsorption** to infected cells. Flaviviruses and togaviruses hemagglutinate similar species of RBCs, but the conditions of pH and temperature required for maximum hemagglutination (or hemadsorption) differ between the viruses of the two families.

Immunologic Characteristics

The **E glycoprotein,** which is on the surface of virions and infected cells, is the **major antigenic structure** of the virion possessing distinct but separate neutralization and hemagglutination-inhibition epitopes. The E protein has antigenic determinants that demonstrate virus-, subgroup-, and flavivirus-specific reactivities. The M and C proteins are also antigenic but do not play a critical role in either pathogenesis or viral classification. Of practical concern for prolonged immunity and artificial immunization is the finding that some flaviviruses isolated at different times (e.g., Murray Valley encephalitis viruses from 1956 and 1969) may be antigenically distinguishable.

Like togaviruses, most flaviviruses remain immunogenic after formalin inactivation. With yellow fever and dengue viruses, however, this treatment markedly impairs their capacity to elicit neutralizing Abs. Accordingly, preparation of successful vaccines required the development of attenuated viruses (see Prevention and Control, below). The existence of only a single known immunologic type of yellow fever virus accounts in part for the effectiveness of the vaccine.

The general immune response to flavivirus infection is similar to that of togaviruses (described previously). Indeed, the considerable cross-reactivity between members of the same subgroup, or even different subgroups, and the marked evidence of "original antigenic sin" response is critical for immunity and immunization procedures.

Host Range

Flaviviruses multiply in the same wide range of vertebrates, arthropods, and cell cultures as togaviruses. In addition to their prolonged replication in mosquitoes and ticks, some flaviviruses can survive in ticks for many months by **transovarian transmission** (e.g., Russian spring–summer encephalitis virus), and through periods of molting and metamorphosis, without apparent injury to the host; this survival during periods of poor transmission furnishes a possible mechanism of overwintering.

Multiplication

The initial stages of infection (i.e., attachment, entry, and uncoating) for flaviviruses are similar to those of togaviruses. Flaviviruses also replicate in the cytoplasm. Although replication of the viral genome is similar in both viral families, the production of the viral proteins is markedly different. Thus, the entire flavivirus genome is translated into a single polyprotein (like picornaviruses; see Chap. 55), which is processed by proteolytic cleavage into the virion structural proteins, from the sequences

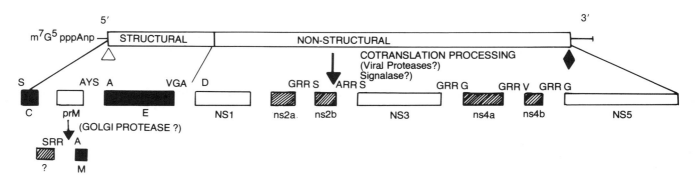

Figure 60–10. Structure of the 17D strain of yellow fever virus. The regions encoding the structural and nonstructural proteins are shown as an open box; the untranslated sequences at the 5' and 3' termini are shown as single lines. The cotranslational processing of the polyprotein encoded in the single open-reading frame is shown (open triangle indicates the initiation codon, AUG; and the solid diamond the termination codon, UGA); structural proteins, identified nonstructural proteins, and hypothesized (based on potential protease cleavage sites) nonstructural proteins are shown as solid, open, and hatched boxes, respectively. (Modified from Strauss JH, Strauss EG, Hahn CS, Rice CM: In Rowlands DJ, Mayo M, Mahy BWJ [eds]: The Molecular Biology of the Positive Strand RNA Viruses, p 75. New York, Academic Press, 1987)

encoded in the 5′ quarter of the genomic plus-strand RNA, and the nonstructural proteins from the remainder (see Fig. 60–10). The processing of the polyprotein occurs rapidly and efficiently, essentially as the RNA is translated (**cotranslational processing**). The protease, which accomplishes the processing, is probably virally encoded but has not yet been identified.

Assembly of the virions characteristically occurs by budding into cytoplasmic vacuoles. Large numbers of virions accumulate in prominent vacuoles from which they are released through narrow canaliculi connecting the endoplasmic reticulum with the plasma membrane.

PATHOGENESIS

The initial insult—the vector bite and injection of virus, the spread of virus from site of local viral multiplication to lymph nodes, and spread of virus through lymphatics and blood into a variety of tissues, including the target organs—is not dissimilar from that of togaviruses (see preceding discussion).

Flaviviruses produce three types of clinical syndromes (see Table 60–2): **central nervous system disease,** manifested mainly as encephalitis (St. Louis, Japanese B, Murray Valley, Ilheus, and Russian spring–summer encephalitis viruses), with pathologic features similar to those produced by togaviruses; **severe systemic disease** involving viscera such as liver and kidneys (yellow fever virus); and **milder systemic disease** characterized by severe muscle pains and a rash that may be hemorrhagic (West Nile, dengue, and some of the tick-borne viruses). Despite the anxieties induced by the appearance of a flavivirus, most of these viruses, except for yellow fever and dengue, predominantly produce subclinical or mild infections that can be recognized only by laboratory diagnostic procedures.

The severity and extent of the **encephalitides** due to flaviviruses vary with the etiologic agent. For example, **St. Louis encephalitis** virus produces mild lesions, low mortality, and few residua compared with **Japanese B encephalitis** virus, which has a mortality rate of about 8% and causes neurologic residua in more than 30% and persistent mental disturbances in about 10% of clinically diagnosed infections.

The pathogenesis of **yellow fever** differs from that of other flavivirus infections: after primary multiplication in lymph nodes extensive secondary viral multiplication, with cell destruction, occurs in many viscera, including the liver, spleen, kidneys, bone marrow, and lymph nodes. The organs involved and the severity of the lesions vary with the infecting strain. Yellow fever may be a grave disease, with a mortality of about 10%.

The pathology of yellow fever is characterized by degenerative lesions of the liver, kidney, and heart, accompanied by hemorrhages and bile staining of tissues. The most distinctive lesions occur in the liver, where a pathognomonic midzonal, hyaline necrosis develops, with preservation of the basic liver architecture and without inflammatory reaction.

In uncomplicated **dengue fever** deaths are rare; hence, the pathologic lesions are not well known. Biopsies of the characteristic skin lesions reveal endothelial swelling, perivascular edema, and mononuclear infiltration in and about small vessels. In some epidemics, particularly in tropical Asia, hemorrhagic fever and shock syndrome may be prominent. **Hemorrhagic dengue** (characterized by high fever, hemorrhagic manifestations, shock, and a mortality as high as 15%) is most often seen in children sequentially infected with different immunologic types of virus during a limited period (about 2 years), suggesting a hypersensitivity or immune-complex type of disease. The pathologic findings and the marked depression of complement components (particularly C3) support this concept, suggesting that Ag–Ab complexes activate the complement system, with subsequent release of vasoactive peptides. However, primary cases have been reported, and their pathogenic mechanism must be explained.

LABORATORY DIAGNOSIS

Definitive diagnosis depends on viral isolation and serologic assays. Isolation of virus from blood is usually only successful in yellow fever, dengue, and tick-borne encephalitis; viral isolation is usually made during the first 2 to 4 days of the disease. For other flavivirus infections, viral isolation is most effectively made from brain tissue. Specific identification of the virus is best made using neutralization titrations, although hemagglutination-inhibition titrations with standard sera can also be employed to identify the immunologic group. Because of the marked cross-reactivity among flaviviruses, and particularly within the same subgroup, considerable care is required to identify the virus.

Serologic diagnosis can be made as described for togaviruses.

EPIDEMIOLOGY

The epidemiologic patterns of flavivirus infections are more varied than those of togaviruses. These patterns will be summarized for a few of the most important diseases.

St. Louis encephalitis* is the major flavivirus infection in the United States; despite its name, it occurs throughout the country. The most severe epidemic to occur since reporting was initiated took place in 1975,

---

* The first epidemic due to this virus was recognized in St. Louis, Mo, in 1933.

affecting the central part of the country; 1815 cases with 140 deaths were recorded. The epidemiologic pattern is similar to that described for WEE virus infections. Wild birds are the major reservoir of the virus, and *Culex taralis* and the *C. pipiens* complex are the most common mosquito vectors. Man is an accidental, dead-end host (see Fig. 60–9).

The epidemiology of **Murray Valley encephalitis, Japanese B encephalitis,** and **West Nile fever** is basically the same as that of St. Louis encephalitis except for the species of mosquito vectors and the avian reservoirs. Serologic surveys indicate that for each case of clinical disease produced by these viruses several hundred **inapparent infections** are also induced.

Yellow fever presents another complex ecologic situation.* Two distinct epidemiologic types exist, **urban** and **jungle yellow fever.** Each has a different cycle, but they may interact. In its simplest form the epidemiologic pattern of urban yellow fever simply involves man and the domestic mosquito, *Aedes aegypti:*

Viremia in man begins 1 or 2 days before and persists for 2 to 4 days after the onset of clinical illness. Viremia is greatest during this period and mosquitoes taking a blood meal may be infected. A 10- to 12-day period of viral multiplication in the cells lining the mosquito's intestinal tract is then required (the **extrinsic incubation period**) until sufficient virus accumulates in the salivary glands to permit transmission to man.

Jungle yellow fever is transmitted by various jungle mosquitoes, primarily to monkeys. Man becomes an accidental host when he enters the animals' domain.

Haemagogus spegazzinii
Monkey   Man
Aedes simpsoni

Infection of man may initiate a cycle of urban yellow fever.

Dengue fever resembles yellow fever epidemiologically. There is an urban cycle (man ⇌ *Aedes aegypti* mosquitoes), and probably a jungle cycle with the monkey as the mammalian host. *Aedes albopictus,* which has recently been introduced into the United States, can also serve as a vector. This disease is still prevalent in the Caribbean islands; more distant subtropical areas are also affected (see Table 60–2). A severe type 2 epidemic occurred in Cuba in 1981.

The **tick-borne** complex of viral infections introduces several unique features into the epidemiology of flavi-

* The monograph *Yellow Fever*, GK Strode (ed), New York, McGraw-Hill, 1951, reviews in exciting detail many of the important facts personally discovered by its authors.

virus infections: ticks may serve as reservoirs by **transovarian transmission** of virus; in addition to being transmitted by ticks (*Ixodes*), some of these viruses (e.g., **Russian spring–summer encephalitis virus**) may also be transmitted to man from the goat by milk instead of by an arthropod.

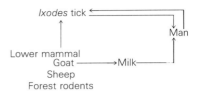

Omsk hemorrhagic fever virus is transmitted by *Dermacentor marginatus,* probably from muskrats, and **Powassan virus** (isolated in Canada and the United States) is transmitted by *Ixodes* ticks from small mammals, probably squirrels.

PREVENTION AND CONTROL

Yellow fever has played a special role in the development of concepts and methods for the control of insect-borne diseases. Reducing the population of the vector, *A. aegypti,* proved effective soon after Reed and his colleagues demonstrated the causative agent and the vector requirement. The use of modern insecticides has facilitated this control measure. It should be noted, however, that *A. aegypti* mosquitoes and other possible vectors are still present in many parts of the world, including the southeastern United States. Moreover, mosquito control measures cannot eliminate jungle yellow fever.

As noted earlier, the loss of immunogenicity of yellow fever virus upon inactivation made it necessary to seek a vaccine containing infectious virus. Theiler and his co-workers attenuated a mouse-adapted yellow fever virus by serial passage in tissue cultures, first of embryonic mouse tissues, then of embryonic chicks, and finally of embryonic chicks without brain or cord. For the most widely used vaccine, the attenuated virus (17D yellow fever virus) is propagated in embryonated chicken eggs. The 17D strain produces few reactions and has an excellent safety record. The duration of protection afforded by the 17D vaccine is not known, but Abs have been detected at least 6 years after immunization.

Eradication of yellow fever in the United States, and substantial reduction of its incidence in South America and other parts of the world, initially suggested that this disease would be eliminated throughout the world. However, the reservoir of jungle yellow fever was later discovered, and yellow fever has increased in incidence in parts of Central and South America and has been creeping northward. With this nidus present the danger of epidemics is real.

A **Japanese B encephalitis vaccine,** prepared by for-

TABLE 60–4. Properties of Bunyaviridae and Arenaviridae

Property	Bunyaviridae	Arenaviridae
Morphology	Spherical, enveloped, projecting spikes Diameter 80–110 nm Helical nucleocapsid	Spherical, pleomorphic; enveloped Diameter 110–130 nm average Beaded, circular nucleocapsid
Nucleic acid*	RNA, single-stranded; 3 (L, M, and S) segments: 2.1–3.1, 1.1–2.3, and 0.28–0.51 × 10^6 daltons (S segment of Phlebotomus viruses: 0.7 – 0.8 × 10^6 daltons)	RNA, single-stranded; two (L and S) segments, virus-specific 2–3.2 and 1.1 to 1.6 × 10^6 daltons (three species host RNA also in virions)
Virion proteins*	Two glycoproteins: (1) 75–120 and (2) 30–60 × 10^3 daltons; two nonglycosylated proteins: nucleocapsid (N) protein 19–60 × 10^3 daltons and RNA polymerase (L) protein 145–200 × 10^3 daltons	Two glycoproteins: (1) 42–72X and (2) 34–40 × 10^3 daltons; two nonglycosylated proteins: (N) 60–70 × and (L) 200 × 10^3 daltons
Effect of lipid solvents	Inactivate	Inactivate
Stability	Unstable below pH 7.0	Unstable
Hemagglutination	1-day chick or goose RBCs	None
Best animal host	Suckling mice; chickens	Various rodents

* Molecular weights vary with different viral species.

malin inactivation of virus propagated in chick embryos, has been employed with apparent success in children in Japan. However, a similar vaccine was ineffective in U.S. Army personnel stationed in Japan and the Far East. An effective **dengue vaccine** containing infectious, attenuated virus has not yet been developed, although efforts continue, particularly with temperature-sensitive mutants. In addition, the development of live virus expression vectors (e.g., a vaccinia virus vector) containing the cDNA encoding the dengue virus E glycoprotein is under active investigation and offers considerable promise. However, until we explicitly understand the role of viral Ag–Ab complexes in the pathogenesis of dengue hemorrhagic fever and shock syndrome such vaccines should be used with caution.

The marked antigenic cross-reactivity of flaviviruses may prove useful for immunization purposes. For example, in experimental animal studies, immunization with an infectious attenuated virus, such as yellow fever virus, and subsequent injection of one or more inactivated or live attenuated heterologous viruses, resulted in a broad immunologic response that protected against a variety of flaviviruses.

Bunyaviruses

On the basis of morphologic, chemical, and structural features, more than 200 of the so-called arboviruses have been grouped into a single family, Bunyaviridae (see Table 60–2), the largest family of RNA-containing viruses. Based on immunologic and genetic differences, this large family has been further divided into four genera: *Bunyavirus, Phlebovirus, Nairovirus,* and *Unkuvirus.* Phle-

boviruses, however, appear to have a unique genome and therefore a different replication strategy (discussed later), and these viruses will probably be classified in a separate family. The large *Bunyavirus** genus, which consists of 124 viruses, includes the California encephalitis, Bunyamwera, and C subgroups, which are the major ones affecting humans (additional members of the genus, the Bwamba, Capim, Guama, Koongol, Patois, Simbu, and Tete virus groups, have similar properties). Sandfly (or *Phlebotomus*) fever virus and Rift Valley fever virus (genus *Phlebovirus*), Crimean-Congo hemorrhagic fever virus (genus *Nairovirus*), and Hantaan virus (unassigned) are other members of the family associated with human epidemics.

PROPERTIES

The **structures** of all viruses studied show marked similarities to each other (Table 60–4) and clear distinctions from togaviruses and flaviviruses. The virions are spherical particles, clearly larger than togaviruses. Their envelopes are unit membranes covered with surface projections, which are indistinct for most bunyaviruses but arranged in a regular lattice on the phleboviruses and Uukuniemi virus. These spikes consist of two glycoproteins, G_1 and G_2, present in equimolar amounts; they have hemagglutinating activity and induce hemagglutination-inhibiting and neutralizing Abs. The envelope does not possess a matrix protein. The nucleocapsid, when released from the virion, appears to have helical symmetry and is present in three distinct segments

* Named for Bunyamwera, Uganda, where the type species virus was isolated.

(each consists of a common N protein and a unique RNA molecule), often seen in circular forms. However, the three species of RNA are linear. The circles are formed by pairing of complementary segments at the free 3' and 5' ends of the RNA molecules: in fact, free RNAs of the Uukuniemi virus are circular under nondenaturing conditions and can reform circles on annealing after denaturation. The viral RNAs are **negative strands,** their 3' ends are not polyadenylated, their 5' ends are not capped, and they are not infectious. The virions contain an RNA-dependent RNA polymerase, probably the transcriptase.

Immunologic Characteristics

Bunyaviridae, like togaviruses and flaviviruses, can be studied by hemagglutination-inhibition, CF, and neutralization titrations. In contrast to these viruses, however, the cross-reactivity of bunyaviruses, except for phleboviruses, is maximal in CF rather than in hemagglutination-inhibition titrations. For example, CF titrations with one or two standard sera can identify an agent as a group C virus, and specific viral identification can be accomplished by hemagglutination-inhibition and neutralization titrations. Thus CF titrations show all bunyaviruses to be antigenically related through the cross-reacting nucleocapsid protein, but neutralization and hemagglutination-inhibition titrations can subdivide members of the *Bunyavirus* genus into 13 subgroups (serogroups). Similarly, the viruses of each of the three other genera are immunologically related through the nucleocapsid N protein, and members of each genus are divided into serogroups (see Table 60–2). Viruses of one genus are immunologically unrelated to members of the other genera.

Host Range

Suckling and newly weaned mice are the laboratory animals of choice for isolation of the viruses by intracerebral inoculation. These viruses also multiply and produce cytopathic changes and plaques in a variety of cultured cells; among the most useful are continuous human (e.g., HeLa) lines and baby hamster kidney (**BHK21**) cell lines.

Multiplication

Only a few members of the Bunyaviridae family have been studied in detail, but the data obtained indicate that these viruses replicate like orthomyxoviruses (see Chap. 56), except for the phleboviruses (see later). After attachment to susceptible cells via the virion surface glycoproteins, the particles are engulfed and uncoated, and the nucleocapsids enter the cytoplasm. RNAs are immediately transcribed, and, like influenza viruses, the mRNAs attain their 5' caps by cannibalizing host cellular mRNAs (see Chaps. 48 and 56). These viral mRNAs are translated, after which the genome RNA is transcribed to make a full-length positive-sense strand (Fig. 60–11).

The multiplication of a single phlebovirus, the Punta Toro virus, has been extensively investigated, but these studies imply that members of this genus have unusual genomes. The nucleotide sequence of the S RNA of Punta Toro virus indicates that the 5' position of the RNA directly encodes the NS protein and is therefore positive-sense RNA. However, the 3' region, from which the nucleocapsid N protein is derived, is negative-sense RNA,

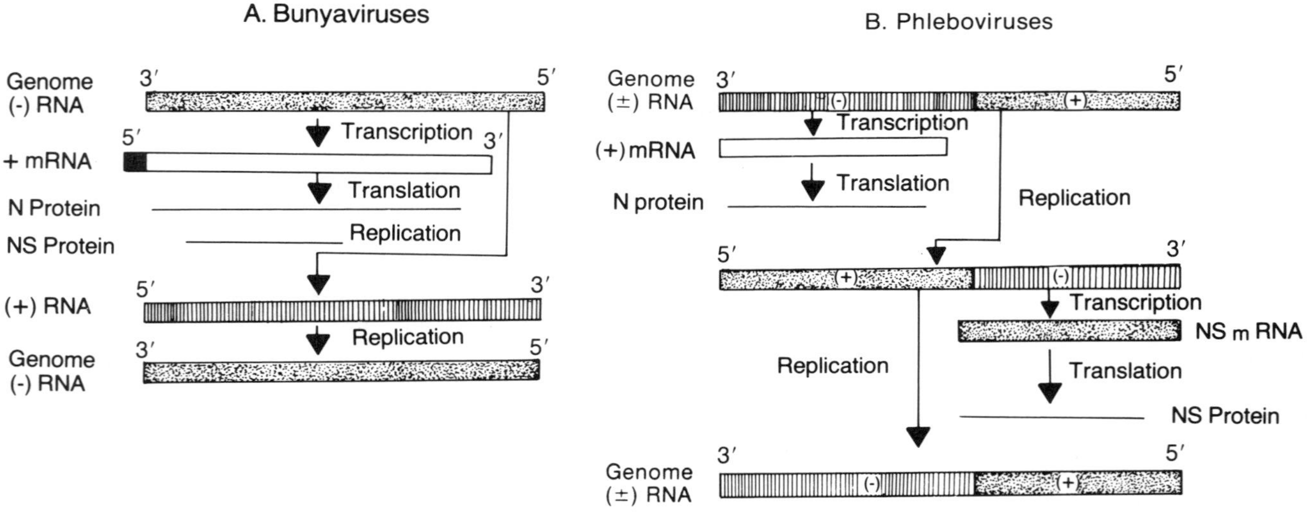

Figure 60–11. Replication strategy of the bunyaviruses (*A*) and phleboviruses (*B*). The models, which include transcription, translation, and replication of the segment of genome RNA, depend on the nucleotide sequences of the small (*S*) segment of the snowball hare bunyavirus (*A*) and the ambisense genome of the Punta Toro phlebovirus (*B*). (Bishop DHL et al: Nucleic Acids Res 10:3703, 1982; Ihara T et al: Virology 136:293, 1984)

like other bunyaviruses. Thus, the proteins are not encoded in overlapping reading frames as in other bunyaviruses, and therefore it must undergo unique replication as shown in Figure 60–11. The phleboviruses are said to have "ambisense" polarity. Whereas the Punta Toro virus S RNA is about twice the molecular weight of the bunyavirus S RNA, the L and M RNA segments are similar in size to those of other bunyaviruses, and it is assumed that their replication follows similar reactions.

Viruses of the Bunyaviridae family have an additional unusual feature: morphogenesis of the virions is accomplished by budding into the vacuoles of the Golgi apparatus (Fig. 60–12) rather than at the plasma membrane or into undifferentiated cytoplasmic vacuoles, like flaviviruses.

PATHOGENESIS AND EPIDEMIOLOGY

Bunyamwera virus, which is the prototype virus from which the family and genus names were derived, was first isolated from a mixed pool of *Aedes* mosquitoes trapped in Uganda. Eighteen immunologically related but distinct viruses have been recognized, including strains isolated in the United States (Florida, Virginia, Colorado, Illinois, and New Mexico). Disease attributed to the Bunyamwera group (see Table 60–2) is rare and usually mild. Like togaviruses and flaviviruses, the virus enters the body when the infected vector bites and partakes of a blood meal. Viremia occurs, and the resulting disease depends on the organs susceptible to a particular virus.

The **California serogroup encephalitis viruses,** which were initially isolated from mosquitoes in the San Joaquin Valley of California, are widely distributed. They produce prominent clinical illnesses, manifested by fever, headache, and mild or severe central nervous system involvement, particularly in children. Recovery is usually complete, although mild residua and even rare deaths have been recorded. The LaCrosse encephalitis virus is the most common arthropod-borne virus to cause human encephalitis in the United States. The snowshoe hare and Jamestown Canyon viruses are other members of this serogroup to cause encephalitis in North America. Clinical disease has been reported from 13 states in all regions of the United States as well as in the countries noted in Table 60–2. The natural reservoir of these viruses is unknown, but the agents have been found in the blood of rabbits, squirrels, and field mice in titers adequate to infect mosquitoes. Although California encephalitis viruses have been isolated from several *Aedes* and *Culicine* species, *Aedes triseriatus* appears to be the principal vector. Moreover, transovarian passage of the virus in this vector may furnish an important mechanism for its maintenance during the winter.

Among a large number of viruses isolated from experi-

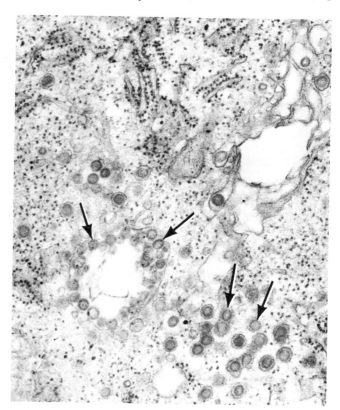

Figure 60–12. Development of Bunyamwera virus in the cytoplasm of a neuron in a mouse brain. Viral particles are present in the cytoplasm and budding into Golgi vacuoles and cisternae of the endoplasmic reticulum. Virions have a mean diameter of 90 nm to 100 nm, a nucleocapsid core of 60 nm to 70 nm in diameter, and an envelope of 15 nm to 20 nm thick. (Original magnification ×39,500; Murphy FA et al: J Virol 2:1315, 1968)

mental (sentinel) monkeys caged in the forested Belem area of Brazil seven different viruses were recognized as being immunologically related to each other but distinct from togaviruses and flaviviruses. Several other immunologically distinguishable but related viruses have been isolated in the Florida Everglades and Central America. These viruses, previously called **group C arboviruses,** but now recognized as a division of the *Bunyavirus* genus, produce mild disease in man, consisting of headache and fever; recovery is complete.

The natural reservoir of C viruses appears to be in monkeys and other forest mammals (e.g., opossums, rats, sloths). The specific mosquito vector has not been established, but culicine and sabethine mosquitoes are likely candidates.

Among the many other viruses of this family some species of the phlebovirus genus, the phlebotomus fever (or sandfly fever) viruses, and Rift Valley fever virus are known to produce disease in man. Most of the viruses were isolated from insects (particularly mosquitoes and ticks) and animals captured in the wild.

The **Congo-Crimean hemorrhagic fever virus,** a **Nairovirus** transmitted by *omma* ticks, initially was considered to cause serious illness that was often fatal. As it became more widely recognized, however, it was realized that the virus did not always cause severe disease but that it caused widespread zoonosis wherever its vector existed (from South Africa through the Middle East and Asia into China).

Arenaviruses

On the basis of morphologic, immunologic, and clinical characterizations the seemingly disparate Lassa virus, lymphocytic choriomeningitis (LCM) virus, and Tacaribe group of viruses, previously considered to be arboviruses, are now classified into a single family called **Arenaviridae** (vernacular, **arenaviruses;** L. *arena*, sand), a term derived from the unique electron microscopic appearance of the virions (Fig. 60–13).

Complement fixation reveals the immunologic relatedness of the arenaviruses, owing to the marked cross-reactivity of the nucleocapsid (N) protein; the Tacaribe viruses, however, are more closely related to each other than to Lassa or LCM virus. Neutralization titrations show the immunologic specificity of each arenavirus.

Although arenaviruses were initially grouped together because of their immunologic relations and similar morphology, they also have comparable epidemiologic, ecologic, and pathogenic patterns. Tacaribe group, LCM, and Lassa viruses do not require arthropods for spread; the natural hosts of all arenaviruses appear to be rodents, in which they often produce chronic infections.

PROPERTIES

The viral particles (see Table 60–4) are spherical or pleomorphic. As viewed in thin sections the virions consist of a dense, well-defined envelope with prominent, closely spaced projections and an unstructured interior containing a varying number of electron-dense granules, probably host ribosomes (see Fig. 60–13), which cause the unique pebbly appearance from which the viruses gained their name. Negative-contrast electron micrographs also show spherical or pleomorphic virions with an envelope having pronounced and regularly spaced club-shaped surface projections.

Within the virion are several species of ribonucleoproteins, which are both virus specific and host cell derived (see Table 60–4). The host cell ribonucleoproteins have characteristics of ribosomes, with 28S and 18S RNAs. The two viral RNA segments are single-stranded RNAs of "ambisense" polarity with the same characteristics as the S RNA of phleboviruses (see Fig. 60–11). The virions contain two glycoproteins associated with the envelope and two nonglycosylated proteins, the N protein (a component of the nucleocapsid) and a large (L) minor protein

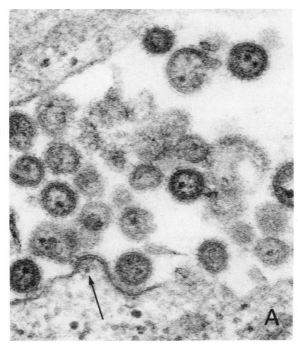
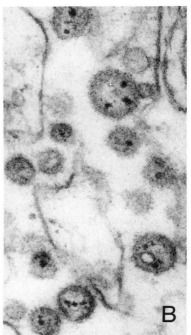

Figure 60–13. (*A*) Machupo virus (a Tacaribe group virus) particle budding (*arrow*) from the plasma membrane of an infected cell; the thickened membrane of the budding particle is prominent compared with the neighboring membrane. Many extracellular mature virions (mean diameter of 11 nm to 13 nm) are present: Their prominent surface projections and internal, ribosome-like particles are readily seen. (Original magnification ×114,000). (*B*) Lymphocytic choriomeningitis (LCM) virus particles in an infected culture of mouse macrophages. The morphology is strikingly similar to that of the Tacaribe group virus in *A*. (Original magnification ×82,000; Murphy FA et al: J Virol 4:535, 1969)

(probably the transcriptase; see Table 60–4). Epitopes on the G_1 glycoprotein are the major antigenic determinants for neutralizing Abs. The S viral RNA segment encodes the N protein in its negative-sense 3' half and the glycoprotein (GPC)—the precursor to the G_1 and G_2 glycoproteins—in its positive-sense 5' half. The intergenic region is GC rich, forms a hairpin structure, and appears to serve as a termination signal for both transcriptions.

The biochemical events of **viral multiplication** follow the patterns described for phleboviruses. The appearance of ribosomal RNAs in the virions is inhibited by dactinomycin (actinomycin D) (0.15 μg/ml), but neither synthesis of viral RNAs nor production of infectious virus is affected. Electron microscopic studies show that virions are formed by budding, chiefly from plasma membrane (see Fig. 60–13). At the sites of budding, the host cell membrane becomes thickened, more clearly bilamellar, and covered with projections. The ribosome-like particles are present within the budding particles before separation from the infected cell. (Their significance is unknown.)

PATHOGENESIS AND EPIDEMIOLOGY

Arenaviruses commonly produce chronic carrier states in their natural hosts, possibly because viral multiplication in animals, as in cell culture, is not associated with extensive cell damage; rather, cell death results from cytotoxic T cell attack as in LCM infections in mice (see Chap. 51). The virus may be isolated from the animals' urine, as well as from their blood and internal organs.

The **Tacaribe group of viruses** (Tacaribe, Machupo, Junin, Tamiami, and five others) has been isolated principally from bats and cricetid rodents in the Western Hemisphere. Junin and Machupo viruses have been frequently isolated from cases of Argentinian and Bolivian hemorrhagic fevers, respectively, and appear to be the etiologic agents of these severe illnesses. Thus, the Tacaribe group viruses, similar to LCM virus of mice, appear to be **spread to humans in excretions of the naturally infected rodents.** Except for Lassa virus, there is no evidence of viral spread from patient to patient.

LCM virus rarely infects humans. When it does, it is usually under conditions in which the indigenously infected mouse population is very dense or from contact with infected hamsters. The disease is generally mild and is manifested most often as a lymphocytic form of meningitis or an influenzalike illness, but occasionally as a meningoencephalitis. Leukopenia and thrombocytopenia frequently develop. Very rarely, LCM virus produces severe and even fatal illnesses associated with hemorrhagic manifestations.

Lassa virus, first isolated in 1969 from an American missionary working in Nigeria, has attracted considerable interest because it is **highly contagious** and produces a **serious febrile illness.** The disease is characterized by severe generalized myalgia, marked malaise, and sore throat accompanied by patchy or ulcerative pharyngeal lesions. Fatal cases also develop myocarditis, pneumonia with pleural effusion, encephalopathy, and hemorrhagic lesions. The virus persists in the blood for 1 to 2 weeks, and during this period it can be isolated from urine, pleural fluid, and throat washings. It is more stable than togaviruses and flaviviruses in body fluids, which probably permits its person-to-person contagion and accounts for the hazard it presents for laboratory isolation or study. Arthropods collected in Nigeria, the only known locale of natural infections, have not yielded virus, and insect cell cultures are insusceptible to viral propagation. The only cycle of Lassa virus transmission outside humans has been detected in the wild rodent *Mastomys natalensis*, which suggests that rodent control may limit viral transmission to man. Lassa virus produces an infection in mice similar to LCM virus infection, and chronic latent infections can be established.

General Remarks

Only a few of the known viruses that multiply in arthropods and vertebrates have been discussed in this chapter. The characteristics of only a few of the viruses have been studied in detail (e.g., Sindbis, yellow fever, dengue viruses), and the properties of most of these agents, and even their clinical and epidemiologic behavior, are known in only a fragmentary fashion. Some, such as the phlebotomus (sandfly) fever virus, transmitted by the bite of the female sandfly *Phlebotomus papatasii*, assumed importance to the U.S. Armed Forces during World War II, when the disease (which is not serious) appeared in military personnel in the Mediterranean area.

Many of the arthropod-borne viruses, including some that can cause serious disease, also produce a very much larger number of inapparent infections in endemic areas; hence, the native human population carries a high level of immunity but the insect population is still highly infectious because of the viral reservoir in lower animals. Such diseases could increase dramatically in quantitative significance when ecologic alterations cause development of a dense population of infected athropods next to a nonimmune human population or when a large, immunologically virginal human population (e.g., military personnel) moves into an endemic area.

Because of their close antigenic relation to human pathogens, those arthropod-borne viruses that have not been associated with human disease cannot be ignored by medical investigators, however esoteric they may seem. Several such viruses have been isolated in the United States or Canada—for example, the Rio Bravo virus (flavivirus group, U.S.), California encephalitis and

Trivittatus viruses (California group, U.S.), and Silverwater virus (Bunyaviridae ungrouped, Canada). Since a change in either the host reservoir, the vector, or the genetics of the viral population might permit these agents to infect man, they remain a potential hazard.

The comforting realization that as many as 200 arthropod-borne viruses of seemingly diverse immunologic groupings, or even without obvious relatives, may be segregated into one family, the Bunyaviridae, on the basis of physical and chemical characteristics, indicates that order is appearing in what previously seemed unmanageable. Thus, the ecologic–epidemiologic classification of otherwise disparate viruses as "arboviruses" is being replaced by classification on a broader base into several new families. In addition, a few viruses initially isolated from arthropods have been placed into well-established families (reoviruses, rhabdoviruses, and picornaviruses), and the arenaviruses have been shown not to be associated with arthropod vectors.

Selected Reading

BOOKS AND REVIEW ARTICLES

Bishop HL, Shope RE: Bunyviidae. Compr Virol 14:1, 1979

Lehmann-Grube F: Portraits of viruses: Arenaviruses. Intervirology 22:121, 1984

Matthews REF: Classification and nomenclature of viruses (Third Report of the ICTV). Togaviridae. Intervirology 17:1, 1982

Schlesinger RW (ed): The Togaviruses. New York, Academic Press, 1980

Schlesinger RW: Dengue Viruses. Vienna, Springer-Verlag, 1977

Schlesinger S, Schlesinger M (eds): The Togaviruses and Flaviviruses. New York, Plenum Press, 1986

Strauss JH, Strauss EG, Hahn CS, Rice CM: The genomes of alphaviruses and flaviviruses. In Rowlands DJ, Mayo M, Mahy BWJ (eds): The Molecular Biology of the Positive Strand RNA Viruses, p 75. New York, Academic Press, 1987

SPECIFIC ARTICLES

Hewlett MJ, Pettersson RE, Baltimore D: Circular forms of Unkuniemi virion RNA: An electron microscopic study. J Virol 21:1085, 1977

Ihara T, Akashi H, Bishop DHL: Novel coding strategy (ambisense genomic RNA) revealed by sequence analysis of Punta Toro phlebovirus S RNA. Virology 136:293, 1984

Iroegbu CV, Pringle CR: Genetic interactions among viruses of the Bunyamwera complex. J Virol 37:383, 1981

Pardigon N, Vialot P, Girard M, Buloy M: Panhandles and hairpin structures at the termini of Germiston virus (Bunyavirus). Virology 122:191, 1982

Reed W, Carroll J, Agramonte A, Lazear JW: The etiology of yellow fever: A preliminary note. Philadelphia Med J 6:790, 1900

Rice CM, Strauss JH: Nucleotide sequence of yellow fever virus: Implications for flavivirus gene expression and evolution. Science 29:726, 1985

Harold S. Ginsberg

Rubella Virus

Dr. George Maton in 1814 realized that a family epidemic of an acute febrile illness with rash, previously considered to be scarlatina (scarlet fever) or rubeola (measles), was a unique disease, which in 1866 was finally given the name *rubella.* This disease resembles measles but is milder and does not have the serious consequences often seen with measles in the very young. Because rubella seems to be such a harmless disease, it did not receive much attention earlier, although its probable viral etiology was demonstrated in experiments with human volunteers in 1938. However, interest in rubella was much increased when Gregg, an Australian ophthalmologist, noted in 1941 that women contracting rubella during the first trimester of pregnancy frequently gave birth to babies with congenital defects. Nevertheless, the cultivation of rubella virus was not achieved until 1962, when Parkman, Buescher, and Artenstein detected the virus through its interference with type 11 echovirus in primary grivet monkey kidney cultures, and Weller and Neva demonstrated unique cytopathic changes (Fig. 61–1) in infected primary human amnion cultures.

Rubella virus is classified as a **togavirus** (genus *Rubivirus*) on the basis of its physical and chemical characteristics (see Chap. 60). However, owing to the worldwide importance of German measles in humans and its unique clinical features and pathogenesis, a separate chapter is devoted to this virus.

Properties

MORPHOLOGY

The virion is roughly spherical and has an average diameter of about 60 nm, in both thin sections and negatively stained preparations; it consists of a roughly isometric core of 30 nm, covered by a loose envelope (Fig. 61–2).

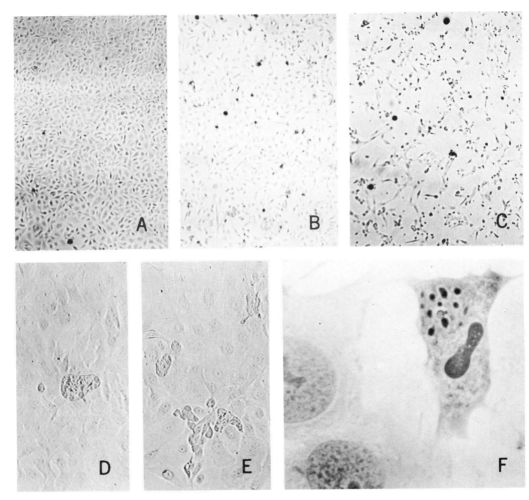

Figure 61–1. Cytopathic effect of rubella virus in human amnion cultures. (*A*) Appearance of normal amnion cell culture as viewed microscopically under low magnification. (Original magnification ×33) (*B*) Rubella-infected culture with estimated 20% destruction of cells on the 14th day after inoculation. (Original magnification ×33) (*C*) Rubella-infected culture with 80% cell destruction on the 28th day after inoculation. (Original magnification ×33) (*D*) Single affected cell with adjacent uninvolved cells on the tenth day after inoculation. (Original magnification ×132) (*E*) Scattered infected cells showing ameboid distortion on the tenth day after inoculation. (Original magnification ×132) (*F*) Infected cell with a large eosinophilic cytoplasmic inclusion and basophilic aggregation of nuclear chromatin, as well as portions of two normal cells. (H&E stain, original magnification ×3500; Neva FA et al: Bacteriol Rev 28:444, 1964)

Negative staining techniques reveal small spikes projecting 5 nm to 6 nm from the envelopes of most particles. Gentle disruption of the envelope with sodium deoxycholate uncovers an angular core; definite symmetry of the nucleocapsid is obscure, but ringlike subunit structures are discernible.

The morphology and growth characteristics of rubella virus resemble those of the alphatogaviruses (see Chap. 60).

CHEMICAL AND PHYSICAL PROPERTIES

The virion contains one molecule of single-stranded RNA with a molecular weight of 3.8×10^6 daltons (11 Kb), significantly smaller than the RNAs of other togaviruses. Two glycoproteins (E_1 and E_2) are present in the envelope, and an arginine-rich nucleocapsid (C) protein is associated with the RNA. Each nucleocapsid capsomer consists of two disulfide-linked C proteins. The virion

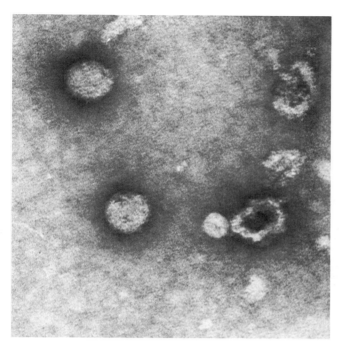

Figure 61–2. Morphology of rubella virus. Note the nucleocapsids within and dissociated from the envelopes. The nucleocapsid cores are somewhat angular and when separate from the envelopes show a subunit structure. (Horzinek MC: Arch Gesamte Virusforsch 33:306, 1971)

RNA is infectious, like that of other togaviruses, and therefore has a positive polarity; it has a 5'-terminal cap and a 3'-terminal poly(A) tract.

Like other enveloped viruses, rubella virus is rapidly inactivated by ether, chloroform, and sodium deoxycholate; it is relatively labile when stored at 4°C and relatively stable at −60° to −70°C.

The **hemagglutinin** and hemolytic activities reside in an integral component of the virion, the E_1 glycoprotein. The hemagglutinin, however, remains biologically active after gentle disruption of the viral particle. It is most effectively assayed with newborn chick, pigeon, goose, and human group O red blood cells (RBCs). As with the other togaviruses, the hemagglutinin does not elute spontaneously from the affected RBCs, and neuraminidase does not render the RBCs inagglutinable (see Mechanism of Hemagglutination, Chap. 44).

IMMUNOLOGIC PROPERTIES

Rubella virus is immunologically distinct from the other togaviruses; only a single antigenic type has been detected. Neutralizing and hemagglutination-inhibiting Abs, which are directed to the E_1 protein, develop during the incubation period of the disease and are commonly present when the rash appears (see Pathogenesis, below); they attain maximal titer during early convalescence and persist (along with immunity) for many years, if not for life. It must be noted, however, that the hemagglutination-inhibiting Abs and the neutralizing Abs are responses to different epitopes of the E_1 protein, since they can vary independently following German measles or immunization. Complement-fixing Abs appear about 6 to 10 days after onset, begin to diminish after 4 to 6 months, and disappear after a few years.

MULTIPLICATION

Rubella virus can replicate in a variety of primary, continuous, and diploid cell cultures of monkey, rabbit, and human origin. When cells are initially infected, the eclipse period is 10 to 12 hours. The viral titer reaches a maximum 30 to 40 hours after infection and may remain high for weeks (indeed, carrier cultures are readily established; see Chap. 51).

As with other togaviruses, the infectious virion RNA initiates viral replication by serving as an mRNA for viral protein synthesis. Viral multiplication is confined to the cytoplasm, and its RNA synthesis proceeds through intermediate replicative forms. Similar to other togaviruses, a 24S RNA as well as 40S RNA is made. The 24S RNA encodes the structural proteins with the gene order NH_2-C-E_2-E_1-COOH (see Synthesis of RNA Viruses, Chap. 48, and Viral Multiplication, Chap. 60).

Morphogenesis of the virions occurs at cell membranes (particularly the plasma membrane), which differentiate by incorporating viral proteins. Nucleocapsids then bud through the thickened vacuolar and surface membranes to form mature viral particles (Fig. 61–3). Unlike other togaviruses, however, nucleocapsids do not accumulate in the cytoplasm. Because viral components, including the hemagglutinin, are incorporated into the cell surface membrane during budding, the infected cells can be detected by hemadsorption.

In many infected cell cultures (e.g., grivet monkey kidney) an increase in viral titer is associated with **increased resistance** to infection with some challenge viruses (e.g., picornaviruses, orthomyxoviruses, measles virus), which provides another procedure for detecting infected cells and isolating viruses. Propagation of rubella virus induces the interference, and it is not consistently associated with interferon production (see Viral Interference, Chap. 49).

Cytopathic changes are not detectable in most rubella-infected cell cultures, but in cultured primary human cells distinctive cellular alterations appear slowly over 2 to 3 weeks. Affected cells are enlarged or rounded, and they often have ameboid pseudopods; staining reveals a disappearance of the nuclear membrane, promi-

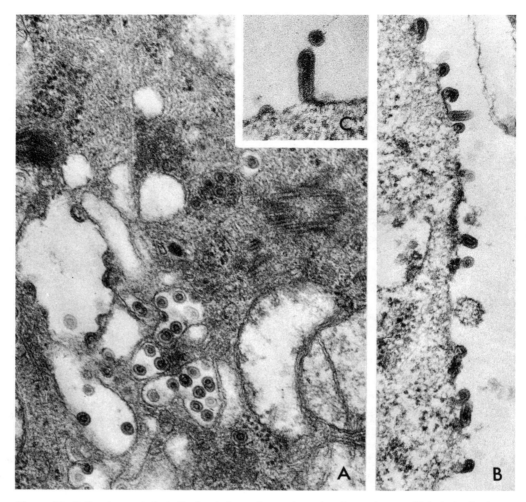

Figure 61–3. Development of rubella virus in the surface and cytoplasmic membranes of infected cell cultures. (*A*) Viral particles budding from cytoplasmic membranes into vacuoles and into the cytoplasm. Numerous mature virions are present with vacuoles. (Original magnification ×60,000) (*B*) Viral particles budding from the surface of an infected cell. (Original magnification ×60,000) (*C*) An elongated form in the budding process. (Original magnification ×84,000; Oshiro LS et al: J Gen Virol 5:205, 1969)

nent clumping of nuclear chromatin, and round or irregular eosinophilic cytoplasmic inclusion bodies (see Fig. 61–1). The cytopathic effects are associated with inhibition of the biosynthesis of host macromolecules and the inability of infected cells to divide. However, the infected cells do not lyse.

In cultures of human diploid cells that are chronically infected (**carrier cultures**) many chromosomal breaks are evident. Such breaks may have a bearing on the pathogenesis of congenital lesions in the infected fetus.

Pathogenesis

The rash appears 14 to 25 days (average 18 days) after infection with rubella virus. During this prolonged incubation period viremia occurs and viral dissemination is widespread throughout the body, including the placenta during pregnancy. Virus multiplies in many organs, but few signs are manifested except for a relatively common arthralgia and arthritis in women, accompanying infection of synovial membranes; leukopenia from viral replication in lymphocytes; occasional thrombocytopenia but uncommon purpuric manifestations; and rare encephalitis. (A chronic progressive disease simulating measles-induced subacute sclerosing panencephalitis infrequently occurs.) The virus can be isolated from nasopharyngeal secretions (and occasionally from feces and urine) as early as 7 days before and as late as 7 days after the appearance of the exanthem. Respiratory secretions are probably the major vehicle for transmitting the virus.

The disease is not unlike measles, except that it is milder, is of shorter duration, and has fewer complications. It is initiated by a 1- to 2-day prodromal period of fever, malaise, mild coryza, and prominent cervical and occipital lymphadenopathy. During the prodromal illness, and for 1 to 2 days after the rash appears, virus can be isolated from the blood. It can also be isolated from the skin lesions.

No characteristic pathologic lesions have been described in rubella, except for the **serious damage induced in infected fetuses.** This damage seems to involve tissues of all germ layers and results from a combination of rapid death of some cells and persistent viral infection in others. The continued infection, in turn, frequently induces chromosomal aberrations and, finally, reduced cell division. The infant infected during the first trimester may be stillborn; if it survives, it may have deafness, cataracts, cardiac abnormalities, microcephaly, motor deficits, or other congenital anomalies in addition to thrombocytopenic purpura, hepatosplenomegaly, icterus, anemia, and low birth weight (the **rubella syndrome**). The greater susceptibility of the early embryo to damage is correlated with the greater placental transmission of virus at that stage: when infection occurs during the first 8 weeks of pregnancy, the virus can be isolated as often from an aborted fetus as from the placenta, whereas in later infections viral isolations are less frequent from the fetus.

PERSISTENCE

When a woman has clinical rubella during the first trimester of pregnancy, the chance that the baby will have a structural abnormality is approximately 30%. Deformed infants born 6 to 8.5 months after intrauterine infection, and even those clinically normal, may still excrete virus in their nasopharyngeal secretions, although high titers of neutralizing IgM and IgG Abs and cell-mediated immunity are present (Fig. 61-4). Viral shedding persists until the clones of infected cells that are still able to divide eventually disappear. Indeed, one infant continued to shed virus in the presence of circulating Abs 2 years after birth, and in another the virus was isolated from cataract tissue at 3 years of age.

The Ab response in congenital rubella, like that noted in congenital cytomegalovirus infections (see Chap. 53) and congenital syphilis (see Chap. 37), indicates that in-

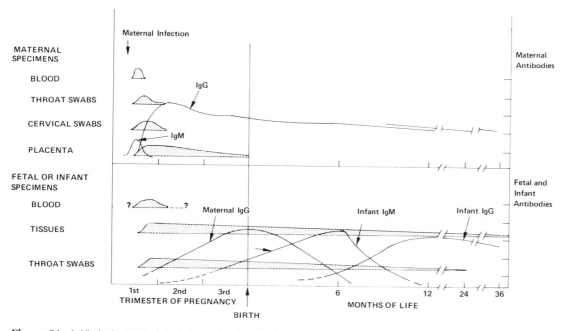

Figure 61—4. Virologic events (*shaded areas*) and antibody response in maternal–fetal rubella infection. (Modified from Meyer HM et al: Am J Clin Pathol 57:803, 1972)

trauterine rubella infections **do not induce immuno-logic tolerance** to rubella virus Ags. However, immunity following congenital rubella shows differences from that following postnatal or childhood infections: some congenitally infected infants fail to develop atypical Ab responses, others lose detectable Abs by 5 years of age, and others show depressed or absent cell-mediated immunity.

Laboratory Diagnosis

Rubella may be clinically confused with measles as well as with infections produced by a number of echoviruses and coxsackieviruses (see Chap. 55). The diagnosis may be confirmed by inoculating infected materials (usually nasopharyngeal secretions) into susceptible cell cultures. The **interference assay,** in primary grivet monkey kidney cultures, is the quickest and most sensitive procedure for initial viral isolation. Infection may also be detected by the development of cytopathic changes in susceptible cell cultures, by immunofluorescence, and by hemadsorption. Serologic diagnosis can be efficiently accomplished by hemagglutination-inhibition titrations, but care must be used to remove lipoprotein nonspecific inhibitors of hemagglutination (e.g., by adsorption of serum with dextran sulfate-CaCl$_2$). The hemagglutination-inhibition titration serves as a standard, since it correlates well with neutralization titrations. Solid-phase immunoenzyme assays (enzyme-linked immunosorbent assays [ELISAs]) are also reliable, and they are now most frequently used for diagnostic purposes because of the ease of performance and rapidity of completion. Complement fixation titrations are now used less often because Abs appear later and decrease earlier than Abs measured with the other techniques.

Evaluation of the immune status is of particular importance for women who are exposed to German measles during the first trimester of pregnancy (even a past history of rubella or previous immunization is not an absolute guarantee of immunity), women of childbearing age who wish to determine whether they should be immunized, and neonates born to a mother who was exposed to rubella during pregnancy or who have suggestive signs of the rubella syndrome.

Epidemiology

Rubella is a highly contagious disease, spread by nasal secretions. Unlike measles or chickenpox, however, **rubella infection is often inapparent,** thus fostering viral dissemination and rendering isolation of patients virtually useless. The ratio of inapparent infections to clinical cases is low (approximately 1 : 1) in children but as high as 9 : 1 in young adults. Patients are most infectious during the prodromal period, owing to the large amount of virus present in the nasopharynx, but communicability may persist as long as 7 days after the appearance of the rash. Transmission usually occurs by direct contact with infected persons, mostly children 5 to 14 years of age. However, **the apparently normal infant excreting virus acquired *in utero* is perhaps the most dangerous carrier:** his infection is unrecognized, and he comes in close contact with nurses, physicians, hospital visitors (including future mothers early in pregnancy), and, later, other children at home.

In urban areas of Europe and the Americas, during the winter and spring months, minor outbreaks are noted every 1 or 2 years. Major epidemics recur every 6 to 9 years, and superepidemics erupt at intervals of up to 30 years. This epidemic pattern succeeds in immunizing 85% of the population up to 15 years of age (the age span of susceptibles is significantly more than that in measles). Infection is almost always followed by long-lasting protection against clinical disease, although reinfections do occur. Most inapparent infections occur in those whose immunity has partially waned; the reinfection induces a secondary immune response (IgG Abs), which probably prevents or reduces the extent of viremia.

Prevention and Control

The extensive epidemic in the United States during 1964 resulted in congenital disabilities in approximately 20,000 infants, causing enormous anguish and an economic loss of well over 1 billion dollars. To avoid these dire consequences, prevention of maternal infection is of utmost importance. Isolation procedures are rarely practical, and passive immunization is of questionable value. However, effective live **attenuated viral vaccines** have been developed and are presently being used effectively to prevent the recurrence of such a devastating experience. The vaccine (RA27/3 strain) currently used in the United States is one isolated in human fetal kidney cells and attenuated by passage in fetal kidney fibroblasts and WI-38 diploid cells at 30°C. The vaccine requires a single injection and is immunogenic in at least 95% of the recipients, but Abs appear later than those following natural infection, and at levels as much as tenfold lower. Nevertheless, immunization effectively protects the recipients from clinical rubella following exposure for at least 15 years, even after extensive exposure during epidemics. Accordingly, vaccination has markedly reduced the incidence of the congenital rubella syndrome (Fig. 61–5), but congenital rubella continues to occur because 10% to 20% of the childbearing-aged population continue to be susceptible. As the presently immune young children increasingly enter the childbearing period, con-

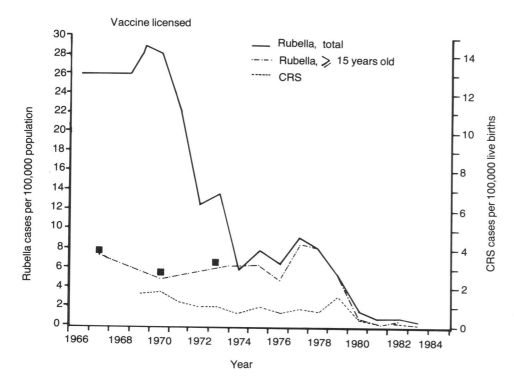

Figure 61–5. Incidence rate of reported rubella cases and congenital rubella cases, United States, 1966–1985. *CRS,* congenital rubella syndrome. (Modified from MMWR 33: 551, 1986)

genital rubella will decrease further, but this process will take 10 to 30 years. Hence, efforts must increase to immunize the susceptibles who are of age to have children.

Though the vaccines employed are highly protective some drawbacks exist. (1) In 2 to 3 weeks after immunization, small amounts of infectious virus appear in the nasopharynx. Viremia is unusual, however, and transmission to susceptible contacts has not been observed. (2) At the time of nasopharyngeal viral shedding, mild arthralgias and occasional arthritis occur in 1% to 2% of children and in 25% to 40% of adult women; adults also occasionally experience mild rash, fever, and lymphadenopathy. (3) The Ab levels attained confer solid immunity, but the long-lasting immunity that follows natural infection may not occur (unfortunately, the most attenuated viruses used in vaccines also are the least immunogenic). (4) A tenfold higher reinfection rate is observed in vaccinees than in those who have had natural infections. (5) An effective herd immunity is not produced, even when as many as 85% of 1- to 12-year-olds are immunized. (6) Despite the marked reduction in incidence of clinical disease, infection is frequently not prevented, even with exposure only 2 to 3 months after immunization. However, the inapparent infection could serve to induce a life-long immunity similar to that following natural disease.

It should be emphasized that immunization has a unique goal: to protect an unborn fetus rather than the recipient of the vaccine. This goal has been approached by two almost opposing plans: immunization of teenage girls, who are the prospective mothers, or immunization of children 1 to 12 years of age, who are the major viral transmitters, thereby indirectly protecting pregnant women. In the United States the latter alternative has been chosen; that is, routine immunization of all children. In addition, immunization is recommended for all women of childbearing age who are without protective Abs. (Birth control must be rigidly practiced for 2 to 3 months following immunization.)

The immunization program pursued in the United States has elicited great concern, since the vaccine may not confer long-lasting immunity and therefore the women may become susceptible during their childbearing years. In Great Britain immunization is given to all girls between 10 and 14 years old and to women of childbearing age who do not have detectable hemagglutination-inhibiting Abs. It is not yet clear which approach is more successful in attaining the goal of preventing fetal damage.

If preventive measures fail, many physicians recommend therapeutic abortion when rubella occurs during the first trimester of pregnancy.

Selected Reading

BOOKS AND REVIEW ARTICLES

Horstman DM: Rubella. In Evans AS (ed): Viral Infections of Humans: Epidemiology and Control, 2nd ed, p 519. New York, Plenum, 1984

Rawls WE: Congenital rubella: The significance of virus persistence. Prog Med Virol 10:238, 1968

Krugman S, Hinman AR, Burke JP (eds): International symposium on the prevention of congenital rubella infection. Rev Infect Dis 7:1, 1985

Preblud SR, Serula MK, Frank JA Jr et al: Rubella vaccination in the United States: A ten year review. Epidemiol Rev 2:171, 1980

SPECIFIC ARTICLES

Davis WJ, Larson HE, Simsarian JP et al: A study of rubella immunity and resistance to infection. JAMA 215:600, 1971

Gregg NM: Congenital cataract following German measles in the mother. Trans Ophthalmol Soc NZ 3:35, 1941

Oker-Bloom C: The gene order for virus structural proteins is NH_2-C-E_2-E_1-COOH. J Virol 51:354, 1984

Sedwick WD, Sokol F: Nucleic acid of rubella virus and its replication in hamster kidney cells. J Virol 5:478, 1970

Vaheri A, Hovi T: Structural proteins and subunits of rubella virus. J Virol 9:10, 1972

62

Harold S. Ginsberg

Reoviruses and Epidemic Acute Gastroenteritis Viruses

Classification and General Characteristics

The term *reovirus* (respiratory enteric orphan virus) refers to a group of RNA viruses that infect both the respiratory and the intestinal tracts, usually without producing disease. Though originally considered members of the echovirus group (and classified as type 10 echovirus), reoviruses are larger and differ in producing characteristic cytoplasmic inclusion bodies. Moreover, these inclusion bodies, which contain specific viral Ags, stain green-yellow with acridine orange, like cellular DNA, rather than red, like the usual single-stranded RNA. This striking observation led to the discovery that **reovirus RNA is double stranded** with a secondary structure similar to that of DNA. This finding was the first indication that such an unusual nucleic acid exists in nature. Viruses with similar chemical and physical properties have since been found to be widely disseminated in vertebrates, invertebrates, and plants. These viruses (more than 150) have been grouped into the family **Reoviridae** (vernacular, **reoviruses).**

All these viruses have segmented, double-stranded RNA genomes. They are similar in morphology (Table 62–1), but differences in structure, antigenicity, stability, and preferred hosts are the bases for dividing Reoviridae into several genera: *Orthoreovirus* (**reovirus**) includes species that infect humans, birds, dogs, and monkeys; *Orbivirus* (Latin *orbis*, ring) comprises members that multi-

TABLE 62–1. Characteristics of Reoviridae That Infect Man

Characteristic	Reoviruses	Rotaviruses	Orbiviruses*
Morphology	Icosahedral; double capsid; no envelope	Icosahedral; double capsid; no envelope	Icosahedral; double capsid—outer is skinlike
Size (diameter)	75 nm; inner capsid 50 nm	70 nm; inner capsid 50 nm	70 nm; inner capsid 50 nm
Nucleic acid	RNA; double-stranded; 10 segments, mol. wt. 0.7–2.7×10^6, total 15×10^6 daltons (46 Kb)	RNA; double-stranded; 11 segments, mol. wt. 0.23–2.04×10^6, total 10×10^6 daltons (31 Kb)	RNA; double-stranded; 10 segments, mol. wt. 0.3–2.7×10^6, total 12×10^6 daltons (37 Kb)
Effect of lipid solvents	Infectivity stable	Infectivity stable	Infectivity stable
Virion polypeptides	8	10	8

* Blue-tongue virus, the best-characterized orbivirus, was used for this comparative analysis. The size range for the genus is 60 nm to 80 nm.

ply in insects, including Colorado tick fever virus, which infects humans; *Rotavirus* (Latin *rota*, wheel) includes some major etiologic agents of infectious infantile diarrhea in humans and several other animals; *Cypovirus* includes the **cytoplasmic polyhedrosis viruses,** which consist of viruses that infect Lepidoptera and Diptera; *Phytoreovirus* contains viruses that infect many different types of plants (e.g., rice dwarf virus and clover wound tumor virus); and *Fijivirus* also contains viruses that infect plants as well as insects (e.g., maize rough dwarf virus). Only the viruses infecting humans will be discussed.

MORPHOLOGY

The virions of all Reoviridae have similar sizes, icosahedral symmetry, and not one but two icosahedral capsids. Fine-structure electron microscopy (Fig. 62–1) reveals that the outer capsid is probably constructed of 32 large capsomers (18 nm in diameter) and that neighboring capsomers share subunits (see Fig. 62–1), which is apparently a unique feature of this family. The inner capsid has 12 prominent projections at its vertices and an undetermined number of intervening capsomers. Each projection is composed of five molecules of λ2 proteins into which the σ1 dimer is anchored (Fig. 62–2). Rotaviruses are slightly smaller than reoviruses (see Table 62–1) but have similar double capsids.

Orbiviruses (e.g., blue-tongue virus of sheep, the type species) show another morphologic variation: the capsomer arrangement of the outer capsid is usually obscure, but exposure of the virions to CsCl below pH 8 removes a thin outer layer, revealing a capsid of 32 large, ring-shaped capsomers (hence, its Latin derivation, *orbis*). Colorado tick fever virus has not been adequately studied.

Reoviridae of different genera do not display any immunologic relatedness.

CHEMICAL AND PHYSICAL CHARACTERISTICS

The virions of Reoviridae are composed of protein and about 15% RNA. The double-stranded nature of the RNA is shown by many properties: complementary base ratios (G = C = 20 mol/dl); sharp thermal denaturation (T_M 90°C to 95°C); pronounced hyperchromicity on denaturation; resistance to a concentration of ribonuclease A that completely degrades single-stranded RNA; and characteristic density in $CsSO_4$. Electron micrography reveals stiff filaments, like those of DNA, and x-ray diffraction patterns are consistent with double-stranded molecules.

The **genomes** of Reoviridae consist of **ten or 11 distinct segments,** which are distributed in **three size classes** (Fig. 62–3). That these are distinct components of the virion rather than products of artificial fragmentation is shown by the following characteristics: the fragment sizes are reproducible within a genus; the different pieces do not cross-hybridize; each fragment contains two free 3'-OH termini and two 5'-terminal diphosphates; and electron micrographs of gently disrupted virions demonstrate molecules of viral nucleic acid of the expected lengths.

Orthoreovirus virions (other genera have not been examined) also contain about 3.7×10^6 daltons of a heterogeneous collection of small, single-stranded oligonucleotides, produced by aborted transcription but whose function is unknown.

The **orthoreovirus** and **orbivirus virions** contain eight species of polypeptides, and those of rotaviruses contain ten polypeptides; their molecular weights are distributed in three size classes. Since the structural proteins (Table 62–2) can be separated only after denaturation of the virion, it is impossible to identify directly their specific functions (i.e., hemagglutination, RNA polymerase, nucleoside phosphohydrolase, group antigenicity, and type-specific antigenicity). However, association of each polypeptide with a virion structure (or a nonstructural protein), the identification of the RNA segment

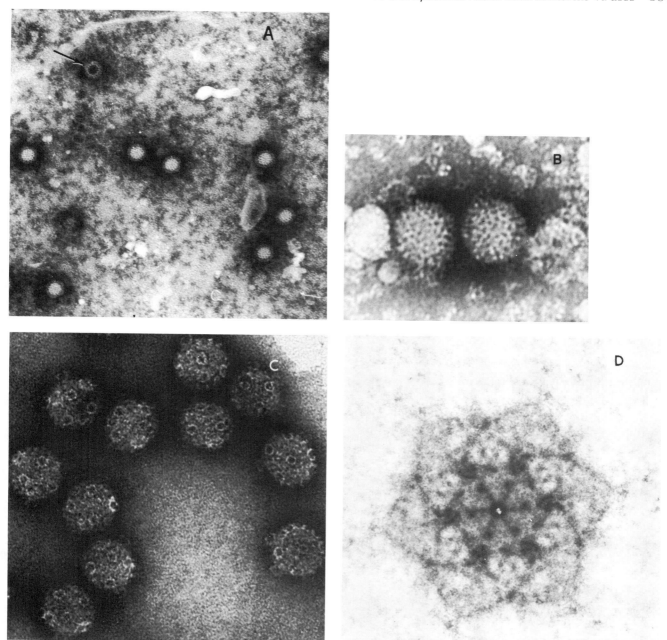

Figure 62—1. Electron micrographs of negatively stained orthoreovirus virions, cores, and capsomers. The virion is 75 nm to 80 nm in diameter. Cubic symmetry, the structure of the capsid, and the absence of an envelope are illustrated by low (*A*) and high (*B*) magnification. Note the empty particle in *A* in which the inner coat that covers the core is apparent (*arrow*). (*A*, original magnification ×75,000; *B*, original magnification ×375,000) (*C*) Cores prepared from purified virions by chymotrypsin degradation of the outer capsid followed by equilibrium centrifugation in a CsCl density gradient. Note the hollow spikes located at the 12 vertices of the icosahedral core. (Original magnification ×280,000) (*D*) Cluster of capsid subunits in which the central capsomer was enhanced by an n = 6 rotation. Rotational enhancement of image detail makes evident the structure of the central capsomer, which is made of six wedge-shaped subunits that exhibit sharing with neighboring capsomers. Note also that each wedge-shaped subunit is also composed of subunits (i.e., polypeptide chains). (Original magnification ×1,200,000; *A, B,* Gomatos PJ et al: Virology 17:441, 1962; *C,* White CK, Zweerink HJ: Virology 70:171, 1976; *D,* Palmer EL, Martin ML: Virology 76:109, 1977)

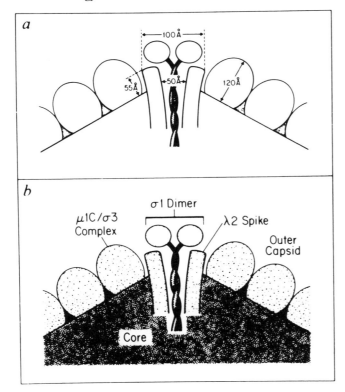

Figure 62–2. Schematic representation of the morphology of the outer capsid of orthoreovirus type 3. (*a*) Dimensions of the virion. (*b*) Orientation of the σ1 protein in the virus. The α-helical region is shown to extend throughout the λ2 channel into the viral core. The globular structure sits on top of the λ2 channel and interacts with host cell receptors. (Bassel-Duby R et al: Nature 315:421, 1985)

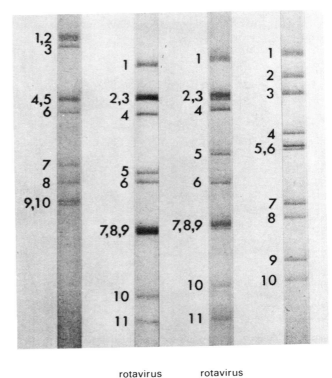

Figure 62–3. Electrophoresis of the virion RNA of orthoreovirus type 3, human rotavirus, calf rotavirus, and an orbivirus on 7.5% polyacrylamide gels (migration was from top to bottom). RNA segments for each virus are distributed in three size classes (L,M,S). (Modified from Schnagl RD, Holmes IH: J Virol 19:267, 1976)

(i.e., the gene) in which each polypeptide is encoded, and the correlation of some polypeptides with function have been possible using genetic and biochemical techniques (see Table 62–2). The outer capsid has a complex structure (see Fig. 62–2) that provides stability to the virion.

The σ1 outer capsid protein is a dimer consisting of the two carboxy-terminal globular receptor-interacting domains and the α-helical coil structure formed from the amino-terminal portions of the polypeptides (see Fig. 62–2). The σ1 protein is also responsible for hemagglutination by all three immunologic types (types 1 and 2 agglutinate human red blood cells (RBCs); type 3 agglutinates bovine RBCs). Orthoreoviruses agglutinate and elute from RBCs, but unlike orthomyxoviruses, they do not destroy the receptor sites on the RBCs, although the reovirus receptors are hydrolyzed by neuraminidase. The RBC receptor is probably a glycoprotein, since either trypsin or periodate inactivates, it and *N*-acetyl-D-glucosamine blocks hemagglutination by binding to the viral capsid.

Orthoreoviruses

PROPERTIES

Immunologic Characteristics

The σ1 outer capsid protein induces specific neutralizing and hemagglutination-inhibiting Abs that distinguish **three immunologic types** of reoviruses. The three types of reoviruses are antigenically related, however, by three or four cross-reacting Ags that can be measured by complement fixation (CF) and immunoprecipitin tests. For example, heterotypic reovirus Abs appear in the serum of about 25% of persons who have primary infections with type 1 reovirus. Type-specific Abs to reoviruses appear 2 to 4 weeks after infection.

Host Range

Reoviruses appear to be ubiquitous in nature: specific viral inhibitors (presumably Abs) have been found in the serum of all mammals tested except the whale; and humans and many other species (including cattle, mice, and monkeys) are naturally susceptible to reoviruses.

Reoviruses and Epidemic Acute Gastroenteritis Viruses 311

TABLE 62–2. Correlation Between Classes of Orthoreovirus* Genomic Segments and Polypeptides in Virions and Infected Cells

Genome				Polypeptides				
Size Class	Segment	Mol. Wt. $(\times 10^{-6})$	Time of Expression	Size Class	Component†	Mol. Wt. $(\times 10^{-3})$	Origin‡	Function§
L	1(L₁)	2.7	Early‖	λ	λ3	135	Core	RNA synthesis
	2(L₂)	2.6	Late		λ2	148	Core	
	3(L₃)	2.5	Late		λ1	155	Core	
M	4(M₁)	1.8	Late	μ	μ2	70	Core	
					μ1	80	Core	
	5(M₂)	1.7	Late		μ1c	72	Outer capsid	
	6(M₃)	1.6	Early‖		μNs	75	Nonstructural	
S	7(S₁)	1.1	Late	σ	σ1	49	Outer capsid	Cell association, hemagglutination, induction neutralizing Abs
					σ1x**	?	?	
	8(S₂)	0.85	Late		σ2	38	Core	RNA synthesis
	9(S₃)	0.76	Early‖		σNs	36	Nonstructural	RNA synthesis
	10(S₄)	0.71	Early‖		σ3	34	Outer capsid	

* Type 3 (Dearing strain) is given as an example. Sizes of the RNA segments and polypeptides are different for each type.

† Identification of a polypeptide with a double-stranded RNA segment was accomplished by *in vitro* translation of specific mRNAs and by intertypic genetic recombination, which takes advantage of the different sizes of RNA segments in different viral types.

‡ Determined by controlled disruption of purified virions and SDS-polyacrylamide gel electrophoresis of infected cell cytoplasmic extracts

§ Identified by genetic studies with temperature-sensitive mutants

‖ Predominant messages transcribed during early stages of infection; mRNAs made in the presence of cycloheximide added at time of infection

\# Processed from μ1

** Translated from second open-reading frame in S₁ RNA; function unknown

Newborn mice are particularly vulnerable to experimental infection, which is often fatal; when infected with type 3 reovirus they occasionally develop a chronic illness similar to runt disease (see Chap. 20).

Cell cultures, including primary cultures of epithelial cells from many animals and various continuous human cell lines, are used to isolate and study reoviruses. The infection causes gross cytopathic changes and permits **hemadsorption** of human **group O RBCs.** Distinctive eosinophilic **inclusion bodies** (Fig. 62–4) are seen in the cytoplasm of infected cells.

Multiplication

The eclipse period is long (6 to 9 hours, depending on viral type and size of inoculum), compared with that of other RNA viruses with icosahedral symmetry (Fig. 62–5; see Chap. 48). Virus then increases exponentially, reaching a maximum titer (from 250 to 2500 plaque-forming units per cell) by approximately 15 hours after infection. Infected cells are not rapidly lysed following viral replication, and release of infectious virus is incomplete.

Owing to the unique structure of the capsid and the novel nucleic acid, replication of orthoreoviruses presents some unusual features. The virus is capable of utilizing the β-adrenergic receptor on the host cell for attachment and cell entry. After **cell penetration,** by receptor-mediated endocytosis, the virions become associated with lysosomes, whose proteolytic enzymes hydrolyze about half of the outer capsid (all of polypeptides σ1 and σ3 and an 8000-dalton piece of polypeptide μ1c), forming a **subviral particle** that is smaller than the virion but larger than the viral core (inner capsid and RNAs). Thus, the core is only partially uncoated and the double-stranded RNA genome segments are not released free into the cytoplasm.

TRANSCRIPTION. The virion RNAs do not function as messengers; but, like DNA, they are first transcribed. However, transcription is **conservative, only the negative (−) strands serve as templates,** and the transcripts are processed into single-stranded mRNAs. This transcription ensues entirely within the subviral particles, which leave the lysosomes after the uncoating process. Initially (beginning 2 to 4 hours after infection), four segments (1, 6, 9, and 10 seen in Fig. 62–3 and Table 62–2) of the viral genome are predominantly transcribed, although all genome segments are copied. An RNA polymerase (transcriptase) contained within the core of the

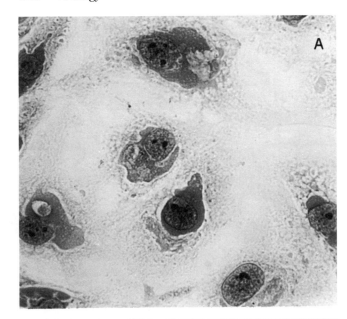

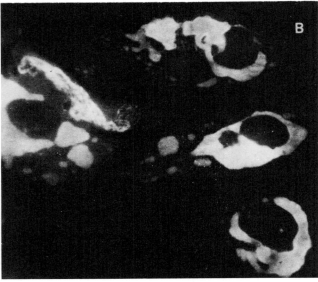

Figure 62–4. Large eosinophilic cytoplasmic inclusion bodies in monkey kidney cells infected with type 1 orthoreovirus. (*A*) H&E stained and viewed with a light microscope. (Original magnification ×1000) (*B*) Stained within fluorescein-conjugated Ab and viewed with ultraviolet optics. (Original magnification ×1500) (Rhim JS et al: Virology 17:342, 1962)

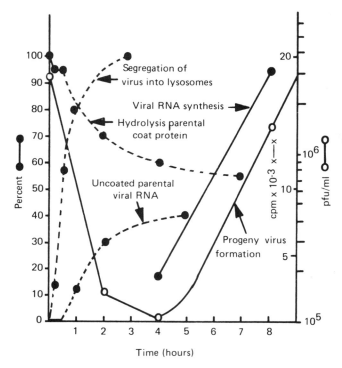

Figure 62–5. Sequential events in the uncoating and replication of orthoreovirus. *cpm*, counts per minute; *pfu*, plaque-forming units. (Silverstein SC, Dales S: J Cell Biol 36:197, 1968)

virion is responsible for this early transcription, which is accomplished without synthesis of new proteins or replication of the viral genome. Four additional enzymatic activities within the virion (nucleotide phosphohydrolase, guanyltransferase, and two methyltransferases) are responsible for synthesizing the caps at the 5′ termini of the mRNAs.

By 6 hours after infection the ten segments of the viral genome are transcribed at comparable rates. The sizes of the mRNAs correspond to those of the virion RNAs (see Fig. 62–3, Table 62–2), and each hybridizes specifically to one strand of the segment of the viral genome of corresponding size. Most of the newly synthesized single-stranded RNA molecules leave the subviral particles, rapidly become associated with polyribosomes, and appear to function as monocistronic mRNAs. These mRNAs are capped, but unexpectedly they are not polyadenylated at their 3′ ends, though the virion contains an oligoadenylate synthetase activity and is relatively rich in poly(A) oligonucleotides.

RNA REPLICATION. Unlike double-stranded DNA molecules (see Chap. 48), orthoreovirus double-stranded RNA is **replicated conservatively.** The negative strand of each segment is copied in great excess into plus single strands (i.e., with the same polarity as the mRNA), and these then serve as templates for synthesis of the minus strands. Free minus strands are not detected because they remain associated with their templates to form the various double-stranded RNA segments.

Replication of the viral RNA requires continuous protein synthesis, apparently to supply the virus-encoded replicase and transcriptase functions. Newly replicated

RNA molecules of all ten size classes are found in infected cells, further strengthening the evidence that the viral genome is indeed segmented. The segmented structure of the genome explains the remarkably high recombination frequency (i.e., 3% to 50%) detected with temperature-sensitive mutants (due to reassortment, as with influenza viruses; see Chap. 56).

TRANSLATION. About 75% of the newly synthesized plus strands become associated with cytoplasmic polyribosomes, which serve as the sites for synthesis of viral proteins. All eight virion structural proteins can be detected in infected cells as early as 3 hours after infection, but only seven of these are primary gene products: μ1c (see Table 62–2) is derived from μ1 by cleavage. In addition, two nonstructural viral polypeptides, μNs and σNs, whose functions are unknown, are also synthesized.

ASSEMBLY. Infectious virions begin to appear 6 to 7 hours after infection (Fig. 62–5), but how the ten genomic segments (which may be likened to chromosomes) are segregated in the appropriate number remains unexplained. Excess viral Ags accumulate and viral particles assemble in close association with the spindle tubules

(Figs. 62–6 and 62–7). However, the mitotic spindle is not essential for viral multiplication, since viral synthesis proceeds unhindered in nondividing cells arrested in metaphase by colchicine, which disaggregates the spindle. Unassembled, newly synthesized double-stranded RNAs and the excess viral Ags accumulate in large masses (see Fig. 62–4), forming the characteristic cytoplasmic inclusion bodies that give green-yellow fluorescence with acridine orange staining.

Orthoreovirus infection inhibits biosynthesis of host cell DNA and proteins within 6 hours, and cell division ceases. In such cells the mitotic index increases more than threefold, but the mitotic sequence is not completed and abnormal mitotic figures form, perhaps because of the viral association with mitotic spindles.

PATHOGENESIS

Orthoreoviruses have frequently been isolated from the feces and respiratory secretions of healthy persons, as well as from patients with a variety of clinical illnesses, particularly minor upper respiratory and gastrointestinal disease. The relation of these viruses to disease is not clear, and human transmission experiments have not

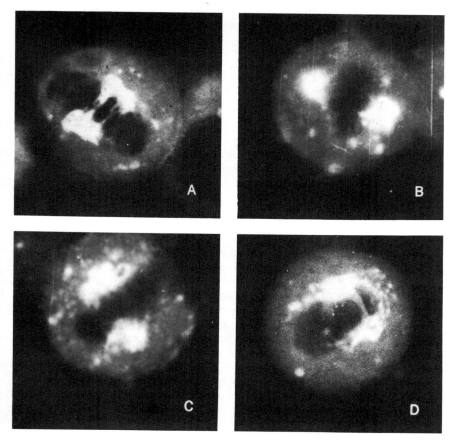

Figure 62–6. Orthoreovirus-infected cells stained with fluorescein-conjugated Ab. Viral Ag is closely associated with the mitotic spindle of virus-infected cells. Cells were examined by dark-field microscopy with ultraviolet illumination. (Original magnification ×750; Spendlove RS et al: J Immunol 90:554, 1963)

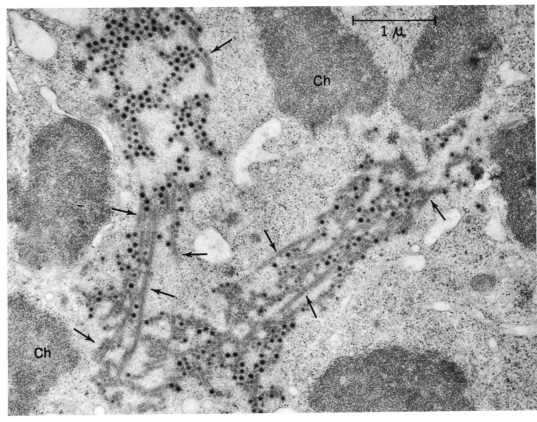

Figure 62–7. Electron micrograph of a cell infected with type 3 reovirus. The section was made through the spindle and chromosomes (*Ch*) of an infected cell in mitosis, showing aggregates of viral particles closely associated with the tubules of the spindle, indicated by arrows). (Original magnification ×15,000; Dales S: Proc Natl Acad Sci USA 50:268, 1963)

been decisive: afebrile respiratory illnesses only occurred in one-third of the volunteers, the symptoms were irregular, and the illnesses were very mild.

In the few deaths that have been attributed to type 3 orthoreoviruses, the pathologic lesions noted (encephalitis, hepatitis, and pneumonia) were similar to lesions found in experimental animals.

LABORATORY DIAGNOSIS

Reoviruses may be isolated from throat washings or fecal specimens by means of cell cultures (e.g., human embryonic or monkey kidney), and they are usually identified as belonging to the reovirus genus by CF or **ELISA** tests. The specific immunologic type can then be identified by hemagglutination-inhibition or neutralization assays.

To permit recognition of Abs in a patient's serum by hemagglutination-inhibition titrations the serum should be pretreated with trypsin or periodate, or adsorbed with kaolin, to remove nonspecific mucoprotein inhibitors.

Since orthoreoviruses are also frequently isolated from healthy persons, an increase in serum Ab titer must be demonstrated before an illness can be assumed to be caused by a reovirus.

EPIDEMIOLOGY AND CONTROL

Though orthoreovirus infections do not seem to be of great clinical importance, studies should be continued to monitor their pathogenetic potential. Both the respiratory and the gastrointestinal tracts may well be sources of their spread. Unrecognized infections are common, for approximately 10% of children by 5 years of age and 65% of young adults in the United States have reovirus Abs in their sera. Antibodies are also frequently found in various wild and domestic animals, but it is not known whether the animals serve as reservoirs for human infections.

Since the meager data assign a limited pathogenicity to these viruses, specific immunization procedures are not warranted.

Rotaviruses

Acute **nonbacterial gastrointestinal infections** (epidemic gastroenteritis and infantile diarrhea) are second only to acute respiratory infections as the cause of illness in families with young children. Hence, the discoveries of viruses that are major causes of these infections encourage the hope that these common illnesses will be better understood and controlled. **Rotaviruses** (initially termed *human reovirus-like agent*) have been identified as a major cause of **sporadic acute enteritis** in infants and young children. In addition, two immunologically distinct viruses that resemble caliciviruses have been shown to be etiologic agents of **epidemic acute gastroenteritis** (discussed separately below).

These new viruses were discovered by direct electron microscopic observation of fecal extracts, duodenal fluid, and duodenal mucosa; none was detected by isolation in a cell culture or animal. A major diagnostic advance, **immune electron microscopy,** was of special utility: sera from convalescent patients were used as a source of specific Abs to agglutinate the virions and make them more easily recognizable and to identify the virus as one that had actually produced an acute infection with an attendant Ab response (Figs. 62–8 and 62–9).

PROPERTIES

The **physical and chemical characteristics** of rotaviruses are similar to those of orthoreoviruses (see above) except that the viral genome consists of 11 double-stranded RNA segments (see Fig. 62–3). The electrophoretic patterns of the 11 segments vary with strains isolated at different times and different locations. These so-called electropherotypes are useful for molecular epidemiologic studies. The physical characteristics of the virion RNAs and the proteins they encode are summarized in Table 62–3. The functions of the individual nonstructural proteins are still unclear, particularly those responsible for transcription and RNA replication.

Rotaviruses isolated from humans belong to **four immunologic serotypes** (1 to 4) based upon neutralization titrations. The outer capsid protein VP7 is the major Ag inducing neutralizing Abs; VP4, which is responsible for hemagglutination, is a minor neutralization Ag. Marked cross-reactivity of human rotaviruses with each other, and with many animal rotaviruses, can be detected by CF, ELISA, and immunofluorescence assays owing to VP6, a major core Ag that makes up about 80% of the virion's protein mass. On the basis of immunologic cross-reactivities (e.g., by CF, ELISA, immunofluorescence, im-

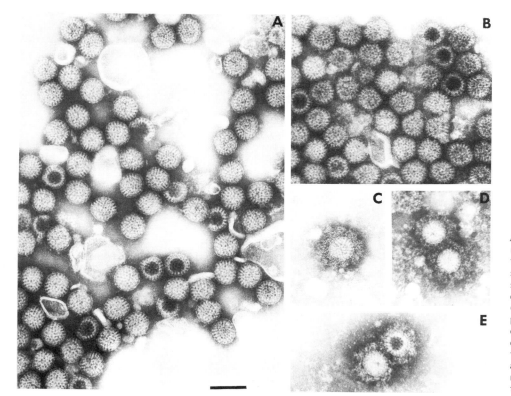

Figure 62–8. Identification of a rotavirus from man by immune electron microscopy. (*A*) Viral particles in stool filtrate. Note the capsomer structure and the appearance of a double capsid in the empty particles. (*B*) Filtrate incubated with acute-phase serum. (*C* through *E*) Viral particles in the same stool preparation incubated with convalescent serum from the same patient. (Kapikian AZ et al: Science 185:1049, 1974. Copyright 1974 by the American Association for the Advancement of Science)

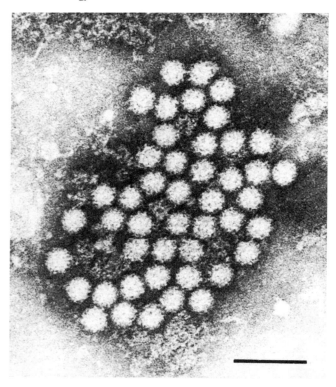

Figure 62–9. Electron micrograph of an aggregate of acute nonbacterial gastroenteritis virus (Norwalk agent) particles in stool from an acutely ill patient. (Original magnification ×200,000; Kapikian AZ et al: J Virol 10:1075, 1972)

mune hemagglutination) all rotaviruses are divided into two serogroups (I and II). A few viruses having the physical characteristics of rotaviruses do not share either group Ag and are called **pararotaviruses.** Serotypes 1 and 2 include only human rotaviruses, but serotypes 3 and 4 include a number of animal rotaviruses. Some animal rotaviruses, however, are distinct and form serotypes 5 to 7. It is noteworthy that the antigenic characteristics of the serotypes are genetically stable and do not correlate with variations of RNA electrophoretic patterns.

Human rotaviruses are fastidious viruses that do not replicate well in cell cultures except after serial passage in the presence of trypsin or when host range mutants are selected by passage in a susceptible animal host (e.g., type 1 multiplies in piglets), or by selecting reassortants after cocultivation with a related animal rotavirus. Thus, although human rotaviruses can now be propagated in cell cultures after adaptation, most data concerning physical and chemical characteristics and multiplication of rotaviruses derive from experiments using animal viruses, particularly SA11, simian rotavirus, which is closely related to human serotype 3 viruses.

Unadapted human rotaviruses induce an abortive in-

fection in cell cultures, and infectious virus is usually not produced or is made in small numbers. However, extensive studies of the simian SA11 virus indicate that replication of rotaviruses follows the same steps in attachment, entry, uncoating, biosynthesis, and assembly as followed by orthoreoviruses (see above). One step in rotavirus morphogenesis, however, is more striking than what is seen with orthoreoviruses: precursor viral particles bud into the cisternae of the endoplasmic reticulum, thus becoming enveloped; but the enveloped stage is transient, and the envelope is stripped to form mature particles. The ability of trypsinization to increase the infectivity of many strains may be due to an enzymatic amplification of the latter step of virion maturation.

PATHOGENESIS

Severe diarrhea and fever, occasionally accompanied by vomiting, is a common syndrome in children less than 2 years old, particularly infants. About half of the most severe illnesses of babies seen in hospitals are caused by the human rotavirus; during the fall and winter the incidence is particularly high. Biopsies of the duodenal mucosa show it to be a principal site of viral multiplication and pathologic lesions, which may account for the symptoms observed. (Whether other parts of the gastrointestinal tract are also affected is uncertain.) Viral excretion in feces is maximal during the first 4 days of illness, but it has been detected as long as 21 days after the onset of diarrhea. In experimental infections of newborn, gnotobiotic calves and piglets, viral replication and pathologic lesions occur in mucosal and submucosal cells of the villous epithelium of the duodenum and spread caudally throughout the entire small intestine, but the large bowel and other organs do not appear to be involved.

LABORATORY DIAGNOSIS

Identification of rotavirus infection requires a laboratory diagnosis because so many agents can produce similar manifestations of gastroenteritis. **Electron microscopic examination** of a negatively stained stool specimen is the quickest, most sensitive, and most reliable method (see Fig. 62–8); it also detects protorotaviruses, which do not share Ags with rotaviruses. However, many laboratories have neither the equipment nor the expertise to use electron microscopy, and therefore the **ELISA** technique has become the mainstay in most laboratories. RNA dot hybridization is a reliable and rapid method to detect rotaviruses in stool specimens and rectal swabs. Isolation of virus from stools can be accomplished; but the isolation rate is only about 75%, and considerable time is required as compared with the preceding techniques. Immunofluorescence permits detection of infected cells more rapidly.

TABLE 62–3. Rotavirus Genomic RNAs and Encoded Proteins*

Genome		Polypeptide			
RNA Segment	Mol. Wt. ($\times 10^{-6}$)	Designation	Mol. Wt. ($\times 10^{-3}$)	Location in Virion	Function
1	2.05	VP1	125	Core	
2	1.68	VP2†	94	Core	
		VP2‡	88		
		VP4‡	84		
3	1.60	NSVP1(?)	Not identified		
4	1.60	VP3†	88	Outer capsid	Hemagglutinin
		VP5‡	60		Minor neutralization
		VP8‡	28		
5	0.98	NSVP2	53		
6	0.81	VP6	41	Core	
7	0.5	NSVP3	34		
8	0.5	NSVP4	35		
9	0.5	VP7	32	Outer capsid	Major neutralization Ag (? glycoprotein; maturation)
10	0.3	NSVP5	20		
11	0.2	VP9	26	Outer capsid	

* Data obtained from the best-studied rotavirus, simian SA11

† Precursor protein proteolytically cleaved

‡ Designates polypeptide as a cleavage product

Ab response is a common and economic means to diagnose rotavirus infections. Owing to the acquisition of Abs at an early age, the illness must induce a fourfold or greater increase in Abs to make a positive diagnosis. There are many serologic techniques to diagnose rotavirus infections; CF, ELISA, immunofluorescence, and viral neutralization titrations (in primary African green or cynomolgus monkey kidney cell cultures) are the most commonly employed.

EPIDEMIOLOGY AND CONTROL

Acute gastroenteritis caused by a rotavirus is a worldwide, sporadic disease, found primarily in young children 6 to 24 months of age. It is a leading cause of childhood deaths in developing countries. Rotaviruses are the third most common cause of gastroenteritis. Their prevalence is demonstrated serologically: over 90% of children in the Washington, DC, area were found to have CF Abs by age 3 years. Nosocomial infections are common in newborns, although clinical illness is rare. The basis for this unexpected finding is unknown. Although the virus is probably transmitted by the fecal–oral route, infection is most common during the cooler part of the year, unlike bacterial diarrheas and dysenteries. The virus is spread successfully because it is present in large amounts in feces (e.g., 10^{11} viral particles per gram of feces), and it is very stable on ordinary surfaces at room temperature.

Outbreaks of rotavirus infections do occur in adults, but subclinical infections are the rule. Indeed, in a small study at least one parent in the family had an inapparent infection at the time illness began in the baby, suggesting that the child's source of virus might literally have been the hand that fed it.

Prevention of this major disease in young children is an important goal whose attainment seems possible but requires additional immunologic and epidemiologic data. The major issues that require resolution before immunoprophylaxis can be generally successful are the following: (1) Does homotypic immunity prevent reinfection and disease? (2) How effective is heterotypic immunity, and how many viral serotypes or type-specific Ags that induce neutralizing Abs must be included in the vaccine? (3) If immunity is homotypic, must the vaccine contain the genes (RNA 9 and RNA 4) for the major and minor neutralizing Ags, respectively? The findings that rotaviruses are genetically stable and that immunologic variants do not continuously arise—as with influenza viruses (see Chap. 56)—despite their similar segmented genomes and ease of genetic reassortment, imply that immunoprophylaxis through vaccination should be successful. Several of the following approaches to develop a rotavirus vaccine are being explored, any or all of which might be effective: **Live attenuated viral vaccine,** which could consist of attenuated human viruses or an animal rotavirus serologically related to human serotypes (e.g., rhesus rotavirus similar to human type 3). **Cloned ro-**

tavirus genes encoding the neutralizing antigens and inserted into either an animal virus such as adenovirus or vaccinia virus or a bacterial vector that transiently colonizes the small intestine. **Synthetic peptide vaccine** can be readily made, since the amino acid sequences of the major and minor neutralizing Ags are known. **Reassortant viral vaccine,** taking advantage of the genes encoding the neutralizing Ags from human rotaviruses and the ability to propagate animal rotaviruses to high titers.

A recent field trial of the rhesus rotavirus vaccine in 247 infants 1 to 10 months old showed that this vaccine protected 68% of infants against type 3 rotavirus infections during a 1-year period. This study also suggested that immunity is homotypic.

Colorado Tick Fever Virus (Orbivirus)

Colorado tick fever is the only tick-borne viral disease recognized in the United States, though Powassan virus, a tick-borne flavivirus (see Chap. 60), had been isolated in Canada. The virions morphologically are classified as orbiviruses (see Table 62–1) having an outer indistinct capsid and an inner capsid. Unlike other orbiviruses, however, the genome of the Colorado tick fever virus consists of 12 segments whose electrophoretic pattern differs from those of other orbiviruses. Its genome is larger than that of any other member of the Reoviridae family. The virus multiplies readily in hamsters, suckling and adult mice, and some continuous human cell lines. Virions replicate, in large numbers, free in the cytoplasm and unassociated with cell membranes or mitotic spindles.

Colorado tick fever virus consists of two serotypes and neither is immunologically related to reoviruses and rotaviruses. The virus is partially inactivated by ether, probably owing to its skinlike outer coat; but it is not inactivated by sodium deoxycholate (in contrast to viruses with classic envelopes, such as togaviruses, which are also transmitted by insects).

Colorado tick fever is an acute, febrile, nonexanthematous infection characterized by acute onset of fever, chills, headache, and severe pains in the muscles of the back and legs. The course of the disease usually consists of two febrile periods of 2 to 3 days separated by a short, afebrile remission. Recovery appears to be complete but infectious virus often persists in RBCs as long as 4 months. Although the disease is uncommon, this viral persistence provides a problem in selecting blood donors in the locale where Colorado tick fever virus infections occur. Infection induces long-lasting, probably lifelong, immunity.

The disease occurs in the western United States, where its major vector, the tick *Dermacentor andersoni*, is found. The virus has also been isolated from the tick *D. variabilis*, collected on Long Island, NY, but no human infections have been reported from this locality. The golden ground squirrel appears to be the major animal reservoir; the virus has also been isolated from chipmunks, other squirrels, and a deer mouse. Infection of man is only incidental and is a dead end in the chain of transmission.

Prevention is directed primarily toward avoiding ticks, either by not entering infested areas or by wearing suitable clothing and using arthropod repellents.

Other Epidemic Acute Gastroenteritis Viruses

NORWALK VIRUS GROUP
Properties

Epidemiologic studies in families and human volunteer experiments suggested that two immunologically distinct viruses cause acute nonbacterial gastroenteritis, one of the most common illnesses in the United States, and second only to acute respiratory illnesses in families. The value of **immune electron microscopy** to detect and characterize rotaviruses (see Fig. 62–8) soon led to the detection of the second group of viruses; the agent initially isolated was termed *Norwalk virus** (see Fig. 62–9). Although these viruses are unrelated to viruses of the family Reoviridae, they will be described in this chapter, since their taxonomic niche is uncertain, and their pathogenesis and epidemiology are so similar to rotavirus infections (which, however, are generally more severe).

The Norwalk virus is the prototype of a group of related viruses that are extremely fastidious and have not been successfully isolated in cell cultures or animals. Electron microscopy has been the primary technique used to characterize these viruses. The virions are nonenveloped, 23-nm to 34-nm, spherical particles with suggestions of surface indentations (see Fig. 62–9). On the basis of morphology, buoyant density in CsCl, resistance to acid and heat inactivation, and failure to lose infectivity when extracted with ether, the viruses were originally considered to be parvovirus-like (see Chap. 52). However, the virions are slightly larger than parvoviruses, the nature of their nucleic acid is unknown because of the inability to obtain adequate quantities of purified virions from fecal specimens, and they contain a single protein of about 39,000 daltons. This latter characteristic distinguishes the Norwalk viruses from parvoviruses, which contain three structural proteins ranging from 60,000 daltons to 85,000 daltons. Thus, the Norwalk and related viruses are not parvoviruses but are similar to **cali-**

---

* The first viruses were identified during an epidemic in Norwalk, Ohio, and were termed the *Norwalk agents*.

civiruses (see below), which also contain only a single structural protein of about 65,000 daltons.

Pathogenesis and Epidemiology

The disease occurs in outbreaks and is characterized by a combination of nausea, vomiting, diarrhea, low-grade fever, and abdominal pain. The illness is self-limited, usually lasting 1 to 2 days; it is commonly found in families or even in community-wide outbreaks. In contrast to the sporadic form of acute gastroenteritis caused in infants by rotaviruses, acute epidemic gastroenteritis mainly affects **school-aged children and adults** and often spreads to family contacts. Both forms of viral gastroenteritis occur most frequently from September to March.

Virions appear in stools with the onset of disease and are shed in greatest number during the first 24 hours of illness, and for not more than 72 hours after onset. Virus has not been detected during the relatively short incubation period (about 48 hours). Antibodies develop following infection, but their persistence and effect on resistance to subsequent infection are still unknown. Only 50% of persons 40 to 60 years of age have detectable Abs to Norwalk virus, whereas 90% of the same persons have rotavirus Abs. **Three distinct immunologic types** have been identified by immune electron microscopy. However, the frequent occurrence of disease in older children and adults may be due to the existence of many different immunologic types of viruses (like rhinoviruses) rather than to failure of Abs to persist or to produce resistance to infection.

Diagnosis of Norwalk virus group infection can be made by radioimmunoassay as well as by immune electron microscopy. The former technique is the more useful, since it is more economical and requires less equipment and expertise.

CALICIVIRIDAE

The virions possess a characteristic six-pointed starlike shape whose surfaces have cup-shaped (chalice) indentations, a morphology that suggested the family name. Most caliciviruses infect animals other than humans (swine, horses, and even sea lions), but one member of the family causes gastroenteritis in children. Unfortunately, the caliciviruses infecting humans cannot be propagated in cell cultures or in animals, but the caliciviruses of lower animals replicate readily *in vitro*. The classical caliciviruses are about 31 nm to 35 nm in diameter and have genomes consisting of a linear single-stranded RNA, and the virion contains a single, major structural protein.

The epidemiologic and clinical features of calicivirus infections are similar to those of rotaviruses, but the occurrence of family and nosocomial infections is considerably less. Epidemics have been studied in Scandinavia, Canada, Japan, and England, but infections have not been reported in the United States. Like rotaviruses, Abs are acquired between 6 and 24 months of age, and over 90% of adults show immunologic evidence of having been infected.

Selected Reading

BOOKS AND REVIEW ARTICLES

Cukor G, Blacklow NR: Human viral gastroenteritis. Microbiol Rev 48:157, 1984

Estes MK, Graham DY, Dimitrov DN: The molecular epidemiology of rotavirus gastroenteritis. Prog Med Virol 29:1, 1984

Joklik, WK (ed). The Reoviridae. New York, Plenum, 1983

Kapikian AZ, Chanock RM: Rotaviruses. In Fields BN et al (eds): Virology. New York, Raven Press, 1985

Kapikian AZ, Hashino Y, Flores J et al: Alternative approaches to the development of a rotavirus vaccine. In XIth Nobel Conference: Development of Vaccines and Drugs Against Diarrheal Diseases, Stockholm, 1985

SPECIFIC ARTICLES

Bassel-Duby R, Jayasuriya A, Chatterjee D et al: Sequence of reovirus hemagglutinin predicts a coiled-coil structure. Nature 315:421, 1985

Fields BN, Green MI: Genetic and molecular mechanisms of viral pathogenesis: Implications for prevention and treatment. Nature 300:12, 1982

Gomatos PJ, Tamm I: The secondary structure of reovirus RNA. Proc Natl Acad Sci USA 49:707, 1963

McCrae MA, Joklik WK: The nature of the polypeptide encoded by each of the 10 double-stranded segments of reovirus type 3. Virology 89:578, 1978

Mustoe TA, Ramig RF, Sharpe AH, Fields BN: Genetics of reovirus: Identification of ds RNA segments encoding the polypeptides of the μ and σ size classes. Virology 89:594, 1978

Smith RE, Zweerink HJ, Joklik WK: Polypeptide components of virions, top component and cores of reovirus type 3. Virology 39:791, 1969

Harold S. Ginsberg

Hepatitis Viruses

The infectious nature of hepatitis was long unrecognized because the disease tends to occur sporadically and because jaundice, a prominent clinical sign, has many diverse causes. Since the time of Virchow the disease was believed to result from obstruction of the common bile duct by a plug of mucus, and it was known as **acute catarrhal jaundice.** In 1942 Voeght first transmitted hepatitis by feeding a patient's duodenal contents to volunteers. Subsequently, it was found that the etiologic agents are filterable and that the disease may be transmitted in two ways: by the intestinal–oral route (**infectious hepatitis**) or by the injection of infected blood or its products (**serum hepatitis**). However, the differences in transmission are not absolute, since the virus of so-called infectious hepatitis can also produce disease when inoculated parenterally, and experimentally, the virus of serum hepatitis has been transmitted orally. The viruses causing these two types of hepatitis showed other differences, particularly the absence of cross-immunity in human transmission experiments. Accordingly, new names were assigned: **hepatitis A virus (HAV)** and **hepatitis B virus (HBV)** for infectious and serum hepatitis viruses, respectively. Epidemiologic and serologic data indicated that there are at least three additional hepatitis viruses.

Evidence for the existence of more than one hepatitis virus evolved from the serendipitous discovery of the so-called **Australia antigen** (also called **hepatitis-associated Ag [HAA]** or **SH Ag**), which appears only in patients with hepatitis B (serum hepatitis) and is identified as the HBV surface Ag (HB$_s$Ag). This Ag was first detected by Blumberg in the serum of an Australian aborigine. In a search for new serum alloantigens he happened to employ test serum from two hemophiliacs who had received multiple blood transfusions. The sera from these subjects, which contained Abs to the so-called Australia

Ag, were then found to react with serum from a variety of patients who had received multiple transfusions or who resided in institutions (e.g., for mental defectives or for lepers) in which the inmates had a high incidence of hepatitis. The detection of Australia Ag was then recognized as signaling the presence of active or inactive serum hepatitis. The discovery of this novel Ag has permitted further characterization of hepatitis B virus and has allowed diagnostic differentiation of the clinical disease.

Of these two viruses only HAV has been propagated in cell cultures, but both can produce infection in chimpanzees and monkeys, and their distinct morphologies can be identified by electron microscopy. Moreover, the identification of hepatitis A and B viruses has led to the recognition that there are at least two other hepatitis viruses, which are unrelated to HAV or HBV. These viruses, termed **non-A, non-B hepatitis viruses,** are at present the major etiologic agents of hepatitis following transfusions in the United States. The continued high incidence of hepatitis (e.g., in the United States approximately 53,000 cases were reported in 1978) and its great morbidity have stimulated increasing interest in these viruses.

Although the hepatitis viruses are taxonomically dis-tinct, their clinical and epidemiologic associations merit their discussion in a single chapter. Similarly, the **delta virus,** which is a defective virus whose replication and transmission requires the help of HBV and which in return makes the HBV disease more severe, will also be included. The properties of these viruses will be described separately, but the similarities of the diseases they produce indicate that their pathogenesis, epidemiology, and control be discussed and compared together.

Properties of Hepatitis A Virus

A 27-nm viral particle (Fig. 63–1) present in the stools of patients with hepatitis A is clearly the etiologic agent of the disease: the particles (recognizable by electron microscopy) are found in the stools of patients with clinical hepatitis A during the peak of the disease and its associated biochemical changes; these particles are not found in patients with serologically identified hepatitis B or non-A, non-B hepatitis; hepatitis A patients develop Abs that react with the viral particles (recognized by neutralization, immune adherence, and complement fixation [CF]); and the particles have been detected, by immune

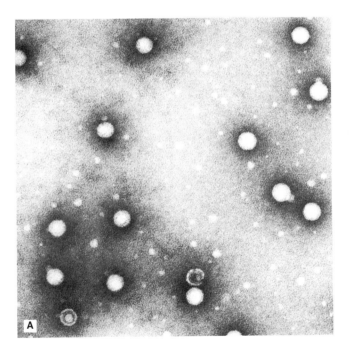
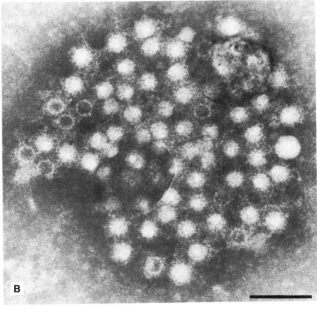

Figure 63–1. Electron micrographs of hepatitis A virus (HAV) particles negatively stained with phosphotungstic acid. (*A*) Virions purified from human stools. Note the fuzzy margins of the virions and the lattice of Ab molecules connecting the virions. (Original magnification ×150,000) (*B*) Immune electron microscopy of HAV virions aggregated by serum from a late-convalescent patient with naturally acquired hepatitis. (Original magnification ×230,000; *A*, courtesy of Dr. John Gerin, Oak Ridge National Laboratory; *B*, Kapikian AZ et al: In Pollard M [ed]: Perspectives in Virology, Vol 9, p 9. San Francisco, Academic Press, 1975)

electron microscopy, in the serum, liver, and feces of marmosets and chimpanzees to whom the disease was transmitted with feces or blood from patients with hepatitis A.

In **morphology** the virions are similar to other picornaviruses (see Chap. 55), with an average diameter of 27 nm, icosahedral symmetry, and no envelope (see Fig. 63–1; see "Properties of Hepatitis A Virus [HAV]"). The **physical and chemical characteristics** of purified hepatitis A virions are like those of enteroviruses: a single-stranded RNA genome, of positive polarity with a 5'-terminal protein (VPg) and a 3' poly(A) tail. The capsid consists of four structural proteins and is even more stable to acid and heat than other enteroviruses (see "Properties of Hepatitis A Virus [HAV]"). The chemical and physical characteristics have led to its formal classification as **enterovirus type 72.** As with most viruses, including HBV, infectivity is inactivated by ultraviolet irradiation, formalin (1:400 dilution) for 3 days at 37°C, and heat for 5 minutes at 100°C (these properties are important for vaccine development). The virions are relatively resistant to common disinfectants, however, and therefore special cleansing procedures must be used to prevent hospital spread.

Human transmission studies and extensive epidemiologic data suggest that the virus exists as a **single immunologic type** and that long-lasting immunity follows infection. Antibodies appear shortly after the onset of clinical disease (CF Abs being detectable earliest); Ab levels rise slowly and persist for years (Fig. 63–2).

Initially, only marmoset-adapted HAV could be propagated in explant cultures of marmoset livers and in a fetal monkey kidney cell line. Presently, however, **viral multiplication** can be carried out in African green monkey and fetal rhesus monkey cells, Vero cells, human diploid lung cells, and a human liver hepatoma cell line. Although viral titers as high as 10^8 TCID$_{50}$ per milliliter are attained in cell culture, cytopathic effects do not occur. Despite the ability to propagate virus *in vitro*, biochem-

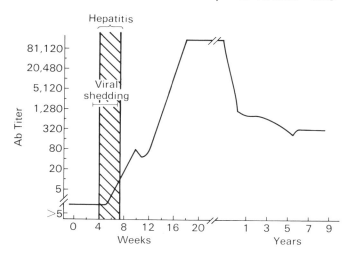

Figure 63–2. Sequential relationship of viral shedding to clinical hepatitis, and the development of virus-specific Abs during HAV infection. (Modified from Krugman S, Friedman H, Lattimer C: N Engl J Med 292:1141, 1975

ical studies of viral synthesis have been difficult to carry out. Nevertheless, it is possible to isolate HAV directly from patients' feces in cell cultures, and the replication of virus *in vitro* has made the production of a vaccine possible (see later).

Although direct biochemical studies of viral replication have not been done, the entire genome has been sequenced from cloned cDNAs. These studies showed that the genome is organized like other picornaviruses, that it can encode a polyprotein with the viral structural proteins at the amino-terminal portion and the RNA polymerase in the remainder.

Properties of Hepatitis B Virus

MORPHOLOGY

Serum from patients with clinical hepatitis B commonly contains three distinct structures (Fig. 63–3) that possess the hepatitis B surface antigen (HB$_s$Ag; see Antigenic Structure, later). The **Dane particle** (named for its first observer) is the least common form, but it alone has the structure attributed to viruses and appears to be infectious. This particle is a complex sphere, 42 nm in diameter with an electron-dense core 28 nm in diameter (Fig. 63–3) surrounded by an envelope. Negative stain can penetrate the Dane particles, revealing the 7-nm outer layer or envelope. Occasional particles lack the core and have a translucent center. The core structures can be isolated by reacting Dane particles with sodium dodecyl sulfate. **Spherical particles** with an average diameter of about 22 nm (range, 16 nm to 25 nm) are most numerous.

PROPERTIES OF HEPATITIS A VIRUS (HAV)

Size: 27–32 nm
Morphology: Icosahedral; naked
Genome: Single-stranded RNA; 2.25–2.28 × 10^6 mol. wt. Positive polarity
Structural proteins:
    VP1—30–33 Kd
    VP2—24–27 Kd
    VP3—21–23 Kd
    VP4— 7–14 Kd
    VPg—
Stability
    Ether: Stable
    Acid: Stable, pH 3.0
    Heat: Stable (60°C/60 min)

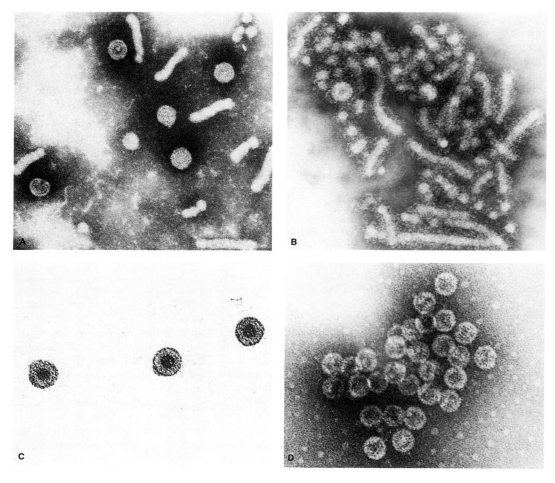

Figure 63–3. Electron micrographs of negatively stained forms of hepatitis B virus (HBV). (*A*) Dane particles, filaments, and 22-nm HB$_s$Ag particles present in the serum of a patient with active infection. (Original magnification ×150,000) (*B*) Immune electron microscopy of Dane particles, filaments, and 22-nm HB$_s$Ag particles aggregated by serum from a late-convalescent patient. (Original magnification ×150,000) (*C*) Purified Dane particles stained with uranyl acetate to show the DNA core. (Original magnification ×200,000). (*D*) Cores of HBV isolated from the liver of a patient with active hepatitis. (Original magnification ×200,000) (*A–C*, courtesy of Dr. John Gerin, Oak Ridge National Laboratory; *D*, Kaplan PM et al: J Virol 17:885, 1976)

These particles are unlike any known virus. Their surfaces appear to have a subunit structure similar to that of Dane particles, but no uniform symmetry is discernible. **Filamentous forms,** about 22 nm in diameter and ranging in length from 50 nm to greater than 230 nm, are also plentiful and have a similar surface structure. When the filaments are mixed with a nonionic detergent, they form spheres that are morphologically and antigenically similar to the 22-nm particles, suggesting that the filaments are aggregates of the spherical particles.

The ease with which the preceding forms can be observed is a reflection of their amazing abundance in patients' sera. In one serum sample, for example, 3×10^{13} 22-nm particles per milliliter were counted, with 1/15 as many filaments and 1/1500 as many Dane particles. A serum may be infectious for humans in a 10^{-7} dilution.

PHYSICAL AND CHEMICAL CHARACTERISTICS

The abundance of viral structures in blood (see Fig. 63–3; Fig. 63–4) has made physical and chemical analyses possible despite the inability to propagate the virus in cell cultures (Table 63–1).

Virions (Dane particles) and isolated 28-nm cores (nucleocapsids) contain the **viral genome, a circular, partially double-stranded DNA molecule,** which consists of a long (L) strand of constant length with a free 3′ terminus and a 5′ terminus that is covalently linked to a

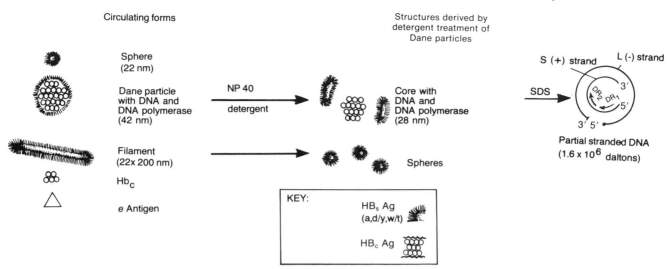

Figure 63–4. Schematic diagram of forms of HBV and the DNA structure. (Modified from Robinson WS, Lutwick LI: N Engl J Med 295:1232, 1976)

TABLE 63–1. *Physical and Chemical Characteristics of Hepatitis B Viral Forms*

Characteristic	22-nm Spheres	Filamentous Forms	Dane Particles (42-nm Spheres)	Core Particles
Buoyant density (CsCl)	1.20 g/ml³	1.20 g/ml³	1.25 g/ml³	1.36 g/ml³
Sedimentation coefficient	39–54S		58.5S	110S
Protein	+		+	+
Glycoproteins	3		3	0
Lipid	+		+	0
Nucleic acid	0		+	0
			Circular, interrupted double-stranded DNA	Circular, interrupted double-stranded DNA

small protein and a short (S) strand that varies from 15% to 60% of the circle length (see Fig. 63–4; Fig. 63–5). The positions of the 5' termini of the L(−) and S(+) strands are fixed, whereas the 3' end of the S(+) strand is variable. The 5' ends of both strands are base-paired, which assures the circular form of the double-stranded DNA. In addition, at both sides of the cohesive 5' ends there is an 11-base-pair direct repeat (5'TTCACCTCTGC); similar direct repeats (DR1 and DR2) are found in all the hepadnaviruses,* which suggests that they play an important role. This DNA is smaller than that of any known animal virus containing double-stranded DNA. The nucleocapsid consists of the core Ag (HB$_c$Ag), a DNA polymerase, and a protein kinase. The DNA polymerase of the virion is dependent on the four deoxynucleoside triphosphates and Mg^{2+}; it completes the single-stranded regions without any added DNA or RNA primer, producing covalently closed, double-stranded circular DNA. The polymerase also serves as a reverse transcriptase (see below).

The viral envelope contains three glycoproteins, termed the *major, middle,* and *large proteins* (see Fig. 63–4). The **hepatitis B surface Ag (HB$_s$Ag),** which is the predominant component of the 22-nm particles found in the blood of patients with active disease and carriers, is a disulfide-bonded dimer of two molecules of the major protein. In the middle protein a 55 amino acid sequence, which is the region distinct from the major protein, probably contains the receptor for polymerized human serum albumin that appears to mediate the attachment

* Viruses with the same morphologic, chemical, and physical characteristics as HBV have also been found in woodchucks, ground squirrels, and Peking ducks. Indeed, hepatitis and hepatocarcinoma are associated with the infection of woodchucks. Accordingly, these viruses have been taxonomically grouped into the family **Hepadnaviridae.**

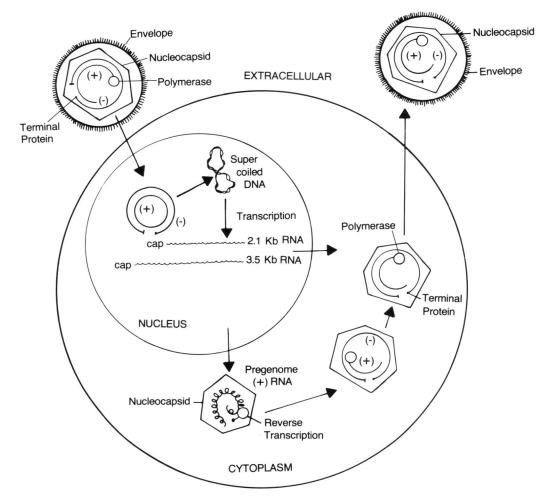

Figure 63–5. Multiplication of hepatitis B virus in hepatocytes.

of the virus to hepatocytes. The large protein consists of the middle protein plus about 128 amino acids added to its N terminus; this exposed additional sequence appears also to be important for attachment to hepatocytes.

A virion contains 300 to 400 molecules of major proteins and 40 to 80 of the middle and large proteins; filaments have the same protein composition as virions. However, the spherical 22-nm particles are different: in chronic carriers in which active viral replication ensues, they contain the same relative numbers of major and middle proteins but 20 times fewer large proteins; in the absence of viral replication they contain mostly the major protein, less than 1% of middle protein, and no large protein.

The **nucleocapsid** contains only a single protein, the **core protein,** which possesses **protein kinase activity**

capable of phosphorylating itself. The core protein is expressed as the **HB$_c$Ag,** and when the capsid is disrupted *in vivo*, proteolytic cleavage uncovers the **HB$_c$Ag** determinant. The core protein also interacts with the HBV genome through its arginine-rich C-terminal region.

ANTIGENIC STRUCTURE

The 42-nm enveloped virion containing DNA has two distinct major Ags, the **surface Ag (HB$_s$Ag)** and the **core Ag (HB$_c$Ag).** The HB$_s$Ag is the only Ag on the two other hepatitis B particles circulating in the blood, that is, the 22-nm spheres and the filamentous particles (see Fig. 65–4).

The HB$_s$Ag contains several antigenic determinants: a **group-specific determinant, a,** which is present in all HB$_s$Ag-containing material, and two sets of generally mutually exclusive type-specific determinants, **d** or **y** and **w** or **r**, which represent sets of alleles of two independent genetic loci. In addition, four variants of the **w** allelic determinant have been observed. Thus, a number of phenotypic combinations or subtypes are possible (Table 63–2). A few sera containing surface antigenic reactivity of mixed subtype (**adyw, adyr, awr, adwr,** and **adywr**) have also been reported; whether these arise from mixed HBV infection or unusual phenotypic mixing is unclear.

The subtypes of HBsAg are not associated with different biological activities of the viruses, and no clinical differences have been observed in the infections caused by the various subtypes. These subtypes are valuable as markers for studies of the epidemiologic behavior of the virus.

Chemical reduction and alkylation of 22-nm particles and purified HBsAg markedly reduces its antigenicity, which emphasizes the critical importance of its disulfide-linked dimeric structure. Deglycosylation also greatly diminishes this reactivity of the HBsAg.

The **HBcAg** is of a **single antigenic type** and is found only in the core of the 42-nm Dane particle and in free core particles present in the nuclei of hepatocytes from infected livers. Hence, to detect the HBcAg in plasma one must disrupt the enveloped 42-nm particles with lipid solvents to expose the core Ag, whereas the HBsAg can be readily detected in serum during acute hepatitis B and in chronic carriers. Some HBsAg-positive sera contain an additional specific Ag, designated **e (HBeAg),** which is considerably smaller than either the HBsAg or the HBcAg particle (about 12S); it is also distinct from the viral DNA polymerase. The HBeAG is specific for hepatitis B virus infection. Two subdeterminants (HBeAg/1 and HBeAg/2) have been detected. The HBeAg, in part bound to immunoglobulins, appears during the incubation period of acute hepatitis B, just after the appearance of HBsAg and prior to clinically apparent liver injury. Antibodies to HBeAg occur frequently in HBsAg-containing serum and even in serum having anti-HBsAbs. The function of the e

Ag is unknown, but epidemiologic evidence indicates that the presence of this viral gene product in serum is associated with active viral replication and liver pathology. Patients with HBeAg are those most likely to have active disease and to be efficient transmitters of infection.

IMMUNITY

Antibodies directed against the surface and core Ags appear at different times and show different patterns of disappearance (Fig. 63–6). Antibodies to HBcAg increase rapidly during the early phase of clinical disease; their appearance is usually associated with reduced viral replication and beginning resolution of disease. However, these Abs are also consistently present in chronic carriers of HBsAg, and they may be a sensitive indicator of continued viral replication. The presence of anti-HBcAbs does not uniformly indicate a good prognosis, and serum from such patients may even be infectious.

Cell-mediated immunity to HBsAg appears near the end of the acute phase of hepatitis, and its increase is correlated with the disappearance of the circulating Ag. In contrast, Abs to HBsAg do not appear until months after termination of the clinical illness (Fig. 63–6). It is also noteworthy that in chronic HBsAg carriers specific cell-mediated immunity is generally decreased and circulating anti-HBsAbs are not demonstrable, but electron microscopic examination of serum detects HBsAg–Ab complexes.

An increase in Abs to HBsAg is clearly correlated with immunity. Thus, human γ-globulin containing these Abs is of value in preventing hepatitis B; and immunization in humans with purified HBsAg induces formation of Abs to HBsAg, accompanied by immunity to challenge. In addition, immunization of chimpanzees with purified HBsAg results in resistance to challenge with HBV-infected serum (see Prevention and Control).

MULTIPLICATION

Without a susceptible cell in which HBV can replicate, data on the steps in viral multiplication were derived entirely from identification of viral structures detected in hepatocytes from patients with acute and chronic HBV hepatitis and from ducks and ground squirrels infected with the hepatocytes. These studies led to a unique model of biosynthesis of a DNA virus (see Fig. 63–5). Following entry and uncoating of the virion into susceptible cells, the viral DNA reaches the nucleus and utilizing the virion DNA polymerase forms complete double-stranded circles that attain a supercoiled configuration. Host cell DNA-dependent RNA polymerase then transcribes the DNA minus-strand to make multiple complete RNA cop-

TABLE 63–2. Phenotypic Combinations or Subtypes of HBs Antigens Observed in Patients

Group Determinant	Subtype Determinant	
	d/y	w/r
a	y	w_1
a	y	w_2
a	y	w_3
a	y	w_4
a	y	r
a	d	r
a	d	w_2
a	d	w_4

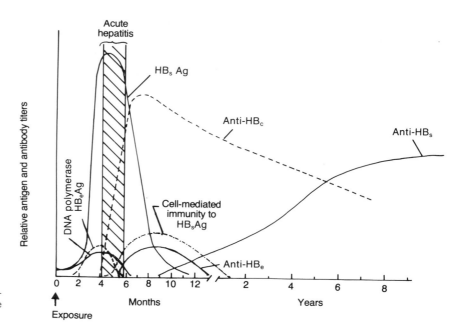

Figure 63–6. Diagram of the time course of circulating HBV Ags and the immune response. Note that HB$_c$Ag is not found free in the plasma.

ies termed the **pregenome RNA,** as well as a smaller capped RNA (see Chaps. 48 and 65). The RNAs are transported into the cytoplasm where the core protein is made and rapidly encapsidates the pregenome RNA as well as the newly synthesized viral polymerase and the DNA terminal protein. There follows a unique reaction for DNA viruses, reverse transcription of the pregenome RNA, with the DNA-linked terminal protein probably serving as a primer to initiate synthesis of the DNA minus-strand. As viral DNA is synthesized, the pregenome RNA is degraded, except for a small fragment of about 20 base pairs derived from the 5′ end that appears to act as a primer for synthesis of the DNA plus-strand from the newly made minus-strand template. The progeny particles usually contain only partially double-stranded viral DNA. They obtain HB$_s$Ag-containing envelopes, probably by budding from the cell's plasma membrane, and become infectious virions.

Thus, hepadnaviruses employ a strategy similar to that of retroviruses for genome replication. However, hepadnaviruses employ the strategy of initiating and terminating the synthetic cycle with DNA rather than RNA genomes. Hepadnaviruses differ from retroviruses in one other characteristic: whereas the retrovirus RNA is always transcribed from an integrated DNA molecule, hepadnavirus can transcribe the pregenome RNA from free as well as integrated viral DNA. It is striking to note that there is significant homology in nucleotide sequence (particularly in the **pol** gene) between the genomes of HBV and several retroviruses.

Properties of Non-A, Non-B Hepatitis Viruses

The sensitive immunologic and electron microscopic methods used to detect the viruses and Ags of hepatitis A and B viruses have made it possible to identify at least two other viruses. One of these has been transmitted to chimpanzees, and the virus has been identified by immune electron microscopy. The virions are nonenveloped icosahedrons about 27 nm in diameter, which, like HAV, resemble picornaviruses in morphologic and some physical characteristics. A second virus, which can also infect chimpanzees, is larger and inactivated by lipid solvents. The non-A, non-B hepatitis viruses do not cross-react immunologically with Ags of hepatitis A or B viruses.

Properties of Delta Hepatitis (D) Virus

In 1977 a new Ag termed *delta Ag* was noted in nuclei of hepatocytes in patients with chronic hepatitis B, but neither the HB$_c$Ag nor other HBV Ags were detected in the cells containing the delta Ag (δ Ag). Indeed, the delta Ag is found only in patients with chronic and often severe HBV hepatitis; it is not noted in patients with non-A, non-B, or HAV hepatitis. The delta virus genome is a 1.75-Kb single-stranded RNA that does not have any homology with the HBV genome. However, the nucleotide sequence of the delta virus RNA shows extensive homology with a common viroid (see Chap. 51). The genome of extracellular delta virus found in blood is associated with δ Ag and encoated by the HB$_s$Ag to form particles of 35 nm to 37 nm.

Studies in chimpanzees, which are susceptible to infection with δ virus as well as HBV, indicate that the δ virus is defective, and its replication requires coinfection with HBV to act as a helper virus. Moreover, δ virus replication appears to depress replication of HBV, just as coinfection of cells with the defective adeno-associated virus (AAV) and an adenovirus results in multiplication of AAV and a decreased yield of its helper adenovirus (see Chap. 52). The HB$_s$Ag covers the δ virus particle and probably serves to protect the δ virus RNA genome as well as to permit the particles' attachment to and entry of the host hepatocytes.

Pathogenesis

The response of humans to infection with hepatitis viruses ranges from inapparent infection and nonicteric hepatitis to severe jaundice, liver degeneration, and death; the disease is often debilitating and convalescence is prolonged. Though clinical differences in type A and type B hepatitis have been described, the acute disease may be clinically indistinguishable, so that the two infections are often differentiated (Table 63–3) mainly by the routes of infection, the length of the incubation periods, and the laboratory identification of specific viruses, Ags, and Abs.

Hepatitis A virus (HAV) usually enters by the oral route and multiplies in the gastrointestinal tract (probably in the epithelium, although, as in poliovirus infection, mesenteric lymph nodes may also be involved). Viremia eventually occurs, and virus spreads to cells of the liver, kidney, and spleen. Virus can be detected in the feces and duodenal contents and in the blood and urine during the preicteric and the initial portion of the icteric phases (Fig. 63–7). When quantitative immune electron microscopy is used, HAV is first detected during the **preicteric period** and attains its highest concentration in the feces prior to the appearance of jaundice (see Fig. 63–7). Indeed, the onset of jaundice usually heralds the approaching termination of viral shedding. More sensitive human transmission studies, however, indicate that virus may be shed in the feces for slightly longer periods. When virus is present in the feces it is also found in liver cells and in the bile. Antibodies to HAV appear as the viral titer decreases and liver damage becomes apparent (see Fig. 63–7). This timing suggests that liver damage may be effected, at least in part, by immunologic mechanisms (although HAV–Ab complexes have not been detected in either chimpanzee or man, and the appearance of specific cytotoxic T cells has not been studied).

TABLE 63–3. Differentiating Characteristics of Hepatitis Types A and B

Property	Type A	Type B
Usual transmission	Fecal–oral	Parenteral inoculation*
Characteristic incubation period†	15–40 days	60–160 days
Type of onset	Acute	Insidious
Fever >38°C	Common	Uncommon
Seasonal incidence	Autumn and winter	Year-round
Age incidence	Most common in children and young adults	All ages; most common in adults
Size of virus	27 nm	42 nm
Virus in feces	Incubation period and acute phase	Not demonstrated
Virus in blood	End of incubation period and briefly during acute phase	Incubation period and acute phase; may persist for years
Appearance of HB$_s$Ag	Absent	14–50 days after infection
Detection of HB$_s$Ag		Blood (less often in feces, urine, semen, and bile)
Duration of HB$_s$Ag		60 days to years
Prophylactic value of γ-globulin	Good	Good if titer of anti-HB$_s$Ag is high

* Parenteral injection is probably not the predominant mode of transmission in developing countries where the means of spread is unknown.

† Considerable overlapping in the duration of incubation periods for types A and B has been noted in volunteers as well as in patients during epidemics, that is, as long as 85 days for type A hepatitis and as short as 20 days for type B hepatitis.

Hepatitis B virus (HBV) enters predominantly by the parenteral route, but its primary site of replication is unknown. However, HB$_c$Ag is present in the nuclei of hepatocytes as early as 2 weeks after experimental infection in chimpanzees, who develop a disease closely akin to that in man. During the latter half of the incubation period (see Fig. 63–6), 5 to 8 weeks after infection, the blood becomes infectious (a danger if used for transfusion), and it contains detectable HB$_s$Ag and viral DNA polymerase. The HB$_s$Ag has also been identified in bile, urine, semen, feces, and nasopharyngeal secretions. It is striking that the HB$_s$Ag and DNA polymerase usually begin to decline in the blood before acute symptoms disappear and are not detectable by the time liver functions return to normal.

Whether immune reactions play a role in the pathogenesis of hepatitis B is still unclear but, as noted earlier, anti-HB$_c$Abs appear at the onset of disease; HB$_s$Ags are present in large quantity weeks before anti-HB$_s$Abs appear; and cell-mediated immunity to HB$_s$Ag becomes detectable during the active disease, and its increasing de-

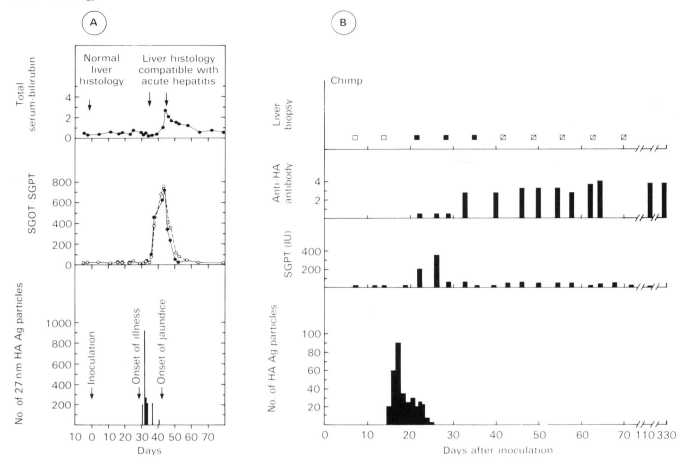

Figure 63–7. Temporal relationship of clinical illness, liver pathology, serum enzyme alterations, fecal shedding of HAV Ag, and Ab response in experimental HAV infection (*A*) in man and (*B*) in the chimpanzee. *SGOT* and *SGPT,* serum glutamic–oxaloacetic and pyruvic transaminases. Liver histology: □, normal; ■, acute hepatitis; ◪, resolving hepatitis. (Dienstag JL et al: Lancet 1:765, 1975; Dienstag JL et al: J Infect Dis 132:532, 1975)

velopment correlates with the fall in HB$_s$Ag titer (see Fig. 63–6).

Both HAV and HBV infections primarily affect the liver. When jaundice occurs, it is usually preceded by anorexia, malaise, nausea, diarrhea, abdominal discomfort, fever, and chilliness. This preicteric phase may last from 2 days to 3 weeks. The icteric stage of hepatitis A usually has an abrupt onset with a sharp rise in temperature, whereas that of hepatitis B characteristically appears more insidiously and with less fever. Nevertheless, **hepatitis B is ordinarily a more severe disease,** the fatality rate sometimes being as high as 50%, whereas in hepatitis A fatalities rarely exceed 1% (although convalescence is often prolonged). These differences, however, may not reflect only the characteristics of the viruses, for persons receiving blood or plasma transfusions are usually ill at the time of inoculation.

Liver biopsies obtained during the course of the illness have revealed early cloudy swelling and fatty metamorphosis at the time clinical symptoms begin and diffuse parenchymal destruction by the time jaundice has developed. No inclusion bodies are seen in affected cells; intracellular HB$_s$Ag, however, can be detected by immunofluorescence early in the course of type B hepatitis, and occasionally 20-nm to 30-nm virus-like particles (probably the nucleocapsids or cores) are seen in liver biopsies. The degeneration of cells is not localized to any one part of the liver lobule, in contrast to the findings in yellow fever (see Chap. 60) or chemical hepatitis. With recovery, hepatic cells regenerate; scar tissue (cirrhosis) develops only after extensive or long-standing destruction of cells, which is primarily associated with hepatitis B. In fatal infections, the liver parenchyma is often almost completely destroyed (**acute yellow atrophy).**

In 10% to 20% of adult patients and about 35% of children HB$_s$Ag persists in the blood for extended periods, but less than half of these become **chronic Ag carriers.** More than half of the chronic carriers continue to manifest biochemical and pathologic evidence of chronic liver disease. Although the mechanism of persistent infection with antigenemia is unexplained, defects in cell-mediated immunity (e.g., that directed at HB$_s$Ag) or other forms of immunodeficiency seem likely. The delta virus produces its most deleterious effects when it infects chronic HB$_s$Ag carriers. Simultaneous infection with HBV and δ virus does not produce a disease different from that induced by HBV alone.

It has been proposed that HBV infection is causally related to hepatocarcinoma based on the evidence that there is a close correlation between the geographic distribution of hepatitis B and the incidence of hepatocarcinoma, HB$_s$Ag is detected in patients with hepatocarcinoma with an unusually high frequency, and integrated and free viral DNA is found in many, but not all, hepatocellular carcinomas. The striking association between the woodchuck hepatitis virus infections and hepatocarcinomas in these animals adds support to the hypothesis that HBV is an oncogenic virus.

Laboratory Diagnosis

Immunologic techniques provide the most sensitive and economical approach for detecting viral Ags and Abs. Immune electron microscopy can be used to identify HAV in feces and HBV in blood or liver biopsy tissue. If a liver biopsy is obtained, both viruses can be identified by immunofluorescent techniques. The HAV Ags or Abs can be easily assayed by CF, immune adherence, enzyme-linked immunosorbent assay (ELISA), or radioimmunoassay (RIA). The CF Abs are detectable first. The ELISA and solid-phase RIA are the most sensitive of these techniques, but the immune adherence titration (which requires purified HAV Ag or standardized serum, guinea pig complement, and human group O red blood cells [RBCs] is simpler, less costly, and faster, as well as being quite sensitive and specific. Since Abs are usually present at the onset of illness (see Fig. 63–7), serologic diagnosis requires detection of HAV-specific IgM Abs. The Abs in serum from clinical cases also neutralize HAV in cell culture, as well as in chimpanzees and marmosets, but this procedure is too expensive and time-consuming.

A number of tests have been devised to detect and measure HBV Ags, particularly the HB$_s$Ag. The most sensitive techniques now available are RIAs and ELISAs with either a micro-solid-phase method or a double-Ab immunoprecipitation procedure. The passive hemagglutination assay (using RBCs coated with either HB$_s$Ag or anti-HB$_s$Abs) is also an accurate and very sensitive method for quantitating either Ag or Ab. Agar gel diffusion and countercurrent immunoelectrophoresis are of considerable value for identifying the subtype Ags and the e Ag. Since HB$_c$Ag does not circulate in the blood, its measurement is not useful for clinical diagnosis. Detection of Abs to HB$_c$Ag is valuable, however, since their presence appears to reflect viral replication. Similarly, HB$_e$Ag has been associated with viral infectivity and therefore signifies specific HBV infection.

Epidemiology

Predominantly, the **type A hepatitis virus** (of the **short-incubation disease** previously called *infectious hepatitis*) is spread by ingestion (particularly in epidemics), and **type B hepatitis virus** (of the **long-incubation disease** formerly termed *serum hepatitis*) is disseminated by parenteral inoculation. However, the viruses of both diseases can be rarely transmitted by both the oral and the parenteral route. The diseases produced by **non-A, non-B hepatitis viruses** may have a relatively short incubation period, but their major spread is also by parenteral inoculation. However, an epidemic form of non-A, non-B hepatitis, probably spread by the fecal–oral route, has been observed in several areas of Southeast Asia and North Africa. Epidemic non-A, non-B hepatitis, which predominantly affects adults, has not been detected in the United States or Western Europe, but it is not known whether the etiologic agent(s) exists in these regions.

HEPATITIS A

Hepatitis A mimics poliomyelitis in many of its epidemiologic features. When environmental factors favor widespread intestinal–oral transmission, the disease is **endemic,** and infection (usually mild or inapparent) occurs in the very young. Under these conditions the disease in adults is uncommon and epidemics are rare. On the other hand, under good sanitary conditions, spread of the virus is restricted, and adulthood is frequently attained without immunity. In such nonimmune populations, especially in military groups, **epidemics** are likely, and the source of virus can usually be traced to **contamination of water or food** by infected humans.

The danger of HAV dissemination from an infected person is greatest during the latter part of the incubation period, when viral shedding in the feces is greatest but is unrecognized because jaundice is not yet present (see Fig. 63–7). Long-term intestinal carriers of the virus have not been detected. Because **subclinical infections,** particularly in children, often predominate, secondary person-to-person spread and contamination of food and drink are common. Such transmission is particularly favored by the unusual stability of hepatitis A virus and its notable resistance to disinfectants, such as chlorine at ordinary concentrations in water. Not only contaminated water but also **shellfish** that live in it (and concentrate the virus) may be sources of infection: for example, raw oysters and clams obtained from polluted waters have been the origin of numerous epidemics throughout the world. Hepatitis A virus is not a common cause of transfusion hepatitis because viremia is usually brief and chronic viremia occurs rarely, if ever.

Subhuman primates, particularly chimpanzees, are the only known natural nonhuman hosts. Hepatitis has developed among handlers 3 to 6 weeks after infected animals arrived in the United States from Africa. Apparently, the young animals were infected by man after capture and subsequently excreted HAV in their feces, and hence infected their handlers. (Some animals showed clinical manifestations of disease.)

HEPATITIS B

Hepatitis B is readily distinguishable from hepatitis A not only by morphologic and immunologic properties but also on epidemiologic evidence, such as the long incubation period, parenteral transmission, increasing incidence with age, and nonepidemicity (see Table 63–3).

Despite sensitive immunologic techniques to detect viral carriers by screening the blood of donors for HB_sAg, blood and its products continue to be a potential source of hepatitis in the United States. However, only a portion of these cases are caused by undetected hepatitis B virus. Indeed, about 90% of the cases now transmitted by **transfusion**[*] in the United States are caused by **non-A, non-B hepatitis viruses.** The unrecognizable chronic carriers of hepatitis B or a non-A, non-B virus are the pernicious sources of infection, spreading virus via transfusions of infected blood, plasma, or convalescent serum; via contaminated fibrinogen; and via inadequately sterilized syringes, needles, or instruments (medical and dental) containing traces of contaminated blood. The last source, which includes common stylets for blood counts and syringes for drawing blood, has been essentially eliminated in many parts of the world by the use of disposable instruments. However, needles used in tattooing and communal equipment used by drug addicts are still a common means of viral dissemination. Injection of as little as 10^{-6} ml to 10^{-7} ml of contaminated blood may transmit infection, as predicted by the electron microscopic finding of as many as 10^7 Dane particles per milliliter of blood.

The large amount of virus and its Ags in blood and the demonstration of HB_sAg (and the probable presence of

virus) in other body fluids as well also explain why type B hepatitis is an **occupational disease** of health personnel—dentists, physicians, nurses, and ward personnel who are frequently exposed to the unrecognized chronic carriers; laboratory workers who handle blood; and technicians who process human plasma and blood products.

Health personnel who work in hemodialysis units and in cancer therapy wards, where the chronic carrier rates among patients are high, appear to be particularly vulnerable. It is striking that infected professionals serving in renal dialysis units, for example, develop acute hepatitis, whereas the patients, who have various forms of immunodeficiencies, do not manifest clinical disease. The mechanism by which virus is transmitted from carriers to these healthy workers is not entirely clear. Many infected workers do not recall any accidental parenteral injections, although possible viral entrance through skin abrasions cannot be ignored. It should be recalled that HB_sAg is detectable in the saliva, urine, and feces of chronic carriers, as well as in blood and its products, so that entrance of the virus through membranes of the eye or mouth must be considered a plausible route of infection.

Additional epidemiologic observations provide further evidence that transmission of HBV by means other than parenteral injection is also likely when chronic infection exists. It is now clear that HBV has disseminated and persists throughout the world, even in remote and insular localities where medical care is primitive and blood and its products are not commonly used for therapeutic or prophylactic purposes. Indeed, there is evidence suggesting that all biological fluids from HBV-infected persons are infectious and potentially able to transmit the virus. Hence, nonparenteral transmission of virus must occur. Furthermore, family clusters of type B hepatitis are being observed with increasing frequency; these cases are grouped around an index case with whom family members have had close person-to-person contact but no known exchange of blood. As noted previously, experimental oral transmission has been demonstrated, and sexual transmission is also probable. Neonates may be infected during gestation by placental transmission (cord blood often contains HB_sAg), at the time of delivery, or in the postnatal period.

Although hepatitis B is usually sporadic, epidemics may occur when many samples of serum or plasma are pooled. For example, in the early 1940s more than 28,000 cases of serum hepatitis in American military personnel resulted from the use of yellow fever vaccine containing contaminated human serum to stabilize the live virus.

NON-A, NON-B HEPATITIS

Sensitive immunologic techniques used to screen blood donors for HBV Ags and to establish the diagnosis of viral

---

[*] Overt hepatitis follows about 1% of blood transfusions in the United States. On the assumption that even more inapparent infections are produced, it has been estimated that 2% to 3% of the adult population carries a hepatitis virus.

hepatitis have sharply reduced the incidence of hepatitis B infections, but they have also revealed another agent (or agents) as a cause of posttransfusion hepatitis. **Non-A, non-B hepatitis** is now the major form of posttransfusion hepatitis in the United States. The incubation period is usually 6 to 8 weeks; however, periods as short as 10 days and as long as 11 weeks have been recorded (which further implies that more than one virus is responsible for non-A, non-B hepatitis). Indeed, in large outbreaks that were epidemiologically similar to those produced by HAV, virions morphologically akin to a picornavirus were detected in feces by means of immune electron microscopic techniques.

Prevention and Control

No specific therapy is available for any of the viral hepatitides, although some antiviral drugs (e.g., acyclovir and adenine arabinoside) have been tried experimentally. In all proved or suspected cases of viral hepatitis great care should be taken in the disposal of feces and of all syringes, needles, plastic tubing, and other equipment used for blood sampling and parenteral therapy. Whenever possible, disposable equipment (including needles and plastic syringes and even thermometers) should be used in hospital and office practice. A syringe, once used, should not be reused with a fresh needle, even merely to obtain a blood specimen. Nondisposable equipment and supplies (e.g., dishes and bed clothing) should be autoclaved at 15 lb pressure (121°C), boiled in water for at least 20 minutes, or heated to 180°C for 1 hour in a sterilizing oven.

Subjects giving a history of jaundice or with detectable HB$_s$Ag in their blood should not be used as blood donors. Blood HB$_s$Ag has been reported in about 0.3% of blood donors in New York and 0.2% to 1.2% of different groups of blood donors in Tokyo; the carrier rate among commercial donors, particularly drug addicts, is 3 to 10 times higher than that among volunteers. However, the incidence of hepatitis non-A, non-B Ags in the population is unknown. Because minute amounts of contaminated plasma can initiate infection, the practice of pooling plasmas should be avoided: then plasma from an infected individual, used unwittingly, will infect only one person.

The protection of individuals exposed to patients, carriers, or contaminated blood requires additional consideration. Since hepatitis A patients usually shed virus only briefly after jaundice appears (see Fig. 63–2), and viremia is transient, these patients in hospitals and at home constitute a hazard for their contacts for only a short period. In contrast, hepatitis B (and probably non-A, non-B hepatitis) patients are potential sources of virus for transmission over a prolonged period, whether they are suffering from acute icteric hepatitis or are chronic carriers without clinical liver disease. Accordingly, all close contacts of carriers or possible carriers of HBV must take every precaution to prevent exposure to blood (and to objects potentially contaminated with blood—e.g., toothbrushes, razors) and to body excreta. These precautions are of greatest significance for personnel working in hemodialysis units, intensive care units, and custodial mental institutions; for dentists; for technicians in clinical laboratories and blood processing facilities; and for close family contacts (particularly spouses).

Type A hepatitis may be prevented by **passive immunization.** Pooled human γ-globulin* reduces the incidence of iceteric disease (but not of infection) when given early in the incubation period. Initially, inconsistent results were obtained in preventing hepatitis B with pooled γ-globulin. However, the advent of assays for anti-HB$_s$Abs has made it apparent that immune γ-globulin containing a high titer of anti-HB$_s$Abs is partially effective; that is, in different recipients it prevents disease, decreases the severity of hepatitis, or markedly prolongs the incubation period. Therefore, γ-globulin containing anti-HB$_s$Abs is recommended for prophylaxis in exposed persons; however, for those who are at a high risk of acquiring HBV infection owing to their occupation, newly developed vaccines are more effective. For the newborn of a mother positive for HB$_s$Ag and HB$_c$Ag, a regimen combining one dose of HB$_s$Ab-positive γ-globulin at birth with an HB$_s$Ag vaccine soon after birth is 85% to 90% effective in preventing infection.

The ability to propagate virus in cell cultures is leading to the development of an attenuated virus HAV vaccine. Trials in marmosets, chimpanzees, and humans are in progress. The potential is also great for the production of a subunit Ag vaccine in *Escherichia coli* through the use of a plasmid vector or in animal cells by means of an animal virus vector, such as vaccinia virus or adenovirus (see Chap. 50). For hepatitis B the relatively large amounts of HB$_s$Ag in the serum of chronic carriers make it possible to purify the Ags that are essential for inducing neutralizing Abs, and allow the use of this material for immunization after inactivation of any infectious virus present. This vaccine has been licensed and is now successfully used. A purified HB$_s$Ag prepared in yeast by a recombinant DNA-vector technique has also proved to be effective experimentally and is now available for general use. An experimental vaccine with infectious vacci-

---

* Infection with hepatitis A virus is so widespread that the serum of many adults contains anti-HAV Abs, and a relatively high titer is present in concentrated γ-globulin pools. HAV and HBV have been eliminated from these pools, along with the fibrinogen, in the usual cold ethanol fractionation of plasma. The recommended dose of γ-globulin is 0.02 ml/kg of body weight, administered by intramuscular injection as soon after exposure as possible.

nia virus as a vector has been shown to produce immunity in chimpanzees, and one with an adenovirus vector in vaccines is under study.

Selected Reading

Dienstag JL, Alter HJ: Non-A, non-B hepatitis: Evolving epidemiologic and clinical perspective. Semin Liver Dis 6:67, 1986

Feinstone SM: Hepatitis A. Prog Liver Dis 8:299, 1986

Hilleman MR: Newer directions in vaccine development and utilization. J Infect Dis 151:407, 1985

Immunization Practices Advisory Committee: Recommendations for protection against viral hepatitis. Morbidity and Mortality Weekly Report, CDC 34:313, 1985

Miller RH, Robinson WS: Common evolutionary origin of hepatitis B virus and retroviruses. Proc Natl Acad Sci USA 83:2531, 1986

Purcell RH, Rizzetto M, Gerin JL: Hepatitis delta virus infection of the liver. Semin Liver Dis 4:340, 1984

Rizzetto M, Verone G, Gerin JL, Purcell RH: Hepatitis delta virus disease. Prog Liver Dis 8:417, 1986

Seeger C, Ganem D, Varmus HE: Biochemical and genetic evidence for the hepatitis B virus replication strategy. Science 232:477, 1986

Ticehurst JR: Hepatitis A virus: Clones, cultures, and vaccines. Semin Liver Dis 6:46, 1986

Tiollais P, Pourcel L, Dejian A: The hepatitis B virus. Nature 317:489, 1985

Zuckerman AJ: The history of viral hepatitis from antiquity to present. In Deinhardt F, Deinhardt J (eds): Viral Hepatitis: Laboratory and Clinical Science, p 3. New York, Marcel Dekker, 1983

64

Renato Dulbecco

Oncogenic Viruses I: DNA-Containing Viruses

Unity of Oncogenic Viruses

The first tumor-producing (oncogenic) virus was discovered in 1908 by Ellerman and Bang, who demonstrated that seemingly spontaneous leukemias of chickens could be transmitted to other chickens by cell-free filtrates. Later (1911) Rous found that a chicken sarcoma, a solid tumor, can be similarly transmitted. The viruses responsible turned out to be retroviruses (see Chap. 65). These virus-induced tumors were then considered by many as a biological curiosity, either not true cancers or perhaps a peculiarity of the avian species. These notions, however, were shaken when DNA-containing viruses were shown to produce a cutaneous fibroma and a papilloma of wild rabbits (by Shope, in 1932), and the renal adenocarcinoma of the frog (Lucke, 1934).

The later discovery of virus-induced tumors in mice provided a particularly suitable system for experimental work. In 1936 Bittner demonstrated that a spontaneously occurring mouse adenocarcinoma is caused by a virus transmitted from the mother to the progeny through the milk, and in 1951 Gross discovered the first of many retrovirus-induced murine leukemias. These studies revealed that the viral etiology of a cancer can easily go unrecognized for several reasons: cancer can be caused as a rare effect by ubiquitous viruses, which may easily be considered as innocuous bystanders; with some oncogenic viruses the viral particles are heterogeneous, and most infect cells without inducing cancer; the disease may not develop until long after infection; and the cancers do not seem contagious, because the method of transmission of the virus is not apparent (e.g., through the embryo or the milk).

Further impetus to investigations of tumor viruses arose from the later discovery of a new group of DNA-containing viruses that cause cancer in mice and other rodents: polyoma virus and simian virus 40 (SV40) (as a passenger virus in cultures of rhesus monkey kidney cells). Finally, the **human** adenoviruses, papilloma viruses, and herpesviruses were also shown to have oncogenic activity in rodents. Moreover, by this time the study of bacterial lysogeny had clearly shown that the genetic material of viruses can become permanently integrated with that of the host. These realizations led to an explosive development of interest in viral carcinogenesis.

Soon the oncogenic effect of several viruses was demonstrated also in tissue cultures, in the form of **cell transformation** (see Chap. 47). Studies in this model system led shortly to a shattering conclusion: **a virus that has induced cancer is often no longer recognizable in the culture by its infectivity,** or by antigenicity. Traces could, however, be found in the form of viral DNA, RNA, and new Ags, in ways reminiscent of lysogeny. It thus became clear that time-tested techniques and approaches for the identification of viral agents of disease may not be adequate in the search for viral agents of human cancer.

The discovery of many new oncogenic viruses in the 1960s revealed a puzzling distribution. Most classes of DNA-containing viruses were found to produce tumors in animals, whereas only one family of RNA-containing viruses, the group now called **retroviruses,** did so (Table 64–1). The replication of retroviruses also displayed a sensitivity, peculiar for RNA viruses, to agents that interfere with DNA replication or transcription. This property was explained when Temin and Baltimore independently discovered that the oncogenic retroviruses, unlike other RNA viruses, replicate through a DNA-containing intermediate, made by a **reverse transcriptase** (RNA-dependent DNA polymerase). This discovery brought unity into the field of oncogenic viruses, suggesting that **oncogenesis is an attribute of viral DNA.** This unification has become stronger more recently with the demonstration that all oncogenic viruses cause cancer and transformation through **oncogenes** they carry or activate.

These developments, and the problems they raise, will be analyzed in this and the next chapter by examining the characteristics of cell transformation induced by several viruses. Many of the findings in animals will be analyzed in detail as the basis for results more recently obtained in humans.

TABLE 64–1. Distribution of Oncogenic Viruses Among Animal Virus Families

Nucleic Acid in Virions	Viral Group (or Family)	Oncogenic Viruses
RNA	Picornaviruses	None
	Togaviruses	None
	Orthomyxoviruses	None
	Paramyxoviruses	None
	Rhabdoviruses	None
	Coronaviruses	None
	Arenaviruses	None
	Retroviruses*	Oncoviruses: leukosis viruses, sarcoma viruses
	Reoviruses	None
DNA	Adenoviruses	Many types
	Papovaviruses	Polyoma virus, SV40, SV40-like human viruses, papilloma viruses
	Herpesviruses	Virus of neurolymphomatosis of chickens (Marek disease), Lucke's virus of frog renal adenocarcinoma, herpes simplex virus (cell transformation),† Epstein-Barr virus (Burkitt lymphoma, nasopharyngeal carcinoma), cytomegalovirus,† primate herpesviruses
	Hepadnaviruses	Hepatitis B virus
	Poxviruses	Fibroma virus
	Parvoviruses	None

* See Chapter 65.

† Transformation *in vitro* after ultraviolet irradiation

Oncogenes

ONCOGENES AND PROTO-ONCOGENES

The study of oncogenic viruses has allowed the identification of a set of genes, also expressed in seemingly spontaneous animal and human cancers, which initiate the cancer process. Many of them play an important role, not only in cancer but also in normal cell growth and differentiation. The illegitimate expression of these genes is responsible for the processes of transformation and neoplasia. This fundamental discovery was made studying the transforming activity of an oncogenic retrovirus, the Rous sarcoma virus (RSV; see Chap. 65). Genetic studies by Martin and Vogt first identified in this virus the *src* gene, responsible for transformation; then M. Bishop and collaborators, probing the genome of normal chicken cells with probes made from the *src* gene, identified a very similar gene, the **cellular *src*** gene (or *c-src*). The normal gene is known as *src* **proto-oncogene,** and the related gene present in the virus as **src oncogene** or *v-src*. *v-src* is incorporated in the RSV genome as a consequence of the recombination of a nontransforming retrovirus with *c-src*. The *c-src* is highly conserved in animal species, suggesting that it performs some important function.

THE v-src ONCOGENE. The possible function of *v-src* was clarified after Erikson identified its protein product by translating *in vitro* RSV RNA: it is a phosphoprotein of 60 Kd, known as **pp60$^{v\text{-}src}$.** He then showed it to be a **protein kinase.** Hunter discovered that it has the unusual property of **phosphorylating tyrosine** in proteins, whereas most other kinases phosphorylate serine or threonine. This finding suggests that the function of *c-src* is somehow related to cell growth regulation, because tyrosine phosphorylation is carried out by many growth factor receptors (see Chap. 47).

v-src VERSUS c-src. The protein expressed in normal cells by the *src* proto-oncogene is similar but not identical to that expressed by *v-src,* and does not cause transformation. This difference might be attributed to two circumstances: one is that the *v-src* gene differs from the *c-src* gene in its 3′ end and lacks a phosphorylation site (tyr-527) apparently crucial for normal control; the other is that the viral gene is more strongly transcribed. Transfection experiments with a cloned *c-src* gene show that the alteration is more important than the overexpression. Apparently the altered gene escapes its normal regulatory restraints, becoming a transforming gene.

OTHER ONCOGENES. After *v-src* many other oncogenes were discovered in oncogenic viruses; and DNA extracted from human cancers was also shown to contain oncogenes able to transform NIH 3T3 cells. Some of these cancer-related oncogenes are similar to those previously recognized in viruses. More than 30 oncogenes of various origins are now known and are listed in Table 64–2.

ONCOGENES OF DNA VIRUSES. Whereas the oncogenes present in oncogenic retroviruses are altered cellular proto-oncogenes that confer no advantage on the virus except that of being selected by the experimenter, the oncogenes present in oncogenic DNA viruses lack sequence homology with known cellular proto-oncogenes and do perform essential functions for the virus. These viruses multiply in resting cells, which lack the enzymes needed for the replication of the viral DNA. The viral oncogenes cause the production of the enzymes by activating the machinery for cellular DNA replication. In all likelihood the growth stimulation is responsible for transformation. Whether or not the oncogenes of DNA

TABLE 64–2. Oncogenes

Oncogene	Location	Function
ONCOGENES PRESENT IN DNA VIRUSES		
E1A	Nucleus, cytoplasm	Regulates transcription
E1B		
PV-ST	Cytoplasm	
PV-MT	Plasma membrane	Binds and stimulates pp60$^{c\text{-}src}$ and pp62$^{c\text{-}yes}$
PV-LT	Nucleus	Initiates DNA synthesis and regulates transcription
SV40-ST	Cytoplasm	
SV40-LT	Nucleus, plasma membrane	Initiates DNA synthesis, regulates transcription and binds p53
ONCOGENES PRESENT IN RETROVIRUSES (see Chap. 65)		
abl	Plasma membrane?	Tyrosine protein kinase
erb A	Cytoplasm	Thyroid hormone receptor
erb B	Plasma and other membranes	EGF receptor Tyrosine protein kinase
ets	Nucleus	
fes	Plasma membrane	Tyrosine protein kinase
fgr	id	id
fms	id	CSF-1 receptor (tyrosine protein kinase)
fos	Nucleus	
fps (see *fes*)		
kit	Membranes	Tyrosine protein kinase
mil/raf	Cytoplasm	Serine/threonine protein kinase
mos	id	Serine protein kinase
myb	Nucleus	
myc	Nucleus	
ras	Plasma membrane	GTP-binding protein
raf (see *mil*)		
rel	Cytoplasm	
ros	Cytoplasm	Tyrosine protein kinase
sis	Cytoplasm and secreted	PDGF subunit
src	Plasma membrane	Tyrosine protein kinase
ski	Nucleus	
yes		Tyrosine protein kinase
ONCOGENES NOT PRESENT IN VIRUSES		

bcl—human follicular lymphoma
bcr—human chronic myelogenic leukemia
int-1, 2, 3, and *4*—breast cancer in rodent
met—chemically transformed human cell line
neu—rat neuroglioblastoma (similar to *erb B*)
p53—active in transformed cells
ret—human lymphoma DNA
rho—similar to *ras*

viruses derive from cellular proto-oncogenes, they have evolved independently for the sake of the virus.

MODE OF ACTION OF ONCOGENES. Most of the known oncogenes encode **proteins important in cell growth control** (see Chap. 47): some are related to growth factor subunits (platelet-derived growth factor [PDGF] by *v-sis*), others to growth factor receptors (an amputated epithelial growth factor [EGF] receptor by *v-erb B*). The trunca-tion of the external domain of these receptors presum-ably keeps the cytoplasmic domain active all the time, in the absence of activation from the outside. Many onco-gene products are protein kinases with tyrosine specific-ity (the *src* family), two (oncogenes *mos* and *raf/mil*) are protein kinases with threonine-serine specificity, others (the *ras* family) are plasma membrane guanosine tri-phosphate (GTP)–binding proteins implicated in signal transduction. Several oncogenes encode nuclear pro-teins (*myc, myb, fos*, E1A, LT) some of which (E1A, LT) are known to regulate gene transcription.

The normal counterparts of oncogenes, the proto-on-cogenes, appear therefore to specify proteins that are parts of regulatory pathways that initiate at the cell sur-face and terminate at the genes, activating cell growth. The oncogenes with nuclear products, which have no known relationship to proteins regulating growth, may represent the proximal end, close to the genes, of these pathways. The important differences in base sequences of proto-oncogenes performing the same function (e.g., tyrosine protein kinase) suggest that there are multiple similar but distinct regulatory pathways that are active in different cell types or that respond to different signals. In normal cells these pathways function only occasionally, in response to well-defined stimuli; the continuous activ-ity of a pathway when a proto-oncogene becomes an oncogene would cause the cell to grow without restraint and would alter the activity of its genes, causing it to be transformed.

This model of transformation is supported by the fol-lowing observations. (1) In normal cells some cellular proto-oncogenes (e.g., *c-fos*) are expressed only in re-sponse to a growth stimulus, in a transient way; (2) nor-mally, each cellular oncogene is strongly expressed only in some cell type (e.g., *c-src* in brain cells, *c-fos* in differ-entiating macrophages) and at different stages of devel-opment (*c-fos* in expressed in placenta); (3) cells trans-formed by the oncogene *v-sis* encoding a PDGF subunit revert to normality when the medium contains antibod-ies to the PDGF. The latter result suggests that the con-tinued action of PDGF produced by the cells on the cells' own PDGF receptors maintains the state of transforma-tion.

ACTIVATION OF CELLULAR ONCOGENES IN VIRUSES AND CANCERS.

The principal method by which a proto-onco-gene is converted into an oncogene with unregulated function is by structural alterations of its coding or con-trol sequences. The coding sequences are altered in the majority of cases. The changes may be single-base muta-tions, causing the replacements of critical amino acids, frame shift mutations, deletions, or insertions. In retro-viruses, fusion of parts of a proto-oncogene with a viral gene may also contribute to the activity of the oncogene. As a result of the alterations, the oncogene may have unregulated transcription (e.g., *v-myc*) or translation (e.g., *v-fos*), or its protein may have lost targets for regulatory modifications (e.g., *v-src*). Proto-oncogenes may become deregulated by interaction with other cellular constitu-ents; an example is the deregulation of *c-src* by complex-ing with polyoma virus MT protein (see Association of T protein With Other Oncogene Proteins, below).

An oncogene may be generated without viral interven-tion when a proto-oncogene is altered by physical or chemical carcinogenic agents or by translocation. Impor-tant in all cases is a high expression by an active en-hancer, either viral or cellular.

Enhancers

Enhancers are DNA sequences activating transcription, which are present in both DNA viruses and retroviruses as well as in cellular genes (e.g., for immunoglobulins, insulin, chymotrypsin). Although the enhancers of differ-ent genes share common properties such as repeats, they vary greatly in sequences. Best known is the SV40 enhancer, which is made up of two 72-base-pair repeats; it increases transcription of viral or of some heterologous genes 10 or more times, in comparison with enhancer-free genomes. An enhancer is usually close to the pro-moter it controls; it is remarkable, however, that they continue to act on the promoter if displaced to positions several thousand base pairs removed in either direction, although with decreasing efficiency at great distances, and they can be turned back-to-front without losing ac-tivity. An enhancer acts only on some promoters: for in-stance, the SV40 enhancer acts strongly on the β-globulin promoter but weakly on the α-globulin promoter.

Enhancers are specifically **activated** by *trans*-acting factors, which may be of viral or cellular origin, and therefore confer cell or tissue specificity on the expres-sion of genes they control. For instance, the SV40 en-hancer does not increase transcription by the herpes-virus promoter in human cells unless the LT protein, an SV40 regulator, is present. Embryonal cells (e.g., undiffer-entiated embryonal carcinoma cells) produce transact-ing factors that **inhibit** some viral enhancers. Viral mu-tants capable of replicating in these cells have altered enhancers (see "Papovaviruses: Polyoma and Related Vi-ruses," below).

Enhancers appear to operate by facilitating access of the transcriptional machinery to DNA. They are in fact recognized in SV40 minichromosomes as segments de-void of nucleosomes (nucleosome gaps), which are highly sensitive to attack by many nucleases. This prop-erty may be associated with the presence of Z-DNA se-quences, which cannot fold to form nucleosomes and bind special proteins.

Cooperation of Oncogenes

Various oncogenes have two main effects on cultures of embryonic rat cells: **immortalization** and **transformation.** Immortalized cells do not undergo *in vitro* senescence, have reduced requirements for growth factors, may be unable to undergo terminal differentiation, and may be morphologically transformed. However, they do not grow in suspension in agar and do not produce tumors in syngeneic animals or in immunodeprived nude mice. Transformed cells are capable of growing in agar and forming tumors but may undergo senescence. For complete transformation and induction of tumor formation both immortalization and transformation are needed in this cell system. Typical immortalizing oncogenes are *v-myc* and *v-fos* among retroviral genes and *E1B* and *Py-LT* among those of DNA viruses; they encode nuclear proteins that may act directly on cellular genes. Among transforming oncogenes are the retroviral *v-ras* and the DNA virus *E1B* and *Py-MT;* they specify membrane proteins that, through intracellular mediators, may alter the regulation of genes.

Studies with other cells show that the distinction between the two classes of oncogenes is not absolute: the effects depend on many variables, such as the species and the type of cell used and the strength of expression of the oncogene. It is likely that the effects of an oncogene are influenced by the balance of expressed cellular genes. But even with these limitations, it is clear that there are different classes of oncogenes, although perhaps not as rigidly defined, and oncogenes of different classes can cooperate to produce a more pronounced oncogenic effect than either alone. This concept is corroborated by observations with oncogenic viruses, to be detailed later, and in the following chapter, which show that neoplasia is a multistep process and that different oncogenes can carry out different steps. Permanent cell lines, in which the effects of oncogenes are much more pronounced than in short-term cultures, have probably undergone one or more of these steps.

Oncogenic DNA Viruses

PAPOVAVIRUSES: POLYOMA AND RELATED VIRUSES

Prominent among DNA-containing viruses for understanding the mechanism of carcinogenesis are papovaviruses. The first virus, isolated by Stewart and Eddy, was found to produce various kinds of neoplasia when injected into newborn mice; it was therefore named **polyoma virus** (PyV), that is, agent of many tumors. The virus is widespread in mouse populations, both wild and in the laboratory; it is normally transmitted to animals after birth, through excretions and secretions. The structurally similar **SV40** was discovered by Sweet and Hilleman as an agent that multiplies silently in rhesus monkey kidney cultures (used for propagating poliovirus) but was found to produce cytopathic changes in similar cultures from African green monkeys *(Cercopithecus aethiops).* It was later shown to produce sarcomas after injection into newborn hamsters.

Subsequently, **human papovaviruses** were isolated, first from the brain of a patient with progressive multifocal leukoencephalopathy (PML); next, others from patients with Wiskott–Aldrich syndrome (defects in cellular and humoral immunity, and reticulum cell sarcomas, due to an X-linked recessive allele); and then, frequently, from the urine of immunosuppressed individuals. These SV40-like viruses are designated by the initials of the persons from whom they were isolated (e.g., JC virus, from a PML patient; BK virus, from urine). These viruses produce tumors in hamsters and transform hamster cells *in vitro.* Since 70% of humans have Abs to the SV40-like viruses, they may be the human equivalent of the widespread SV40 of rhesus cultures and polyomavirus of mice.

Properties

The virions of PyV, SV40, and SV40-like human viruses are small, naked icosahedrons with a diameter of 45 nm and with 72 capsomers (see Chap. 44). They contain a minichromosome made up of a **cyclic double-stranded** DNA, about 5 Kb long, associated with octamers of cellular histones (H2a, H2b, H3, H4) to form nucleosomes similar to those present in cellular chromatin. Removal of the proteins upon extraction causes a deficit in intrinsic helical turns, which is compensated by twisting of the double helix (**supercoiling**). The SV40 genome is very similar in sequences to those of the SV40-like viruses. It also has significant, although lower, homology to the PyV genome, suggesting a common origin in evolution.

All papovaviruses are very resistant to inactivation by heat or formalin; hence, SV40 (from rhesus kidney cultures) survived in some early batches of formalin-killed poliovirus vaccine.

Multiplication in Cell Cultures

These viruses can produce either a **productive** or a **nonproductive** infection. The outcome depends on the species of the cells and their physiological state, because viruses with such small genomes depend heavily on cellular functions. Thus, cells of certain **permissive** species (Table 64–3) are killed by infection and yield virus, whereas those of **nonpermissive** species are **transformed** without virus production. In **semipermissive** cultures some cells are transformed while others are killed and yield virus (probably depending on the state of

TABLE 64–3. Some Permissive and Nonpermissive Cells Used With Polyoma Virus and SV40

Cell	Virus	Type of Culture
Permissive	Polyoma virus	Secondary cultures of mouse embryo cells
		Primary cultures of mouse kidney cells
		3T3 and 3T6 cell lines (mouse subcutaneous tissue)
	SV40	Primary cultures of African green monkey kidney cells
		BSC-1 CV-1 Vero } cell lines (African green monkey kidney)
Nonpermissive or semipermissive	Polyoma virus	Secondary cultures of hamster embryo cells
		Secondary cultures of rat embryo cells
		BHK cell line (baby hamster kidney)
	SV40	Secondary cultures of hamster, rat, or mouse embryo cells
		3T3 cell line (mouse subcutaneous tissue)

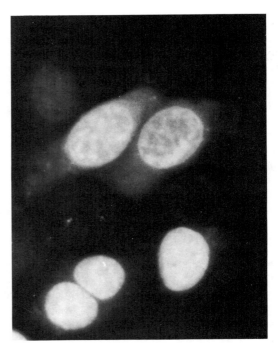

Figure 64–1. Fluorescence photomicrograph of cultured mouse kidney cells productively infected by polyoma virus. The accumulation of capsid Ag in the nuclei is revealed by its combination with fluorescein-conjugated Abs, which emit green fluorescence under ultraviolet light.

the cells). Permissive cells can be transformed by virus when viral multiplication is prevented by viral mutations or by damage in the viral DNA.

PRODUCTIVE INFECTION. As with most DNA viruses, viral DNA replication and capsid assembly occur in the cell nucleus (capsid Ag is detected there by immunofluorescence [Fig. 64–1] and virions by electron microscopy); each nucleus can produce up to 10^8 viral particles. As with other naked viruses, release depends upon **disintegration of the cells.**

Transcription of the supercoiled DNA is carried out by the host polymerase II. In the **early transcription** (Fig. 64–2) genes for DNA replication are transcribed from the "early" DNA strand; in the **late transcription,** which starts after DNA replication has begun, genes for virion proteins are transcribed from the opposite, "late" DNA strand, at a much higher rate than the persisting early strand transcription. Early transcription is **autoregulated** because it is inhibited by the large T Ag (see Viral Proteins, below).

Both early and late transcription initiate in the **control region,** which includes the origin of DNA replication (see Fig. 64–2), between the early and the late regions of the genome, and proceed in divergent directions. The control region regulates all phases of transcription as well as DNA replication. In SV40 (Fig. 64–3) it contains a series of repeats with different functional significance. Three 21-base-pair repeats, each containing

two GC-containing hexamers, act as **promoters** for early transcription. At the downstream end of the repeats is a **TATA box;** at the upstream end two 72-base-pair repeats constitute the **enhancer.** Each control sequence (the 21-base-pair repeats, the TATA box, the enhancer) binds activation factors specified by cellular genes, the presence of which makes the cells permissive for the virus. Especially important is the enhancer: although SV40 does not multiply in lymphoid cells, a construct of SV40 DNA associated with the enhancer from an immunoglobulin gene will. The three regulatory elements operate in concert and are somewhat redundant: deletion of several hexamers of the 21-base-pair repeats reduces only slightly the efficiency of viral multiplication. Artificially changing the distances between the elements has worse effects, suggesting that regulation depends on interactions among the proteins.

The regulatory elements control both early and late transcription. Early transcription is dependent on all three types of elements. It begins at several different sites, which change as infection proceeds (see Fig. 64–3). Late transcription depends on two of them: the 21-base-pair repeats and the enhancer; it also has heterogeneous starting points. The switch from early to late transcription is brought about by the binding of the large T Ag (see

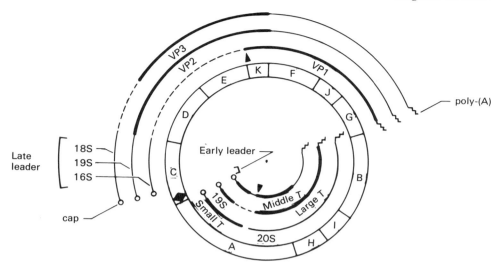

Figure 64–2. Polyoma virus mRNAs and proteins. Heavy lines indicate the translated parts of the mRNAs; wavy lines are untranslated parts; dashed lines (introns) are parts removed by splicing. The middle T is known only for polyoma virus. The two black triangles indicate shifts of the reading frame.

later) to specific sites in the control region and also by the replicating state of viral DNA.

In PyV the general organization is the same, but the different regulatory elements partially overlap with each other and the binding sites for T Ag. The PyV enhancer is made up of two domains, one homologous to the SV40 enhancer, the other to the enhancer of the E1A region of adenovirus (see below), showing the evolutionary relatedness of DNA oncogenic viruses. The function of PyV enhancer is blocked by E1A-encoded peptides, explaining its inactivity in cells (such as murine embryonal carcinoma cells) that express an E1A-like function. The activity is restored by viral mutations that modify the enhancer, changing the host range of the virus.

VIRAL PROTEINS (see Fig. 64–2). During the early infection with PyV, three **early proteins** are synthesized, collectively known as T (tumor) antigens. They are detected by immunofluorescence or immunoprecipitation of infected cell extracts with the serum of an animal bearing a large virus-induced tumor. They are the nuclear **large T (LT)** Ag, the cytoplasmic as well as nuclear **small T (ST)** Ag, and the membrane-bound **middle T (MT)** Ag. Both LT and MT have the properties of oncogenes. These proteins are specified by three different early mRNAs with a common leader sequence and different splices: parts of the PyV LT and MT mRNAs derive from the same DNA segment, but in different phases. In contrast, during SV40 infection only two T Ags are made: LT and ST. A function

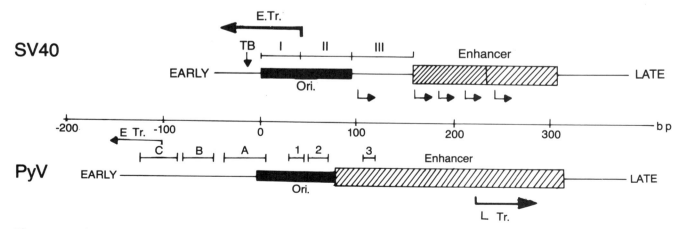

Figure 64–3. Organization of the control regions of SV40 and PyV. The early region of the genome is on the left, the late region on the right. *TB,* TATA box; *E.Tr,* beginning of early transcription; *L.Tr,* beginning of late transcription (at multiple sites in SV40); *Ori.,* origin of replication; *I, II, III, 1, 2, 3, A, B, C,* binding sites for LT. Distances are in base pairs from the origin. (Data from Frisque RJ, Bream GL, Cannella MT: J Virol 51:458, 1984 for SV40; Cowie A, Kamen R: J Virol 52:750, 1984 for PyV)

equivalent to that of PyV MT is carried out by a fraction of the SV40 LT, which, like MT, is present in the cell plasma membrane.

In both viruses LT performs **several different functions** on viral multiplication and transformation. During multiplication, by binding to different sites in the control region of the viral DNA, it causes the initiation of viral DNA replication, blocks early transcription, and promotes late transcription. In transformation it acts like an immortalizing oncogene. In addition, the C terminus of SV40 LT has a function absent in PyV LT, the **helper function** for adenovirus (see Viruses With Hybrid DNA, below).

Of the four **late proteins,** three are in the virions: the main capsid protein (virion protein 1, VP1) and two minor proteins, VP2 and VP3 (the sequence of VP3 being contained in that of VP2). The three capsid proteins are synthesized on three different late mRNAs with a common untranslated leader and different splices (see Fig. 64–2). The VP2 and VP3 peptides end at a common terminator about halfway through the late region; VP1 is read on a different frame. The fourth protein (**agnoprotein,** i.e., of unknown function), small and very basic, is encoded in the mRNA leader; it probably participates in the assembly of virions. Thus, like small bacteriophages (see Chap. 45), the small genome of these papovaviruses expresses several distinct proteins from the same DNA segment in both its early and late region.

Viral **DNA replication** begins after an unusually long lag of 10 to 12 hours. During this time **cellular genes essential for viral replication are activated.** In crowded quiescent cultures infection stimulates the **synthesis of cellular DNA,** mRNAs, and histones to a level comparable to that of growing cultures. It also stimulates formation of **enzymes involved in DNA synthesis:** thymidine kinase, deoxycytidylate deaminase, and DNA polymerase α. The induced thymidine kinase is that normally expressed in growing cells, showing that the cells are converted by the virus to a growing state. This conversion, probably caused by the MT protein with PyV and by the equivalent function with SV40, may be related to the changes the virus induces in transformation; its obvious advantage for the virus would then explain why these viruses have transforming genes.

DNA replication takes place in supercoiled replicative intermediates (Fig. 64–4); it initiates at a palindrome at the viral origin (see Fig. 64–3) after LT binds to the origin region, especially sites I and II in SV40, and then proceeds bidirectionally. Replication is carried out by a complex of host enzymes, including a primase, DNA polymerase α and β, topoisomerases I and II, RNAse H. Some of the viral DNA molecules generated are recycled back to replication; others become associated with histones and are then encapsidated into virions.

MUTATIONS AFFECTING PRODUCTIVE INFECTION. The analysis of mutants (**temperature sensitive [ts], host range, plaque size, and deletions**) identifies **four regions** in the PyV or SV40 genome. Two regions are expressed **early:** (1) Mutations in region A affect LT: they make it heat labile, prevent viral DNA replication, and cause overproduction of early mRNAs by removing the negative regulation of LT. (2) PyV hr-t mutations (for the host range-transformation), which are not temperature sensitive, affect ST and MT Ag and restrict viral multiplication to special cell types. Two genes are expressed **late:** mutations in region B/C affect VP1, producing either heat-sensitive virions or small plaques; mutations in region D affect VP2 and VP3, preventing penetration.

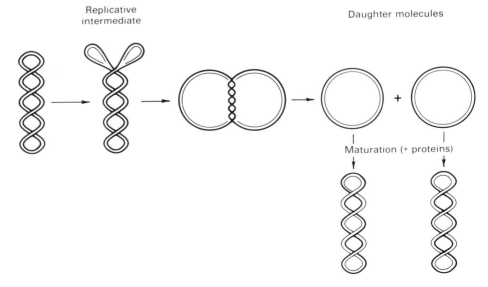

Figure 64–4. Replication and maturation of polyoma virus DNA. The replicative intermediate is partly supercoiled, partly relaxed, and therefore has intermediate buoyant density in CsCl-ethidium bromide gradients. When replication is almost complete, it yields two daughter molecules, each with a gap in the new strand at the replication terminus. These molecules are then sealed and, in conjunction with cellular histones, are converted to chromatin containing covalently closed DNA supercoils. Supercoiling is caused by unwinding of the helix by topoisomerase after it binds the histones.

Replicative intermediate

Daughter molecules

Maturation (+ proteins)

Cell Transformation

After nonpermissive rodent cells are exposed to the virus, transformed clones are usually identified by their distinct morphology (Fig. 64–5), ability to form foci overgrowing a monolayer of untransformed cells, or ability to grow in suspension in agarose or methylcellulose. The various procedures select for different subsets of transformed cells, with somewhat different properties. The number of stable transformed clones generated by a viral sample is much less (10^{-3} to 10^{-5}, depending on the cell type) than the number of plaques on permissive cells. Yet transformation of a cell is caused by a single virion, because its frequency is proportional to the viral titer (see Dose Response Curve of the Plaque Assay in the appendix to Chap. 44).

Isolation of transformed clones in soft agar allows a distinction between cells undergoing **stable transformation,** which form rare large colonies, and those undergoing **abortive transformation,** which form small but much more frequent colonies. In abortive transformation the cells return to normality after four to six generations and stop growing in suspension; however, they continue to grow if transferred to a dish with liquid medium. Abortive transformation is due to failure of integration (see later).

INTEGRATION OF THE VIRAL DNA INTO THE CELLULAR DNA. Cells permanently transformed by PyV or SV40 contain one or a few viral genomes per cell, as shown by the kinetics of hybridization of their DNA with labeled viral DNA. During fractionation the viral DNA sequences are always found associated with the cellular DNA, even when completely denatured to single strands (by sedimentation in an alkaline sucrose gradient). Hence, the viral DNA is integrated, that is, covalently bound to the cellular DNA as the prophage of lysogenic cells (see Chap. 46). The integrated viral genomes are often duplicated in tandem (i.e., head-to-tail) and often have rearrangements or deletions. A fraction of the early region at its 5' end is always present, because the function of the MT Ag (in PyV) or the N-end of the LT Ag (in SV40) is required for the expression of the transformed state. Integration occurs by a nonhomologous recombination between the cyclic viral DNA and the cellular DNA and needs the activity of the LT Ag: ts A mutants do not integrate at nonpermissive temperature, causing abortive transformation.

To locate the **viral and the cellular integration sites,** fragments containing viral DNA generated by restriction endonucleases from the DNA of transformed cells or from the virus are compared with each other. Evidently, only the hybrid linker fragments (containing both cellular and viral DNA), which derive from the two ends of the integrated genome, differ from those obtained from the free viral DNA, showing in which fragment the viral DNA has been opened at integration. The lengths of the hybrid fragments depend mostly on the integration site in the cellular DNA; hence, these sites can be compared in different clones of transformed cells.

The results show that **neither the viral nor the cellular sites of integration are constant** for the same integrated genome in cell clones that derive from independent transformation events; hence, it is unlikely that integration causes transformation by inserting the viral genome in a specific cellular gene.

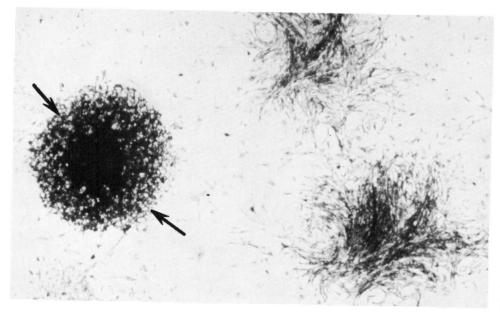

Figure 64–5. Colonies of the BHK line (hamster kidney) infected by polyoma virus. One colony is transformed (*arrows*) and is recognizable by its considerable thickness and the random orientation of its cells. The untransformed colonies are thin and contain cells that tend to orient parallel to one another. (Courtesy of M. Stoker)

TRANSCRIPTION OF THE INTEGRATED GENOME. Only the early region of the integrated viral DNA or the part of it that is essential for maintaining transformation is transcribed in nonpermissive cells. The restriction of transcription is caused by the structure of the chromatin rather than by the specificity of the host transcriptase. Thus, with transcription *in vitro* by *Escherichia coli* transcriptase, chromatin extracted from SV40-transformed cells yields the same RNA as is synthesized *in vivo*.

ROLES OF VIRAL PROTEINS IN TRANSFORMATION. All T Ags are normally expressed in cells transformed by PyV or SV40. The function of the various proteins is dissected by studying mutants and by transfecting nonpermissive cells with vectors that express selected parts of the early regions.

The best understanding is obtained with **polyoma virus,** which specifies three early proteins. With cells of **permanent rat lines** neither LT nor ST alone causes transformation or appreciable changes, whereas MT causes transformation and abolishes requirement for growth factors. With **primary rat embryo cultures** MT causes transformation if strongly expressed; LT only immortalizes the cells. The effect of MT is graded: in cells containing a vector with a promoter controlled by dexamethasone, which can be stimulated to different degrees, the effect varies from morphologic transformation to focus formation, anchorage-independent growth, or tumorigenesis, depending on the degree of MT expression. Mutations, such as hr-t, that alter MT, abolish transformation. The ts A mutations, which affect LT, prevent the onset of transformation at the nonpermissive temperature; once transformation has taken place at the permissive temperature, it may or may not revert at the higher temperature, depending on the location of the mutation. The ST Ag has an accessory function in transformation by both PyV and SV40, especially in primary cultures; in some types of cell it confers a more extreme transformed phenotype. It acts especially on actin components of the cytoskeleton. The cooperation of the three proteins, as in infection with complete PyV, causes the most pronounced transformation of primary cultures.

With **SV40,** LT performs the functions of both PyV LT and MT; the majority of the molecules go to the nucleus and cause immortalization, whereas a small proportion, after fatty acid acylation, go to the plasma membrane as transmembrane proteins, performing functions similar to those of MT. In fact, mutants blocked in nuclear transport still transform cells of permanent lines (an MT function), but not of primary cultures because they lack the immortalization function (an LT function). The transforming functions reside at the 5' end of the LT gene: mutations affecting the other half of the gene do not interfere with transformation, although they may interfere with viral multiplication in permissive cells.

Replication of the viral DNA is not needed for transformation, for many mutations dissociate the two functions. Moreover, SV40 genomes in which the origin of replication is deleted efficiently transform permissive human cells because they do not kill them for lack of late proteins.

ASSOCIATION OF T PROTEINS WITH OTHER ONCOGENE PROTEINS. In infected cells the PyV MT is associated with the phosphoprotein specified by the proto-oncogene *c-src* (pp60^{c-src}, see above) and *c-yes* (pp62^{c-yes}); the complex is precipitated from cell extracts by antibodies to either protein. The precipitate has strong protein kinase activity with tyrosine specificity, which belongs to the pp60^{c-src} and pp62^{c-yes}, and is strongly enhanced by the association with MT; MT itself is phosphorylated at tyrosine by this activity.

Formation of this complex is very important for transformation, for in MT mutants the ability to transform and to activate the protein kinase is similarly affected. Association with MT enhances the kinase activity of the oncogene proteins. In pp60^{c-src} this is accompanied by a change of the site of phosphorylation from tyr-527 to tyr-416, as in the *v-src* present in Rous sarcoma virus. Transformation by PyV and by RSV, two unrelated viruses, appears therefore to share a common mechanism, the activation, by different mechanisms, of a cellular oncogene product. The SV40 LT is associated with another cellular phosphoprotein, p-53, which is strongly expressed in embryonic and neoplastic cells and has the properties of a proto-oncogene. The SV40 LT-p-53 complex also has protein kinase activity, but with specificity for serine and threonine. It is likely that the association is important for transformation.

COMPLEXITY OF TRANSFORMATION. Transformation requires a complex of factors. It involves the expression of two or three oncogenes or their equivalent, the formation of complexes between the proteins encoded by these genes and other cellular proteins. It is affected by the strength of the expression of the oncogenes and by the nature and species of the cells in which they are expressed. All these requirements evidently have the role of minimizing the danger of transformation for cells.

REVERSION OF TRANSFORMATION. Reversion, whereby the cells again acquire normal characteristics, can occur by three mechanisms: (1) **Alterations of the integrated viral genome,** especially deletions resulting from recombination. (2) **Excision and loss** of the integrated genome: in the presence of LT function an integrated viral genome containing the origin of replication can repeatedly undergo local replication. Recombination within the replicated segment may cause excision of a complete viral genome, leading to viral replication in permissive or semipermissive cells (**induction**). This event may lead to death of the cell. In nonpermissive cells, excision of the integrated genome may give rise to revertants if the sequences required for transformation are completely excised. (3) **Changes of cellular genes.** Reversion may be caused by changes of cellular genes that control the expression of the integrated viral genome. Either an "antioncogene" becomes activated or a gene producing a needed transacting factor becomes inactivated.

Changes of cellular genes play an important role in transformation in other ways. They are responsible for the increase in malignancy (**progression**) during prolonged growth of the transformed cells either in culture or in the animal.

IMMUNOLOGIC CONSEQUENCES OF TRANSFORMATION. PyV- or SV40-transformed cells display new antigenic determinants (**transplantation antigens**) that make them foreign in an isogeneic host, and therefore liable to attack by cytotoxic T and other immune cells (see Chap. 50). New determinants are contributed by the PyV MT or SV40 LT present at the cell surface.

SV40 Genome in Transgenic Mice

The microinjection of plasmids containing SV40 DNA associated with a suitable promoter–enhancer combination into the male pronucleus of a fertilized mouse oocyte gives rise to a transgenic mouse with SV40 sequences in its genome. Such animals usually develop tumors localized in cells in which the viral control region can be expressed. If the SV40 genome is connected to its own control region, **intracranial choroid plexus tumors** usually develop, because the virus has tropisms for these cells (see below). If the genome is connected to the control region of an insulin gene, insulinomas are produced. Cells containing and expressing the viral genomes (usually as LT) may remain normal, but if they are kept in culture for a long time they may become transformed. Apparently the expression of the viral genome induces changes in cellular genes that then cause the transformed phenotype.

Role of SV40 and SV40-like Viruses in Human Pathology

SV40, although able to grow in human cells, does not transform them: the cultures grow for many generations until they undergo a **crisis** and die. The SV40-infected cells, however, can be transformed by infecting them with a murine sarcoma virus (see Chap. 65), implying collaboration of oncogenes. The virus is not oncogenic for humans: this was shown in the 1950s when large numbers of people were accidentally inoculated by SV40

present, unrecognized, in early batches of poliovirus vaccine. Epidemiologic studies have since shown no oncogenic effect.

SV40 has a tropism for the brain. A persistent infection is sometimes responsible for the chronic human brain disease **progressive multifocal leukoencephalopathy (PML).** All affected cells have nuclear LT, but few produce virus; they contain nonintegrated viral DNA and express only early functions. The cells are stimulated to multiply, and display many chromosomal abnormalities. These are probably produced by the ability of LT to induce local and generalized DNA replication and nonhomologous recombination.

JCV and BKV are very similar to SV40 in general genome organization. The main differences are in the control regions, especially the enhancers, which determine the type of human cells in which they grow. BKV is restricted to the kidney and multiplies in primary cultures of human embryonic kidney cells. Many adults have a latent kidney infection with BKV, which can be activated by immunodepression. Like SV40, JCV has tropism for the brain and can give rise to PML. The virus is restricted to glial cells, and multiplies in primary cultures of human fetal brain rich in spongioblasts, or in a line of human fetal glia cells (astroglia) immortalized by infection with origin-defected SV40. Transgenic mice that express the T Ag of JCV in the brain have impaired myelin formation with destruction of oligodendrocytes, as in PML.

Both JCV and BKV are oncogenic in some animal hosts. JCV induces a variety of tumors in hamsters and in primates, mostly of neural type. BKV, with the exception of some mutants, is not oncogenic in hamsters, but it collaborates with oncogene Ha-*ras* in transforming human embryonic kidney cells.

An oncogenic simian papovavirus called **B-lymphotropic papovavirus** isolated from a B-lymphoblastoid line derived from an African green monkey is related to SV40 and BKV. It transforms hamster embryo cells *in vitro*. Antibodies to this virus are frequent among humans and primates, suggesting a widespread infection with it or some antigenically related virus.

Viruses With Hybrid DNA

SV40 DNA can integrate in viral DNAs, with which it has very little homology, as it does in cellular DNA. Recombinants with adenovirus DNA enclosed in adenovirus capsids (**adeno-SV40 hybrids**) have been powerful tools for dissecting the function of the SV40 genome in transformation. They were observed in adenovirus grown in monkey cells, in which occult SV40 helps adenovirus multiplication. The recombination occurs in the adenovirus E_3 region, which is nonessential for its multiplication (see Chap. 52). The SV40 **helper function,** which resides in the last 113 amino acids at the C-terminal portion of LT, is expressed in the hybrids and provides an initiation factor required for synthesis of adenovirus late proteins. All hybrids may have various defects in other SV40 functions. Those expressing early SV40 proteins transform cells, which are indistinguishable from cells transformed by SV40 alone.

PAPOVAVIRUS: PAPILLOMA VIRUSES

The first papilloma virus was discovered in rabbits by Shope, who produced warts in the skin of either wild or domestic rabbits *(Oryctolagus cuniculus)* by inoculating extracts of warts of wild cottontail rabbits *(Sylvilagus floridanus)*. The papilloma virions are structurally similar to polyoma virus virions, but somewhat larger (55 nm); they also contain a cyclic double-stranded DNA (about 8 Kb in length). Papilloma viruses are, however, unrelated to polyoma viruses in DNA sequences. Papilloma viruses infect many animal species, producing **benign warts,** called **papillomas,** and probably malignant tumors (e.g., cervical carcinoma in humans). About 50 types are known for the human virus, at least six for the bovine viruses. The types are defined on the basis not of serology, as for other viruses, but of DNA homology: different types have less than 50% homology. The human types form 12 groups on the basis of residual DNA homology (Table 64–4). The genomes of the various types have the same general organization but may differ considerably in length, DNA sequences, and diseases they produce.

Multiplication and Pathogenesis

In most species papilloma viruses induce papillomas in the skin and some mucous membranes in individuals of the same or related species, but cattle virus can also produce mesenchymal tumors in horses and hamsters. Bioassay, based on wart formation in the skin, shows that the virus infects with extremely low efficiency. The ratio of physical particles to infectious units ranges between 10^5 and 10^8. These viruses do not propagate in tissue cultures; the main sources of virus are skin papillomas. Nonproductive infection, however, may occur *in vitro*: DNA of human papilloma type 5 (HPV5) transfected into a line of mouse cells (C127) causes them to grow in agar and form tumors in nude mice; and other HPVs transform human epithelial cells grafted to nude mice. Bovine papilloma virus (BPV) has an especially broad host range: it replicates in both epithelial and fibroblastic cells, causes tumors in hamsters, and transforms the C127 and other murine cells as well as epithelial cells of bovine origin. For these properties it is especially useful in experimental work. Human viruses can produce a latent infection in cervical and laryngeal epithelium.

GENE ORGANIZATION AND TRANSCRIPTION (Fig. 64–6).

All papilloma viruses have the same organization. The genome contains ten open reading frames, eight (E_1 to E_8) in the **early region,** which encompasses 70% of the genome, and two (L_1 and L_2) in the **late region.** Transcription takes place on only one strand. In productive infection, as in cottontail warts, the whole genome is transcribed; in unproductive infection, as in domestic rabbit warts, only the early region is transcribed. Beginning and ending of transcription are separated by a non-coding area of approximately 1000 base pairs, which contains sequences that regulate DNA transcription and replications. A trans-activator of transcription acting on one such sequence is encoded by gene E_2. Transcripts are spliced in different ways, generating several messengers.

TABLE 64–4. Human Papilloma Virus Homology Groups

I	II	III	IV	V A	V B	V C	VI	VII	VIII	IX	X	XI	XII
1	2	6	4	5	9	24	7	16	18	30	34	35	41
	3	11		8	15			31	32				
	10	13		12	17			33	40				
	26	27		14	37			34	42				
	27	27		19	38								
	28	32		20									
	29			21									
				22									
				23									
				25									
				36									

Underlined groups, associated with carcinomas

A

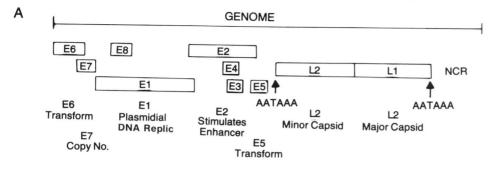

B

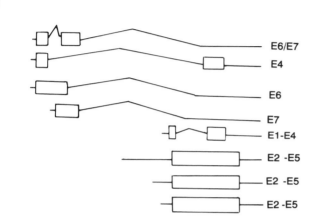

Figure 64–6. (*A*) The genome of bovine papilloma virus, showing the open reading frames (*boxed*) with their functions (below, two polyadenylation sites [AATAAA]). *NCR*, noncoding region with the origin of DNA replication (as cyclic plasmid), enhancer, and promoter. (*B*) Transcripts of the early region found in transformed cells. Coding sequences are boxed. Slanted lines, spliced-out sequences. (Chen EY, Howley PM, Levinson AD, Seeburg PH: Nature 299:529, 1982; Yang Y, Okayama H, Howley PM: Proc Natl Acad Sci USA 82:1030, 1985)

DNA REPLICATION. The viral DNA is present in the nucleus of papilloma cells or transformed culture cells as cyclic, highly methylated molecules that replicate autonomously but in synchrony with the cellular DNA. The progeny molecules segregate equally to the daughter cells, which thus maintain a **constant copy number** per cell (**plasmidial state**), usually 100 to 200 but as much as 10,000 for BPV in terminally differentiated keratinocytes. Autonomous replication and maintenance of copy number depend on the function of viral genes. Replication requires the activity of the E_1 gene, which has some homology with the SV40 sequences for LT Ag; copy number depends on gene E_7. In the absence of the E_1 function (e.g., in cancer cells; see below) the viral DNA cannot replicate autonomously and persists in the cells only if it becomes integrated.

Oncogenic Activity—Papillomas

Skin papillomas begin as a proliferation of the dermal connective tissue, followed by proliferation and hyperkeratinization of the epidermis. In warts of wild cottontail rabbits the nuclei of the keratohyaline and keratinized layer contain viral capsid Ag (recognizable by immunofluorescence) and viral particles (recognizable by electron microscopy and infectivity). The proliferating

connective tissue and basal epidermis, in contrast, contain neither Ag nor virions (Fig. 64–7), but they do contain viral DNA, which is probably responsible for the stimulus to proliferate.

Papillomas, including human warts, often **regress spontaneously.** If a rabbit has several separate papillomas, they all regress simultaneously, suggesting an **immunologic mechanism.** In contrast, the papilloma-derived carcinomas in rabbits usually do not regress and can be serially transplanted. Presumably they are less immunogenic, or their high growth potential overcomes the allograft reaction (see Chap. 17).

CARCINOMAS. It is likely that papilloma virus infection does not itself cause the cells to become cancerous, but this happens as the result of **additional events.** In fact, papillomas of cottontail rabbits do not regularly evolve to malignant cancers, and then only after a period of 1 to 2 years. The progression is much more frequent in domestic rabbits, showing the role of genetic factors; it is also favored by exposure to carcinogenic polycyclic hydrocarbons. Papillomas produced with bovine papilloma virus type 4 in the alimentary canal frequently evolve into carcinomas in cattle that eat bracken. This fern contains carcinogens that cooperate with the virus, inducing pro-

Figure 64–7. Fluorescence photomicrograph of a frozen section of wild rabbit papilloma stained with fluorescent antiviral Abs. Capsid Ag is present in nuclei and therefore appears as small, discrete, bright areas; it is restricted to the keratohyaline (*H*) and keratinized (*K*) layers of the epidermis. There is no capsid Ag in the cells of the proliferating basal layer (*P*). (Noyes WF, Mellors RC: J Exp Med 104:555, 1957)

gression. The rabbit cancers retain the viral genome in the form of plasmids, whereas the bovine ones have only **integrated** viral genomes; they often lack the viral genome altogether, showing that it is required for **initiating** the cancer but not for progressing to the malignant state.

Role of Papilloma Viruses in Human Tumors

The tumorigenic activity of papilloma viruses is demonstrated experimentally by the transformation of cells in culture or in grafts and by the evolution of both rabbit and bovine papillomas to carcinomas.

Several kinds of human tumors are associated with the various types of human papilloma viruses (HPV; see Table 64–3). They include various forms of warts, premalignant lesions, and malignant cancers of various origins, which have viral DNA sequences in their cells. Viral DNA is also frequently detected in anogenital tumors, such as the common genital wart (**condylomata acuminata**) of the cervix, vulva, and anus. Those with irregular mitoses may evolve into **squamous carcinomas.** Regularly associated with HPV DNA is **epidermodysplasia verruciformis,** a rare, lifelong disease characterized by disseminated flat warts and macular lesions, which may develop into cancers after several years. A third HPV-associated disease is the **juvenile-onset laryngeal papillomatosis;** the papillomas are benign but resist treatment, recur stubbornly, and tend to spread through the respiratory tract. Certain types of HPV are typically associated with carcinomas (see Table 64–4), especially types 16 and 18; their presence in cervical dysplasias carries a high risk of progression to cancer. Malignant cancers tend to contain integrated HPV genomes, whereas premalignant lesions

always contain genomes in plasmidial form; the significance of this difference is unknown.

ETIOLOGIC CONSIDERATIONS. The regular association of certain HPV types with several kinds of human tumors, although not proving an etiologic role of the virus, suggests it strongly, especially in view of the recognized transforming and tumor-inducing ability of the virus. For the squamous carcinomas of the uterine cervix this concept is supported by its high frequency in women having intercourse with multiple partners, who themselves also have multiple partners. Such promiscuity appears to favor the spread of genital viruses such as HPV, which is present in human semen.

MECHANISM OF TRANSFORMATION. The mere presence of viral DNA integrated in cells does not entail a neoplastic transformation, for HPV16 DNA is commonly found in the normal epithelium surrounding a cervical carcinoma. The finding suggests that among a population of DNA-carrying cells, some become neoplastic, owing to the interplay of additional events. As to the viral role in transformation, studies with mutant or deleted papilloma virus genomes of animals, especially bovine, as well as transfection of cells with cloned DNA fragments, have identified two genomic regions that are essential for transformation of cultured cells or tumor induction in animals. They are the E_6–E_7 and the E_5 regions of the early part of the genome. E_6 encodes a peptide present both in the nucleus and in membranes, whereas E_5 encodes a small hydrophobic peptide, probably membrane-bound. Each region alone can transform mouse cells, but together they act synergistically and produce a more complete transformation, a situation analogous to that encountered with polyoma viruses. In transformed or tumor cells the two regions are transcribed by two separate mRNAs from which the E_1 region is spliced out. The absence of this gene makes integration the only means for persistence of the viral genome in the cancer cells. In human cervical carcinomas, which contain only integrated viral DNA, the integration is such as to allow expression limited to the E_6–E_7 region. Genes included in these regions may be related to oncogenes. The function of these transforming genes is necessary and sufficient for initiating cell transformation or the formation of benign warts in lower animals, but insufficient for producing recognizable changes in human cells. These viral functions are insufficient for the progression of the cells to the malignant state in any species.

HUMAN ADENOVIRUSES

Human adenoviruses cause a productive infection of human cells and kill the cells (see Chap. 52). Many of these viruses, however, transform nonpermissive hamster or rat fibroblasts, in which infection is nonproductive, and cause tumors when injected into newborn hamsters. Tumors are induced at high frequency and rapidly by viruses of serologic group A (**highly oncogenic**); those of group B are **weakly oncogenic;** the others are **nononcogenic.** Viruses of all groups transform hamster cells *in vitro.* Permissive human cells can be transformed by subgenomic viral DNA fragments, which lack genes necessary for DNA replication and cell killing.

The virus-transformed cells contain integrated fragments of the viral DNA, usually multiple and in nonequimolar amounts, corresponding altogether to a fraction of the genome, sometimes to its entirety. Cells transformed by viruses of any group contain at least **four tumor (T) Ags:** two specified by the E1A region of the genome and two by the E1B regions. They are revealed by Abs present in tumor-carrying animals.

MECHANISM OF TRANSFORMATION. The four genes that specify the tumor Ags are located in the early region at the extreme left segment of the viral DNA (see Chap. 52); DNA fragments containing that segment of the genome can transform cells. These and other early genes are transcribed in nonpermissive transformed cells; late genes either are absent in the proviral DNA or are not transcribed, ensuring cell survival. The cooperation of the four genes has results comparable to those of the polyoma virus early genes. Like PyV LT, E1A has an immortalizing function; like PyV MT+ST, E1B contributes a complete transforming function. E1A, like other immortalizing oncogenes, can cooperate with a *ras* oncogene in transforming primary cells.

The E1A region specifies a transactivator of transcription of viral genes; it also has profound effects on the cellular genome. It initiates cellular DNA replication, often in a disorderly way, causing irregularities in chromosome numbers and chromosomal aberrations, and has contrasting effects on transcription of cellular genes. It enhances the activity of cellular regulatory factors that activate growth-stimulatory genes; it inhibits the function of some viral enhancers, such as those of PyV and SV40. The E1A region of highly oncogenic adenoviruses inhibits transcription of the gene for the heavy chain of the class I major histocompatibility (MHC) antigen in both rodent and human cells, with important consequences for the tumorigenicity of the cells they transform (see below). With the latter effect the E1A region mimics the function of certain regulatory cellular genes expressed in early embryonic cells, which also prevent expression of class I MHC genes. Nononcogenic adenoviruses (such as Ad 5) do not inhibit transcription of MHC-I genes; on the contrary, the introduction of the E1A region of Ad 5 into cells transformed by Ad 12 restores MHC-I transcription.

Cells morphologically transformed (an incomplete transformation) by nononcogenic adenoviruses, upon continuous cultivation, acquire the ability to grow in agar and to produce tumors in syngeneic animals. This evolution to full oncogenicity is comparable to the **progression** of malignant cancers; it may be caused by the chromosomal rearrangements induced by the virus. A similar mechanism may explain how some rodent tumors induced by highly oncogenic adenoviruses retain the tumorigenic property upon loss of the viral genome.

ROLE OF MHC EXPRESSION. Highly oncogenic adenoviruses produce tumors that grow indefinitely in immunologically competent syngeneic animals, showing no sign of rejection. Resistance to rejection is apparently due to lack of expression of class I MHC Ags on the tumor cells, which prevents the MHC-restricted activity of cytotoxic T (CTL) lymphocytes (see Chap. 50). Lack of expression is due to the modulation of transcription of cellular genes by the products of the EIa gene, which activates certain promoters while inhibiting others. Introducing an MHC gene by transfection restores MHC expression and abolishes tumorigenicity. Cells transformed *in vitro* by highly oncogenic adenoviruses behave similarly when transplanted into immunocompetent animals. In contrast, cells transformed *in vitro* by nononcogenic adenoviruses express the MHC Ags and form tumors only in immunocompromised animals, such as nude mice, that lack a CTL response. Suppression of class I MHC genes is therefore important for tumor formation in animals but not for transformation *in vitro.*

ONCOGENIC HERPESVIRUSES

Members of all three subclasses (α, β, and γ) of the herpesvirus (HSV) family have a demonstrated or suspected oncogenic activity. They are herpes simplex types 1 and 2 in the α subclass; cytomegalovirus in the β subclass; and Epstein-Barr, herpesvirus saimiri, and the Marek disease virus of chickens in the γ subclass.

All these viruses are able either to immortalize or to transform cells *in vitro.* Epstein-Barr virus, herpesvirus saimiri, and Marek disease virus produce malignant tumors in animals. Moreover, the human viruses are implicated in some human malignancies by epidemiologic observations.

Alphaherpesviruses

UV-irradiated human herpesvirus of either type 1 or type 2 transforms hamster embryo cell cultures *in vitro.* The transformed cells produce tumors when injected into weanling hamsters. Some fragments of viral DNA immortalize the cells, which then become tumorigenic upon serial passages. With intact virus, transformation cannot be seen because the cells are killed. The rationale for UV treatment is that in some virions the radiation, with its random effects, inactivates cell-killing genes but not transforming genes. Using transfection of cloned viral DNA fragments, the transforming region has been located at map units 0.31 to 0.42 for HSV-1 and 0.58 to 0.62 for HSV-2 (see Chap. 53). A characteristic of these transformants is that the infecting viral DNA fragment is unstable and can be lost. Lack of viral DNA sequences, although the cells remain transformed, suggests involvement of cellular genes in transformation.

Human herpesvirus type 2 has been implicated in **squamous cervical and vulvar cancer** in women. However, although the cancer cells often contain herpesvirus proteins, recognizable serologically, and herpesvirus DNA, a prospective epidemiologic study did not confirm an etiologic herpes 2 role. Given the frequent occurrence of herpes 2 infection in promiscuous women, the virus is likely to be an occasional passenger in the cancer cells.

Betaherpesviruses

Like HSV-1 or -2, UV-irradiated cytomegalovirus (HCMV), as well as some cloned fragments of the viral DNA, transform hamster or rat embryo cells in cultures, making them tumorigenic. The fragments assign the transforming sequences to the immediate early region of the genome. However, there is no clear indication that HCMV is oncogenic in humans or animals.

Gammaherpesviruses

Epstein-Barr virus (EBV) was discovered in electron microscopic sections of the **Burkitt lymphoma,** a neoplasm of B-lymphocytes that affects the bones of the jaws and abdominal viscera. The lymphoma cells have a characteristic cytologic appearance and display at their surface immunoglobulins (Igs), all of the same kind in a given tumor; that is, the tumors are **monoclonal.** Subsequently, the virus was recognized in other human cancers, such as the **nasopharyngeal carcinoma (NPC),** some cases of **primary intracerebral lymphoma** and of **lymphoepithelioma-like carcinomas of the thymus,** as well as in **polyclonal lymphoproliferative disorders in immunodeficient persons.** The virus induces a lymphoma in New World monkeys such as marmoset monkeys and cotton-top tamarins. EBV is also the agent of **infectious mononucleosis (IM),** an extensive but self-limiting lymphoid proliferation (see Chap. 53).

STATE IN TUMOR CELLS. Cell cultures derived from BLs usually do not produce EBV virus, but some **producer** lines do, in small amounts. In both kinds of cultures EBV is present in all or most cells as a **latent infection.** The cells contain viral DNA, of which one or two copies are randomly **integrated** in the host DNA; the others (50 to 100 per cell) are present extrachromosomally, as nuclear **plasmids** (as papilloma virus DNA in papillomas). Of the

large EBV genome only a fraction is transcribed, generating three mRNAs, which correspond to two regions: one specifying a nuclear Ag, the other a membrane antigen.

The **nuclear Ag** (called EBNA, from EB virus nuclear Ag) is a soluble complement-fixing agent recognized by antibodies in the patient's serum. It contains five proteins. Of these, one is free in the nucleus; another, containing repeats of the dipeptide glycine-alanine, binds to metaphase chromosomes, probably to a periodic structure, such as chromatin. In this way it may regulate the host genome. The **surface Ag,** present in the cell plasma membrane, contains two proteins; it is recognized by immunofluorescence by means of an antiserum to a protein made in bacteria from a cloned EBV DNA fragment. This Ag elicits a cell-mediated immune response against the infected cells. One of the surface Ags confers tumorigenicity on cells to which it is transfected. It therefore appears to be a **transforming gene,** possibly equivalent to PyV MT.

Except for producer lines, no other viral gene is expressed in BL cells; EBV DNA is replicated as a plasmid by cellular enzymes. As with papilloma virus, the replication is **synchronous** with that of the host DNA and maintains a **constant copy number** per cell. Maintenance appears to depend on the action of the EBNA-1 protein on the origin of replication of the viral DNA. EBNA-1 may therefore be equivalent to the PyV LT.

INDUCTION. Viral multiplication can be induced in latently infected cell lines by tumor promoters (perhaps through activation of protein kinase C; see Chap. 47) or butyrate, or, more significantly, by introducing into the cells certain cloned fragments of viral DNA. The expression of two viral genes (EB1 and EB2) generates a *trans*-activating factor, the absence of which determines latency. The first new gene products appearing after induction, before replication of the viral DNA, are several polypeptides of the **early Ag.** In some cell lines no other genes are activated, giving rise to an **abortive infection.** In more permissive lines the early Ag is followed by the **viral capsid Ag** and a new DNA polymerase, which makes **linear viral DNA molecules,** equal to those present in virions; this enzyme, in contrast to the cellular enzyme, is **inhibited by phosphonoacetic acid or acyclovir** (see Chap. 49). These cells release progeny virions. In producer cell lines spontaneous induction frequently takes place in rare cells.

INFECTIVITY OF RELEASED VIRUS. Virus released through induction of BL cells is infectious for human cells carrying **receptors** for the C3d–CR2 component of complement, which are also receptors for the virus. They are present on resting human B-lymphocytes and epithelial cells of oral and nasopharyngeal squamous epithelia, and the prickle cell layer of the tongue. Infected human cord blood lymphocytes are induced to differentiate into Ig-secreting blast cells, capable of indefinite multiplication; they give rise to permanent **lymphoblastoid cell lines,** which are immortalized but not transformed. Cells lacking receptors can be infected by circumventing their need, either by transfection with purified viral DNA or by exposure to the DNA enclosed in the envelope of another virus. Resistant cells can also be made sensitive by the injection of the receptor gene; receptors then appear on the surface. These approaches permit the infection of a variety of cell types.

The ability of the virus to immortalize B-lymphocytes shows that it performs in human cells one of the functions needed for transformation. In New World monkeys it displays a complete oncogenic potential by inducing lymphomas.

VIRAL TRANSMISSION AND CONSEQUENCES OF INFECTION IN HUMANS. The virus is transmitted horizontally. In underdeveloped countries, because of poor hygiene and overcrowding, essentially all children are infected by the age of 3 years. In developed countries infection occurs later, and about 20% of the people escape infection altogether. Primary infection takes place through the throat. The virus productively infects epithelial cells around the oropharynx, which are recognized as being EBNA-positive; they give rise to abundant viral shedding. A small number of B-lymphocytes are also infected, but latently, and are immortalized. These lymphocytes express different Igs at their surfaces (**polyclonal reaction**). Because these cells also express the viral surface Ag, they are killed by the cytotoxic T-cell response of the organism; the infection remains clinically silent. Primary infection in adolescents or young adults through kissing, in contrast, frequently generates infectious mononucleosis (IM). Again, this polyclonal B-cell proliferation is brought under control by a strong T-cell reaction. Most circulating lymphocytes present during the acute phase of the disease are cytotoxic T cells specific for the viral surface Ag. In both subclinical infection and IM the infected person is finally seroconverted, produces Abs to a variety of EBV Ags, and becomes solidly immune to a new infection. Virus production persists for life in the buccal cavity and is the main source of viral transmission. The persistence of the infected epithelial cells may be attributed to their failure to express or to present adequately the viral surface Ag, thus avoiding the cellular immune response of the host.

CONSEQUENCES OF INFECTION IN IMMUNODEFICIENT PERSONS. Immunodeficient persons include organ transplant patients made immunodeficient for safeguarding the transplant and persons with genetic or acquired defects of the immune system. In these persons, infected B cells persist, giving rise to a massive polyclonal B-cell proliferation, which itself can be lethal. In some cases the disease evolves to a monoclonal lymphoma owing to the occurrence of an additional, rare event of unknown nature. Immunodeficiency may also favor the development of a BL (see later).

BURKITT LYMPHOMA. BL is a cancer that has the greatest incidence in certain areas of the world, especially central Africa and New Guinea (**endemic BL),** where it affects children from 2 to 16 years of age; in other places it is very rare (**sporadic BL).** Cells derived from endemic BL regularly contain many copies of the viral genome, but those derived from sporadic BL usually lack them. The discrepancy suggests that the virus performs a function, probably immortalization, that can also be supplied in other ways. An important factor in the development of BL is the activation of the proto-oncogene *c-myc* (located on chromosome 8) by characteristic chromosome translocations, which are regularly present in BLs. The translocations bring the *c-myc* gene close to one of the immunoglobulin loci. Most frequently the translocation involves the heavy-chain Ig locus (chromosome 14), more rarely the locus for the κ light chain (chromosome 2) or for the λ light chain (chromosome 22). The Ig produced by a particular tumor is usually specified by the gene involved in the translocation: BLs with an 8–2 translocation produce a κ-chain Ig; those with the 8–22 translocation produce a λ-chain Ig. This suggests that the *c-myc* is activated only if translocated near an active Ig gene. There is great variability in the localization of the break point in respect to both the *c-myc* and the Ig genes, but the 5' end of *c-myc* is always altered; this alteration may be the cause of *c-myc* activation. The translocation and *c-myc* activation abrogate the immune response against the surface Ag, probably by reducing its expression; the mechanism is unknown.

The endemic forms of BL occur in areas with high incidence of ***Plasmodium falciparum* malaria.** The suppression of T-cell activity and stimulation of B-cell proliferation caused by the parasite probably favor the development of the tumor. Malaria immunosuppression may also contribute to the early age of primary EBV infection in endemic areas.

BL is therefore produced by three functions. The virus performs the immortalizing function in endemic areas, where infection in children is widespread. The activation of *c-myc* contributes the transforming function; and a favoring function is caused by malaria, probably by allowing the formation of a large population of immortalized B cells susceptible to transformation. In nonendemic areas some other unknown, less frequent event provides the immortalizing function. It is interesting that in BL cells the activated *myc* gene acts as a transforming oncogene as it does in avian cells infected by the MC29 virus (see Chap. 65); in contrast, in rodent primary cells *myc* acts as an immortalizing oncogene. These differences point to the importance of the species of the cells, and therefore the state of the rest of the genome, in determining the effect of an oncogene.

NASOPHARYNGEAL CARCINOMA (NPC).

NPC is one of the most frequent cancers in males in certain ethnic groups in southern China, where a high consumption of Cantonese-style salted fish during childhood appears to provide a necessary cofactor. The tumors contain proliferating epithelial cells as well as lymphocytes: EBV DNA is present in the epithelial cells in multiple (100 to 150) copies as plasmids. As in the abortive infection of B-lymphocytes, only the early Ag is produced. The patients have Abs of G and A isotypes to this Ag. The Abs are useful for early diagnosis because their appearance precedes by 1 to 2 years the onset of the tumor and because they are easily screened; during this time the mucosa displays characteristic hyperplasia and atypia. The titers of the Abs are related to tumor burden, so they are a good indicator of the course of disease.

HERPESVIRUS SAIMIRI.

Herpesvirus saimiri naturally infects squirrel monkeys without producing any clinical symptoms. Virus can be isolated from normal animals and propagated in cultures of owl monkey kidney cells. In New World monkeys, especially marmosets, the virus is highly oncogenic, rapidly producing T-cell lymphomas and acute leukemias. *In vitro* the virus immortalizes owl monkey T cells, making them independent of the interleukin-2 growth factor (see Chap. 47). Immortalized cells as well as tumor cells contain large numbers of viral DNA molecules in stable plasmidial form; it is not known whether the cells also have integrated DNA. This virus is therefore similar to EBV except that it affects T cells rather than B cells. A small proportion of the genome, extremely variable among strains, is responsible for oncogenesis; these sequences are not needed for viral multiplication in owl cells.

MAREK DISEASE VIRUS.

Marek disease virus (MDV), a chicken herpesvirus, causes both a **productive** infection in the epithelium of feather follicles and an **abortive** infection, with neoplastic transformation, in lymphoid T cells; the tumor cells infiltrate many visceral organs and the peripheral nerves. Most lymphoma cells contain little or no infectious virus, but their inoculation into healthy chickens transmits the disease. The involvement of T cells in the neoplasia, and the formation of suppressor cells, leads to severe **immunosuppression,** which favors development of the disease by decreasing the immunologic defenses.

A successful live vaccine against MDV infection of chickens has been developed, using a virus adapted to the turkey that is altered in the presumptive transforming region of the genome. The vaccine has markedly decreased the incidence of lymphoma.

HEPATITIS B VIRUS

Infection with **hepatitis B virus** (HBV; see Chap. 63) is intimately associated with **primary hepatocellular carcinoma (PHC)** in humans, which is the leading cause of death from cancer worldwide, and in some animals, such as the woodchuck, the ground squirrel, and Peking ducks. The virus has wide distribution: more than 10% of individuals are infected worldwide. It multiplies only in cultures of human hepatocellular carcinoma cells and does not transform any type of cells *in vitro*. The virions contain a cyclic, partially double-stranded molecule, which in the cells is replicated by reverse transcription through an RNA intermediate (see Chaps. 48 and 65).

In the hepatocellular carcinomas of man and the woodchuck and in lines derived from them, the HBV genome is found **integrated.** The opening of the DNA circle for integration appears to occur at various places, but the sequences for the surface (s) Ag are usually uninterrupted, suggesting an important role in cancer induction. Often multiple copies (five to seven) are integrated, scattered over the cellular genome. In some cases short direct repeats (11 base pairs to 12 base pairs) flank the genome, as is the case with transposons and retroviruses, to which the virus is related by the use of reverse transcription for DNA replication (see Chap. 8). In cell lines derived from human hepatocellular carcinomas the integrated viral genome and the host flanking sequences are often heavily rearranged, but they may be less so in the cancer cells themselves. The cancer cells may express the surface (s) Ag (earlier termed *Australian Ag*), but not core (c) Ags. The gene for this Ag is in highly methylated form in the cancer cells; in culture, 5-azacytidine, which prevents methylation of the newly made DNA, elicits the production of c Ag. Some cancers do not express any viral antigen.

Experimental evidence for an etiologic role of HBV in PHC is based on the experimental induction in the woodchuck. In humans such a role is supported by a striking correlation between the incidence of PHC and that of chronic HBV infection. Both are common in many parts of Africa and Asia but rare in North America and Europe. In the south coast provinces of China, in which PHC constitutes 50% of all cancer deaths in males and 25% in females, 86% of the patients are serologically positive for s Ag. An important factor for cancer development seems to be infection at young age, frequently from the mother; the infected children end up as virus carriers with chronic hepatitis and develop PHC 30 to 40 years later. A prospective epidemiologic study in Taiwan showed that the risk of PHC is 200 or 300 times higher in **chronic carriers** than in noncarriers. Alcoholic liver cirrhosis does not provide a background for this cancer. In the carriers the HBV DNA is not integrated in the liver cells, whereas it is in the cancers.

The mechanism of cancer induction by HBV is unclear. The virus does not seem to contain any oncogene, nor does the integration activate neighboring cellular proto-oncogenes, as happens with retroviruses (see Chap. 65), because it occurs at random sites. The tumors are monoclonal in respect to the mode of integration, suggesting the need of other events for cancer formation. Rearrangement of cellular DNA induced by the virus may contribute the additional event through alteration of cellular genes. In all likelihood several factors concur in generating the cancer, including environmental factors, such as mycotoxins. The prevalence of the cancer in males may be related to the presence of hormone-responsive control elements in the viral genome.

DNA TUMOR VIRUSES AS CLONING VECTORS

Many viral genomes of DNA tumor viruses have been converted into cloning vectors (see Gene Manipulation, Chap. 8). **SV40 DNA** was the first to be used and is present in several currently used vectors. They contain the basic SV40 replicon, that is, its origin of replication and the LT gene, the rest being foreign DNA. In cells coinfected with a ts A virus as helper maintained at 41°C the hybrid DNA replicates and is encapsidated by the capsid proteins of the helper. The SV40 replicon is also incorporated in plasmids (such as pBR322; see Chap. 8), together with selective markers, converting them into **shuttle vectors,** which can replicate in either animal cells (using the SV40 replicon) or in bacteria (using the plasmidial replicon). The LT gene can be dispensed with if the vector is grown in **cos cells** in which LT is generated by the expression of an integrated SV40 genome.

Bovine papilloma virus (BPV) DNA is useful as a vector, owing to its persistence in cells, because extrachromosomal plasmids allow easy purification. Sixty-nine percent of the genome is required for replication and maintenance. It can accept at least 16 Kb of foreign DNA; with larger inserts it tends to integrate into the host DNA. Fused to pBR322 it constitutes a useful shuttle vector. Foreign genes introduced at the BVP–pBR322 junction are properly regulated in animal cells.

Also useful as cloning vectors are the EB virus, herpes simplex virus, and adenovirus.

Selected Reading

Baker CC, Howley PM: Differential promoter utilization by the bovine papillomavirus in transformed cells and productively infected wart tissues. EMBO J G: 1027, 1987

Bishop JM: Viral oncogenes. Cell 42:23, 1985

Brinster RL, Chen HY, Messing A et al: Transgenic mice harboring SV40 T-antigen genes develop characteristic brain tumors. Cell 37:367, 1984

Das GC, Niyogi SK, Salzman NP: SV40 promoters and their regulation. Nucleic Acids Res Mol Biol 32:218, 1985

Fluck MM, Staneloni RJ, Benjamin T: Hr-t and ts-a: Two early gene functions of polyoma virus. Virology 77:610, 1977

Greenfield C, Fowler MJF: Hepatitis B virus and primary liver cell carcinoma. Mol Biol Med 3:301, 1986

Hanahan D: Heritable formation of pancreatic β-cell tumours in transgenic mice expressing recombinant insulin/simian virus 40 oncogenes. Nature 315:115, 1985

Hunter T, Cooper JA: Protein tyrosine kinases. Ann Rev Biochem 54:897, 1985

Kato S, Hirai K: Marek's disease virus. Adv Virus Res 30:225, 1985

Khoury G, Gruss P: Enhancer elements. Cell 33:313, 1983

Kingston RE, Baldwin AS, Sharp PA: Transcription control by oncogenes. Cell 41:3, 1985

Klein G, Giovanella BC, Lindhal T et al: Direct evidence for the presence of Epstein-Barr virus DNA and nuclear antigen in malignant epithelial cells from patients with poorly differentiated carcinoma of the nasopharynx. Proc Natl Acad Sci USA 71:4737, 1974

Kovesdi I, Reichel R, Nevins JR: Identification of a cellular transcription factor involved in E1A *trans*-activation. Cell 45:219, 1986

Lewis AM, Cook JL: The interface between adenovirus-transformed cells and cellular immune response in the challenged host. Curr Topics Microbiol Immunol 110:1, 1984

Li JJ, Peden KWC, Dixon RAF, Kelly T: Functional organization of the Simian Virus 40 origin of DNA replication. Mol Cell Biol 6:1117, 1986

Lupton S, Levine AJ: Mapping genetic elements of Epstein-Barr virus that facilitate extrachromosomal persistence of Epstein-Barr virus–derived plasmids in human cells. Mol Cell Biol 5:2533, 1985

Manos MM, Gluzman Y: Simian Virus 40 large T-antigen point mutants that are defective in viral DNA replication but competent in oncogenic transformation. Mol Cell Biol 4:1125, 1984

McKnight S, Tjian R: Transcriptional selectivity of viral genes in mammalian cells. Cell 46:795, 1986

Mounts P, Shah KV: Respiratory papillomatosis: Etiological relation to genital tract papilloma viruses. Progr Med Virol 29:90, 1984

Palmiter RD, Brinster RL: Transgenic mice. Cell 41:343, 1985

Rassoulzadegan M, Cowie A, Carr A et al: The roles of individual polyoma virus early proteins in oncogenic transformation. Nature 300:713, 1982

Ratner L, Josephs SF, Wong-Staal F: Oncogenes: Their role in neoplastic transformation. Am Rev Microbiol 39:419, 1985

Sefton BM: The viral tyrosine protein kinases. Curr Topics Microbiol Immunol 123:39, 1986

Spector DH, Spector SA: The oncogenic potential of human cytomegalovirus. Progr Med Virol 29:45, 1984

Tiollais P, Pourcel C, Dejean A: The hepatitis B virus. Nature 317:489, 1985

Tosato G, Blaese RM: Epstein-Barr Virus infection and immunoregulation in man. Adv Immunol 37:99, 1985

van Beveren C, Verma IM: Homology among oncogenes. Curr Topics Microbiol Immunol 123:1, 1986

van der Ebs AJ, Bernards R: Transformation and oncogenicity by adenoviruses. In Doerfler W (ed): The Molecular Biology of Adenovirus 2. New York, Springer-Verlag, 1984

Wang D, Liebowitz D, Kieff E: An EBV membrane protein expressed in immortalized lymphocytes transforms established rodent cells. Cell 43:831, 1985

Yang Y-C, Okayama H, Howley PM: Bovine papillomavirus contains multiple transforming genes. Proc Natl Acad Sci USA 82:1030, 1985

Zuckerman AJ: Prevention of hepatocellular carcinoma by immunization against hepatitis B. Int Rev Exp Pathol 27:60, 1985

65

Renato Dulbecco

Oncogenic Viruses II: RNA-Containing Viruses (Retroviruses)

The family of **Retroviridae** (L. *retro*, backward; vernacular, **retroviruses**) is characterized by the presence of a reverse transcriptase in the virions. It includes several genera (Table 65–1), some of which are not oncogenic. The tumorigenic retroviruses are members of the three genera of **oncoviruses** (Gr. *oncos*, tumor). They induce sarcomas, leukemias, lymphomas, and mammary carcinomas. Retroviruses exist in the most diverse species, from fish to humans. They can be transmitted both horizontally and vertically; occasionally they incorporate and transmit sequences of the host DNA.

The virions are enveloped and ether-sensitive, about 100 nm in diameter; the capsid, probably icosahedral, encloses the single-stranded RNA genome.

Retroviruses have an unusual method of multiplication. Within the cells the viral RNA, released from the envelope, gives rise to a double-stranded DNA copy by **reverse transcription.** This copy moves to the nucleus and becomes **integrated** in the cellular DNA as **provirus.** The provirus is **transcribed** by the DNA-dependent RNA polymerase II, generating RNA copies, some of which are the genomes of progeny virions; others are processed to mRNAs.

Components of Virions

The viral RNA is a positive-sense, single-stranded molecule of about 9 Kb in oncovirus, 10 Kb in lentivirus, and

TABLE 65–1. Retroviridae

Genus	Subgenus	Species
Cisternavirus A		
Mice, hamster, guinea pig		
Oncovirus B		
Mammary carcinomas in mice		Mouse mammary tumor viruses: MMTV-S (Bittner's virus), MMTV-P (GR virus), MMTV-L
Oncovirus C	Human	Human lymphotropic viruses I and II (HTLV-I and II)
	Avian	Rous sarcoma virus (RSV)
		Rous-associated viruses (RAV)
		Other chicken sarcoma viruses
		Leukosis viruses (ALV)
		Reticuloendotheliosis viruses
		Pheasant viruses
	Mammalian	Murine sarcoma viruses (MSV)
		Murine leukosis virus G (Gross or AKR virus)
		Murine leukosis viruses (MLV)-F,M,R (Friend, Moloney, Rauscher viruses)
		Murine radiation leukemia virus
		Murine endogenous viruses
		Rat leukosis virus
		Feline leukosis viruses
		Feline sarcoma virus
		Feline endogenous virus (RD114)
		Hamster leukosis virus
		Porcine leukosis virus
		Bovine leukosis virus
		Primate sarcoma viruses (woolly monkey; gibbon ape)
		Primate sarcoma-associated virus
		Primate endogenous viruses: baboon endogenous virus (BaEV), stumptail monkey virus (MAC-1), owl monkey virus (OMC-1)
	Reptilian	Viper virus
Oncovirus D		
Primates		Mason-Pfizer monkey virus (MPMV)
		Langur virus
		Squirrel monkey virus
Lentivirus		Human immunodeficiency virus (HIV)
		Visna virus of sheep
		Caprine arthritis-encephalitis virus
		Equine infectious anemia
Spumavirus F		Foamy viruses of primates, felines, humans, and bovines

shorter in some defective viruses (see later). Like cellular mRNAs, it has a poly(A) tail at the 3′ end, a cap at the 5′ end. Electron micrographs of the RNA extracted from several type C viruses show that **each virion contains two RNA molecules** held together by a dimer linkage structure near the 5′ end. The RNA dimer (70S) separates upon denaturation into the two genetically identical molecules; hence, the virion is **diploid.** The virions also contain some cellular RNAs of low molecular weight; among them are tRNAs, which, as seen below, perform an essential function.

As shown in Fig. 65–1, the viral RNA contains three basic genes: *gag, pol,* and *env* in 5′ to 3′ direction. Human oncoviruses have an additional gene, and lentiviruses have five (see below). The two ends of the RNA have distinctive features, which, as will be seen below, are very important for the functions of the virus. The 5′ end contains, in 5′ to 3′ direction, the **cap; a terminal redundancy (R)** of from 10 to 80 nucleotides, depending on the virus; the **U5 sequence** (meaning unique 5′); and the **primer binding site** (PBS) of 16 to 18 nucleotides complementary to the 3′ end of a tRNA, which is bound to it. Avian viruses have tryptophan tRNA; murine and feline viruses have proline tRNA; the mouse mammary tumor virus and lentiviruses have lysine tRNA. The bound tRNA is the **primer** for reverse transcription. The AUG codon at beginning of the **gag** protein is several hundred nucleotides beyond the cap site. In this interval there are two important sequences: the **dimer linkage site (DLS),** where the two copies of the RNA present in virions are held together, and the **packaging signal** (ψ), which allows the packaging of the RNA in the virions.

The 3′ end contains, beyond the end of the *env* gene: the **+ strand primer region** (+P) of 12 bases rich in purines, which is important in reverse transcription; the **unique (U3) sequence,** which contains important signals for the transcription of the provirus; the **terminal redundancy (R),** identical to that at the 5′ end; and a **poly(A)** chain. The U3 sequence is quite long and variable in different viruses. It has 150 to 170 bases in type C avian oncoviruses, but over 1200 in mouse mammary tumor virus (type B), which has the only U3 sequence with an open reading frame capable of encoding a small protein.

The general organization of the genome, with its terminal repeats enclosing the genes, is similar to that of bacterial and eukaryotic transposons (see Chap. 8). Additional similarities will emerge later.

Multiplication (Fig. 65–2)

Most retroviruses are not cytopathic and do not appreciably alter the metabolism of the cells they infect. In cultures, infected cells continue to multiply while releasing progeny particles. The adsorption of virions (through

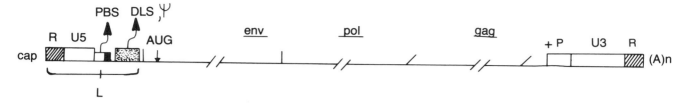

Figure 65–1. Features of a retroviral genome. *R,* terminal redundancy; *U5,* unique 5′ end region; *PBS,* primer binding site; *DLS,* dimer linkage site; ψ, packaging signal; *AUG,* initiation codon for *gag* protein synthesis; *+P,* plus-strand primer region; *U3,* unique 3′ end region; *L,* leader region; *(A)n,* poly(A). *gag, pol,* and *env* are the three genes.

their glycoproteins to specific cell receptors) and penetration take place as with other enveloped viruses (see Chap. 48).

The outstanding feature of oncovirus multiplication is the **DNA intermediate** in the replication of the viral RNA. Such an intermediate was predicted by Temin from the extraordinary sensitivity of oncornavirus multiplication to agents that inhibit DNA replication or transcription (BUdR, FUdR, or actinomycin D) and from the presence of sequences complementary to the viral RNA in the DNA of infected cells. The ability of this DNA to infect other cells, which then produce virus, later supplied direct evidence for a viral DNA.

Less than 1 hour after infection of growing chicken cells with an avian oncovirus, viral DNA synthesis begins in the cytoplasm. A continuous **minus strand** (complementary to the viral RNA) is made by the viral reverse transcriptase (Fig. 65–3), and even before it is completed, the synthesis of the **plus DNA** strand (complementary to the minus strand) begins. When the minus strand is completed, the part of the viral RNA still paired to it is degraded by RNase H.

Initially, the viral DNA is a double-stranded **linear** molecule containing gaps, with a continuous minus strand and a discontinuous plus strand. Between 6 and 9 hours after infection the gaps are filled, and the DNA moves to the nucleus and becomes **cyclic;** by 24 hours several DNA molecules have become integrated in the cellular DNA as **proviruses** at different random sites in each cell. The **linear, cyclic,** and **proviral forms** of the viral DNA contain a complete genome because they are **all infectious.** The viral DNA does not undergo independent replication either in linear or in cyclic form, because it is not a complete replicon. Progeny RNA is generated by regular transcription of the integrated provirus.

As a consequence of this method of replication the **double-stranded viral DNA is different from the virions' RNA,** having increased its length by 500 to 600 nucleotides. The differences are at the two ends, which have become equal (Fig. 65–4); they are known as **long terminal repeats (LTRs).** Each LTR consists, in the 5′ to 3′ direction, of a U5, R, and U3 region. Each LTR contains

a signal for cap addition, a TATA and a CAT site determining the beginning of transcription, as well as a site for addition of poly(A). The provirus utilizes only the initiation signals of the upstream (5′) LTR because the advancing transcription interferes with initiation at the downstream LTR. The 3′ LTR contributes the termination signals but can initiate transcription if the 5′ LTR is missing. Like the insertion sequences present at each end of bacterial transposons, each LTR has at the two ends short direct repeats of the cellular integration site (five to 13 base pairs in different proviruses).

A salient feature of LTRs is a **transcriptional enhancer** (see Chap. 64), consisting of a tandem repeat of an 85-base-pair (in a murine virus) sequence, which upon interaction with trans-acting factors of both viral and cellular origin, controls the rate of transcription in *cis.* Enhancers have cell-specific activity: the range is broad for some oncoviruses that are transcribed in many kinds of cells, narrow for others. In this way the LTRs contribute to determining the viral host range.

INTEGRATION

The LTRs are crucial in the integration of the viral DNA into the cellular DNA. Only covalently closed cyclic molecules, with the two LTRs connected end to end in a direct repeat, are integrated (Fig. 65–5). The integrase cuts the two strands separately at the U5–U3 junction; it makes two staggered cuts in the cellular DNA and inserts the viral DNA into it. After the gaps are repaired, the inserted viral DNA, or **provirus,** is four base pairs shorter than the cyclic DNA (they will be replaced at the next reverse transcription) and is flanked by the short repeats of cellular DNA, which are the hallmark of integrated transposons (see Chap. 8). Regular transcription, initiating at the 5′ LTR and ending at the 3′ LTR, generates the viral RNA present in virions. These proviruses are rarely excised by recombination between the LTRs.

The complex method of retroviral genome replication is similar to that of other eukaryotic transposons, such as the Ty elements of yeast and the *copia*-like elements of *Drosophila.* It also has some similarity to the mechanism

(Text continues on p 360.)

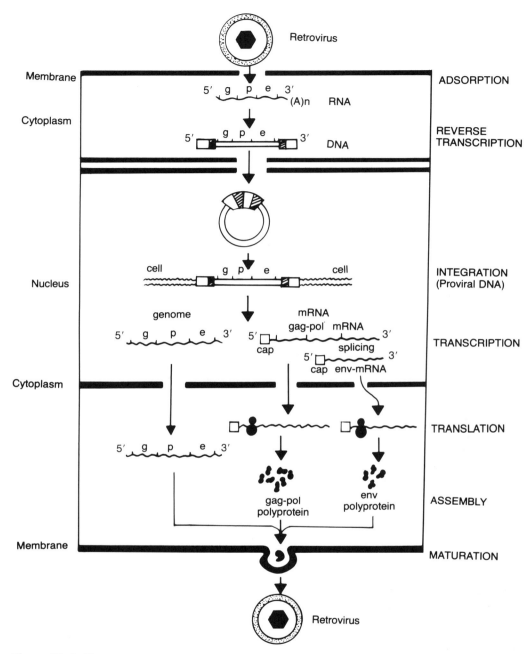

Figure 65–2. The retroviral life cycle, from entry of the infecting virions (*above*) to release of progeny virions (*below*). (Modified from Verma I: In Becker Y [ed]: Replication of viral and cellular genomes, p 275. Boston, Martinus Nijhoff, 1983)

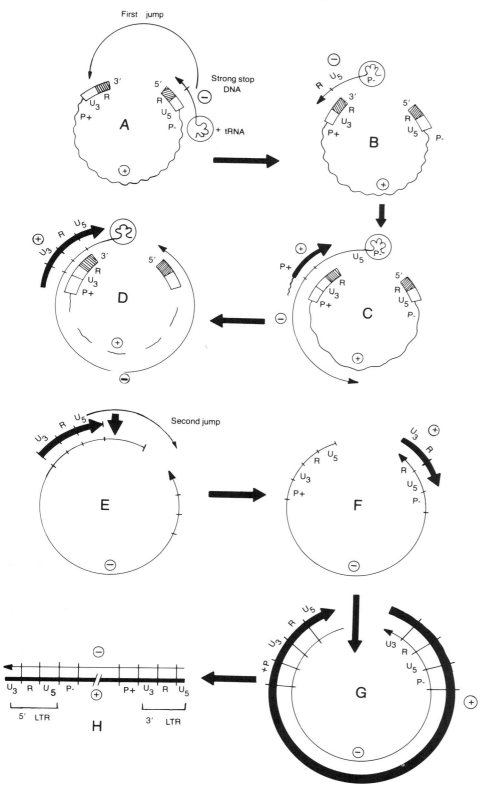

Figure 65–3. Model of reverse transcription of a retrovirus RNA. (*A*) Synthesis of the minus(−)-strand DNA initiates at the tRNA primer bound to the primer binding site (*P−*) close to the 5′ end of the RNA and continues to the 5′ end. This step yields the **strong stop DNA** attached to the tRNA primer. (*B*) The strong stop DNA (−) has "jumped" to the terminal redundancy (*R*) at the 3′ end of the RNA. (*C*) The minus DNA strand grows toward the 5′ end of the RNA. At the same time plus-strand synthesis begins at an RNA primer connected to the viral DNA minus strand at the plus primer binding region (*P+*) in proximity to its 5′ end. The primer is isolated by the RNase H function of the reverse transcriptase from the viral RNA. (*D*) The plus DNA strand grows toward the tRNA primer and stops at its beginning. (*E*) The tRNA primer and the viral RNA are degraded by RNase H. (*F*) The plus DNA has "jumped" to the other (3′) end of the minus DNA strand. (*G*) Both plus and minus strands are completed, generating a double-stranded DNA molecule bounded by two LTRs (*H*). The "jumps" must be considered exchanges between the two ends, which are held in close proximity by proteins, including the reverse transcriptase. The exchanges may occur between ends of the same molecule or between those of the two molecules forming a dimer in the virions. (Modified from Verma I: In Becker Y [ed]: Replication of Viral and Cellular Genomes, p 275. Boston, Martinus Nijhoff, 1983)

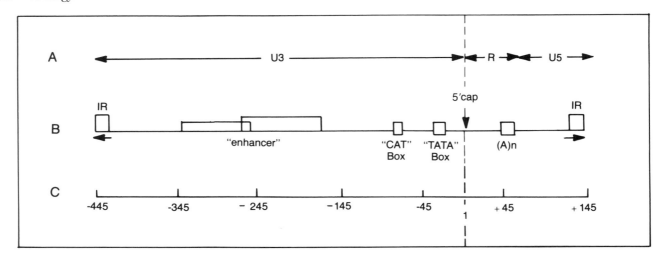

Figure 65–4. Organization of one of the LTRs flanking an Mo-MuLV provirus. (*A*) The main LTR regions (symbols are as in Fig. 65–1). (*B*) Map of the LTR. *IR,* inverted repeats of 11 base pairs in the integrated LTR. The unintegrated DNA contains two additional A-T pairs at each end. *"TATA" Box* is the polymerase II initiation site; *"CAT" Box* is a sequence associated with polymerase II activity; (*A*)*n* is the signal for poly(A) addition; *5' cap* is the site for cap addition (*arrow*). When the provirus is transcribed, the cap site (nucleotide #1) marks the beginning of the RNA. (*C*) Distances in base pairs. (Modified from Verma I: In Becker Y [ed]: Replication of Viral and Cellular Genomes, p 275. Boston, Martinus Nijhoff, 1983)

of integration of temperate bacteriophages, such as lambda (see Chap. 46). In fact, the site created by the joining of the two LTRs in the cyclic molecules is referred to as the **att** site, and the integrase is referred to as the **int** enzyme.

REVERSE TRANSCRIPTASE

The crucial protein reverse transcriptase, specified by gene *pol,* is a multifunctional enzyme with the following activities: (1) **RNA-dependent DNA polymerase** that builds a DNA strand complementary to the **template** plus RNA strand as an extension of the tRNA bound to the template acting as primer. The two strands remain together in a **hybrid** (RNA–DNA) **helix.** (2) **DNA-dependent DNA polymerase,** which builds a plus DNA strand complementary to the newly built strand, generating a double-stranded DNA helix. (3) **RNase H,** which degrades the RNA strand in RNA–DNA hybrids, leaving an oligonucleotide as primer for the synthesis of the plus DNA strand, and cuts off the primer tRNA. (4) **Integrase,** which integrates the double-stranded viral DNA into the cellular DNA. The polymerase, RNase H, and integrase functions are carried out by separate parts of the protein. Of the various functions only the integrase is virus-specific; the enzyme of one virus performs the other functions on the RNAs of all other viruses and on nonviral RNAs.

SYNTHESIS OF VIRAL PROTEINS (Fig. 65–6)

Synthesis of viral proteins begins at the same time as that of the viral DNA and takes place on two main messen-

gers. A 35S messenger, probably identical to the RNA that ends up in virions, is the template for synthesis of the *gag* protein precursor. The adjacent *pol* gene is out of frame with *gag* and is read 5% of the time by frame-shifting ribosomes, generating a *gag–pol* polyprotein. A second 24S messenger, spliced from the 35S RNA, is the template for the *env* precursor, which is also read on a different frame. The various precursors (*gag* polyprotein, *gag–pol* polyprotein, *env* polyprotein) are later cleaved into the final products (see Fig. 65–6). Processing is carried out by a viral protease recognizing cysteine, which is specified by a sequence between the *gag* and *pol* sequences.

MATURATION (see Fig. 65–2; Fig. 65–7)

The *env* polyprotein enters the tubes of the endoplasmic reticulum during synthesis and moves to the Golgi apparatus, where it is glycosylated. From the Golgi it moves to the plasma membrane. Most of the *gag* polyprotein remains in the cytosol, but a fraction follows the same pathway of the *env* polyprotein, is glycosylated, and reaches the outer side of the plasma membrane. About 8 hours after infection the *gag* and *gag–pol* precursors, together with viral RNA, start to assemble nucleocapsids under the cell plasma membrane while budding of the virions takes place (Fig. 65–8); the polyproteins are cleaved in the process. The nucleocapsids bind to *env* polyprotein, which is then cleaved into the transmembrane matrix protein p15E and the external gp 70. The latter remains bound to p15E by S–S bonds.

The packaging sequence ψ of the viral RNA is essential for nucleocapsid assembly. If it is deleted, no virions are assembled, but the virus can supply the proteins to an

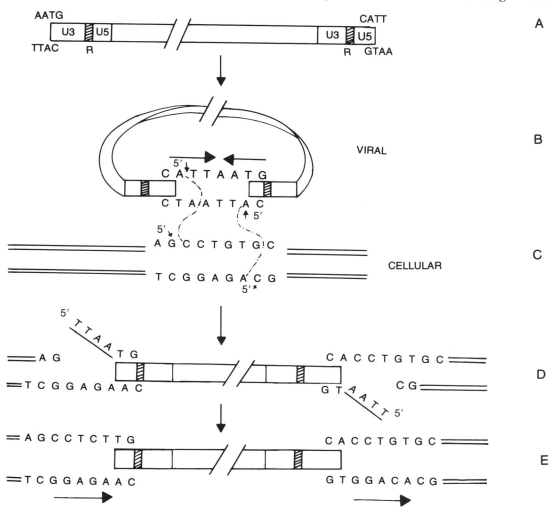

Figure 65–5. Integration of proviral DNA. (*A*) Linear form of proviral DNA. (*A*) It closes covalently to form a circle (*small arrows*). (*B* and *C*) The integrase function of the polymerase makes two staggered cuts in the cellular DNA and two in the viral DNA (*large arrows*). The 3′ ends of the cut viral DNA join the 5′ ends of the cut cellular DNA. (*D*) Four bases of the proviral DNA are removed at each end (*underlined*). (*E*) The gaps are filled by copying the two single-stranded cellular sequences, generating two direct repeats (*arrows*). (Modified from Panganiban AT: Cell 42:5, 1985)

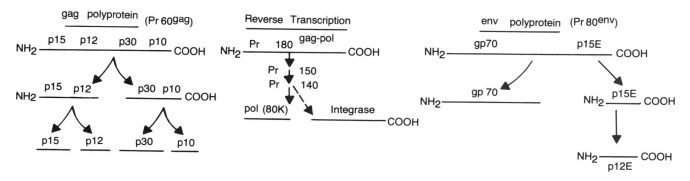

Figure 65–6. Processing of the polyproteins of a murine retrovirus. *Pr,* precursor protein.

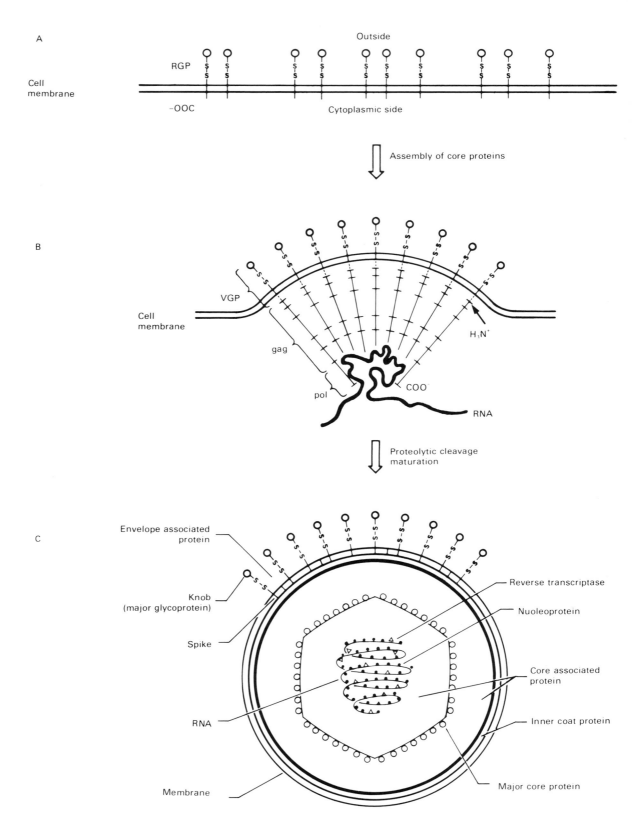

Figure 65-7. Assembly and maturation of a leukosis virus. (A) Incorporation of the viral glycoproteins (VGP) into the cellular membrane. (B) Assembly of the core polyprotein under the cellular membrane; interaction with VGP and viral RNA. (C) Mature virions. (Modified from Bolognesi DP et al: Science 199:183, 1978. Copyright 1978 by the American Association for the Advancement of Science)

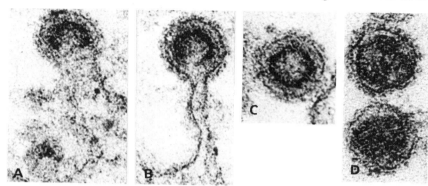

Figure 65–8. Thin sections of type C particles from cells producing a mouse leukosis virus. (*A* and *B*) Two phases of budding of the virions at the cell surface. (*C*) Detached immature virion, with an electron-lucent nucleoid. (*D*) Two mature particles with dense nucleoids. (Courtesy of L. Dmochowski)

intact genome, which will then be assembled. The packaging sequence is present in the 35S RNA but not in the *env* mRNA, from which it is removed with the intron; lacking the sequence, this messenger is not packaged. This explains why only genomic RNA is present in virions. Moving the packaging sequence to the 3′ end of the genome by *in vitro* DNA recombination allows the packaging of *env* messengers.

PHENOTYPIC MIXING

In cells infected by more than one kind of oncovirus some progeny virions contain glycoproteins not specified by the enclosed genome. In this way defective genomes, incapable of specifying virions' proteins, can form virions in a cell that is also infected by a normal genome, which acts as **helper.** The glycoproteins may even derive from an **unrelated virus,** such as vesicular stomatitis virus (VSV). Such **pseudotypes** have a **host range determined by the envelope,** which may be broader than the host range of the genotype. Thus, a murine sarcoma virus genome in an envelope derived from a feline or primate virus can induce tumors in the usual host of the latter virus.

Genetics

Many markers are available for genetic studies in both avian and murine viruses: **temperature-sensitive (ts) mutations** affecting the reverse transcriptase or the cell-transforming ability; differences in the **host range** or **antigenicity** determined by the glycoproteins; differences in the **electrophoretic mobility** of individual proteins; **restriction endonuclease fingerprints** of the various forms of the viral DNA; and, for Rous sarcoma virus (RSV), morph mutants (see Fig. 65–11), which produce **transformed cells of fusiform,** rather than round, morphology.

Oncoviruses have a **high mutation frequency,** suggesting frequent errors by the reverse transcriptase. For example, an avian sarcoma virus unable to infect duck cells generates mutants that are able to do so, with a frequency of 10^{-4} to 10^{-5} per generation.

In genetic crosses with avian viruses **the frequency of recombination is very high,** 10% to 20% between *env* and *pol* and between *env* and *src*. Indeed, the progeny may have undergone two to five crossovers per molecule. Vogt has proposed that this high frequency, far exceeding that of any other animal virus (DNA or RNA), is due to the diploidy of the virions. It can be shown that in crosses between two strains with different markers **heterozygous virions** are produced during the first multiplication cycle. In the next cycle they generate recombinants, possibly because the synthesis of the minus DNA strand frequently switches from one RNA molecule to its partner (after the initial switch required for replication; see Fig. 65–3). Recombination has helped in **mapping** the viral genes, establishing their order as shown in Figure 65–1.

Viral Assay

IN VITRO. Virus can be identified and assayed by several characteristics. In fibroblastic cultures oncogene-carrying viruses transform cells, which then form colonies (**foci**) over the normal resting cells. **Syncytial plaques** are produced by the fusion of XC cells (rat cells transformed by RSV) plated together with cells infected by murine leukosis viruses. **Immunofluorescence plaques,** revealed by fluorescent Abs to a viral Ag, are produced by many viruses. Virions in sufficient concentrations are assayed by measuring the **reverse transcriptase activity.**

IN VIVO. Some assays use as endpoints the induction of leukemias or tumors; the production of spleen foci is used for Friend virus (Fig. 65–9), and an increase in spleen weight is used for Rauscher virus.

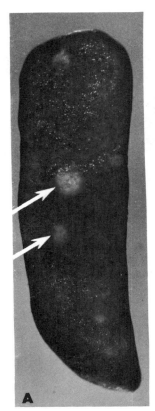

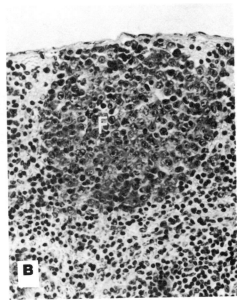

Figure 65–9. Foci in the spleen of C3H mice inoculated with Friend leukemia virus. (*A*) Macroscopic appearance of the foci (*arrows*) in whole spleen. (*B*) Microscopic appearance at low magnification of a section of spleen with one focus (*F*). The focus contains large cells, characteristic of Friend's leukemia. (Axelrod AA, Steeves RA: Virology 24:513, 1964)

Immunologic Reactions of Virion Proteins

The virion proteins possess various kinds of antigenic sites that are useful for classifying the viruses. Abs elicited against the envelope glycoproteins react only with those of the same virus (**type-specific** Abs); Abs elicited against core proteins (**gs Ag**) also react with proteins of other viruses of the same viral species (**group-specific** Abs); some Abs react with viruses of other species as well (**interspecies** Abs). In the mammalian viruses two specificities can be distinguished in the same polypeptide chain of the major capsid protein: gs^1 is common to all viruses of a given species but does not cross-react with viruses of other species, while gs^3 is common to all mammalian (but not avian) oncoviruses, suggesting a **common origin** for all mammalian type C oncoviruses. The reverse transcriptase too is antigenic and contains type, group, and interspecies determinants.

Antibodies to various determinants may have different effects, for example, immunoprecipitation or neutralization. In the presence of complement, antibodies against envelope proteins are cytotoxic for infected or transformed cells carrying these proteins in their plasma membranes (**complement-dependent cytotoxicity;** see

Chap. 50). In the absence of other signs of infection such surface Ags have in the past been mistaken for genuine cellular Ags. Thus, GIX, originally recognized as a thymocyte differentiation Ag, was later shown to be a type-specific determinant of the main envelope glycoprotein (gp 70) of AKR virus. Antibodies against glycoproteins are useful for **preventing viremia** in cats and, consequently, preventing the horizontal spreading of infection. Heterologous Abs, which are active especially toward interspecies Ags, can prevent tumor formation by murine or feline viruses even if administered several days after the virus.

Antibodies against viral glycoproteins in chronically infected mice (especially of the NZB strain) are correlated with an **autoimmune disease.** The same mechanism may be involved in the pathogenesis of human **lupus erythematosus,** because in this disorder the kidneys contain Ags that cross-react with the group-specific Ags of mammalian type C viruses.

Oncoviruses

Oncoviruses are classified (Table 65–2) according to their appearance in electron micrographs of thin sections:

TABLE 65-2. Comparative Morphologic and Biochemical Properties of Infectious Type A, B, C, and D Oncoviruses

| Characteristic | Type A | Type B | Type C | | Type D |
			Avian	Mammalian	
Prototypes	M432*	MMTV	RSV RAV	MuLV SSV BaEV	MPMV
Presence of intracytoplasmic A particles	–	+	–	–	+
Complete nucleoid at budding	+	+	–	–	+
Preferred cation for DNA polymerase activity	Mg^{2+}	Mg^{2+}	Mg^{2+}	Mn^{2+}	Mg^{2+}
Mature particle					
Nucleoid morphology	Centric	Eccentric	Centric	Centric	Eccentric
Envelope spikes	NA	Long, with knobs	Short	Short	Short
Hormone responsiveness of virus	No	Yes	No	No	No

* M432 is an endogenous virus from *Mus cervicolor* containing extensive sequence homology with type A intracisternal A particle (IAP) genes (4). There is no extracellular form of the IAP; budding taking place into endoplasmic reticulum.

NA, not applicable

(Chin IM, Callahan R, Tronick SR et al: Science 223:364, 1984)

mature **B particles** have an acentric core; **C particles** have a central core; **D particles** have a morphology intermediate between **B** and **C** particles. **A particles,** found only within cells, have a double shell with an electronlucent center.

CLASSIFICATION OF TYPE C ONCOVIRUSES

Avian type C viruses have seven subgroups (A to G); murine viruses have three: G (AKR virus), FMR (Friend, Moloney, and Rauscher viruses), and NZB xenotropic; feline viruses also have three (A, B, C). This classification is based on serology, interference, and host range. These properties depend mostly on the envelope glycoprotein.

Serology depends on antigenic differences of the glycoproteins. **Interference** is the resistance to further infection of cells already harboring an identical or related virus whose envelope glycoproteins block the cell receptors for the superinfecting virus. The block is at the adsorption–penetration step.

HOST RANGE. With **avian** viruses susceptibility of different breeds of chicken (and their cultured cells) is controlled by several dominant cellular genes that specify the cell surface receptors; susceptibility is dominant. The viruses are classified according to the genotype of the cells they infect. In contrast, with **murine** viruses, resistance of **mouse cells** takes place at a stage between penetration and integration. It is mainly controlled by the cellular Fv-1 gene, which exists in the *n* or *b* alleles (so named from prototype mouse strains). Virus strains that grow in Fv-1^{nn} cells (e.g., Swiss NIH mice) are called N-tropic; those that grow in Fv-1^{bb} cells (e.g., BALB/c mice) are B-tropic. Heterozygous strains of mice (Fv-1^{nb}) are resistant to both viruses; hence, in contrast to the avian case, resistance is dominant. The tropism of the virus is determined by a *gag* protein; restriction can be overcome by viral mutations generating NB-tropic viruses, or by treating cells with glucocorticoids. Another mouse gene, **FV-2^r,** prevents focus formation by the Friend virus; susceptibility is dominant.

CATEGORIES OF TYPE C ONCOVIRUSES OF LOWER ANIMALS

Oncoviruses Not Carrying Oncogenes

Exogenous oncoviruses, which are acquired by animals through infection, were discovered in mice and chickens because they induce various forms of leukemias after a **long latent period** but do not transform cells *in vitro*. Among the exogenous **murine leukosis viruses (MuLVs)** the Moloney MuLV induces T-cell lymphomas; the Rauscher and Friend MuLVs cause erythroblastosis, and then erythroleukemia, with characteristic foci in the spleen. **Avian leukosis viruses (ALVs)** and **feline leukosis viruses (FeLVs)** produce B-cell lymphomas. We will see below that these neoplasias do not result from the activity of viral genes, but from the activation of proto-oncogenes present in the cells. Some retroviruses of this class, however, can directly produce changes in the host: an example is the **osteopetrosis** induced by an avian retrovirus, a proliferative disease of the bone caused by abnormal growth and differentiation of osteoblasts (the bone-forming cells).

Endogenous oncoviruses, which are produced by proviruses present in the cells, were recognized through observations in inbred mouse strains. Some strains were selected for a high incidence of leukemia at a young age (e.g., AKR, C58), others developed the disease infrequently and at an older age (e.g., BALB/c, C57BL, C3H/He), and some (e.g., NIH Swiss) were essentially leukemia free. Gross found that filtered extracts of AKR leukemic cells could transmit leukemia to newborn low-leukemia (C3H) animals; a virus (called **AKR** or **Gross virus**) was identified as the agent.

Gross also showed that animals of a high-leukemia strain did not need postnatal infection in order to develop leukemia, because their fertilized ova, when implanted into the uterus of a female of a low-leukemia strain, developed into mice with a high incidence of leukemia—evidently **the virus is transmitted congenitally (vertically).** Another murine virus of this category, isolated by Kaplan from x-irradiated mice, produces a low incidence of thymic lymphomas (**radiation leukemia virus**).

Several groups of **avian leukosis viruses** are transmitted vertically. Most induce lymphatic leukemia, with B cells as targets; some do not induce neoplasia of any sort.

ORIGIN OF ENDOGENOUS VIRUSES. The demonstration that murine endogenous viruses are vertically transmitted shows that they are not ordinary infectious agents. In fact, in crosses between high- and low-leukemia mouse strains, transmission follows Mendelian genetics, as expected of integrated proviruses. In the AKR strain all mice have such a provirus, Akv, but individual mice may have several, all evidently copies of Akv. They are probably derived from infection of oocytes by virus generated after activation of the original Akv provirus (see below). This view is corroborated by the production, in the laboratory, of mice with Moloney MuLV proviruses by infecting oocytes with Mo MuLV virus. With viruses that produce pronounced viremia in their host, establishment of a new provirus in the germ line occurs frequently (once every 10 to 30 years of inbreeding). This mechanism also explains the presence in cats of the RD114 provirus, which is closely related to a provirus present in baboons.

Endogenous proviruses (as well as **provirus-related sequences**) are common in the DNA of many animal species, up to 100 or so proviruses per haploid genome. Because detection depends on suitable hybridization probes, many may still be undiscovered.

EXPRESSION OF ENDOGENOUS PROVIRUSES. A few proviruses are capable of generating virus; the majority are not, because they are defective or are not expressed. Expression depends on the state of the cellular genome. Thus, cellular genes determine whether chicken cells containing a RAV-0* provirus will make infectious RAV-0 or just its envelope proteins. Also important for expression is the stage of differentiation of the cells that affects the state of the chromatin at the site of integration: only proviruses located in active chromatin and with undermethylated enhancers are expressed. Expression of proviruses with highly methylated enhancers is promoted by 5-azacytidine, which inhibits methylase activity. Demethylation probably favors the interaction of the enhancer with activating factors, but it is not itself sufficient for expression. Spontaneous induction of proviruses frequently occurs in the genital tract (ovary, placenta, testes, prostate) or in the embryo.

A competent but inactive provirus can be **induced** to generate virus, using agents that also induce prophages (see Chap. 46), such as halogenated pyrimidines, chemical carcinogens, radiations, or inhibitors of protein synthesis. Induction may also occur after infection with other oncoviruses. Each procedure may induce some proviruses but not others in the same cell.

VARIOUS KINDS OF MURINE PROVIRUSES. Each animal generally contains different kinds of endogenous proviruses; the mouse has three kinds. Upon induction, **ecotropic** proviruses, probably of recent origin, generate virus able to infect mouse cells. **Xenotropic** proviruses generate virus that does not infect mouse cells, but cells of other species; the difference is in the *env* gene and the LTRs. **Amphotropic** proviruses—found in wild mice—generate virus that infects both murine and nonmurine cells. The origin and evolutionary state of the latter two classes of proviruses are not known.

CONSEQUENCES OF ENDOGENOUS PROVIRUSES. Endogenous proviruses can have important consequences for cells: (1) Generation of **leukemia** or **lymphoma,** as is the case for Akv. (2) Contributing genetic information to other viruses by **recombination** (see Recombinant Viruses later). (3) **Replication help** for defective viruses in the form of virion proteins or polymerase. (4) **Inactivation of host genes** inside which they are integrated. In the mouse a provirus inserted into a gene for coat color generates the **dilute** mutation. Excision of the provirus (by recombination between the LTRs) causes reversion of the mutation. (5) Insertion of the proviral glycoprotein into the cell membrane may confer **new antigenic specificity** on the cells, or **resistance to infection** by related retroviruses by blocking their receptors (interference). This is the mechanism of a mouse mutation (Fv-4) that renders cells

---

* *RAV*, Rous-associated virus. Such viruses are isolated from stocks of defective Rous sarcoma virus, in which they act as helper. Each one is identified by a number.

resistant to many ecotropic murine leukemia viruses. (6) Evolutionary spreading of the provirus through the host genome by reverse transcription like other eukaryotic transposons (see Transposons, Chap. 8).

BASES OF ONCOGENICITY. Neoplasia develops with a lag of 6 months or longer after the body of an animal is invaded by a replication-competent type C oncovirus generated by either **infection** with an exogenous virus or spontaneous **induction** of an endogenous provirus. The long time is required either for the integration of a new provirus at a specific site or for the production of special recombinants.

INSERTIONAL ACTIVATION OF ONCOGENE. The most general mechanism of oncogenicity is the insertional activation of the oncogene. It was discovered by Hayward and collaborators in the generation of bursal lymphomas in chickens by ALV. In the neoplastic cells a new provirus, absent in other cells, is localized in the proximity of the *c-myc* proto-oncogene, most frequently in the control area at the 5' end of the gene, increasing its rate of transcription. The provirus is usually defective but always contains one LTR. These characteristics suggest that *c-myc* is activated by the enhancer function of the LTR, which replaces the normal controls; structural changes caused by the insertion or occurring independently also contribute to the activation. Another example is the activation of *c-erb B* by ALVs to induce erythroblastosis in chickens. In other neoplasias the provirus is localized at a number of constant sites in the tumor cells, activating new presumptive oncogenes, the role of which in tumor induction is in some cases supported by their involvement in translocations characteristic of some nonviral neoplasias.

Insertional oncogene activation generates **monoclonal neoplasias** in which all malignant cells (i.e., capable of producing tumors in isogeneic hosts) have the provirus at the same localization. This stage is often preceded by a stage of **polyclonal** proliferation of cells that are not malignant (they do not produce tumors in isogeneic hosts) and that contain proviruses integrated at various locations. The large number of infected cells produced at this stage increases the chance of provirus insertion near the oncogene, which generates the malignant cell. Such a cell overgrows all others, giving rise to the monoclonal neoplasia.

RECOMBINANT VIRUSES (See Fig. 65–9; Fig. 65–10). In murine leukoses two types of recombinant viruses are important intermediates in oncogenesis. They are formed when an ecotropic oncovirus, of either exogenous or endogenous origin, induces endogenous xenotropic proviral sequences and recombines with them. The recombinants are of two kinds: the mink cell focus-forming viruses and the spleen focus-forming viruses.

The genomes of **mink cell focus-forming viruses (MCFV)** are mostly ecotropic, with part of the *env* gene and of the U3 region of the LTR of xenotropic origin. They usually result from separate acts of recombination with different endogenous proviruses. The recombinants are **replication competent** and **dualtropic,** owing to the xenotropic component; in addition to infecting mouse cells, they produce **necrotic foci** on a line of mink lung cells. MCFVs were first observed during the spontaneous development of **thymic lymphomas in AKR mice,** in which the Akv ecotropic provirus is spontaneously induced in various organs after birth, causing viral spread through the organism. Soon MCFVs appear in the thymus, where lymphomas are formed later, at the age of about 6 months. MCFVs persist in the lymphoma cells. Similar recombinants were subsequently recognized in leukemias induced by exogenous MuLVs.

MCFVs in oncogenesis are classified by their properties. They have a **special host range** that enables them to efficiently infect certain cell types not efficiently infected by the parental ecotropic virus. For instance, some will infect only T-lymphocytes, others B-lymphocytes. This property depends on the enhancer in the U3 region: replacing the U3 region with that of another virus may change the host range.

MCFVs have a direct tumorigenic effect: inoculation of MCFV into newborn AKR mice accelerates the production of lymphomas, whereas inoculation of ecotropic

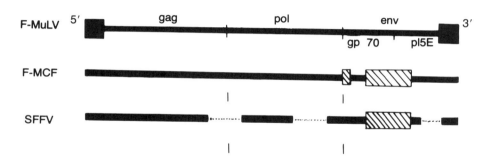

Figure 65–10. Main features of two recombinant viral genomes (FMCF and SFFV) derived from Friend MuLV, compared with the original genome. Xenotropic sequences are shown as hatched segments, ecotropic sequences as solid lines, and deletions as dotted lines. (Gonda MA, Kaminchick J, Oliff A et al: J Virol 51:306, 1984)

AKR virus does not. Clearly, the injected MCFV bypasses the need for *ex novo* formation from AKR. That MCFVs are **required** for oncogenicity is shown by abrogation of lymphoma formation by a mouse mutation that prevents the replication of MCFV but not of the ecotropic AKR virus. The mechanism of oncogenesis is oncogene activation by promoter insertion. In AKR lymphomas MCF proviruses are often integrated at several distinct **specific sites.** One is within (or in the proximity) of the *c-myc* oncogene; others are near presumptive oncogenes designated variously, for instance, *Mlvi*, *Pim*. Activation of the oncogene is caused by structural changes, increased transcription, or messenger stabilization. As in other cases of insertional activation, the onset of clonal neoplasia is preceded by a polyclonal cell proliferation of non-tumorigenic cells, which lasts until a provirus becomes localized near an oncogene; then this cell initiates the monoclonal tumor. The sequential activation of several presumptive or recognized oncogenes such as *Pim-1*, *Pim-2*, and *myc* in the same cell, appears to bring about the evolution of the lymphoma to its malignant state by synergistic action.

Spleen focus-forming viruses (**SFFV**) are probably formed by recombination between ecotropic and MCFV genomes in mice infected by the erythroleukemia-inducing Friend or Rauscher MuLVs. They have a structure similar to MCFVs, but with deletions in the *env* region, which is largely of MCFV origin, and are therefore defective. They replicate using the parental MuLV as helper; therefore, each SFFV can exist only in a complex with its helper. SFFV is required for oncogenicity: injected into an **adult** animal the complex produces rapid erythroleukemia, which is at first reversible but then progresses to malignancy through additional steps. In contrast, the replication-competent Friend or Rauscher viruses, which act as helper, do not produce erythroleukemia in adult mice but do produce it when they are inoculated into **newborns;** they must first generate the SFFV.

Owing to the *env* deletion, SFFVs make an abnormal, short glycoprotein, of 52 Kd to 55 Kd instead of 70 Kd, which is incorporated into the cell membrane but not in the virions; in the progeny, SFFV genomes have envelopes containing helper glycoproteins. The altered glycoprotein gene confers on SFFVs two crucial properties: tropism for erythroblasts and a growth-promoting activity for these cells. These changes are essential for tumorigenicity, which is abolished if the recombinant *env* gene is mutated.

The role of the SFFV is to generate the initial polyclonal nontumorigenic erythroid proliferation in which the clones of tumorigenic cells will later appear. For this purpose the SFFV must reach a high titer in the thymus. In fact, an intrathymic injection of helper-free SFFV has no effect; several injections are needed to induce a tran-

sient splenomegaly with polyclonal proliferation of growth-factor-independent erythroid cells. A malignant leukemia is only obtained by repeating the injections for many days. In the natural infection a sustained SFFV replication is maintained by the helper; in some of the SFFV-infected cells an additional event leading to malignancy will then occur. The nature of these additional events is unknown. In rats, however, SFFVs are not required: the Friend MuLV by itself causes erythroleukemia, without production of SFFV. (Rat cells do not have equivalent xenotropic proviruses.) The initial polyclonal proliferation is probably produced by the MuLV itself.

In infected animals, a large number of different MCFVs or SFFVs are formed. Only those with sequences especially suitable for performing the specific growth-promoting effect are important for carcinogenesis.

FELINE LEUKEMIA VIRUSES. Cats have two main types of **endogenous** viral genomes. One type is represented by the RD114 provirus, which is homologous to a baboon endogenous virus. The RD114 provirus is present in only some of the *Felidae* species, showing that it was acquired at some point during their evolution, probably by cross-species infection from baboons. Cats have some 20 proviruses of this kind, mostly defective; one or two are replication competent. They are expressed, producing *gag* proteins, in tumors and during embryonic life and are induced, producing infectious virus, in mixed cultures of feline cells with heterologous cells. Other endogenous genomes are related to the exogenous FeLVs and are defective.

The **exogenous** FeLVs belong to three serologically distinct groups: A, B, and C. Whereas viruses of group A occur alone, those of groups B and C occur in conjunction with A and may be generated by it. Oral or nasal infection with exogenous FeLV through saliva occurs frequently in cats; 2% of all cats become persistently infected. Transplacental transmission also occurs. In the cats the virus replicates in hematopoietic cells. The animals generate antibodies to FOCMA (feline oncornavirus associated membrane antigen, a product of the *gag* gene) and to the glycoproteins of the virions. The latter Abs are neutralizing. There is a persistent infection of the bone marrow revealed by appearance of the *gag* protein p27 in serum or by relapsing viremia.

The viruses produce leukemias of both lymphoid and myeloid type, probably by insertional activation of an oncogene; the *c-myc* gene is activated in about one-third of the cases. The viruses, especially those of group C, also produce anemia and **immunosuppression,** through the action of structural proteins of the virions—especially p15E—on immune and erythroid cells. For these multiple effects FeLVs are useful as a model for immunosuppression by human retroviruses (see below).

Immunologic Effect. The virions elicit the formation of neutralizing Abs, especially after removal of the carbohydrate side chain of the envelope glycoprotein. Some of the Abs are cross-reactive among most strains. Abs to the envelope Ags can protect cats against the development of lymphomas. Subunit vaccines reproducing parts of envelope glycoprotein prepared in yeast seem effective in protecting cats against FeLV infection.

Oncogene-Containing Oncoviruses

Oncogene-containing oncoviruses are called **acutely transforming viruses (ATVs)** because upon injection into a suitable host they produce tumors within a few months. They also transform fibroblastic cells *in vitro*, producing characteristic **foci** (Fig. 65–11) that are useful for assay. Both properties are due to the presence of an activated oncogene (see Oncogenes, Chap. 64) in the viral genome. Peyton Rous isolated the first strain of ATV, which is known as **Rous sarcoma virus (RSV),** from a chicken with a spontaneous sarcoma. Many other strains were subsequently isolated in various species.

ATVs are all **defective,** with the exception of two RSV strains (Prague C, Schmidt-Ruppin), and all contain one or more oncogenes in replacement for normal viral sequences (Fig. 65–12). The two nondefective RSV strains, which also contain an oncogene, were probably generated from a defective genome by recombination with other oncoviruses. The oncogenes were captured by rare illegitimate recombination with the normal proto-oncogenes of the cells, in which both viral genes and the proto-oncogene were altered. Deletions of the viral genome explain the defectiveness of ATVs; alteration of the proto-oncogene explains its activation (see Oncogenes, Chap. 64).

The deleted genomes of ATVs, being **replication defective,** require a **helper** for their multiplication; ATVs are then generated as **pseudotypes** with helper envelope. **Nonproducer** transformed cells, which do not contain helper, can be obtained by transforming cells at low multiplicity with the ATV–helper mixture—to obtain cells infected by ATV alone—or, in the murine system, using helper-free ATV. Such virus is obtained from cells harboring a Ψ-free helper, which helps the production of ATV but is not itself packaged into virions.

AVIAN SARCOMA VIRUSES (Table 65–3). The prototype is the Rous sarcoma virus, but the group contains many other viruses unrelated to it. RSV contains the oncogenic *v-src*, which is inserted in replacement of *env* sequences, and is transcribed by a special short mRNA in which it is connected directly to the leader sequence.

ROLE OF v-src. The *v-src* oncogene encodes a tysosine-specific protein kinase, the phosphoprotein pp60^{v-src}, which is endowed with much higher activity than the product of *c-src*. Transformation is caused by the activity of the kinase, for when cells transformed by ts mutants in the *src* gene are shifted to a nonpermissive temperature, they lose the transformed phenotype within an hour, as soon as the *v-src* protein is denatured. Deletions within the *v-src* gene give rise to **transformation-defective (Td)** strains. When injected into chickens, some of them

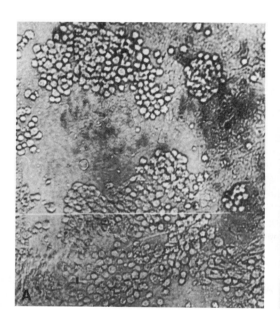
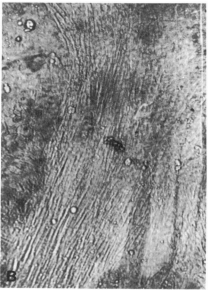

Figure 65–11. Transformation of cultures of the iris of the chick embryo by either wild-type Rous sarcoma virus (*A*) or its morph^f mutant (*B*). Cells transformed by the wild-type virus are round, whereas those transformed by morph^f virus are fusiform and form bundles. The untransformed cells form an epitheliumlike sheet and can be recognized because they contain various amounts of black pigment, whereas both kinds of transformed cells are almost colorless. (Courtesy of B. Ephrussi and H. Temin)

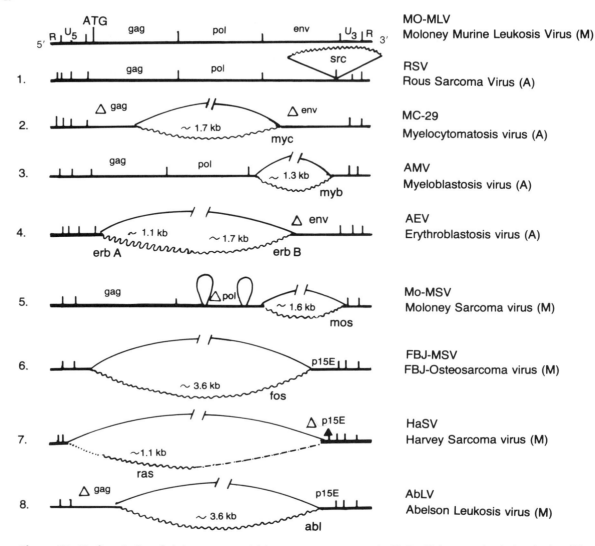

Figure 65–12. Organization of viral genomes containing oncogenes, compared with the Moloney murine leukemia virus (*Mo-MuLV, top*). Wavy lines indicate oncogenes, with attached denominations. Dotted lines at *7* indicate sequences derived from a rat endogenous retrovirus; thin lines indicate deleted MuLV sequences; heavy lines indicate retained sequences. A triangle preceding a gene indicates that the gene is truncated. AEV (*4*) has two oncogenes.

unexpectedly produce tumors, which, in contrast to those induced by regular RSV, arise after a lag period of many months, and far from the site of inoculation. Hana-fusa isolated oncogenic virus from the unusual tumors and showed that it is derived by recombination of the Td RSV genome with the cellular *src* proto-oncogene. Rescue only takes place if the 3' end of the *v-src* oncogene persists in the Td mutants; these sequences are essential for conferring a high tumorigenic activity on *v-src*. Complete replacement of *v-src* with the *c-src* proto-oncogene abolishes the transforming activity; a single amino acid replacement in *c-src* can restore it if the gene is strongly

expressed. Strong expression of *v-src* and of other onco-genes as well, however, can be lethal for the cells, showing that transformation depends on a delicate balance of genes.

For transformation, pp60$^{v\text{-}src}$ must be connected to the cell plasma membrane through a fatty acid (myristic acid) bound to its N-terminal glycine. Replacing glycine with another amino acid, which prevents myristylation, prevents transformation. Mutants in which the protein is myristylated but does not bind to the membrane also fail to transform.

Presumably pp60$^{v\text{-}src}$ causes transformation by phos-

TABLE 65–3. Avian Sarcoma Viruses

Virus	Genome Size (Kb)	Oncogene	Oncogene Protein
Rous sarcoma virus (RSV)	Varies	v-src	pp60^{src}
Fujinami sarcoma virus (FSV)	4.5	v-fps	p140$^{gag-fps}$
PR6 II	4.0	v-fps	p105$^{gag-fps}$
Y73	4.8	v-yes	p90$^{gag-yes}$
UR-2	3.3	v-ros	p68$^{gag-ros}$
ASV-17	3.5	v-jun	p55$^{gag-jun}$
Myelocytomatosis virus MC-29	5.6	v-myc	p110$^{gag-myc}$
OK10	8.2	v-myc	p200$^{gag-pol-myc}$ p57$^{gag-myc}$
CM II	6.0	v-myc	p90$^{gag-myc}$
MH2	6.0	v-myc v-mil	p100$^{gag-mil}$
Reticular endoteliosis virus	5.5	v-rel	p56$^{env-rel}$
Avian erythroblastosis virus (AEV)	3.0	erb A erb B	p75$^{gag-erb A}$ p44$^{erb B}$
Avian myeloblastosis virus (AMB)	7.0	myb	p48^{myb}
E26	5.7	myb ets	p135$^{gag-myb-ets}$
Sloan-Kettering viruses	5.7–8.9	ski	p110$^{gag-ski-pol}$ p125$^{gag-ski}$
S13	8.5		gp 155

phorylating cellular proteins. Several such proteins are known, but it is not clear whether any of them are directly involved in transformation. Gene activation by promoter insertion does not play any role in transformation, for the site of localization of the RSV provirus in the host genome is different in different tumors. The localization is important, however, in mammalian cells transformed *in vitro* by RSV. In these cells the strength of viral transcription and the degree of transformation depend on the state of the chromatin around the provirus, as detected by nuclease sensitivity, and the degree of methylation of the enhancer DNA.

Tumors induced by RSV in chickens tend to regress spontaneously because the cells are killed by the cell-mediated immune response of the organism to the helper glycoprotein present in the cell membrane.

Other avian acutely transforming viruses produce sarcomas and other tumors different from sarcomas: MC-29 produces myelocytomatosis, endotheliomas, and carcinomas; avian erythroblastosis virus (AEV) produces an erythroblastic leukemia; and avian myeloblastosis virus (AMV) causes a myeloblastic leukemia. These viruses carry a variety of oncogenes (see Table 65–3), which specify proteins of different groups: some are tyrosine-protein kinases, at least in structure if not in function (*src, fps, yes, ros, fms, kit,* and *erb B*), and are associated with the cell plasma membrane. One (*mil*) is a cytoplasmic serine/threonine-specific protein kinase. *Myc, fos,* and *myb* have nuclear localization; *erb A* and *rel* have

cytoplasmic localization. In most viruses the oncogene replaces part of the *gag* gene (see Fig. 65–12) and is expressed in the same messenger. The oncogene protein is then a **gag–onc fusion protein.** Whereas most ASVs have one oncogene, some have two: MH-2 (related to MC29) has *v-myc* and *v-mil*; avian erythroblastosis virus has *v-erb A* and *v-erb B*; E-26 (related to avian myeloblastosis virus) has *v-myb* and *v-ets*. In these cases the added oncogene confers on the virus either more extreme transforming power (as in AEV) or the ability of y to cause different tumors (as in E-26, which transforms cells of the erythroid lineage), showing the importance of cooperation between oncogenes. Cooperation can occur in various ways. In MH-2, which transforms macrophages, *v-myc* is the transforming gene, whereas *v-mil* causes cell immortalization by inducing the production of a growth factor needed by the macrophages themselves (**autocrine growth**). With the reticuloendotheliosis virus, the single oncogene *rel* cooperates in the tumors with *c-myc*, which is activated by the helper virus integrated within its regulatory sequences. This is one of the few cases in which the helper virus participates in causing cell transformation. Other significant helpers are those of avian myeloblastosis virus (called MAVs: myeloblastosis-associated viruses), which by themselves produce kidney and bone tumors.

The **murine sarcoma viruses (MuSVs;** Table 65–4) were all derived from Mo-MuLV during passages in mice or rats. They contain several types of oncogenes; their

TABLE 65–4. Murine Sarcoma Viruses

Virus	Genome Size (Kb)	Oncogene	Oncogene Protein
Harvey-MuSV	5.5	Ha-ras	p21
Kirsten-MuSV	4.5	Ki-ras	p21
BALB-MuSV	6.8	bas	p21
Mo-MuSV	5.8	mos	p37
Myeloproliferative sarcoma virus (MPSV)	7.0	mos	p37
Abelson leukemia virus	4.0	abl	p130$^{gag-abl}$
Rat SV	6.7	ras	p29
FBJ osteosarcoma virus ⎫ FBR osteosarcoma virus ⎬		fos	p55 p75$^{gag-fos}$
3611 MSV		raf	p75$^{gag-raf}$

TABLE 65–5. Feline Sarcoma Viruses

Virus	Oncogene	Oncogene Protein
Snyder-Theilen FeSV	fes/fps	p85$^{gag-fes}$
Gardner-Arnstein FeSV	fes/fps	p110$^{gag-fes}$
Hardy-Zuckerman-1	fes/fps	p95$^{gag-fes}$
Susan McDonough	fms	p160$^{gag-fms}$
Gardner-Rasheed	fgr	p70$^{gag-actin-fgr}$
Parodi-Irgens	sis	p75$^{gag-sis}$
Hardy-Zuckerman-2	abl	p95$^{gag-abl}$
Hardy-Zuckerman-4	kit	p80$^{gag-kit}$

proteins are expressed by themselves, except for that of the Abelson murine leukosis virus, which is expressed as a *gag-abl* fusion protein. Some of the oncogenes originate from the mouse genome, others from the rat genome. In the case of Harvey sarcoma virus, the oncogene is enclosed in sequences of a rat leukemia virus, which presumably captured it first.

Transformation by MuSV occurs in steps, possibly produced by cooperation among oncogenes. In fact, in Abelson MuLV-transformed cells that contain the *v-abl* oncogene, *c-myc* is often amplified and strongly expressed; transformation is decreased when there is lack of expression of p53, a presumptive oncogene protein, or of the EGF receptors, which mediates the effect of tumor growth factor α produced by the same cells. In contrast, in highly malignant cells the *v-abl* oncogene may be absent, suggesting that other genes maintain the malignant state.

In addition to transforming fibroblasts, some of the murine sarcoma viruses affect hemopoietic cells. The myeloproliferative sarcoma virus stimulates the growth of hematopoietic stem cells and of myeloid and erythroid precursors; the special host range is determined by the U3 sequence. Ha-MSV and Ki-MuSV stimulate the growth of erythroid cells; Ki-MuSV, in addition, transforms human keratinocytes immortalized by adeno-SV40 hybrid virus, again showing the importance of cooperation between oncogenes. The Abelson virus causes lymphoid and mast cell tumors.

FELINE SARCOMA VIRUSES (FeSV; Table 65-5). Defective, oncogene-carrying derivatives of FeLV cause sarcomas in FeLV-infected young cats. Virus obtained from the tumors produces fatal tumors in young cats; in older cats cell-mediated immunity causes tumor rejection. Ten different strains of feline sarcoma viruses have been identified. They carry a variety of oncogenes: *fms, fes, fgr, sis, abl,* and *kit.* In addition, T-cell lymphomas are produced by a defective FeLV derivative containing the *myc* gene. In all cases the oncogenes are expressed as *gag–onc* fusion proteins. A puzzling finding is that one third of the naturally occurring lymphoid tumors in cats contain neither infectious FeLV nor its genome (tracked through its U3 region), yet epidemiologic evidence shows that they are associated with exposure to FeLV. Perhaps the virus induces secondary changes that maintain the neoplastic state and is then lost.

ROLE OF ONCOVIRUSES IN CHEMICALLY AND PHYSICALLY INDUCED NEOPLASIA AND CELL TRANSFORMATIONS

Though mutagens and viruses have long appeared to provide different mechanisms for inducing cell transformation, they are now understood to act in a similar way: by activating oncogenes. The various agents can cooperate, because many viruses produce transformed foci in cell cultures in conjunction with carcinogens but not without them. Presumably, just as different viruses can cooperate by introducing or activating different oncogenes, so can oncogenes activated by carcinogens cooperate with the viral ones.

After the radiation leukemia virus was obtained from an x-ray-induced thymoma, viruses seemed the answer to physical or chemical carcinogenesis. A viral origin of tumors induced by these agents is not general, however, because many radiation-induced thymomas can be shown not to be virus mediated. Moreover, chemically induced AKR thymic lymphomas lack the integrated MCFs that are regularly present in tumors induced by virus. The unifying factor is probably represented by oncogenes.

Role of Type C Oncoviruses of Lower Animals in Human Neoplasia

Since the viral etiology of leukemias and related tumors in many animal species has been recognized, a possible role of these viruses in human leukemias has been suspected. The viruses of cats have especially been scrutinized, because feline leukemia and sarcoma viruses are shed in the saliva of infected cats and can infect human cells in culture. However, in epidemiologic studies disease in pets does not correlate with human infection.

PRIMATE AND HUMAN TYPE C ONCOVIRUSES
Primate Leukemia Viruses

Type C viruses are present in many primates, such as baboons, Old World monkeys, and great apes. **Endogenous** viruses have been recognized in some species: baboons have about 100 proviruses related to the **baboon endogenous virus (BaEV),** of which only some can be induced to multiply and release infectious virus; colobus and other Old World monkeys have 50 to 70 related copies of other proviruses. **Exogenous** viruses belong to the **gibbon ape leukemia virus (GaLV)** group, which were isolated from various kinds of tumors or leukemias and induce the same neoplasias in young gibbon apes. To this group belongs the defective **simian sarcoma virus (SSV),** the only known oncogene-carrying virus among the primate viruses: it carries the *sis* oncogene, which is expressed in a p28 protein corresponding to the B chain of the platelet-derived growth factor (PDGF). Binding of p28, expressed at the cell surface, to the PDGF receptors of the same or other cells is required for transformation, which is therefore based on self-stimulation of growth. In fact, *v-sis*–transformed fibroblasts behave *in vitro* like normal fibroblasts stimulated by PDGF. Additional steps are presumably required for attaining the malignant phenotype in the sarcomas.

Primate leukemia viruses do not show extensive homology with other type C viruses, but share with them some common antigenic determinants of the reverse transcriptase and the p30 *gag* protein. They share gp-70 determinants with D-type primate viruses. T-lymphotropic simian viruses are discussed later, together with human viruses.

Human Oncoviruses

Several types of **endogenous** proviruses have been identified by hybridizing probes from simian viruses to human DNA under conditions, allowing detection of moderate homology. Some are in multiple copies (50 to 100) in the human genome; others are as a single copy. Some have the length and organization of a complete oncoviral genome, but none can give rise to infectious virus, owing to termination codons and frame shifts; others are highly defective. Evidently there has been evolutionary restraint to their expression, which has caused their divergence. These proviruses are related to other mammalian C-type viruses, but some are recombinants with mouse mammary tumor virus (type B) or D-type viruses. The similarity of some of these proviruses to endogenous simian proviruses shows that they originated before the separation of man from other primates.

Human endogenous viruses, although not expressed in adult tissues, are expressed in placentas in which antigens of p30 *gag* are recognizable, and type C particles—probably not infectious—are frequently seen budding from cells of the trophoblastic syncytium.

Exogenous human type C oncoviruses are of a single family, which includes the **human T-lymphotropic viruses (HTLV-I and -II);** they are unrelated to human endogenous retroviruses. HTLVs are similar in organization to **bovine leukemia viruses** and some **simian T-lymphotropic viruses** and differ from other oncoviruses in several properties. (1) They have a very restricted host range, limited to helper T-lymphocytes and other cells expressing the OkT4—also known as Cd4—antigen; this Ag is the receptor for viral adsorption. (2) They induce the formation of multinucleated syncytia in certain indicator cell lines, a useful diagnostic tool. (3) Their *gag* gene is shorter and specifies only three proteins; it lacks the equivalent of the MuLV p12. (4) They have additional sequences that encode a *trans*-acting activator of the viral promoter.

HTLV-I and -II. HTLV-I is associated with a special form of leukemia, **adult T-cell leukemia (ATL),** also known as mycosis fungoides or Zésary cutaneous T-cell leukemia. The closely related HTLV-II has been associated with two cases of hairy cell leukemia (also of T cells). HTLV-I was discovered by Gallo by growing T cells from patients with ATL *in vitro* in the presence of interleukin-2 (IL-2; T-cell growth factor). The cells then release HTLV-I into the medium. Clonal cultures of helper T cells infected by the virus have abnormal properties. They are nonspecifically activated: they have IL-2 and transferrin receptors and provide the helper function to B cells in the absence of antigen, causing polyclonal production of antibodies (see Chap. 16); but some have lost helper activity or are alloreactive. These cells express on their surfaces altered major histocompatibility (HLA) antigens, together with HTLV-I *env* products, gp 61, and gp 45.

RELATION OF HTLV-I TO ATL. An important role of the virus seems likely: the virus immortalizes normal human T cells *in vitro;* and in leukemic individuals the leukemic cells, but not the normal cells, contain the HTLV-I provirus. The leukemias are **clonal,** having the provirus at the same site in all cells of the same individual, although the site varies in different individuals. Clonality shows that a leukemia arises from a single cell already harboring the provirus, excluding a secondary infection after the leukemia was established. It is likely, however, that additional changes are required for generating the leukemic clone.

HTLV-I does not have an oncogene and must therefore act by changing the expression of cellular gene. It does not act in *cis* on a cellular oncogene, because the provirus does not have unique localizations. It probably promotes proliferation of quiescent T cells by changing the expression of cellular genes in *trans* through the expression of the *tat* gene (also present in **bovine leukemia virus**) which specifies a *trans*-activator of transcription. The main effect is the activation of genes for the IL-2 receptor and for synthesis of IL-2, which allow the cells to multiply autonomously. The *tat* gene, located at the 3' end of *env* and extending into the U3 region, is highly conserved among the various isolates of the HTLV-I virus; it is expressed by a double-spliced mRNA, which is not produced by other oncoviruses (Fig. 65–13), and encodes a nuclear 40-Kd protein that acts on enhancerlike sequences present in the viral LTR, stimulating viral transcription. Gene *tat* may be comparable to the adenovirus E1A gene (see Chap. 64). The regular preservation of the gene in the leukemic cells harboring a defective provirus suggests that it plays an important role in leukemogenesis.

EPIDEMIOLOGY. Antibodies against viral *gag* and *env* antigens are present in ATL patients and in some unaffected persons in endemic areas in the southwestern United States, the Caribbean islands, parts of South America and south Italy, many parts of Africa, and the southern islands of Japan. The virus and the disease persist in migrant populations from these areas. Not all ATL cases, however, are HTLV-I positive. The virus is also implicated in a neurologic disease (tropical spastic paraparesis) that is prevalent in areas where HTLV-I is endemic.

OTHER ONCOVIRUSES
Oncovirus B

MOUSE MAMMARY TUMOR VIRUSES (MMTV). Most mice, both wild and inbred, have **endogenous MMTVs.** These proviruses are related to each other but vary in location, number, and degree of completeness. They were probably acquired by different and fairly recent infections of the germ line. More distantly related sequences are either precursors of these proviruses or products of their divergence. The proviruses can be expressed to various degrees, probably depending on their location in the mouse genome. Except for the GR mouse, these endogenous MMTVs seem to play little or no oncogenic role.

Exogenous viruses are important for oncogenesis. The prototypes are the **Bittner virus,** derived from the C3H mouse strain, and a related virus derived from the RIII strain. Both are **high-cancer strains,** in which mammary cancers appear in 90% of the animals by 1 year of age. In contrast, **low-cancer strains,** such as BALB/c, have a 20% to 50% incidence of mammary tumors by 2 years of age.

The Bittner virus was discovered as a result of questioning the seemingly obvious genetic basis of the high incidence of mammary cancers in C3H mice. In 1936 Bittner showed that newborns of a low-cancer strain nursed by females of a high-cancer strain acquire a high incidence and vice versa. This simple experiment revealed that the cancer is induced by a transmissible agent present in the milk (the **milk agent),** later recognized as a virus. The virus is generated in the lactating mammary gland by induction of an endogenous provirus, MTV-1, which in the absence of exogenous infection gives rise to a low frequency of mammary cancers by spontaneous induction later in life (18 to 24 months).

PROPERTIES. The virions are generally similar to those of the C type described earlier, but they differ in some details, such as the presence of spikes (Fig. 65–14) and the eccentricity of the cores (Fig. 65–15). One major difference is the organization of the LTRs, which contain an open reading frame of 1.1 Kb, coding for a 36-Kd protein. Other differences include the molecular weights of the peptides, and a Mg^{2+} preference of the reverse transcriptase (containing two subunits) compared to a Mn^{2+} preference for type C viruses.

Viral replication in lines of rat or mink cells is similar to that of type C viruses. However, the **viral yield** in cultures is markedly **increased by glucocorticoids** and, in some cells, by progesterone. A 120-base-pair sequence within the LTR is responsible for this effect. Either hormone binds to its receptor in the cytoplasm; the complex moves to the nucleus where it binds to the LTR enhancer, increasing both viral multiplication and oncogenicity. Hormonal activation of LTRs is responsible for the predominant localization of the effects of the virus to the mammary gland.

ASSAY. A slow and qualitative bioassay uses as endpoint the development of breast cancer in mice. Physical, biochemical, and serologic methods are also available.

Figure 65–13. Genome organization of HTLV-I. Underneath is the constitution of the double-spliced mRNA that expresses the *tat* gene. Solid lines, messenger sequences; dashed lines, spliced-out sequences.

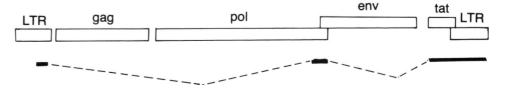

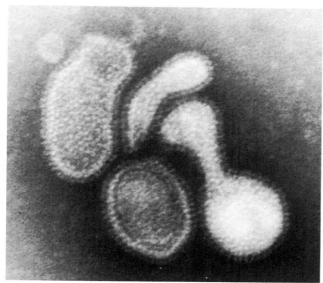

Figure 65–14. Electron micrograph of the purified mouse mammary tumor virus (MMTV) with negative staining. Note the envelope, covered by spikes, which surrounds an internal component. (Lyons MJ, Moore DH: JNCI 35:549, 1965)

PATHOGENESIS. Suckling animals infected through the milk with the C3H or the RIII virus develop **hyperplastic alveolar nodules (HANs)** and **adenocarcinomas.** The cancers induced by the RIII virus are dependent on ovarian hormones for growth, whereas those induced by the C3H viruses are independent.

The hyperplastic nodules, which are constituted of normal-looking, milk-producing mammary tissue (Fig. 65–16) containing large amounts of virus, are **preneoplastic lesions** (i.e., have a high tendency to evolve into cancer). Thus, pieces of infected mammary gland transplanted to gland-free mammary fat pads of isologous,

virgin, virus-free females undergo neoplastic transformation more frequently if they contain nodules. In the GR mouse, virus produced by spontaneous induction of the MTV-2 provirus under hormonal stimulation during pregnancy also proliferates in lactating glands, causing the formation of **plaques,** benign tumors that regress at the end of pregnancy (not to be confused with those used for virus assay in cultures; see Chap. 44). However, after several pregnancies they tend to generate pregnancy-independent carcinomas. In all cases, therefore, MMTVs induce hyperplastic lesions (nodules, plaques) whose progression to cancer requires other events.

The exogenous Bittner virus causes HANs and mammary cancers in the same way as the avian leukosis viruses: by activating cellular proto-oncogenes. In the cancers the MMTV proviruses are localized almost always near one of the oncogenes *int-1*, *int-2*, *int-3*, or *int-4*, which are normally silent, and cause their transcription. *Int-1* is normally expressed only in precursors of male germ cells and in the embryonic nervous system, but stimulates cell growth upon transfection into normal mammary cells.

HOST AND HORMONAL FACTORS IN CANCER INDUCTION. The importance of hormonal factors is evident in the pregnancy dependence of the plaques induced in GR mice by the MTV-2 virus. Moreover, with the Bittner virus the cancer incidence is low in virgin infected females but high in infected females that are force-bred (i.e., made to bear litters in quick succession without nursing them) or given estrogen. It is even more striking that when virus-infected males are castrated and are injected with estradiol and deoxycorticosterone over a long period, they frequently develop mammary cancer, whereas normal males have no mammary cancer at all. All these effects can be attributed to the hormonal activation of the LTRs, which allow extensive viral multiplication, and subsequent occasional provirus insertion near an *int* gene.

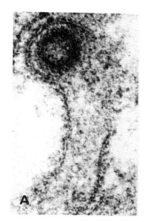
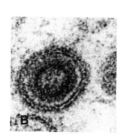
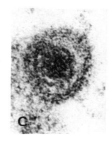

Figure 65–15. Electron micrographs of thin sections of a mammary tumor producing mouse mammary tumor virus. (*A*) Budding of the virions at the cell membrane. (*B*) Immature virion with an electron-lucent core. (*C*) Mature B-type particle with a dense, eccentric core. (Courtesy of L. Dmochowski)

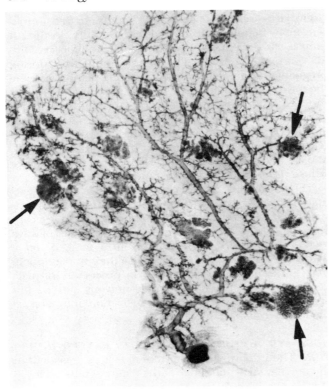

Figure 65–16. Hyperplastic alveolar nodules in the mammary gland of a multiparous C3H female mouse bearing a mammary tumor. The nodules (some indicated by *arrows*) are filled with milk. (Whole mount of gland, hematoxylin staining, original magnification ×6) (Nandi S et al: JNCI 24:883, 1960)

GENETIC FACTORS. The expression of an MMTV is strongly affected by the genetic constitution of the host, which determines the time of tumor appearance. For instance, resistance of the C57BL strain to the Bittner virus (of which they are normally free) is determined by several genes that affect viral multiplication. Moreover, the development of cancers in the presence of the virus is controlled by two additional genes. Various H2 alleles also affect the incidence of tumor formation.

RELEVANCE IN HUMAN BREAST CANCER. It is not known whether viruses similar to MMTVs are involved in the genesis of human breast cancer. Particles containing 70S RNA dimers and a reverse transcriptase have been isolated from neoplastic tissues, which sometimes contain Ags related to the viral core Ag of MMTV. Nucleic acid homology and serologic specificity show that the particles bear some relatedness to MMTV. Cell cultures from human breast cancers shed a protein immunologically related to the MMTV glycoprotein.

Oncovirus D

The retroviruses of the oncovirus D group differ from others in virion morphology, nucleic acid sequences, and antigenicity. The virions resemble those of B oncoviruses in the budding mechanism and in the preference of their reverse transcriptase for Mg^{2+}; they resemble type C virions in morphology and in the sizes of their peptides. Two **endogenous** type D viruses have been isolated from Old World monkeys, one from cells of a spectacled langur and the other from those of a squirrel monkey.

Among **exogenous** viruses, **Mason-Pfizer monkey virus (MPMV)** was isolated from a mammary carcinoma and normal tissues from rhesus monkeys. Its genome seems to have evolved from a recombinant between a murine B oncovirus and a primate C oncovirus related to the baboon endogenous virus (BaEV). The virus is propagated in rhesus or human cell lines and transforms rhesus foreskin cultures; as with MMTV, the yield is increased by dexamethasone. A syncytial plaque assay uses cells of a human glioma line transformed by RSV. In rhesus monkeys the virus is not oncogenic but induces **T-cell deficiency** resulting in a wasting disease with opportunistic infections, lymphoadenopathy, and thymic atrophy. Similar viruses have been isolated from cases of spontaneous immunodeficiency syndrome in rhesus monkeys kept in several regional primate centers in the United States (see simian HIV-related viruses, later).

Cisternavirus A

Two kinds of intracisternal A particles (IAPs) with a double shell and a clear core are observed by electron microscopy in the cisternae of the endoplasmic reticulum in rodent tissues: large ones are present in oocytes and in various kinds of tumors, especially plasmocytomas; small particles occur in the cells of early embryos. Both are noninfectious. Their genomes are present in about a thousand copies per cell in the DNA of all mouse species, in Syrian hamsters, and in rats. In the mouse they are highly heterogeneous, the largest being 7.3 Kb. The genomes, flanked by LTRs of 300 to 400 base pairs, contain the three oncovirus genes, and have considerable homology to parts of the oncovirus B and C genomes. The products are a *gag*-like protein, p73, the main component of the IAP inner shell, and a Mn^+-dependent reverse transcriptase; many stop codons prevent *env* expression. Variants of the *gag* protein are secreted by some cells as IgE-binding factors, which regulate IgE synthesis by B-lymphocytes. The viral genomes are, like transposons, mobile within the host genome, giving rise to **activation or inactivation of genes** by inserting near or within them. Insertional activation of the *c-mos* proto-oncogenes has been observed in a mouse myeloma and of the IL-3 gene (see Chap. 47) in a myelomonocytic leukemia. Insertional inactivation has been observed in an im-

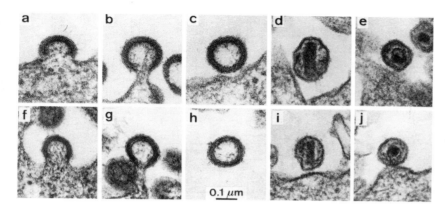

Figure 65–17. Electron micrographs of thin secretions of cells in various stages of the release of HIV (*a* to *e*) or the Visna virus of sheep (*f* to *j*). Note the similar bar-shaped nucleoids in *d* and *i*. (Original magnification ×100,000; Gonda M et al: Science 227:173, 1985)

munoglobulin gene. IAP genomes may represent a gene pool important for retroviral evolution.

Lentiviruses

Lentiviruses give rise to slowly developing diseases. One of them, the human immunodeficiency virus (HIV) is responsible for the human **acquired immunodeficiency syndrome (AIDS).** Other viruses cause slow diseases in animals: the **visna-maedi virus,** which causes an encephalitis (**visna**) or pneumonia (**maedi**) of sheep; the **caprine arthritis-encephalitis virus;** and the **equine infectious anemia virus.** A **feline lentivirus (FIV)** causes an immunodeficiency syndrome in cats. The diseases are characterized by a long incubation period and protracted course. The viruses cause an infection, mostly latent, of monocytes and macrophages, and from them spread to other cells.

Lentiviruses have the following distinctive characteristics: The virions have a bar-shaped rather than spherical nucleoid and contain proteins with an arrangement similar to other retroviruses, but of different sizes. The larger genome (approximately 10 Kb) contains genes not present in other retroviruses. It uses a tRNA^{lys} as primer for negative-strand synthesis, whereas most other infectious mammalian retroviruses use tRNA^{pro}. Genes *pol* and *env*, which overlap in oncoviruses, are separated in HIV by a region that includes sequences for other genes (see below). The genomes of lentiviruses have homology with each other but not with other retroviruses. The main targets of lentiviruses are mononuclear phagocytes (monocytes and tissue macrophages). Their precursor cells in the bone marrow undergo latent infection, which becomes productive when the cells reach the various tissues and mature to macrophages. Release of lymphokines may then give rise to inflammation and disease. A factor contributing to the protracted course of disease is the high mutability of the viral genome, with production of mutants capable of evading the immune response.

HUMAN IMMUNODEFICIENCY VIRUS*

Human immunodeficiency virus (HIV) occurs in many strains, some of which are given special names. HIV-1 is the collective denomination of strains isolated in Europe, America, or Central Africa; HIV-2 is a strain isolated from prostitutes in West Africa. Both cause the lethal AIDS disease, although HIV-2 tends to produce milder forms. The disease is characterized by severe **immunodepression,** with large reduction of the ratio of CD4+ to CD8+ cells, and **opportunistic infections** of various kinds: fungal (pneumonia from *Pneumocystis carinii*, oral or esophageal *candidiasis*), bacterial (salmonellosis, tuberculosis), and viral (cytomegalovirus, SV-40 like JC virus). The disease is also accompanied by secondary neoplasias, such as **Kaposi's sarcoma** and **B-cell lymphomas. Brain lesions** with dementia are frequently present and may be the only manifestation. The viruses can also cause less severe symptoms, of which the **lymphoadenopathy syndrome** is the most common, sometimes as a prelude to the immunodepression.

The virus was isolated by Montagnier from patients affected by the lymphoadenopathy syndrome by cultivating lectin-activated peripheral blood lymphocytes (PBLs) with IL-2. He observed expression of reverse transcriptase and production of viral particles, which, like those of other lentiviruses, have a **bar-shaped nucleoid.** Subsequently many other isolations of similar viruses followed. Like HTLV-I and -II, all strains have strong tropism for CD4+ (helper) lymphocytes.

GENETIC PROPERTIES (FIG. 65–18 A AND B). The genome of HIV (approximately 10 kb) contains nine genes, with important differences between HIV-1 and -2. Like other retroviruses both viruses contain the *gag, pol,* and *env* genes. The products of the *gag* gene is synthesized on an unspliced messenger, that of the *env* gene on

---

* The virus is also known as LAV (lymphoadenopathy virus), HTLV-III (for its tropism toward CD4+ cells), or ARV (AIDS-related virus).

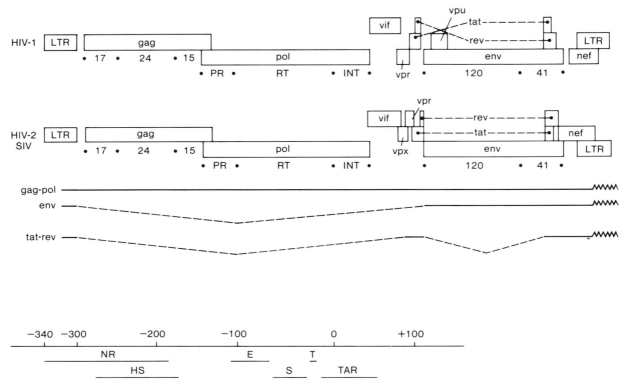

Figure 65–18. *(A)* The genomes of the HIV-1 and -2 proviruses. They differ in genes *vpu* and *vpx,* as well as in the frames of the various exons. In both viruses, genes *tat* and *rev* have two separate exons. 17, 24, 15: proteins p17, p24, p15 derived from processing of the *gag* polyprotein. PR, RT, INT: protease, reverse transcriptase, integrase, resulting from processing of the *pol* polyprotein. 120, 41; glycoproteins gp120 and gp41 derived from processing of the gp160 *env* polyprotein. Below the genomes are the main splicing patterns of the transcript, which lead to expression of *gag-pol, env,* and *tat-rev.* *(B)* The regulatory region of HIV-1. Numbers: base pairs upstream (−) or downstream (+) to the cap site (0). NR: negative regulatory region; HS: region homologous to IL-2 and IL-2 receptor genes; E: enhancer; S: SP1 binding region; T: TATA box; TAR: *tat*-responsiveness region.

a single-spliced messenger. The products are polyproteins; the *gag-pol* product is processed by a viral protease which is part of the *pol* polyprotein, whereas the 160-Kd product of the *env* gene is processed by a cellular protease into a gp-41 transmembrane glycoprotein and a gp-120 external protein. Both viruses have five other small genes: *tat, rev, nef, vif,* and *vpr*; in addition, HIV-1 has *vpu,* whereas HIV-2 has *vpx.* Genes *tat, rev,* and *nef* are synthesized on a double-spliced messenger. Mutations in genes *tat* and *rev* strongly inhibit viral multiplication, whereas those in *nef* enhance it. These genes specify proteins that act in trans, regulating expression of other viral genes.

Gene *tat* specifies a 15.5-Kd **transactivator** that, in collaboration with a cellular protein, enhances expression of all viral genes by increasing the production and translation of active messengers. It acts on an RNA hairpin loop present in the TAR region of the genome, increasing 100-fold the synthesis of viral proteins. The increase affects both transcription initiation (20-fold) and the amount of protein synthesized per mRNA (5-fold). The mechanisms of these changes are not known. By its

strong effect on the production of viral proteins, the *tat* gene acts as a replication trigger, once transcription is activated at the 5′-LTR (see below, Regulation of Provirus Transcription).

The 19-Kd *rev* protein is required for the expression of the *gag, pol,* and *env* genes, but not for *rev* and *tat,* by promoting the formation of unspliced and single-spliced messengers. In the absence of *rev* protein, only the double-spliced mRNAs are made. The *rev* protein promotes the export out of the nucleus of unspliced or single-spliced mRNAs; it does not affect splicing directly. In the absence of *rev* protein these RNAs remain entrapped in the nucleus, probably by cellular factors that are part of the splicing machinery. The *rev* protein acts by binding to a hairpin loop in the 3′-half of the *env* gene, the *rev*-responsive element (RER) facilitating access of the RNA to the export system. The mode of action of *rev* on the *gag* and *pol* genes may be similar.

The combination of the *tat* and *rev* proteins greatly increases viral replication by amplifying the effect resulting from transcriptional activation of the genome (see the following section, Regulation of Provirus Transcription). The *tat* protein acts as replication trigger, whereas the *rev* protein shifts the synthesis from regulatory to structural proteins. By decreasing its own synthesis toward a steady-state level, it regulates synthesis of structural proteins and viral multiplication.

The *nef* protein binds to the NR (negative regulatory) region. This protein has properties similar to those of the p21 product of the *ras* oncogene: it is myristoylated, binds GTP, is phosphorylated by protein kinase C, and has GTP-ase activity. It was originally thought to be a negative regulator of viral growth, but this function is no longer accepted. It has little effect on viral multiplication *in vitro*; it may have unrecognized effects *in vivo*. The *vif, vpu, vpr,* and *vpx* genes probably specify virion proteins that increase the efficiency of infection.

REGULATION OF PROVIRUS TRANSCRIPTION. (FIG. 65–18 B). Transcription of the HIV genome is stimulated or inhibited by the binding of suitable regulatory factors of either viral or cellular origin to the 5′ LTR. Thus the inhibitory NR region responds to the *nef* protein and to a cellular protein, Ppt-1, expressed in nonactivated T cells, which also inhibits expression of the IL-2 receptor gene. Several regions (the SP1 binding sites, the enhancer, the TATA box) increase transcription in response to many cellular factors, such as NFk-B, which also acts on an immunoglobulin gene. These factors are produced in T lymphocytes activated by Ag or mitogens and act on cellular genes. In this way cell activation serves also to activate a resident latent HIV provirus. Transcription is also activated by the products of some viral genomes, such as HTLV-I and -II (gene *tax*), EBV, cytomegalovirus, herpes simplex virus type 1, and adenoviruses (gene E1A), which interact with the enhancer.

CELL INFECTION. The virus infects cells expressing at their surface the 58 Kd CD4 protein, which is the viral receptor. Among these cells are T helper lymphocytes, antigen-presenting cells, (such as macrophages, their monocyte precursors, and the dendritic cells of lymph nodes), astroglia, B-lymphocytes, especially if immortalized by EBV, colonic epithelia in culture. Any human cell transfected with the CD4 gene and expressing the protein at its surface can be infected. In contrast, mouse cells similarly transfected are not infectible, although they can produce virus if transfected with the cloned HIV provirus. Thus other molecules besides CD4 are also necessary for infection. Among animals, some primates and rabbits can be infected, offering possible models for studying the biology of the virus; it seems, however, that they do not develop the AIDS disease. Immunodeficient *scid* mice reconstituted with human hemopoietic cells are susceptible to HIV infection; they may be an important tool for research.

Important in infection are the two cleavage products of gp160, the *env* product, of which gp41 is anchored to the cell membrane, whereas gp120 is connected to gp41 by noncovalent bonds. Gp120 binds to the host CD4 protein, a step blocked by certain monoclonal Abs specific for CD4. Mutational studies show also a requirement for the extracellular N-terminus of gp41, which has a **fusogenic** property (see Chap 48, Initial Steps of Viral Infection). It is likely that the binding of gp120 to CD4 exposes the gp41 terminus, allowing its interaction with an undefined receptor at the cell surface; its penetration into the host cell membrane would cause its fusion to the viral membrane. By this mechanism the viral core would directly enter the cytoplasm of the host cell. By a similar mechanism the *env* protein present at the surface of the infected cells may cause their fusion to CD4+ uninfected cells, with formation of **syncythia** (Fig. 65–19). This phenomenon is pronounced with some cell lines, but is not of general occurrence. HIV virions have also been seen in endosomes, suggesting that the virus can also enter the cells through the endocytosis pathway (see Chap. 48, Initial Steps of Viral Infection). CD4 endocytosis may be caused by its phosphorylation, through the phosphokinase C pathway, following the attachment of HIV virions.

An important consequence of infection is the progressive disappearance of the CD4 protein from the cell surface (**CD4 downregulation**), by a posttranscriptional mechanism, with a profound impairment of the function of the cells. As with other retroviruses (see above, Classification of type C oncoviruses), the mechanism of downregulation is, at least in part, interference, because the *env* protein, by binding the CD4 protein intracellularly, prevents the expression of the latter at the cell surface. Indeed, intracellular complexes of CD4 with gp160 or gp120 can be precipitated from cell extracts by certain MoAbs to CD4. Moreover, introduction into the cells of a vaccinia virus vector that expresses only the *env* protein causes substantial downregulation of CD4. This may not be, however, the only mechanism, because a vaccinia vector expressing only the *nef* gene also causes CD4 downregulation.

The consequence of infection for cell survival may vary. Quiescent T-helper lymphocytes infected *in vitro* may undergo a **latent infection,** with little production of viral proteins and persistence of the viral genome as an integrated provirus. It seems that the viral genome is in some cases in a non-replicating, non-integrated state, whereas in other cases it is integrated. In either case, when the cells are activated by Ag or mitogen, they enter **productive infection,** with a burst of viral replication and release of infectious virus, followed by cell death. Cells of certain permanent human CD4+ cell lines (such as H9) are infected by HIV but are not killed; they are the source of the virus used for experimental work and diagnostic

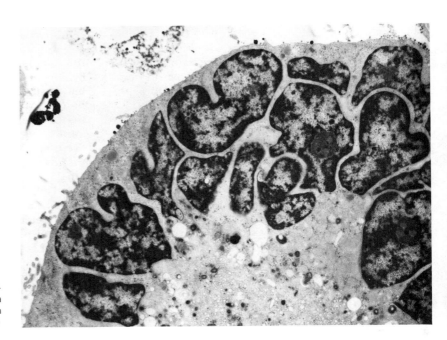

Figure 65–19. Electron micrograph of a giant multinucleated cell of a cultured T4⁺ lymphocyte from a culture infected with HIV (original magnification ×8000). (Klatzmann D et al: Science 225:59, 1984)

assays. Cells of the monocyte lineage tend to undergo a productive infection that can last for a long time, with accumulation of virions in intracellular vacuoles but poor extracellular release.

The death of productively infected cells is not due to the formation of syncytia, as formerly thought. A lethal factor may be the gp41 across the cell membrane, because amputation of its cytoplasmic domain can abolish cell killing, and some gp41 peptides, conjugated to a protein carrier, can slowly kill cells. Its role may be related to that of p15E in FLV infection (see previous discussion of Feline Leukemia Viruses). Another factor in cell death may be the accumulation of unintegrated viral DNA, a characteristic of lentiviruses.

HYPERVARIABILITY. A characteristic of HIV-1 and -2, as of other lentiviruses, is a very high genetic variability caused by a reverse transcriptase that causes mutations at a rate of a million-fold higher than that observed in the replication of DNA viruses. In fact, independently isolated viral strains differ from each other owing to mutations, deletions and insertions, and even in the same individual variant strains continue to appear all the time. In the *env* gene, the gp120 part has five hypervariable regions; the *gag* and *pol* genes are less variable, probably because constraints in viral assembly and enzymatic activity limit the range of acceptable mutations. A consequence of the variability is the emergence of strains with **tropism** for certain cell types. Thus, virus isolated from infected macrophages multiplies in macrophages much more readily than in T lymphocytes, and vice versa. An-

other consequence is the rapid emergence, often by a single aminoacid change, of **Ab-resistant variants** during viral growth in the presence of antibodies.

IMMUNE RESPONSE. HIV-infected persons develop a humoral and a cell-mediated response to virion Ags, both internal (such as p24 or the reverse transcriptase) and external (such as gp41 and gp120).

HUMORAL RESPONSE. Seropositivity appears after HIV infection more slowly than in most other viral infections, leaving a fraction of potentially infectious individuals unrecognizable for some time by serological tests. Abs to p24 and to some gp41 epitopes are recognized first but decreased markedly as symptoms appear, whereas those to gp120 remain more constant; the neutralizing titers are relatively low. The epitopes recognized are type-specific, but each serum has a broad specificity for viral strains because it contains Abs to many different epitopes; within one individual the epitopes recognized vary with time.

CELL-MEDIATED RESPONSE. A response of T-helper lymphocytes can be recognized by growth stimulation of PBLs exposed in vitro to cells presenting viral Ags. In infected but symptom-free individuals the response is strong for p24 but weak for gp120. CTLs capable of lysing cells with HIV Ags on their surfaces are present in the peripheral blood and other body fluids. They may be restricted to either class I or class II HLA Ags. Infected macrophages present viral Ag in the context of class I Ag, but in CD4+ cells that bind gp120 to surface CD4 molecules, the complex undergoes endocytosis, and Ags are presented in association with HLA class II Ags. The peripheral blood also contains mononuclear cells with characteristics of Natural Killer cells, which are capable of lysing HIV-infected target cells; there are also lymphocytes that lyse these targets in the presence of specific Abs (ADCC).

The various forms of immune response are highest in infected symptomless individuals, but decline in strength as the disease progresses. The causes for the decline may be multiple: killing of T-helper and antigen-presenting cells by HIV infection; killing of T-helper cells by cell-mediated immunity against adsorbed or processed viral Ags; virus-induced CD4 down-regulation; reduced production of lymphokines, such as IL-2 or IFN-gamma, that are needed for the operation of some effector cells; production of suppressor factors by the infected cells; genetic variation affecting crucial viral epitopes.

PATHOGENESIS. Infection occurs by transfer of body fluids containing either free virus or infected cells. The main vectors are blood and semen, but virus is also present in tears, saliva, bronchial secretion, and milk. Entry of the virus or virus-infected cells into the body is followed by infection of antigen-presenting cells and CD4+ lymphocytes. A long latent phase follows, during which the HIV provirus is recognized in a small fraction of PBLs by DNA hybridization using viral probes, but immunocytology fails to reveal viral proteins. These appear in a small proportion of PBLs in the phase of productive infection, in CD4+ cells that are activated. Activation may be caused initially by alloantigens introduced with the infecting inoculum (blood cells, spermatozoa) and subsequently by antigens introduced by infecting agents. These activated cells transmit the infection to new susceptible cells. The infected persons remain as **asymptomatic carriers** for an undetermined period of time, some perhaps indefinitely. It seems that many, perhaps most of the infected persons will ultimately develop the disease; after a large proportion of CD4+ cells and antigen-presenting cells have disappeared or become nonfunctional, immunodepression and the AIDS disease develop, months or years following the onset of infection. Profound **perturbations of the immune system** become detectable at this stage by the study of PBLs: a decrease of CD4+ in comparison to CD8+ lymphocytes; reduced in vitro growth response to anamnestic Ags (such as tetanus toxoid, or CMV Ags), reduced response to T-cell mitogens, deficient synthesis of lyphokines important in antimicrobial defense (IL-2, IFN-gamma), reduced expression of HLA class II Ags, and decreased response to IFN-alpha connected to downregulation of

receptors, probably caused by accumulation in the se-'rum of an unusual acid-labile IFN-alpha.

Frequently in HIV-1 infected persons, B cells, although not HIV-infected, undergo a **polyclonal activation** with release of large amounts of Abs of various isotypes into circulation. Stimulation may be provided by HIV virions or by abnormal production of lymphokines. Concurrent infections such as EBV and CMV may also contribute to the lymphoadenopathy. **Brain disease** is accompanied by the presence in the cerebrospinal fluid of virus-specific Abs as well as viral Ags, which are produced by HIV-infected macrophages and glial cells.

The origin of the **neoplasias** developing in some AIDS cases is uncertain. Changes similar to those of Kaposi's sarcoma (KS) were observed in transgenic mice expressing the HIV-1 *tat* gene. In these animals the gene is expressed in the epidermis, but the KS-like lesions are localized in the underlying subcutaneous tissue; they lack *tat* expression. The neoplastic changes may be induced by the *tat* protein, which is released by the infected cells, and can penetrate other cells, in which it acts as a growth factor. Opportunistic infections (EBV, CMV) may be involved in the origin of the lymphomas.

LABORATORY DIAGNOSIS. Infected peripheral blood cells produce viral proteins, demonstrable serologically, or an increase of reverse transcriptase, upon cocultivation with H9 or similar cells. Infection is also revealed by demonstration of viral DNA in PBLs by hybridization or by the polymerase chain reaction technique. ELISA assays are most widely used for the detection of serum Abs that bind to HIV proteins, both core (p24) or surface (gp 120, gp41), or to specific immunodominant and conserved peptides of these proteins. More sensitive is the western blot, based on the detection of electrophoretically separated viral proteins adsorbed on paper, using Abs from the patients. **Seroconversion** becomes detectable after a period of several weeks after infection, depending on the assay used.

EPIDEMIOLOGY. HIV infection has reached pandemic proportions. Virus is transmitted by infected persons, usually asymptomatic carriers who may be even more infectious than AIDS patients, in which viral multiplication is reduced by the disease of the helper and antigen-presenting cells. Transmission through the blood is observed especially among drug addicts who share syringes, in hemophiliacs who receive blood concentrates, and in individuals receiving transfusion after surgery, although the contribution of the latter class is now minimized by screening the donated blood for HIV Abs. Sexual transmission, originally observed among homosexuals, occurs also in heterosexual intercourse, in both directions. Through heterosexual contacts the disease, which was originally prevalent in men, has rapidly spread to women, closing the gap between the two sexes. The practice of rectal sex appears to be particularly responsible for heterosexual transmission. Infection is frequent among prostitutes, especially in central Africa, and in groups with pronounced sexual promiscuity. Transmission also occurs from an infected mother to her child, either transplacentally or perinatally (by contamination with mother's blood). In contrast, nonsexually related persons living together with infected persons, or providing health care to them, have a very low risk of infection, except through rare needle-stick accidents or exposure of wounds or skin abrasions to infected blood. These studies indicate that transmission through contacts not involving blood or sexual secretions is highly exceptional. Exposure to high titer virus in the laboratory may however lead to infection.

CONTROL. The best approach to control is **prevention.** Reduction of sexual promiscuity and adoption of prophylactic measures (such as the use of condoms) can reduce sexual transmission. Transmission through shared needles by drug addicts may be reduced by education. Transmission via blood or blood concentrate may be eliminated or at least markedly reduced by serologic screening of donated blood; the use of coagulation factors produced by recombinant DNA technology will eliminate this risk altogether. Transmission from mother to child can be reduced by recognizing infection and preventing or interrupting pregnancy in infected women. **Vaccination** attempts have been carried out, but the information available is still very limited. In humans a CTL response was obtained in Zaire in a group of healthy volunteers immunized with a vaccinia vector expressing the *env* protein, and later boosted with injection of *env* peptides. Chimpanzees inoculated with viral Ags have responded with Ab production and cell-mediated immune response, but the animals were not protected, because they became infected after challenge with the virus. Best results were obtained with SIV virus (see below, Other HIV Strains and HIV-Related Simian Viruses) in susceptible rhesus monkeys, in which a vaccine made with inactivated virus was capable of preventing infection. The difficulties of vaccination are compounded by the latency of infection, viral transmission upon cell contact, and the great variability of the virus. In chimpanzees, **passive immunization** with high titer *env* Abs did not prevent infection.

Control by **chemotherapy** is of some use, because many **dideoxynucleosides** (among which is 3′-azidothymidine [AZT]; see Chap. 49) have selectivity, that is, they can block DNA synthesis through reverse transcription more effectively than regular DNA replication. These substances have other advantages: they are active orally, have moderate toxicity, and penetrate the blood-brain barrier. AZT has been shown to prolong survival of AIDS patients, but it is not suitable for long term treatment of asymptomatic carriers, owing to bone marrow suppression. **Interferon alpha** has shown some success in a fraction of cases against Kaposi sarcoma.

OTHER HIV STRAINS AND HIV-RELATED SIMIAN VIRUSES. A virus related to HIV (simian immunodeficiency virus, SIV) causes an AIDS-like disease in monkeys (SAIDS, simian AIDS). Various strains were isolated from captive rhesus monkeys (*Macaca mulatta*) in a primate research center in USA, as well as from monkeys caught in the wild; the isolates are more closely related to HIV-2 than to HIV-1. Abs to SIV are found in many monkey species, although they do not develop disease. The precise genealogical connections between human and monkey viruses are not clear.

Selected Reading

Chen ISY: Regulation of AIDS virus expression. Cell 47:1, 1986

Coffin JM: Genetic variation in AIDS viruses. Cell 46:1, 1986

Cullen BR, Greene WC: Regulatory pathways governing HIV-1 replication. Cell 58:423–426, 1989

Evans LH, Cloyd MW: Friend and Moloney murine leukemia viruses specifically recombine with different endogenous retroviral sequences to generate mink cell focus-forming viruses. Proc Natl Acad Sci USA 82:459, 1985

Felber BK, Paskalis H, Kleinman-Ewing C et al: The pX protein of HTLV-I is a transcriptional activator of its long terminal repeats. Science 229:675, 1985

Friend C, Pogo BG-T: The molecular pathology of Friend erythroleukemia virus strains. Biochim Biophys Acta 780:181, 1985

Gallo RC: The AIDS virus. Sci Am 256:46, 1987

Ghysdael J, Bruck C, Kettmann R, Burny A: Bovine leukemia virus. Curr Topics Microbiol Immunol 112:1, 1984

Gartner S, Markovits P, Markovitz DM et al: The role of mononuclear phagocytes in HTLV-III/LAV infection. Science 233:215, 1986

Gonda MA, Kaminchick J, Oliff A et al: Heteroduplex analysis of molecular clones of the pathogenic Friend virus complex: Friend murine leukemia virus, Friend mink cell focus-forming virus, and the polycythemia- and anemia-inducing strains of Friend spleen focus-forming virus. J Virol 51:306, 1984

Guyader M, Emerman M, Sonigo P et al: Genome organization and transactivation of the human immunodeficiency virus type 2. Nature 326:662, 1987

Hahn BH, Shaw GM, Taylor ME et al: Genetic variation in HTLV-III/LAV over time in patients with AIDS or at risk for AIDS. Science 232:1548, 1986

Haseltine WA, Sodrowski JD, Patarca R: Structure and function of the genome of HTLV. Curr Topics Microbiol Immunol 115:177, 1985

Hirsch V, Riedel N, Mullins JI: The genome organization of STLV-3 is similar to that of the AIDS virus except for a truncated transmembrane protein. Cell 49:307, 1987

Hull R, Covey SN: Genome organization and expression of reverse transcribing elements: Variations and a theme. J Gen Virol 67:1751, 1986

Ju G, Cullen R: The role of avian retroviral LTRs in the regulation of gene expression and viral replication. Adv Virus Res 30:180, 1985

Levy JA: Mysteries of HIV: Challenges for therapy and prevention. Nature (London) 333:519, 1988

Neel BG, Hayward WS, Robinson HL et al: Avian leukosis virus-induced tumors have common proviral integration sites and synthesize discrete new RNAs: Oncogenesis by promoter insertion. Cell 23:323, 1981

Panganiban AT: Retroviral DNA integration. Cell 42:5, 1985

Rabson AB, Martin MA: Molecular organization of the AIDS retrovirus. Cell 40:477, 1985

Robert-Guroff M, Markham PD, Popovic M, Gallo RC: Isolation, characterization, and biological effects of the first human retrovirus: The human T-lymphotropic retrovirus family. Curr Topics Microbiol Immunol 115:7, 1985

Rosen CA, Terwilliger E, Dayton A et al: Intragenic cis-acting *art* gene-responsive sequences of the human immunodeficiency virus. Proc Natl Acad Sci USA 85:2071, 1988

Ruscetti S, Wolff L: Spleen-focus forming virus: Relationship of an altered envelope gene to the development of a rapid erythroleukemia. Curr Topics Microbiol Immunol 112:21, 1984

Sharp PA, Marciniak RA: HIV TAR: An RNA enhancer? Cell 59:229–230, 1989

Siliciano RF, Lawton T, Knall C et al: Analysis of host-virus interactions in AIDS with anti-gp120 cell clones: effect of HIV sequence variation and a mechanism for CD4+ cell depletion. Cell 54:561, 1988

Sonigo P, Barker C, Hunter E, Wain-Hobson S: Nucleotide sequence of Mason-Pfizer monkey virus: An immunosuppressive D-type retrovirus. Cell 45:375, 1986

Stevenson M, Meier C, Mann AM et al: Envelope glycoprotein of HIV induces interference and cytolysis resistance in CD4+ cells: sistance in CD4+ cells: mechanism for persistence in AIDS. Cell

Verma IM: Retroviral vectors for gene transfer. In Microbiology—1985, pp 229–232. Washington, American Society for Microbiology, 1985

Vogel J, Hinrichs SH, Reynolds RK et al: The HIV gene *tat* induces dermal lesions resembling Kaposi's sarcoma in transgenic mice. Nature (London) 335:606, 1988

Vogt PK (ed): Leukemia virus. Curr Topics Microbiol Immunol 115, 1985

Weiss R, Teich N, Varmus H, Coffin J (eds): RNA Tumor Viruses. Cold Spring Harbor, NY, Cold Spring Harbor Laboratory, 1984

Whitlock CA, Witte ON: The complexity of virus–cell interactions in Abelson virus infection of lymphoid and other hematopoietic cells. Adv Immunol 37:74, 1985

Wong-Staal F, Gallo RC: Human T-lymphotropic retroviruses. Nature 317:395, 1985

Yamanoto N, Hinuma Y. Viral aetiology of adult T cell leukemia. J Gen Virol 66:1641, 1985

Index

Boldface page numbers indicate figure citations.
Italic page numbers refer to in-text chemistry citations.
Page numbers followed by the letter *n* in italics indicate information is footnote.
Page numbers followed by the letter *t* in italics indicate table citations.